STRUCTURAL MATERIALS

STRUCTURAL MATERIALS

EDITED BY GEORGE WEIDMANN,
PETER LEWIS and NICK REID

The Open University

MATERIALS DEPARTMENT
OPEN UNIVERSITY, MILTON KEYNES, UNITED KINGDOM

Butterworth-Heinemann Ltd
Linacre House, Jordan Hill, Oxford OX2 8DP

A member of the Reed Elsevier group

OXFORD   LONDON   BOSTON
MUNICH   NEW DELHI   SINGAPORE   SYDNEY
TOKYO   TORONTO   WELLINGTON

First published 1990
Reprinted 1994

British Library Cataloguing in Publication Data
Structural materials
  1. Materials
  I. Weidmann, George   II. Lewis, Peter   III. Reid, Nick
  IV. Series
  620.1'1

ISBN 0 7506 1901 5

Library of Congress Cataloguing in Publication Data
Structural materials/edited by George Weidmann, Peter Lewis and Nick Reid
  p. cm. (Materials in action series)
  Includes bibliographical references
  ISBN 0 7506 1901 5
  1. Materials—dynamic testing   I. Weidmann, George
  II. Lewis, Peter   III. Reid, Nick   IV. Series
  TA418.32.S77   89–25456
  620.1'1–dc20   CIP

Designed by the Graphic Design Group of the Open University

Typeset and printed in Great Britain by Alden Press (London and Northampton) Ltd, London

This text forms part of an Open University course. Further information on Open University Courses may be obtained from the Admissions Office, The Open University, PO Box 48, Walton Hall, Milton Keynes MK7 6AB.

Series Preface

The four volumes in this series are part of a set of courses presented by the Materials Department of the Open University. Although each book is self-contained, the first volume, 'Materials Principles and Practice' is an introduction to the ideas, models and theories which are then developed further in the separate areas of the other three. It assumes that you are just starting to study materials and that you are already competent in pre-university mathematics and physical science.

Unlike many introductory texts on the subject, this series covers materials science in the technological context of making and using materials. This approach is founded on a belief that the behaviour of materials should be studied in a comparative way and a conviction that intelligent use of materials requires a sound appreciation of the strong links between product design, manufacturing processes and materials properties.

The interconnected nature of the subject is embodied by the use of two sorts of text. The main theme, or story line, of each chapter is in larger black type. Linked to this are other aspects, such as theoretical derivations, practical techniques, applications and so on, which are printed in colour. These links are flagged in the main text by a reference such as ▼Assessing hardness▲ and the linked text, under this heading, appears nearby. Both sorts of text are important, but this format should enable you to decide your own study route through them.

The books encourage you to 'learn by doing' by providing exercises and self-assessment questions (SAQs). Answers are given at the end of each chapter, together with a set of objectives. The objectives are statements of what you should be able to do after having studied the chapter. They are matched to the self-assessment questions.

This series and the Open University courses of which it is a part are the result of many people's labours. Their names are listed after the prefaces. I should particularly like to thank Professor Michael Ashby of Cambridge University for reading and commenting on drafts of all the books and the group of student 'guinea pigs' who worked through early drafts. Finally, thanks to my colleagues on the course team and our consultants. Without them this project would not have been possible.

Further information on Open University courses may be had from the address on the back of the title page.

Charles Newey
Open University
February 1990

Preface

The focus of this book is on load-bearing materials. Taken literally, this would first include virtually every solid material. A coat of paint, for example, is both protective and aesthetic, but it can also carry loads when things impinge on it or when its substrate is strained. Secondly, liquids and gases, such as the fluid in a shock absorber, or the air in a car tyre, are also load bearing. We shall, however, confine ourselves primarily to solid materials and their use in applications whose principal purpose is to carry loads, such as the materials used in bridge building.

Note that we shall be concerned with both the material *and* the application. We shall also continue to apply the philosophy of the first book in this series: namely, that in the real world of materials applications, the material is inseparable from the design of the product and the manufacturing processes employed — what we have called the 'product, processing, properties' philosophy.

Different classes of material are distinguished by differences in chemical bonding and structural organization. As such, they require different descriptions or models of the relationships between their structures and their mechanical properties. Thus, to emphasize the common features within each class and highlight the distinctions between them, the chapters follow a fairly conventional classification of materials. Six of these (ferrous and nonferrous metals, ceramics and glasses, polymeric and cementitious materials) are based on chemical similarity, while two (fibres and composite materials) are based on form or geometry.

In the real world, however, the different classes of material are not neatly segregated, but are very frequently in competition with one another. For example, silicate glass competes with transparent plastic for spectacle lenses, steel may be replaced by glass-reinforced plastic for car bodies. To reflect this, comparisons between materials occur throughout the book and selection of appropriate materials is a principal feature of the first and the final chapters. There are three appendixes: the first covers units of measurement; the second covers polymeric repeat units and abbreviations; the third is the Periodic Table.

In addition to thanking our authors and consultants, we join the Series Editor both in thanking our external assessor, Professor Michael Ashby, for his perceptive and helpful comments, and in acknowledging the efforts of our student 'guinea pigs'. We should also like to thank our course team colleagues for their penetrating and constructive criticisms, particularly on the early drafts, and Dr Phil Thompson (ex-Portsmouth Polytechnic), Ken Shopland (ex-Gwent College) and Dr Chris Burgoyne (Cambridge University) for their critical reading and zeal in spotting errors. Without the sterling and unstinting work of our publishing editors, Gerald Copp and Allan Jones, our course managers, Ernie Taylor and Andy Harding, our designer Debbie Crouch and Graphic Artist Alison George, photographer Mike Levers and our secretaries, Tracy Bartlett, Lisa Emmington and Anita Sargent, this book could not have happened — by the time you read this we shall have bought them all a drink.

George Weidmann, Peter Lewis and Nick Reid
February 1990

Open University Materials in Action course team

MATERIALS ACADEMICS
Dr Nicholas Braithwaite (Module chair)
Dr Lyndon Edwards (Module co-chair)
Mark Endean (Module co-chair)
Dr Andrew Greasley
Dr Peter Lewis
Professor Charles Newey (Course and module chair)
Professor Nick Reid
Ken Reynolds
Graham Weaver
Dr George Weidmann (Module chair)

TECHNICAL STAFF
Richard Black
Colin Gagg
Jim Moffatt
Naomi Williams

PRODUCTION
Phil Ashby (Producer, BBC)
Gerald Copp (Editor)
Debbie Crouch (Designer)
Alison George (Graphic Artist)
Andy Harding (Course manager)
Caryl Hunter-Brown (Liaison Librarian)
Allan Jones (Editor)
Mike Levers (Photographer)
Carol Russell (Editor)
Ernie Taylor (Course manager)
Pam Taylor (Producer, BBC)
Bob Walters (Producer, BBC)

SECRETARIES
Tracy Bartlett
Lisa Emmington
Angelina Palmiero
Lesley Phelps
Anita Sargent

CONSULTANTS FOR THIS BOOK
Dr John Briggs
Dr Martin Buggy (University of Limerick)
Dr Christina Doyle (Queen Mary College, University of London)
Dr Tom Frank (Roehampton Institute of Higher Education, London)
Dr Nigel Mills (University of Birmingham)
Stan Pugh
Dr Andrew Stevenson (MERL Ltd, Hertford)

EXTERNAL ASSESSOR
Professor Michael Ashby FRS (University of Cambridge)

Contents

Chapter 1
Materials and Mechanics

by Nick Reid and George Weidmann

Chapter 1 Materials and mechanics

1.1 Aims and structure of the book

The focus of this book is on load-bearing materials. In it we aim to develop further a number of important propositions about such materials — propositions that we assume you have already encountered, either explicitly or implicitly, at an introductory level. These are:

1 The properties of a material are primarily a consequence of the nature of the cohesive forces between its constituent atoms and/or molecules.

2 The properties of a material, including combinations of unlike materials, are influenced significantly by changes in the internal structure or 'microstructure' at all levels from 0.1 nm up to 1 cm. Knowledge of the connection between properties and microstructure enables new or modified materials to be *designed*, both in terms of their constituents and their required microstructures, in order to satisfy a given need.

3 The microstructure of a material depends on the way in which the material is made. This means that thermal, chemical and mechanical treatments can have a controlling effect on material properties. Such treatments are frequently involved in the processes used to manufacture a product.

4 The performance of a product is determined by *both* the properties of the materials of which is it made *and* the geometrical form of the product. Product performance can therefore be improved by making changes to the material, the form of the product or both. This requires understanding the response of both the material and the geometrical form of the product to the application of forces.

5 The suitability of a material for a given function depends on its properties. The suitability of different materials can be compared provided that precise criteria involving geometrical and performance requirements can be drawn up. Using these criteria, combinations of materials properties, called merit indices, can be identified and used to aid materials selection. These often involve the unit cost of the material.

6 In general, materials are seldom truly stable under service conditions. This leads to the concept that a product has only a finite lifetime. The changes occurring during service that are of concern are those that cause a deterioration in properties such as stiffness or strength. They may be chemical changes (e.g. corrosion) or mechanical changes (e.g. crack growth) or hybrids of these (e.g. stress corrosion cracking). Many changes that affect the lifetime of a product start at the surface (e.g. corrosion, wear, fatigue).

The structure of the book is based on a conventional classification of materials into ferrous and nonferrous metals, ceramics and glasses, polymeric and cementitious materials, fibres and composite materials. Notice that the first six groupings are based on chemical similarity, while the last pair are based on a geometrical classification. In this way, the materials within each group have enough in common for their mechanical properties to be modelled in the same way. Since, in general, the models required vary greatly from group to group, this structure allows us to concentrate on one set of models at a time.

This segregation of materials into classes contrasts with the usual situation where, in the real world, different types of material often compete with one another for the same application. To remind you of this important fact of life, there are wider comparisons of materials within each chapter, building up to a materials-selection exercise in the last chapter. This also serves as revision of the preceding chapters.

Before exploring the differences between these classes of materials, we should note that, for some purposes, an individual material can be regarded as 'stuff' with a given set of properties. If you know the values of these properties, you can then carry out simple design tasks on load-bearing products using a wide range of materials. An example of this is the subject of the next section.

1.2 A design task

Figure 1.1 A bus shelter

Since the work of Hooke in the seventeenth century (see ▼Robert Hooke (1635–1703)▲), it has been known that all materials deform when they are loaded and that the deformation is approximately proportional to the load, provided the load is not excessive. If such deformations disappear when the load is removed then the deformation is said to be **elastic**. In this section, we shall consider elastic deformation in some detail because in most cases products are designed to deform elastically in service. To give this some point, we shall look at a simple design task and then derive some theory to assist in finalizing the design.

Suppose you wish to design the roof of a road-side bus shelter. The requirements of the roof may be stated briefly as follows: to provide cover from the elements over a given period of time, the designed lifetime of the shelter, at the minimum cost. A full design procedure might involve drawing up a **product design specification** (PDS) containing a multitude of factors. Here, we shall consider only a selected fraction of these, affecting load-bearing performance and cost.

1 *Materials*   These should be readily available from suppliers in the form required and should meet the requirements of the design specification.

2 *Product lifetime*   Take this to be twenty years.

3 *Loading*   The roof should provide *safe* cover to the people underneath. It should not collapse under a heavy fall of snow, when the loading can reach $1500\,\mathrm{N\,m^{-2}}$, and it should resist loads due to winds, which can be large as $2000\,\mathrm{N\,m^{-2}}$.

4 *Environment*   Take it to be that of a typical inland city in the United Kingdom.

5 *Maintenance*   There should be no need for maintenance (e.g. painting) in normal use.

6 *Weight*   This will affect the design of the supporting structure and will require some limit so that the loading due to weight will be no more than, say, 10% of the maximum loading in service ($200\,\mathrm{N\,m^{-2}}$).

7 *Dimensions*   The roof should cover a queue of people two abreast and ten in line, and thus cover an area of $1.5 \times 5\,\mathrm{m^2}$. The roof should be simply supported along its long edges by two parallel walls, 1.5 m apart. It should have a minimum headroom of 2.1 m.

This design specification can be met by a roof that is made of sheet material and is gently sloping, to allow for drainage. The existence of a given lifetime implies that the roof must be *stable* over this period. Absolute stability is not necessary, but any gradual alterations to the roof must be kept within strict limits. Factors such as chemical changes within the material due to its exposure to sunlight or to the atmosphere must be limited if they are accompanied by adverse changes in properties. For example, the roof material should not become embrittled, and corrosion over this period should not reduce the thickness of any section by more than, say, 10%.

▼Robert Hooke (1635–1703)▲

The fact that there is a definite relationship between force and elastic deformation was first perceived by Robert Hooke, who in 1676 published it in the form of a Latin anagram, *ceiiinosssttuv*. Three years later he revealed the answer, *ut tensio sic vis* ('as the extension, so the force') or force is proportional to extension. This proportionality, however, involves the geometry as well as the elastic properties of the deformed material. What we know as Hooke's law had to wait over 100 years, until Thomas Young, in about 1800, realized that expressing the proportionality in terms of stress and strain removed the geometric factors and left a material property — the elastic modulus.

Hooke, like others of his contemporaries, was active in many fields: microscopy (Figure 1.2a) astronomy, chemistry, optics and palaeontology. He worked with Robert Boyle in Oxford on the properties of gases — pV = constant was almost Hooke's other law!

His work on elasticity was stimulated by his interest in the use of springs to power clocks and watches. He was also an architect. After the Great Fire of London in 1666, Hooke contributed to the reconstruction by designing several buildings. Indeed, the city of Milton Keynes is blessed with a fine example (Figure 1.2b).

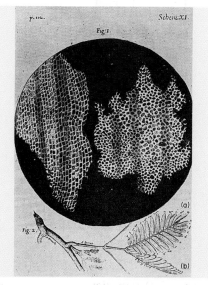

Figure 1.2 (a) Microstructure of cork as observed and drawn by Robert Hooke

Figure 1.2 (b) Willen Church, Milton Keynes, designed by Robert Hooke

Provided it meets the performance specification, the fate of a roof design is likely to be decided on the grounds of the total cost — the cost of materials plus the cost of shaping them, both on and off site, plus any indirect costs, such as the cost of the roof-supporting structure.

Consider five contrasting commonplace materials for this application: aluminium-magnesium alloy (AA), mild steel (MS), poly(vinyl chloride) (PVC), soda-lime silica glass (GL) and glass-fibre reinforced polyester (GFRP). For each of these materials, we must decide what the shape and dimensions of the roof should be in order to meet the specification. To enable you to do this scientifically, rather than by trial-and-error, requires some simple 'continuum' mechanics.

1.3 States of stress

In this section, we shall consider elastic deformation under a variety of common modes of loading, briefly survey stress states and give examples of loading geometries. We shall then focus on two important types of loading: namely, bending and torsion.

1.3.1 Tension or compression

Tension, or more properly, uniaxial tension, is one of the most important stress states, even though there are comparatively few applications, such as in bicycle spokes or ropes, in which materials are subjected to a pure and simple tension. One reason for its importance is that, being 'one dimensional', tension is conceptually straightforward and easy to model. Another is that tensile tests can be performed readily on most types of material and data on the tensile properties of materials therefore tend to be the ones most often available in reference books.

When considering uniaxial loading, it is easy to forget that a material's *response* is rarely uniaxial. In general, it is triaxial as the material deforms in all three dimensions. Two extremes of behaviour can be imagined: either the *proportions*, or *shape*, of the specimen remains constant, in which case its volume must change; or its *volume* stays constant, requiring a change in proportions (Figure 1.3). Examples of these extremes of behaviour are rare, nearly all materials exhibiting different combinations of changes in both shape and volume. For example, in the tensile test, a tensile strain is imposed along the longitudinal axis of the specimen and, in general, this is accompanied by a change in its cross-section. If the material is **isotropic** (i.e. its properties are nondirectional), all directions within the cross-section will change by the same factor and you can express the changes in terms of the ratio

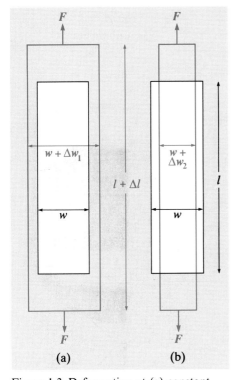

Figure 1.3 Deformation at (a) constant proportions (b) constant volume

$$v = -\frac{\text{width strain}}{\text{axial strain}} = -\frac{\Delta w/w}{\Delta l/l} \tag{1.1}$$

In isotropic, elastic materials this ratio, v, is known as **Poisson's ratio**. Note that when Δw is negative (a contraction), v is positive and vice versa. Together with the Young's, the shear and the bulk moduli (E, G and K), Poisson's ratio is one of the set of four elastic constants, any two of which are sufficient to describe fully the elastic deformation of an *isotropic* material.

What, then, are the limiting values of v?

For the extreme case where the proportions of the specimen do not change

$$\frac{\Delta w}{w} = \frac{\Delta l}{l}$$

and therefore $v = -1$.

For the other extreme of a deformation at constant volume, $\Delta V = 0$ and, since $V = Al$, where A is the area of the cross-section and l is the length, differentiation gives

$$\Delta V = l\Delta A + A\Delta l = 0$$

or

$$\frac{\Delta A}{A} = -\frac{\Delta l}{l}$$

For simplicity, consider a square cross-section of width w, so that $A = w^2$ and, again differentiating, $\Delta A = 2w\Delta w$. Then

$$\frac{\Delta A}{A} = \frac{2\Delta w}{w} = -\frac{\Delta l}{l}$$

Rearranging

$$-\frac{\Delta w/w}{\Delta l/l} = v = \frac{1}{2}$$

Thus, v can range in value from -1 to $\frac{1}{2}$ and a negative value indicates that the area of the cross-section will *increase* on stretching. Not many materials show such eccentric behaviour, however, and iron pyrites is a rare example. At the other extreme, elastomers, such as natural rubber, come close to having $v = \frac{1}{2}$. Most materials have intermediate values of v. Finally, it is worth noting that the strain ratio accompanying the *plastic* deformation of metals is approximately $\frac{1}{2}$, indicating that there is little or no volume change associated with this process.

SAQ1.1 (Objective 1.2)
Show that in an isotropic material, the fractional change in its cross-sectional area associated with an elastic strain ε in the direction of uniaxial loading is $-2v\varepsilon$. Take the section to be square with a width w.

1.3.2 Other common stress states

Some of the common stress states are sketched in Figure 1.4 in terms of the stresses acting on a small sample of material. In a sense these are all abstractions, since, apart from uniaxial tension, it is difficult to achieve a uniform simple stressed state in a body. In many cases, however, the actual stress state within a limited volume of material is sufficiently close to one or other for them to be used as a good approximation. An example of this is in the walls of pressure vessels and pipes. In the wall of a gas pipe under pressure there are three stresses acting — a circumferential tension (the **hoop stress**), a tension parallel to the pipe's axis and a radial compression through the thickness of the wall. If the thickness of the wall is less than about one twentieth of the pipe's diameter, the radial compression becomes negligible compared with the other stress components and the stress state can be modelled as a biaxial tension, a much simpler system (see Figure 1.4c).

The most general state of stress, triaxial stress, has no less than six components: three **direct** stresses acting perpendicular to the faces of a cubic element and three **shear** stresses acting parallel to the faces. In

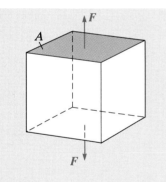

(a) simple uniaxial tension, $\sigma = \dfrac{F}{A}$

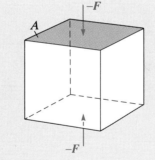

(b) simple compression, $\sigma = \dfrac{-F}{A}$

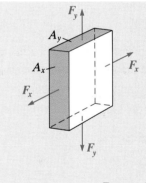

(c) biaxial tension: $\sigma_x = \dfrac{F_x}{A_x}$, $\sigma_y = \dfrac{F_y}{A_y}$

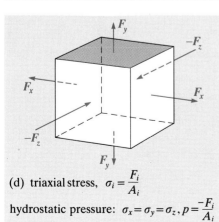

(d) triaxial stress, $\sigma_i = \dfrac{F_i}{A_i}$

hydrostatic pressure: $\sigma_x = \sigma_y = \sigma_z, p = \dfrac{-F_i}{A_i}$

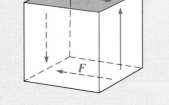

(e) shear, $\tau = \dfrac{F}{A}$

Figure 1.4 Some stress states (a) simple tension (b) simple compression (c) biaxial tension (d) triaxial stress (e) pure shear

such cases, there is always *one* orientation of co-ordinate axes (the **principal** axes) which makes the shear stresses disappear, leaving three principal direct stresses (Figure 1.4d).

Note the convention of treating compression as a negative tension. Hydrostatic pressure is a special case of triaxial stress in which not only are all three principal stresses equal, but the sign convention is reversed so that a positive pressure results from three equal compressive principal stresses. Pure shear is not easily achieved in solid materials, though the torsion of thin-walled shafts gets close to it.

1.4 Structural analysis

Load-bearing structures come in many shapes and sizes and degrees of complexity. To analyse these structures so as to predict their performance when they are loaded, it is usual to break them down into their structural elements. The most important of these, certainly for our purposes, are those elements subjected to tensile, bending and torsional loads.

The essentials of structural analysis lie in obtaining three different relationships:

(a) that between *strain* and *displacement*;
(b) that between *stresses* and *strains* (called the **constitutive equation**);
(c) that between *forces* and *stresses* (the **equilibrium equation**).

Although not done explicitly, these have already been covered for the case of tension in linear elastic, isotropic bodies: namely,

The strain–displacement equation

$$\varepsilon = \frac{\Delta l}{l_0} \tag{1.2}$$

The constitutive equation

$$\sigma = E\varepsilon \tag{1.3}$$

The equilibrium equation

$$\sigma = \frac{F}{A} \tag{1.4}$$

The first two of these should be clear enough, since they come directly from the definitions of tensile strain and Young's modulus, respectively. The second equation is also known as **Hooke's law**.

The third equation is connected, perhaps not so obviously, with the requirements of equilibrium, although it arises from the definition of tensile stress. It can be thought of in terms of Figure 1.4(a). When an external tensile force F acts on a body it generates an internal stress σ. For there to be no net force on the shaded top element of the body, which is the condition for equilibrium, the effects of the external load and the internal stresses must balance. If the internal stresses are evenly distributed over any cross-section of the element, which is the usual assumption, then for the forces to balance

internal stress, σ × area, A = external force, F

This leads to the equilibrium equation above (Equation 1.4).

How can you express the equivalents to these three different relationships for bending and torsional loadings?

1.4.1 Strain-displacement and constitutive relationships

Let us start with bending, or flexure, and consider a length l_0 of a straight beam that has been deformed elastically into a circular arc of radius R by, for example, the application of two pairs of forces F on socket spanners at each end of the beam (Figure 1.5). Since the forces in each pair are offset by a distance d, they exert a rotational effect on the ends of the beam, producing a **bending moment**, M, in the beam which is equal to $F \times d$.

In bending, the stresses vary from tensile on one surface to compressive on the other. This implies that there is an axis or plane between the two surfaces along which the tensile stress is zero and it is known as the **neutral axis** or **neutral surface**. For the cross-sections we shall be considering, it is located symmetrically between the two surfaces: that is, half way between them. As indicated in Figure 1.5, the radius of curvature R is measured to the neutral axis n–n.

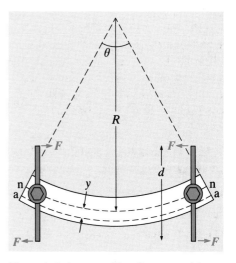

Figure 1.5 A state of bending caused by applying equal and opposite forces at the ends of a beam

Consider the line a–a, at a distance y from the neutral surface. Measured towards the surface that is under tension, y is said to be positive. It is considered negative when measured towards the compressive surface. The line a–a has increased in length by Δl on bending, while the length l_0 of the neutral surface is unchanged. Hence the strain in the surface a–a is

$$\varepsilon = \Delta l / l_0$$

From the geometry of the beam

$$\frac{\Delta l}{l_0} = \frac{(R + y)\theta - R\theta}{R\theta}$$

and therefore

$$\varepsilon = \frac{y}{R} \tag{1.5}$$

This is the strain–displacement relationship for bending in terms of the variable y. The strain varies linearly through the thickness of the beam, with its maximum (positive) value on the tensile surface and its minimum (negative) value on the compressive surface. Since bending involves direct strains (both positive tension and negative compression), the constitutive equation is the same as that for uniaxial tension or compression (Equation 1.3).

In the case of torsion, the analogous relationship is obtained by considering a bar of length l (Figure 1.6). If one end of the bar is twisted through the angle θ with respect to the other, the resulting **shear strain**, γ, is defined as the angle of rotation, measured in radians, of the line AC on the surface of the bar.

The arc AB subtends an angle θ at a distance r away, so its length is equal to $r\theta$, where θ is in radians. AB also subtends an angle γ at a distance of l, so it is also equal to γl. Hence

$$\gamma l = r\theta$$

Therefore

$$\gamma = \frac{r\theta}{l} \qquad (1.6)$$

This is the strain-displacement relationship for torsion and it means that for a given angle of twist per unit length, θ/l, the shear strain, γ, produced is proportional to the distance r from the central axis.

The shear strain produced by a torsion is associated with a shear stress, τ, and the appropriate constitutive equation is the shear version of Hooke's law (Equation 1.3), but in this case involving the shear modulus, G. By comparison with Equation (1.3) we have

$$\tau = G\gamma \qquad (1.7)$$

G can be expressed in terms of Young's modulus and Poisson's ratio as follows

$$G = \frac{E}{2(1 + v)}$$

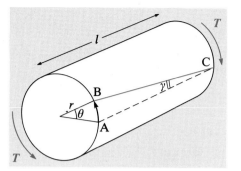

Figure 1.6 The twisting of a cylinder by the application of equal and opposite torques T to its ends

SAQ 1.2 (Objectives 1.1 and 1.3)
An 800 mm long bar, whose circular cross-section has a diameter of 4 mm, was deformed elastically in the following ways:
(a) axial extension of 0.5 mm;
(b) torsional twist of three degrees;
(c) bending into a circular arc of radius 4 m.

In each case, derive an expression for the strain and calculate it, indicating whether the strain is direct or shear and describe how the strain is distributed over the cross-section.

SAQ 1.3 (Objective 1.3)
The bar in SAQ 1.2 is made of a steel with $E = 210\,\mathrm{GN\,m^{-2}}$ and $v = 0.3$. If the deformations are all linearly elastic, calculate the maximum stress for each case given in SAQ 1.2, indicating clearly whether the stress is direct or shear.

1.4.2 Equilibrium equations

Again we shall tackle the bending case first. Since the strain in the beam is proportional to the distance from the neutral axis, so, too, in a material that obeys Hooke's law, is the stress, as shown in Figure 1.7. To derive the equilibrium equation, we need to relate the stress to the bending moment M by taking moments about the neutral surface.

Consider an element of area dA in the cross-section of the beam, a distance y from the neutral surface, with a stress σ acting on it as shown in Figure 1.8. The stress produces a force equal to σdA, which acts at a distance y from the neutral surface. This produces a moment about the neutral surface equal to the force × distance, $y\sigma dA$. The total moment, M, produced over the entire cross-section is found by adding up *all* such moments over the whole area. Mathematically, this is equivalent to integrating with respect to area.

Thus

$$M = \int_A y\sigma\,dA$$

You know that σ varies linearly with y, so σ/y is a constant and can therefore be taken outside the integral, so

$$M = \int_A \left(\frac{\sigma y}{y}\right) y\,dA$$

$$= \frac{\sigma}{y} \int_A y^2\,dA \qquad\qquad (1.8)$$

The integral

$$\int_A y^2\,dA$$

is known as the **second moment of area, I,** of the section and is discussed further in ▼**Moments of area**▲.

Thus, from Equation (1.8), we have the equilibrium equation for bending: namely,

$$\frac{M}{I} = \frac{\sigma}{y} \qquad\qquad (1.9)$$

Comparing this with Equations (1.3) and (1.5) gives

$$\frac{\sigma}{y} = \frac{E\varepsilon}{y} = \frac{E(y/R)}{y} = \frac{E}{R}$$

The double equality

$$\frac{M}{I} = \frac{\sigma}{y} = \frac{E}{R} \qquad\qquad (1.10)$$

is the most important statement of **engineers' bending theory**.

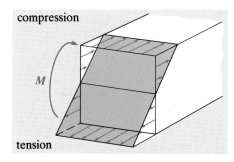

Figure 1.7 The distribution of stress across a section under bending

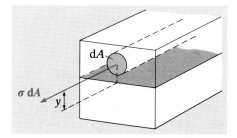

Figure 1.8 A small element of area, dA, situated a distance y from the neutral surface is subject to a small force σdA

▼Moments of area▲

We have seen that I is called the second moment of area. Is there a *first* moment of area? Yes there is and it is used in defining the **centroid**, sometimes called the **centre of gravity**, of an area. The centroid of an area in the xy-plane is defined by the condition that the first moments of area about the centroidal axis are both zero, so that

$$\int x\mathrm{d}A = 0 \quad \text{and} \quad \int y\mathrm{d}A = 0$$

The centroid can be visualized as defining the point at which the whole mass of the area acts. In other words, the area would balance on top of a point support located at the centroid.

The second moment of area I involves the *square* of the distance y. What I does is to describe how material (or an area) is distributed with respect to an axis *in the plane of the area*, while the **polar second moment of area, J**, is similar except that it refers to an axis *perpendicular* to the plane of the area. Since both I and J depend on

the *square* of the distance between the axis and an element of area dA, they are increased by redeploying material at a greater distance from the axis. For tubing of a given material and a given weight per unit length, the tube with the greatest diameter will have the largest values of I and J, in spite of having the smallest wall thickness. Table 1.1 contains expressions for I and J for some simple cross-sections. One special case we shall need is that of the thin-walled cylindrical tube. If the walls are of thickness t, then I and J are approximately $\pi r_0^3 t$ and $2\pi r_0^3 t$.

Table 1.1   Second moments of area for some familiar sections

Section	Dimensions	I	J
solid cylinder		$\dfrac{\pi r^4}{4}$	$\dfrac{\pi r^4}{2}$
cylindrical tube		$\dfrac{\pi}{4}(r_0^4 - r_i^4)$	$\dfrac{\pi}{2}(r_0^4 - r_i^4)$
thin-walled cylindrical tube		$\pi r_0^3 t$	$2\pi r_0^3 t$
solid rectangle		$\dfrac{wt^3}{12}$	$\dfrac{wt}{12}(w^2 + t^2)$
rectangular tube		$\frac{1}{12}(w_0 d_0^3 - w_i d_i^3)$	$\frac{1}{12}[w_0 d_0(w_0^2 + d_0^2) - w_i d_i(w_i^2 + d_i^2)]$
sinusoidal corrugations $s = \dfrac{\lambda}{2}\left[1 + \left(\dfrac{\pi d}{2\lambda}\right)^2\right]$		$\dfrac{wd^2 t}{8}\left(1 - \dfrac{0.81}{1 + 2.5\,(d/2\lambda)^2}\right)$	
I-section (RSJ)		$\left[\dfrac{WD^3}{12} - 2\left(\dfrac{wd^3}{12}\right)\right]$	

For the torsional case, we shall simplify matters by considering the torsion of a thin-walled tube. This is similar to the simplification achieved by assuming that a pressurized gas pipe has thin walls. It means that the shear stresses can be assumed to be uniform throughout the thickness of the wall of the tube (Figure 1.9), with a mean value $\bar{\tau}$. The shear stresses act circumferentially to the tube across the area of the cross-section (Figure 1.10), so

$$\text{shear force} = \text{cross-sectional area} \times \text{shear stress}$$

$$= 2\pi r t \bar{\tau}$$

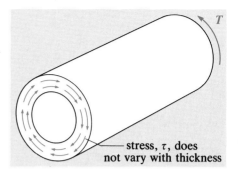

Figure 1.9 The shear stress, τ, in a thin-walled tube under torque T

This force, acting at the radius r, has a moment about the axis of the tube that is in equilibrium with the applied external torque T, so

$$T = \text{force} \times \text{radius}$$

$$= 2\pi r^2 t \bar{\tau}$$

On rearrangement, this yields the equilibrium equation for torsion

$$\frac{T}{J} = \frac{\bar{\tau}}{r}$$

where $J = 2\pi r^3 t$ is the **polar second moment of area** for the thin-walled tube (see 'Moments of area'). From Equations (1.6) and (1.7), this can be rewritten as a double equality, like that for the bending case:

$$\frac{T}{J} = \frac{\tau}{y} = \frac{G\theta}{l} \qquad (1.11)$$

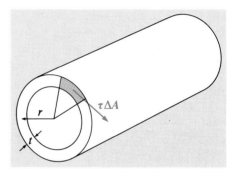

Figure 1.10 The shear force, $\tau \Delta A$, acting on an element of a tube

This equation applies to *any* radially symmetrical shaft (e.g. a solid one) provided the appropriate expression for J is used. (Remember τ is the shear stress at a distance y from the central axis — in solid sections, τ and γ vary linearly from zero along the axis to their maximum values at the surface.)

Thus we have established the strain–displacement, constitutive and equilibrium relationships for three of the more important structural elements. These are summarized in Table 1.2.

Table 1.2   Summary of relationships

Relationship	Uniaxial tension	Bending	Torsion
strain–displacement	$\varepsilon = \dfrac{\Delta l}{l}$	$\varepsilon = \dfrac{y}{R}$	$\gamma = \dfrac{y\theta}{l}$
constitutive	$\sigma = E\varepsilon$	$\sigma = E\varepsilon$	$\tau = G\gamma$
equilibrium	$\sigma = \dfrac{F}{A}$	$\sigma = \dfrac{My}{I}$	$\tau = \dfrac{Ty}{J}$

1.5 Designing the roof

1.5.1 The need for stiffness

The stiffness required of the material for the roof of the bus shelter you are designing will depend on how it is to be supported. If the underlying structure supports the roof continuously, the roofing material is merely a cladding and has little need for much stiffness and therefore does not need to have a large value for Young's modulus. On the other hand, if the roof contains any unsupported spans, the Young's modulus of the material will be an important factor in the design. The product design specification for the bus shelter dictates that the roof should be simply supported along its long edges by two walls, 1.5 m apart, with an unsupported span between them. Such a design requires stiffness. As in the design of most load-bearing structures, during normal service a roof is intended to deform within the elastic range, because such deformations are usually small and are reversible, and therefore cause no permanent distortion of the structure.

The maximum elastic strain in an unsupported roof section should be kept within limits. The strain is proportional to the deflection of the roof — if the roof sags too much under its own weight, it will not drain properly and it will tend to vibrate excessively in the wind. There is a further reason to limit the strain. Excessive compressive strain results in an initially elastic instability called **buckling**, which usually leads to plastic deformation and eventual collapse of the roof.

Both the sag and the buckling load are controlled by the **flexural rigidity**: that is, by the value of EI as Equation (1.10) shows. For a given value of bending moment, M, the radius, R, to which the roof is bent, is proportional to EI. A small sag — that is, large R — requires a large value of EI. To some extent the effect of a low value of Young's modulus can be offset by enlarging the section and, thus, increasing the value of I. Consider a beam of a given material with a square cross-section. How would you increase I without changing the mass per unit length?

The second moment of area, I, may be increased by redeploying material at a greater distance from the neutral axis. For instance, by changing to a rectangular section of the same area, with the longer side perpendicular to the neutral surface. The I-section rolled-steel joist (RSJ) is an example of applying this principle.

The same strategy can be followed with sheet material. To increase I, some of the material can be redeployed further away from the neutral axis. This can be done by attaching stiffening ribs to the sides of the sheet, as shown in Figure 1.11(a). They may also be attached to both sides of the sheet. Alternatively, corrugations can be put into the sheet to increase its sectional depth (see Figure 1.11b). It must be emphasized, however, that the orientation of the corrugations is critical (Figure 1.12). Such stratagems are used to increase the bending stiffness of most sheet materials, even when the modulus of the material is large. For

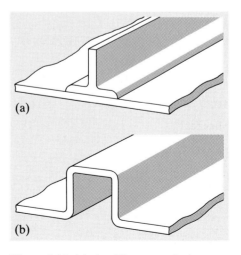

Figure 1.11 (a) A stiffener attached to one side of a sheet. (b) Corrugations for stiffening

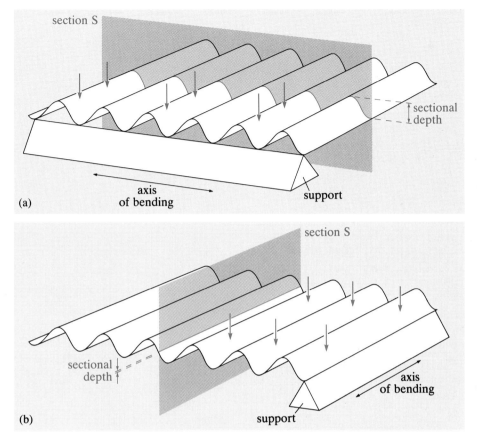

Figure 1.12 (a) In this orientation, corrugations increase the depth of the section, S, within a plane parallel to the axis of bending. (b) In this orientation, the depth of the section is *not* increased

example, steel panels on car bodies often contain one or more corrugations; this is not just an aesthetic whim of the designer, but an important stiffening device.

In principle, I can be manipulated to promote stiffness in any material, provided the material can be readily formed into the shape sought, but for a *given* cross-section, the bending stiffness will be proportional to the Young's modulus, E, of the material. The Young's modulus for each of the materials being considered for the roof of the bus shelter is therefore shown in Table 1.3, together with other design data.

Table 1.3

Material	MS	AA	PVC	GL	GFRP
approximate cost, $£\,kg^{-1}$	0.3	1.3	1.0	0.9	1.7
Young's modulus, $E/GN\,m^{-2}$	210	71	3.0	71	13
proof stress, $\sigma_p/MN\,m^{-2}$	350	200	50*	50	180
density, $\rho/Mg\,m^{-3}$	7.9	2.8	1.45	2.5	1.5
$(I/w)/10^{-9}\,m^4$	35	102	2417	102	558
thickness, t/mm (flat sheet)	7.5	11	31	11	19
depth, d/mm (corrugated sheet)	23	33	95	—	59
maximum stress, $\sigma_{max}/MN\,m^{-2}$					
flat sheet	67	32	4.0	32	11
corrugated sheet	203	100	12	—	33

*Short-term yield stress.

On this criterion mild steel (MS) is the stiffest, while PVC is the least stiff, or most compliant. To compensate for its low stiffness, a section in PVC would need a value of I some 70 times greater than that of mild steel, requiring a much larger volume of material. Also PVC is liable to undergo creep deformation at ambient temperatures, due to its viscoelastic nature, and therefore the *required* value of E depends on the timescale over which the material is loaded. We shall assume that the most severe loading is due to transient wind gusts, so the appropriate value of E is a short-term one.

1.5.2 The need for strength

In the last section we said that, during service, products are usually designed to deform within the elastic range. The range of loading within which deformations remain wholly elastic is revealed by carrying out a tensile test on the material of interest. Figure 1.13 shows graphs of stress against strain for the five materials being considered. For each material, other than PVC, the maximum stress up to which all deformations are elastic is indicated by the point at which the graph, according to some criterion, ceases to be linear. This is known as the **proof stress**. Values of the proof stress for all these materials except PVC appear in Table 1.3. The deformation of PVC is viscoelastic

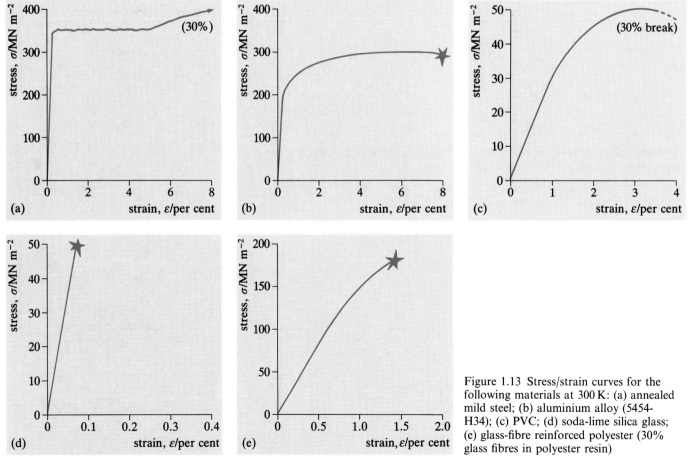

Figure 1.13 Stress/strain curves for the following materials at 300 K: (a) annealed mild steel; (b) aluminium alloy (5454-H34); (c) PVC; (d) soda-lime silica glass; (e) glass-fibre reinforced polyester (30% glass fibres in polyester resin)

whereby the strain increases with time under a constant load and disappears over a period of time on unloading. The viscoelastic stress–strain behaviour of PVC is essentially nonlinear (Figure 1.13c) up to the onset of yielding, so the value of the *short-term yield stress* of PVC, rather than a proof stress, is given in Table 1.3.

1.5.3 Putting in some numbers for stiffness

We shall now use the bending theory given earlier to calculate how thick the roof must be in order to have adequate stiffness. First, a criterion for 'adequate stiffness' must be decided. Suppose that under the maximum loading of $2200\,\mathrm{N\,m^{-2}}$, due to gusts of wind plus its own weight, the roof should not sag by more than, say, 2 cm. Although this might seem a large deflection for a rigid structure, it is actually quite conservative, being only 1/750th of the span.

We are dealing with a *distributed* load of, say, f per unit area, which we shall consider to be acting on a simply supported roof of span s and width w (see Figure 1.14). This loading sets up a bending moment, M, which varies across the span, reaching a maximum value of $fws^2/8$ at the centre (see ▼Bending moments from loads▲).

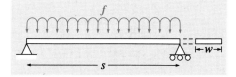

Figure 1.14 A simply supported panel of span s and width w acted on by a uniformly distributed load f

At each cross-section, Equation (1.10) tells us that the radius of bending, R, is equal to EI/M and therefore R also varies across the span. In cases in which R is *constant*, it is an easy matter to calculate the maximum deflection, or sag, from R (see ▼Deflections in bending▲). In this case, where R varies across the span, a complicated integration is required so we shall simply quote the result

$$\Delta_{\max} = \frac{5fws^4}{384EI} \tag{1.12}$$

As expected, $\Delta_{\max}$ is directly proportional to the distributed load, f, and inversely proportional to the flexural rigidity, EI. By rearranging this expression, the value of I required is

$$I = \frac{5fws^4}{384\Delta_{\max}E}$$

For the bus shelter, $f = 2200\,\mathrm{N\,m^{-2}}$; $s = 1.5\,\mathrm{m}$ and $\Delta_{\max} = 0.02\,\mathrm{m}$ so

$$I = \frac{5(2200)(1.5)^4 w}{384(0.02)E}\,\mathrm{m^4}$$

$$= \frac{7251w}{E}\,\mathrm{m^4} \tag{1.13}$$

The value of I/w required for each material being considered is given in Table 1.3. If the roof panel is flat, of width w and thickness t, it has a rectangular section and, therefore, according to Table 1.1, I is given by $wt^3/12$. Putting this into Equation (1.13)

$$t^3 = \frac{12 \times 7251}{E}\,\mathrm{m^3} \tag{1.14}$$

The panel thickness t required in each material is shown in Table 1.3.

The volume of material required can be reduced by redeploying some material away from the neutral axis — the RSJ principle. One common way is to use a corrugated section. With the exception of glass, all the materials are available in corrugated sheets. In glass they are not so readily produced, so a glass roof would have to be flat.

Let us choose a section of sinusoidal corrugations of depth d and 'wave length' $4d$, made from sheet of thickness t_c. In this case, provided t_c is small compared with d, (see Table 1.1)

$$I = wt_c d^2/36$$

If the proportions of the section are fixed at $t_c = d/10$, then $I = wd^3/360$. Putting this into Equation (1.13) you get

$$d^3 = \frac{360 \times 7251}{E} \text{ m}^3 \tag{1.21}$$

The depth, d, of corrugation for each material is also given in Table 1.3.

▼Bending moments from loads▲

Three-point bending

The bending moment, M, acting at a given cross-section of a beam can be derived from a knowledge of the external forces acting on the beam. First consider the case of a beam that is simply supported at its ends and loaded centrally by a force F — as in Figure 1.15(a). This is often called **three-point bending**.

In order to keep the beam in equilibrium the sum of the forces acting upwards on the beam must be equal to the load, F, and since the load acts at the mid-point of the beam, by symmetry, the force at each end support must be $F/2$.

If we take all the forces to one side of a particular cross-section of the beam, the sum of their moments about that section is the **bending moment**. Consider a cross-section S situated at an arbitrary distance x from the left-hand support. To the left of S, there is only one external force, $F/2$, acting on the beam and the moment of this force about S is

moment = vertical force
$\times$ horizontal distance from S

The bending moment, M, is therefore

$$M = \frac{F}{2} x \tag{1.15}$$

Thus the bending moment increases linearly with the distance, x, from the left-hand support, attaining a maximum value at the centre, where $x = s/2$. Hence the maximum bending moment is $Fs/4$.

From the symmetry of the beam, the bending moment must also increase linearly with the distance from the right-hand support, again reaching a maximum of $Fs/4$ at the centre.

This variation in M along the length of the beam gives the triangular distribution shown graphically in the bending-moment diagram (Figure 1.15b).

Distributed load

Now suppose instead of a point load at the centre of the beam you have a load of f per unit area distributed uniformly over the top surface. If the length of the beam is again s and its width is w (see Figure 1.16a), the area of the top surface will be ws and the total load is fws. The forces at each support must therefore be $fws/2$.

To the left of section S there is now an upward force $fws/2$ from the support and a downward force fwx from the

distributed load. Since the downward force, fwx, is uniformly distributed, we shall take it to act mid-way between the left-hand support and S. That is, at a distance $x/2$ from S.

The force $fws/2$ from the left-hand support acts to rotate this portion of the beam clockwise about S, whereas the distributed load fwx acts in the opposite sense. Therefore the net clockwise bending moment is

$$M = \frac{fws}{2} x - fwx \frac{x}{2}$$
$$= \frac{fw}{2}(sx - x^2) \tag{1.16}$$

This is clearly a non-linear distribution which attains a maximum value of $fws^2/8$ at the centre of the beam, where $x = s/2$ — as shown in the bending-moment diagram (Figure 1.16b).

Four-point bending

Finally, consider a simply supported beam that is loaded with two equal forces F, applied at a distance d from each end (Figure 1.17a). This is often called **four-point bending**.

By symmetry, to maintain equilibrium the force acting on the beam at each support must be equal to F.

First, consider cross-sections situated between the left-hand support and the left-hand load. The only force acting to the left of any such section is the supporting force, F, and the bending moment along this part of the beam is therefore

$$M = Fx \qquad (1.17)$$

where x is the distance from the left-hand support.

At the loading point, where $x = d$, the bending moment is Fd.

For sections between the two loading points, the left-hand supporting force has a clockwise moment, Fx, and the left-hand load has an anti-clockwise moment $-[F(x - d)]$. The bending moment, M, between the two supports is the sum of these moments.

$$M = Fx - [F(x - d)]$$
$$= Fd \qquad (1.18)$$

This is independent of x and so the bending moment between the two loading points is constant. Thus, by using this type of loading, a large proportion of a beam may be subjected to a uniform load and, for this reason, four-point loading is often used in laboratory tests.

SAQ 1.4   (Objective 1.4)
A diving board of length L is loaded at its free end by a vertical force F. Write down an expression for the bending moment M at a section of the board a distance x from the free end. Where is M largest and what is the expression for this maximum value?

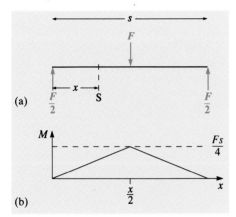

Figure 1.15 A simply supported, *centrally* loaded beam and its bending-moment diagram

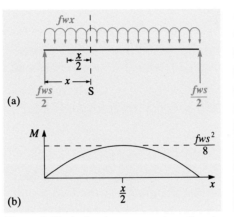

Figure 1.16 A simply supported, *uniformly* loaded beam and its bending-moment diagram

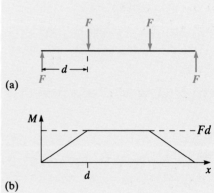

Figure 1.17 A beam in symmetrical four-point bending and its bending-moment diagram

▼Deflections in bending▲

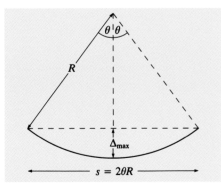

Figure 1.18 A beam of span s bent into a circular arc of radius R

Engineers' bending theory (Equation 1.10) gives the radius, R, of a section of beam that is subjected to a bending moment, M. Often you would prefer to know the vertical deflection Δ of the beam at points along its length. Consider the case of a length s of beam acted on by a uniform

bending moment M. Equation (1.10) shows that the beam is bent to a constant radius R, where

$$R = EI/M \qquad (1.19)$$

The maximum vertical deflection Δ_{max} occurs at the centre and can be related to R by simple geometry (see Figure 1.18).

$$\Delta_{max} = R - R \cos \theta$$

If θ is small, $\cos \theta \approx 1 - \theta^2/2 + \dots$

$$\Delta_{max} = R(1 - \cos s/2R)$$
$$= R(1 - \cos s/2R)$$
$$\simeq R[1 - (1 - (s/2R)^2/2)]$$
$$= s^2/8R$$

From Equation (1.19)

$$\Delta_{max} = \frac{s^2 M}{8EI} \qquad (1.20)$$

Since the bending moment, M, is constant, this expression would apply to the deflection between the supports of a beam in four-point bending, the deflection being measured relative to the loading points.

SAQ 1.5   (Objectives 1.4 and 1.5)
A rectangular glass test-piece ($22 \times 5 \times 2\,mm^3$) is loaded in symmetrical four-point bending through its thickness with two equal loads of 30 N. The supports are 20 mm apart and the loading points are 10 mm apart. Calculate the maximum stress and deflection in the bar, assuming that the deformation is elastic. Take the value of Young's modulus from Table 1.3 and the second moment of area from Table 1.1.

We want now to compare the volumes of the equivalent flat and corrugated roof panels. The volume v_F of a flat panel of thickness t and length L is simply twL. For a corrugated panel of depth d and thickness $d/10$ the volume, v_C, is $0.115dwL$ (see ▼Volume of a corrugated sheet▲). The ratio of these volumes is therefore

$$\frac{v_C}{v_F} = \frac{0.115d}{t}$$

Substituting for t and d from Equations (1.14) and (1.21)

$$\frac{v_C}{v_F} = 0.115 \left(\frac{360}{12}\right)^{1/3}$$

$$= 0.36$$

The volume, and therefore the weight and the cost of material, is reduced by 64%, without loss of stiffness, by changing from a flat to a corrugated shape.

This is a graphic illustration of the fourth proposition in Section 1.1.

1.5.4 Putting in some numbers for strength

The designs we have produced have adequate stiffness, but do they have adequate strength? We need to calculate the maximum stress that would be set up in a flat and a corrugated roof by the worst case of loading ($f = 2200\,\mathrm{N\,m^{-2}}$) and compare this with the strength of the material (Table 1.3). To do this, once again we need Equation (1.10), which for the maximum bending moment is

$$\frac{M_{max}}{I} = \frac{\sigma_{max}}{y_{max}}$$

From 'Bending moments from loads', the maximum bending moment for a distributed load is $M_{max} = fws^2/8$. The maximum stress, σ_{max}, occurs at the surfaces, so the distance from the neutral axis to the surface, y_{max}, is $t/2$ for the flat sheeting and $d/2$ for the corrugated sheeting. Making these substitutions for the flat panel gives

$$\frac{wfs^2/8}{I} = \frac{\sigma_{max}}{t/2}$$

This can be rearranged to give the maximum stress in a flat panel of

$$\sigma_{max(F)} = \frac{wfs^2 t}{16I} \tag{1.22}$$

Taking as an example the values of I/w and t for mild steel that appear in Table 1.3,

$$\sigma_{max(F)} = \frac{(2200)(1.5)^2(7.5 \times 10^{-3})}{16(35 \times 10^{-9})}$$

$$= 67\,\mathrm{MNm^{-2}}$$

▼Volume of a corrugated sheet▲

Consider a corrugated sheet of length L, width w and thickness t. If the length of a half wave is s (see Figure 1.19), the volume of this half wave is simply Lts.

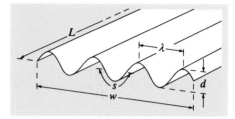

Figure 1.19 If the wave length of a corrugated sheet is λ then in a panel of width w there will be w/λ waves — and $2 \times w/\lambda$ half waves

Since there are $2w/\lambda$ half waves, the total volume of the panel is

$$v_C = Lts \times 2w/\lambda$$

From Table 1.1

$$s = \frac{\lambda}{2}\left[1 + \left(\frac{\pi d}{2\lambda}\right)^2\right]$$

The wave length of the corrugation in the roof panel is four times the depth (i.e. $\lambda = 4d$) and therefore

$$s = 1.15\lambda/2$$

Putting this into the expression for the total volume of the panel

$$v_C = Lt\left(\frac{1.15\lambda}{2}\right)\frac{2w}{\lambda}$$

$$= 1.15Ltw$$

In this case $t = d/10$, so

$$v_C = 0.115Ldw$$

This stress is less than a fifth of the proof stress of mild steel so the strength requirement is met with a large safety factor of 5.2. The maximum stresses in the other materials also lie well below the proof stresses given in Table 1.3.

The maximum stress in the corrugated panel, $\sigma_{max(C)}$, is found by replacing t in Equation (1.22) by d. Thus

$$\sigma_{max(C)} = \frac{wfs^2 d}{16I} \qquad (1.23)$$

Using the values of I/w and d for mild steel given in Table 1.3,

$$\sigma_{max(C)} = \frac{(2200)(1.5)^2(23 \times 10^{-3})}{16(35 \times 10^{-9})}$$

$$= 203\,\text{MNm}^{-2}$$

This value is larger than that of the flat panel because the corrugated panel is deeper and the stress rises linearly with depth, but there is still a reasonable factor of safety (1.7). These comments also apply qualitatively to the corrugated panels in the other materials.

SAQ 1.6 (Objectives 1.4 and 1.5)
For each type of panel (flat and corrugated), in each of the materials (mild steel, aluminium alloy, PVC, glass, GFRP), calculate the maximum sagging deflection Δ_{max} produced when the panel is loaded uniformly to a level that gives a safety factor for strength of two. That is, the material is loaded to a maximum stress equal to 50% of its strength.

SAQ 1.7 (Objective 1.6)
For the corrugated design in each material, calculate the mass and cost per square metre of roof and compare each value with that required in the design specification.

1.6 Implications of the designs

The designs demonstrate the importance of *form*, as well as the material properties, in determining the performance of a product. By changing from a rectangular to a corrugated section, the volume of material required was reduced to about a third. It is important to realize that improvements in performance may be achieved either by changes of material or by changes in form and, in practice, often come from a combination of both.

In this application, there were two independent mechanical requirements — for stiffness and for strength. In the event, stiffness was the critical factor and, in this case, by designing the sections to have sufficient stiffness, you found that they had more than enough strength. This particular design task is 'stiffness limited', but it need not always be the case. Another task may be 'strength limited', but you have to go through the design calculations to find out which requirement is the limiting one.

For the design of the bus shelter, the corrugated panels would be preferred to the flat ones in all the materials except glass because the corrugated design requires much less material. The thicknesses required are feasible. In practice, deeper corrugations than those calculated are often used and, in metal sheets, sometimes they have a profile that is the shape of a trapezium rather than being sinusoidal. The use of a deeper section simply brings a larger factor of safety.

A flat glass panel and a corrugated steel panel would exceed the weight limit prescribed in the design specification, but the panels in the other materials would have weights well within the limit. The GFRP and the aluminium-alloy panels are the lightest. However, the steel panel is the cheapest by far.

For a glass roof panel, flat sheets 11 mm thick would be required. These would usually contain a grid of metal wires to 'reinforce' the glass. Actually, since the wires are situated near the neutral surface of the panel, they do little or nothing to 'reinforce' the glass — that is, to increase its Young's modulus or strength. The purpose of the wires is to hold the pieces of glass together if the panel is cracked.

Provided that the mild-steel panel was galvanized with a thin coating of zinc to protect it against rusting and the PVC contained an ultraviolet stabilizer, all the materials should meet the lifetime requirement of the specification.

1.7 Summing up

In this short chapter, we have introduced the themes and structure of the book, we have developed some aspects of the theory of stress, strain and elasticity, and we have applied some of these ideas to a design task involving a range of disparate materials. This theory will be called upon throughout the book.

Our analysis of the roof has, of necessity, been a slimmed-down and simplified version of the real thing, so it is encouraging to find that it produces answers of a plausible order of magnitude. One simplification was to assume that the roof panel was simply supported, when in practice it would be fixed to its supporting walls. The constraint introduced by fixing could significantly reduce the thickness of material required to achieve a given level of stiffness. The value assumed for the maximum deflection of the roof and the use of the same value for all materials were both arbitrary. A stress-based criterion would yield different deflections, as SAQ 1.6 showed. Finally, we only considered one type of loading — that due to short-term gusts of wind. Others, such as thermal stresses due to changes in temperature, or impact loads due, for example, to vandalism, might well be significant. For a viscoelastic material, like PVC, steady loading due to a thick covering of snow for several days could be the most severe loading. Any design represents a compromise between performance and cost, while aesthetic considerations might predominate, even in a bus shelter.

The next seven chapters are each devoted to a different class of material. Among them are included the five materials considered for the roof of the bus shelter: mild steel (Chapter 3), aluminium alloys (Chapter 2), PVC (Chapter 5), glass (Chapter 4) and GFRP (Chapter 7).

Objectives for Chapter 1

After studying this chapter, you should be able to:

1.1 Give a commonplace example of a body subjected to each of the following types of stress: (a) uniaxial, (b) biaxial, (c) triaxial, (d) hydrostatic.

1.2 Define Poisson's ratio and relate it to the changes in shape and volume that occur during elastic deformation under a uniaxial load.

1.3 Recognize and apply equations of equilibrium and compatibility, and the constitutive equation for a linear elastic material, in each of the following cases of loading: (a) uniaxial, (b) bending, (c) torsion.

1.4 Calculate the bending moment and maximum deflection in a simply supported beam subjected to either a point or a distributed load.

1.5 Define the term 'second moment of area', and explain how it influences the stiffness and strength of a body subjected to bending.

1.6 Evaluate simple designs of loaded products in terms of mass and cost.

Answers to self-assessment questions

SAQ 1.1 The area of the section, $A = w^2$ and, by differentiation, $\Delta A = 2w\Delta w$. Therefore

$$\frac{\Delta A}{A} = \frac{2w\Delta w}{w^2} = \frac{2\Delta w}{w}$$

For an isotropic material

$$\text{Poisson's ratio, } v = -\frac{\Delta w/w}{\Delta l/l}$$

Therefore

$$\frac{\Delta w}{w} = -v\frac{\Delta l}{l}$$

$$= -v\varepsilon$$

Thus

$$\frac{\Delta A}{A} = -2v\varepsilon$$

SAQ 1.2

(a) The axial strain is given by the displacement divided by the length of the bar. Thus

$$\varepsilon = \frac{\Delta l}{l}$$

$$\varepsilon = \frac{0.5\,\text{mm}}{800\,\text{mm}}$$

$$\varepsilon = 0.000625$$

This is a direct strain and it is uniform over the cross-section of the bar.

(b) The shear strain varies linearly from a maximum on the outer surface to zero at the centre. The maximum value is given by

$$\gamma = \frac{r\theta}{l}$$

$$\gamma = \frac{2\,\text{mm}}{800\,\text{mm}} \times \frac{3° \times 2\pi\,\text{rad}}{360°}$$

$$\gamma = 0.000131$$

(c) The bending strain (a direct strain) is zero at the centre of the bar and has equal maximum values, but of opposite signs, at the top and bottom surfaces. The strain, ε, varies linearly with the distance, y, from the neutral surface and is given by

$$\varepsilon = \frac{y}{R}$$

$$\varepsilon = \frac{2\,\text{mm}}{4000\,\text{mm}}$$

$$\varepsilon = 0.0005$$

SAQ 1.3

(a) The strain is uniform across the section and so will be the stress, which is given by

$$\frac{\sigma}{\varepsilon} = E$$

$$\sigma = 210\,\frac{\text{GN}}{\text{m}^2} \times 0.000625$$

$$= 131\,\text{MN}\,\text{m}^{-2}$$

This is a direct stress.

(b) In order to calculate the shear stress a value of shear modulus is required

$$G = \frac{E}{2(1+v)}$$

$$= \frac{210}{2(1.3)}\,\text{GN}\,\text{m}^{-2}$$

$$\simeq 81\,\text{GN}\,\text{m}^{-2}$$

The maximum shear stress is on the outside surface of the bar and is given by

$$\tau = G\gamma$$

$$= 81\,\frac{\text{GN}}{\text{m}^2} \times 0.000131$$

$$\simeq 11\,\text{MN}\,\text{m}^{-2}$$

(c) These are tensile and compressive, that is, direct stresses of equal value.

$$\sigma = E\varepsilon$$

$$= 210\,\frac{\text{GN}}{\text{m}^2} \times 0.0005$$

$$= \pm 105\,\text{MN}\,\text{m}^{-2}$$

SAQ 1.4 Consider an arbitrary section S situated a distance x from the free end of the board (Figure 1.20). To the right of S there is only one force, F, which is acting vertically. The bending moment M at S is

$$M = \text{vertical force} \times \text{horizontal distance}$$

$$= Fx$$

M is a maximum when x is a maximum: that is, at $x = L$, the fixed end of the board. Therefore

$$M_{max} = FL$$

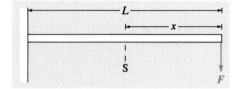

Figure 1.20 The cantilever referred to in SAQ 1.4

SAQ 1.5 The maximum bending moment, M_{max}, is constant over all sections between the two loading points (Figure 1.17) and, from Equation (1.18), it has the value

$$M_{max} = Fd$$

$$= 30 \times 5 \times 10^{-3}\,\text{N}\,\text{m}$$

$$= 150 \times 10^{-3}\,\text{N}\,\text{m}$$

From Equation (1.10)

$$\frac{M_{max}}{I} = \frac{\sigma_{max}}{y_{max}}$$

Rearranging

$$\sigma_{max} = \frac{y_{max}M_{max}}{I}$$

From Table 1.1, $I = wt^3/12$ and therefore

$$\sigma_{max} = \frac{M_{max}t/2}{wt^3/12}$$

$$= \frac{150 \times 10^{-3} \times 2 \times 10^{-3}/2}{5 \times 10^{-3} \times (2 \times 10^{-3})^3/12}$$

$$= 45\,\text{MN}\,\text{m}^{-2}$$

The deflection is given by Equation (1.20)

$$\Delta_{max} = \frac{s^2 M}{8EI}$$

where s is the span over which M is constant. In this case, $s = 10\,\text{mm}$ and therefore

$$\Delta_{max} =$$

$$\frac{(10 \times 10^{-3})^2 \times 150 \times 10^{-3}}{8(71 \times 10^9)(5 \times 10^{-3})(2 \times 10^{-3})^3/12}$$

$$= 0.008\,\text{mm}$$

SAQ 1.6 If the proof stress is σ_p and there is a safety factor for strength of two then the maximum stress in a panel must be $\sigma_p/2$.

Flat panels
Recalling Equation (1.22) and equating the maximum stress to $\sigma_p/2$

$$\sigma_p/2 = \left(\frac{wfs^2}{16I}\right)t$$

From Equation (1.12)

$$\Delta_{max} = \frac{5wfs^4}{384IE}$$

$$= \left(\frac{wfs^2}{16I}\right)\frac{5s^2}{24E}$$

Combining the last two equations, the deflection of the flat panels is

$$\Delta_{max(F)} = \left(\frac{\sigma_p}{2t}\right)\frac{5s^2}{24E}$$

$$= \left(\frac{\sigma_p}{Et}\right)\frac{5(1.5)^2}{48}$$

$$= 0.23\left(\frac{\sigma_p}{Et}\right)$$

Taking the values of σ_p, E and t from Table 1.3, we get the values of $\Delta_{max(F)}$. These are given in Table 1.4.

Under these design conditions, only the glass panel is stiff enough. The other flat panels, particularly those in PVC and GFRP, would vibrate alarmingly in a gale!

Corrugated panels
The line of reasoning is the same as that employed for flat panels. From Equation (1.22)

$$\sigma_p/2 = \left(\frac{wfs^2}{16I}\right)d$$

Combining this with Equation (1.12), as above, gives the maximum deflection for the corrugated panels.

Table 1.4

	$\sigma_p/\mathrm{MN\,m^{-2}}$	$E/\mathrm{GN\,m^{-2}}$	t/mm	d/mm	$\Delta_{max(F)}/\mathrm{mm}$	$\Delta_{max(C)}/\mathrm{mm}$
MS	350	210	7.5	23	51	17
AA	200	71	11	33	61	20
PVC	50	3.0	31	95	124	40
GL	50	71	11	–	15	–
GFRP	180	13	19	59	168	54

Table 1.5

Material	$\rho/\mathrm{Mg\,m^{-3}}$	d_c/mm	$m/\mathrm{kg\,m^{-2}}$	Cost/$\mathrm{£\,m^{-2}}$
MS	7.9	23	20.9	6.3
AA	2.8	33	10.6	13.8
PVC	1.45	95	15.8	15.8
GFRP	1.5	59	10.2	17.3
GL(flat)	2.5	($t = 11$)	25.0	22.5

$$\Delta_{max(C)} = \left(\frac{\sigma_p}{2d}\right)\frac{5s^2}{24E}$$

$$= 0.23\left(\frac{\sigma_p}{Ed}\right)$$

For all the materials, except glass, of course, the values of the maximum deflections in these panels are given in Table 1.4. They are much smaller than those of the flat panels, but the deflections of the PVC and GFRP panels well exceed the value specified (20 mm).

SAQ 1.7 First, you must work out the volume of unit area of corrugated panel. Consider one 'wavelength' λ, of corrugation. The length $2s$ can be found using Table 1.1 and by multiplying this by the sheet thickness t_c, you get the volume of unit length of a corrugation.

$$v = \lambda t_c\left[1 + \left(\frac{\pi d}{2\lambda}\right)^2\right]$$

Unit width of panel will contain $1/\lambda$ such corrugations, so the volume of unit area of panel is

$$v = \left(\frac{1}{\lambda}\right)\lambda t_c\left[1 + \left(\frac{\pi d}{2\lambda}\right)^2\right]$$

The mass m of this panel is obtained by multiplying by the density ρ.

$$m = \rho t_c\left[1 + \left(\frac{\pi d}{2\lambda}\right)^2\right]$$

For the panels considered, $\lambda = 4d$, so

$$m = \rho t_c\left[1 + \left(\frac{\pi}{8}\right)^2\right]$$

$$= 1.15\rho t_c$$

Since $t_c = d_c/10$

$$m = 0.115\rho d_c$$

The values of m for each panel are given in Table 1.5.

The mass of the mild-steel panel just exceeds the limit specified, while the masses of the other corrugated panels are well within the limit. The GFRP panel is the lightest, by a small margin. Note that the mass of the flat glass panel would exceed the limit by a significant amount.

The cost of unit area of panel is obtained simply by multiplying the mass per square meter, m, by the cost per kilogram, which is given in Table 1.3. The values obtained appear in Table 1.5 and show that a corrugated-iron roof would be the cheapest.

Chapter 2
Nonferrous Metals

by Nick Reid   (consultant: Stan Pugh)

Chapter 2 Nonferrous metals

2.1 The pound in your pocket

Barring penury, we are rarely without nonferrous metals — metals and alloys other than those based on iron. We all carry coins around in our pockets and purses, and, in the United Kingdom, these are invariably made from nonferrous metals and have been so for many hundreds of years. Why is this? Is it simply a habit or tradition? Certainly, the first metals to be found and used by man were those few metals existing in 'native' form: gold, copper and possibly silver. The only sources of native iron were the small remnants of meteorites that had survived the burn-up on entering the Earth's atmosphere. Not surprisingly, then, the first coins were based on these nonferrous metals. As recently as the nineteenth and the early twentieth century, gold was used as the basis of international trade and it was South African and, to a lesser extent, Australian gold that provided for the economic viability of the British Empire. When gold was subsequently replaced in economics by the concept of 'confidence', its value fell to low levels, but in recent decades, with loss of faith in international monetary systems, the value of gold has revived, making investment in gold popular again. The fact that metals such as gold do occur as native metals is an indication that they are chemically stable — an important requirement for a coin.

We now want to consider the technical requirements of coinage materials and to see to what extent they are met by the property profile of nonferrous metals.

2.2 Materials for coinage
2.2.1 The design specification of a coin

The properties required of modern coinage may be grouped under five headings.

1 *Cost* Modern coins are debased in that the cost of material is below the face value of the coin. Otherwise, 'entrepreneurs' may melt down the coins to obtain the material from which they are made. On the other hand, the cost of the finished coin must not be too low as this would be a temptation to make counterfeit coins.

2 *Formability* The coin must be capable of being formed by economical processes. The production route used by the Royal Mint at Llantrisant is:

• melting and continuous casting;
• cold rolling, to form sheets the thickness of a coin;
• blanking — punching out coin-sized discs;

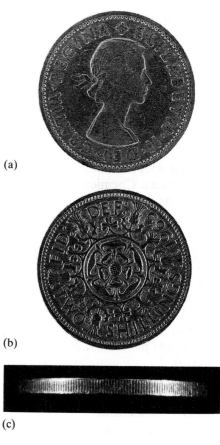

(a)

(b)

(c)

Figure 2.1 A Royal Mint ten-pence piece: (a) obverse face; (b) reverse face; (c) edge

- annealing and pickling, to soften and clean the discs;
- rimming — forming a raised edge on the discs in order to enable coins to be stacked and to protect the faces from damage;
- coining — stamping the faces of the coin between two shaped dies to produce the 'head' and 'tail' in relief.

This route has been followed in principle for hundreds of years; since 1552 in France and since 1662 at the Royal Mint, but with many advances in the speed and quality of production over the intervening period. Coining is the most demanding stage of production for the material. A coining press runs at speeds of up to 400 coins per minute and therefore the material must be capable of shear deformation at high rates without the coins becoming cracked or subjecting the dies to excessive wear. This requires a material that is ductile and not too hard.

Wear of the dies also depends on the die profile used. Products of the Royal Mint have sharply outlined obverse and reverse designs and the dies used have designs and inscriptions with sharp angles. The lives of these dies are in the region of 100 000 pieces. In the USA, coins have softer designs whose inscriptions have rounded corners and the life of a die often exceeds 500 000 pieces.

3 *Durability* Coins should survive twenty-five to thirty years of use, which calls for a material with good resistance to brittle fracture, wear and oxidation. To meet this requirement a range of properties, both mechanical and chemical, must be considered, and there is some conflict here with the previous requirement. To avoid brittle fracture, the material must have a high toughness. To resist wear during handling calls for a high value of hardness (see ▼Wear▲), but this would cause wear of the dies during coining. Compromise is called for on hardness. Normally, the Royal Mint prefers annealed blanks to have a hardness of about 85 H_V from surface to core; this rises to over 100 H_V for the struck pieces. Chemical stability is required under the wide range of conditions that may apply: in high-humidity or coastal atmospheres; in contact with perspiration, body salts or perfumes.

4 *Electrical conductivity* Coins are used to operate automatic vending machines. All such machines contain a device that selects or rejects coins and depends for its operation on coins being made of materials that are good electrical conductors, but at the same time, are not ferromagnetic. When a coin is inserted it rolls down an incline and is made to pass through a magnetic field. As the coin moves through the field it cuts perpendicularly through the lines of magnetic flux. Electric currents, known as eddy currents, are thus induced in the coin. In effect, some of the kinetic energy of the coin is transformed into electrical energy (eddy currents) and, thence, into heat. As the coin loses kinetic energy it slows down.

The speed at which the coin emerges from the detector determines whether the machine accepts or rejects it. This speed depends on the net force acting on the coin; the accelerating force depends on the coin's

(a)

(b)

(c)

Figure 2.2 A US 'quarter': (a) obverse face; (b) reverse face; (c) edge

weight and, hence, the density of the material used, while the retarding force depends on the total eddy current generated, which in turn depends on the material's resistivity.

5 *Appearance* Modern coins, although debased, are made to resemble traditional coins in colour, shape, size and weight, so as to retain the faith of the public in their monetary value. The traditional materials were based on copper (density $9.0 \, \text{Mg m}^{-3}$), silver ($10.5 \, \text{Mg m}^{-3}$) and gold ($19.3 \, \text{Mg m}^{-3}$).

▼Wear▲

'Wear' is normally thought of as the loss of material from surfaces that have been rubbed against one another and it is often measured in terms of the mass lost in a given time under specified conditions. More precisely, wear involves a redistribution of material which adversely alters the surface.

The first point to make is that surfaces are never truly flat. They consist of 'hills' and 'valleys' such as those shown in Figure 2.3, a profile of the surface of shot-blasted

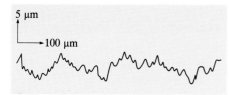

Figure 2.3 The topography of a shot-blasted steel surface obtained using a profilometer — a device with a fine stylus that transverses the surface. Notice the different horizontal and vertical scales

steel. When two such surfaces are brought together, contact occurs only where 'hills' on one surface touch those on the other. If the surfaces are pressed together, the points of contact will deform and, if the force applied is small, the deformations will be elastic. For metal surfaces, as the force increases, the points of contact will become flattened by plastic deformation and the area of contact is also increased. It is found that, for a particular material, if the force acting normally on the surfaces is N and the area of contact is A_c, the average contact pressure, N/A_c, is a constant. It is known as the flow pressure,

P, and is approximately equal to the indentation hardness H_v (see Chapter 3, 'Hardness measurements'). In other words, as N increases, the area of contact A_c increases in proportion and

$$A_c = N/P \qquad (2.1)$$

This is usually much smaller than the *apparent* area of contact, A. To explain friction, it is postulated that some *welding* of the two surfaces occurs within the contact area and that a frictional force, F, is needed in order to fracture the welds, which continually reform as sliding progresses. It follows that the shear force, F, is equal to the shear stress required to break the junctions, τ, multiplied by the total area of the junctions, assumed to be fA_c, a fraction f of the contact area A_c. Thus you have

$$F = \tau f A_c$$

where τ, the shear stress to break the junctions, is called the shear strength.

From Equation (2.1), you can substitute N/P for A_c to get

$$F = \left(\frac{\tau f}{P}\right) N$$

This is regarded as a law of friction with the term in brackets called the coefficient of friction, μ.

The largest value this can have is when the whole *contact* area is welded and $f = 1$. Then taking the shear strength, τ, when this occurs to be half the flow pressure, P, you get $\mu = 1/2$. Very clean metal surfaces may approach this value, but more commonly the existence of surface films of oxide or grease reduces f and lower values of μ are observed.

Adhesive wear

The mechanism of adhesive wear follows directly from the idea of two welded surfaces. When the top surface is made to slide over the bottom surface, material breaks away and does so at the weakest sections. Figure 2.4 shows three cases of surfaces in contact. Case (a) involves two similar metals and fracture occurs in both materials. The junctions at which the surfaces are in contact have been strengthened by work hardening and therefore the fractures take place within the materials, at some distance away from the interfaces between the points of contact. Each surface tears out some material from the other and both surfaces become roughened as they gouge and score one another. Wear is rapid and, for this reason, sliding combinations of similar metals are usually avoided in good engineering practice. In case (b), rupture tends to occur within the softer of the two materials and the harder material picks up some of the soft. Wear is largely confined to the softer material. In case (c) rupture occurs at a weak interface, for example, between oxide films, and little wear or surface damage occurs.

Abrasive wear

In the mechanism for abrasive wear a hard particle in one surface indents, grooves and then cuts material from the other surface. In service, the main cause of abrasion between sliding metals is the presence in one of the two surfaces of particles of hard materials, such as carbide in steels, work-hardened wear fragments or hard oxide films. The particles may also be airborne 'dirt' such as grit. The

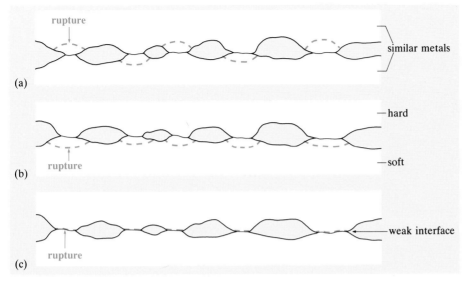

Figure 2.4 A sketch of the contact between two clean metallic surfaces and the fracture paths taken (broken lines) when the junctions are ruptured by sliding. (a) Similar metals in contact — rupture may occur under either surface. (b) Dissimilar metals — rupture occurs beneath the surface of the softer material. (c) A weak interfacial film — rupture occurs at the interface

Figure 2.5 Fatigue cracks on the race surface of a rolling bearing

hardness of the metal in contact with these particles controls the depth of penetration, up to a point. Generally, the harder the surface, the less the penetration and, therefore, the lower the abrasive wear. If the hardness is too high, though, the material may be so brittle that large flakes of it are detached by the abrading particles and wear is rapid.

Wear caused by fatigue

When there is relative motion between two surfaces in contact the state of stress at any given point on or near the surfaces varies with time and this may cause fatigue — the slow growth of cracks. The development of such cracks may eventually detach pieces of material from the surfaces, thereby constituting wear. For example, in a rolling bearing, balls or rollers traversing the surface continually apply a normal force. Fatigue cracks may form below the surface at points where the shear stress is greatest, or where there is some heterogeneity, such as a brittle inclusion particle. These cracks initially grow parallel to the surface, but later they veer towards the surface, detaching a fragment of material (Figure 2.5) and leaving a cavity (Figure 2.6). Fatigue wear is also thought to occur between transiently touching asperities in lubricated sliding.

Chemical or corrosive wear

Finally, there are wear processes associated with chemical effects. The most common example with metals is the repeating cycle of formation, removal and reformation of oxides. The worn fragments here are usually very small, but may appear as compacted flakes consisting of a mixture or 'fudge' of fine oxide and metal particles. Wear may also occur due to corrosion of the metal surface by acid in the lubricating oil.

SAQ 2.1   (Objective 2.3)
A coin of mass 10 g and a flow pressure of $1 \, GN \, m^{-2}$ rests on a flat, glass table top. Make a rough estimate of the actual area of contact. Would you expect the coefficient of friction to have a high or low value?

Figure 2.6 A cavity formed by the spalling of material

2.2.2 The materials used

Nonferrous metals have virtually a world-wide monopoly on coinage. Although this fact may owe something to tradition, it is also a recognition of the suitability of the property profiles of these materials.

Actually, pure metals are never used because they lack the required hardness, which should be of the order of $100\,H_V$. Coins are therefore made of alloys in which other elements have been added to increase the hardness by solution hardening. Additions of 2.5% by weight of zinc and 0.5% by weight of tin to copper form 'coinage bronze' (density $8.9\,Mg\,m^{-3}$) which is used in British copper-coloured coins. An addition of 25% by weight of nickel to copper forms 'cupronickel' in silver-coloured coins (density $9.0\,Mg\,m^{-3}$). These alloys are similar to copper and silver in terms of colour and feel.

These solid-solution alloys, although hardened, still have enough ductility to be formed and enough toughness to resist brittle fracture. They also have acceptable chemical stability. Under normal atmospheric conditions, copper and nickel are protected from further oxidation by passive oxide layers on the surface.

Finally, these alloys are acceptable on the criterion of cost. Their good formabilities enable them to be manufactured by an economical route and their material costs are not too large. In 1989 the cost per kilogram was of the order of £1.7 for copper and £6.9 for nickel. It is not difficult to see why periods of inflation have eliminated silver, at £150 per kilogram, and gold, at £8000 per kilogram, as materials for coinage.

SAQ 2.2   (Objective 2.1)
Based on what you have read so far, give up to two objections to choosing each of the materials in Table 2.1 as a replacement for a silver coinage alloy.

Table 2.1   Properties of potential coinage materials

Material	Density/$Mg\,m^{-3}$	Resistivity/$10^{-8}\,\Omega\,m$	Hardness, H_V	Cost/£ kg^{-1}
mild steel	7.9	15	140	0.3
stainless steel 301 (non-magnetic)	8.1	72	180	1.5
titanium	4.5	55	90	20

Coinage is one of the oldest surviving applications of nonferrous metals and it brings out the virtues of these materials — that, at reasonable cost, they provide an attractive property profile: they are ductile, tough, formable, electrically conducting materials with good chemical stability. As you will soon see, they also have the potential for development into high-strength materials. It is interesting to contemplate the extent to which nonferrous metals, in the form of coinage, lubricate trade and commerce and thereby underpin the very foundations of our society, credit cards and folding money notwithstanding!

2.3 Nonferrous metals and the Periodic Table

Many metals are chemically reactive and, in making chemical bonds with other atoms, they *donate* electrons, thereby becoming positive ions — they are thus said to be electropositive elements. Metals are found in the vertical columns to the left and centre of the periodic table, have high thermal and electrical conductivities, and usually have high ductility and the ability to form alloys of variable composition.

On the basis of this description, about eighty of the elements, a big majority, are metals. Of these, iron is a special case because it accounts for over 90% of the Western World's total output of new metals per year (see Table 2.2).

Iron's importance is recognized by devoting an entire branch of metallurgy to it and this will be the theme of Chapter 3; the other metallic elements are the subject of this chapter. They are all more expensive than iron, so they are used when they have properties, including aesthetic ones, that a cheaper iron alloy would not have: for example, the high electrical and thermal conductivities of copper, silver, gold, aluminium and the alkali metal, sodium, are utilized in electrical and nuclear engineering, electronics and telecommunications; the resistance to corrosion and consequent good appearance of gold, silver and platinum make them attractive for jewellery and for hoarding; the low neutron-capture cross-section (a measure of a material's probability of absorbing neutrons) of beryllium, zirconium, magnesium and aluminium are exploited in nuclear reactors; for their high strength to density ratios, aluminium, magnesium and titanium are used widely by the aerospace industry; so too, are nickel-based alloys, on account of their strength and stability at the service temperatures of gas turbines (up to 1400 K).

In a mere chapter, we haven't the space to consider further the great multiplicity of nonferrous metals, so we shall concentrate on a few representative examples. The factors that have to be considered in assessing nonferrous metals for service are best illustrated by looking at some specific examples of where these materials have been given load-bearing duties. This will also involve considering the principles by which the properties of a nonferrous metal can be improved.

The examples chosen have been selected to cover a range of service conditions. The first considers the choice of material for the fuselage skin of the first jet-propelled commercial airliner, the Comet, while the second considers how the choice is affected if the airliner is to travel supersonically. The third and last example involves the choice of material for the 'hot spot' of a gas turbine engine.

Table 2.2   Average annual output, in tonnes, of new metal by Western World, 1975–85

Fe	Cu	Al	Ni	Ti	U	Ag	Au	Pt
3×10^8	6×10^6	5×10^6	5.5×10^5	3×10^4	2.5×10^4	8000	1250	80

2.4 Choosing material for the skin of the Comet

The Comet, the first turbojet powered airliner to enter airline service, was conceived in 1946 by the de Havilland Company. It was designed to achieve a greatly increased performance by flying at a cruising speed some 50% higher than that of the American and Canadian piston-engined airliners then entering service. At that time, a trans-Atlantic flight took a rather tedious fifteen hours and such an increase in speed, it was believed, would be attractive to both airlines and travellers, a view that was later amply confirmed.

In order to have acceptable fuel economy at high speed, it is necessary to fly at high altitude where the air density and, hence, aerodynamic drag are reduced. The Comet was therefore designed to cruise at a height of 10 700 m, double that of its contemporaries. This is well above the top of Mount Everest and, at this altitude, an extra supply of oxygen would be essential. Rather than make everyone wear an oxygen mask, it was decided that extra air would be pumped into the airliner, resulting in a pressure inside the cabin above that outside. At high altitude, this would produce a pressure difference of $57 \, kN \, m^{-2}$ across the cabin wall, or 'skin', about 50% greater than was then customary, and would be accompanied by a large temperature difference between the interior ($20 \, °C$) and the exterior ($-50 \, °C$). During service, these differences in pressure and temperature would occur in cycles as the aircraft went repeatedly through the various stages of flight — take-off, climb, cruise, descent and landing.

The task of choosing a material for the Comet's skin is similar to the problem that you met in Chapter 1 of selecting a suitable material for the roof of a bus shelter, but the technical specification is very much more demanding. First of all, the structure must be as light as possible for the following reason. The aircraft is accelerated and lifted at frequent intervals and the energy required to do this depends on the mass of the structure. This energy is dissipated, not stored — you

Figure 2.7 The de Havilland Comet

cannot recover it in a reuseable form during the descent. For an aircraft to have a good payload and range, the structure should thus have the minimum mass, or weight, to do its job safely throughout its designed life, which for the Comet was taken to be 15 000 flights. The engines can lift only a limited weight and a kilogram saved in the weight of the structure allows a kilogram more of payload — the very purpose of the machine! In order to minimize structural weight, the skin must be made to 'work hard' by carrying the maximum safe stress in service.

Secondly, the environmental conditions that confront an airliner are far more hostile than those affecting a bus shelter. We have mentioned already the temperature changes, but there are also chemical changes in the atmosphere as aircraft travel to the many airports located by the sea.

Finally, safety must underlie the whole cabin design. Any structural failure may prove fatal for passengers and expensive for the airline; much less is at stake with a bus shelter!

Basically, the cabin of an airliner is a cylindrical tube with closed ends and acts as a 'pressure vessel' when it is loaded by a net internal air pressure. If the thickness of the wall, t, is small compared with the radius of the vessel, R, then for a net internal pressure, p, the state of stress is constant at all places within the wall. At any point in the wall there is a tensile stress $\sigma_\theta = pR/t$ acting in the 'hoop direction' and a tensile stress, σ_z, in the longitudinal direction which is of half this value (see ▼Stresses in the walls of a pressurized vessel▲). We shall now use these relationships to devise a 'merit index' of lightness.

2.4.1 A 'merit index' for lightness

The case for having an airliner with a structure of minimum weight has been made earlier. Clearly, the volume of skin material needed will depend on the cabin size — a design decision — and on the mechanical properties of the material chosen. The weight of the skin depends additionally on the density of the material used. Thus it may be useful to devise a 'merit index', a combination of material properties, that is proportional to the lightness, or 'reciprocal weight', which can be used in guiding the choice of material.

The section on ▼Merit indices▲ presents a general method for deriving a merit index and we shall use this recipe to identify a 'lightness index' for a cylindrical vessel of given length, L, and given diameter, $2R$, that has a net internal pressure, p. The vessel is made of a material with a proof stress σ_p and a density ρ, and it must deform *elastically* under the hoop stress that is set up by pressurizing the cabin.

Step 1   Write down a true statement of the vessel's mass, M.

$$M = \text{volume} \times \text{density} = (2\pi RtL + 2\pi R^2 t)\rho$$

Step 2   Eliminate t, the wall's thickness, because it is neither fixed nor a property of the material. The largest stress, the hoop stress, must not exceed the proof strength. Thus $\sigma_p \geqslant pR/t$ and so $t \geqslant pR/\sigma_p$. Using this to eliminate t

$$M \geqslant 2\pi R(L + R)pR\rho/\sigma_p$$

Step 3 The index is the property term, σ_p/ρ, in the expression for reciprocal mass

$$1/M \leqslant (\sigma_p/\rho)[2\pi R^2 p(L + R)]^{-1}$$

The property σ_p/ρ is often called the **specific strength**, values of which for a range of light structural materials are given in Table 2.3.

When the strain or deflection, rather than the stress, is to be kept within limits then the relevant merit index for lightness is the **specific modulus**, E/ρ. If a material is to be used in a thin section under compression, it will be necessary to keep the stress in the direction of the longitudinal axis below the value that would cause elastic buckling. In this case, the relevant merit index for lightness is either $(E/\rho^2)^{\frac{1}{2}}$ or $(E/\rho^3)^{\frac{1}{3}}$ (see 'Merit indices'). In both these indices, the density is raised to a higher power than the modulus and, therefore, if either is used to measure lightness, *lower-density* materials are favoured.

To derive the relationships for the stresses in the walls of an aircraft cabin, we considered the cabin merely to be a pressurized vessel. The loading on the cabin, however, is not solely due to pressurization. In flight, the cabin is supported by the wings near its centre while the ends of the hull bend downwards under the weight of the structure and the payload. This sets up bending stresses in the top and bottom of the hull, which reach a maximum near the wing and are superimposed on the longitudinal pressurization stress, σ_z. The bending stresses are most severe when the aircraft flies through gusts of upward-moving air and this must be allowed for in the design. Large bending stresses could put the bottom of the cabin into axial compression, whereupon the buckling resistance of the skin would have to be considered.

Table 2.3 Values at room temperature of density, strength and stiffness for some potential airframe materials

Material	Density $\rho/\mathrm{Mg\,m^{-3}}$	Proof or fracture stress $\sigma_p/\mathrm{MN\,m^{-2}}$	Young's modulus $E/\mathrm{GN\,m^{-2}}$	$\sigma_p/\rho^{-1}/\mathrm{MN}$ $\mathrm{m\,Mg^{-1}}$	$E\rho^{-1}/\mathrm{GN}$ $\mathrm{m\,Mg^{-1}}$	$E\rho^{-2}/\mathrm{GN}$ $\mathrm{m^4\,Mg^{-2}}$	$E\rho^{-3}/\mathrm{GN}$ $\mathrm{m^7\,Mg^{-3}}$
spruce	0.45	35	9	78	20	44	99
balsa wood	0.15	7.5	2.4	50	16	107	710
carbon-fibre-reinforced plastic, CFRP (54% fibres by volume, 0°/90° laminate in epoxy resin)	1.50	930	83	620	55	37	25
glass-reinforced plastic, GFRP (60% uniaxial glass by volume in epoxy resin)	1.80	700	40	390	22	12	6.9
aluminium alloy (7075-T6)	2.81	500	71	178	25	9.0	3.2
stainless steel (FSM 1)	7.86	980	185	125	24	3.0	0.38
beryllium	1.48	272	290	184	196	132	89
titanium alloy (6%Al–4%V)	4.46	1100	107	247	24	5.4	1.24
magnesium alloy (AZ 31)	1.78	120	45	67	25	14.2	8.0

▼Stresses in the walls of a pressurized vessel▲

Consider the pressurized cylinder sketched in Figure 2.8. The symmetry of the shape suggests that the most important directions in the body are radial, circumferential and axial. First consider the stress acting in the radial direction: that is, parallel to a radius of the circular cross-section. Apart from the rather feeble force due to atmospheric pressure, p_o, which is about $100\,\mathrm{kN\,m^{-2}}$, the outside surface is free of any forces perpendicular to it. The inside surface feels the larger internal pressure, p_i, and the radial stress in the wall must change continuously from p_o to p_i on traversing the thickness, t.

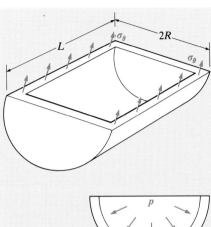

Figure 2.9 A free-body diagram for a pressure vessel

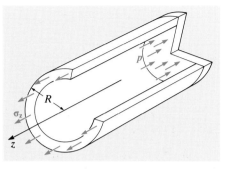

Figure 2.10 Another free-body diagram for a pressure vessel

Figure 2.8 A pressure vessel

To find the stress, σ_θ, acting around a circumference — along a 'hoop' of the wall — you can think of the cylinder as consisting of two halves split longitudinally, as shown in Figure 2.9. This is called a *free-body diagram*. The internal pressure tends to blow the two halves apart, while the 'hoop stress' in the wall tends to hold them together. If the vessel is neither to explode nor implode, there must be equilibrium and these tendencies must balance. When they balance, each half experiences zero net force and remains at rest. The net internal pressure is $p_o - p_i$ and the vertical 'exploding' force is simply $2RLp$ — the

net internal pressure, p, multiplied by the area of the split section $2RL$. (This can be shown rigorously by integrating the vertical component of pressure over all the small elements of area of the wall.) The exploding force must be balanced by the force of attraction between the two halves of the vessel. If the vessel is so long that we can ignore the force carried by the end-caps, then since the cross-sectional area of each split wall is tL, the total attractive force provided by both walls is $\sigma_\theta 2tL$. Equating these forces, you get

$$\sigma_\theta 2tL = 2pRL$$

Hence

$$\sigma_\theta = pR/t \qquad (2.2)$$

Finally, consider the longitudinal or axial stress, σ_z, in the walls. To help, we shall use the free-body diagram shown in Figure 2.10. The principle that the forces must balance if the vessel is in equilibrium

is again applied to each half. The exploding force is $p\pi R^2$, the net internal pressure multiplied by the area of the end cap. The cohesive force is $\sigma_z 2\pi Rt$, the axial stress times the area of the section over which it acts. Balancing these forces, you get

$$\sigma_z 2\pi Rt = p\pi R^2$$

and hence

$$\sigma_z = pR/2t \qquad (2.3)$$

For a thin-walled vessel, the radius is very much greater than the thickness of the walls, so as you can see from Equations (2.2) and (2.3), both σ_θ and σ_z are very much larger than p. Thus the radial stress, which is equal to the net internal pressure, is ignored and the only significant stresses in the walls are the hoop and axial stresses, which both act parallel to the walls. Of these, the hoop stress is twice the size of the axial stress.

These simple equations are the basis of designing high-pressure gas cylinders, which, for reasons of safety, are covered by standard specifications (e.g. BS 5045). In this case, the radius of the cylinder, R, is measured to the middle of the wall.

▼Merit indices▲

A merit index is a combination of certain properties of a material that may be used during the materials selection process to compare different materials under a particular criterion. The properties involved will depend on the criterion. For a given application, it is first necessary to identify the appropriate merit index and to do this the following approach can be adopted.

In general, an application will require a number of unconnected attributes, such as resistance to corrosion, strength, or stiffness, to be considered, but a merit index can address only one of these at a time. Perhaps an example will make this clear. Suppose that for a given application, suspension bars are required of a given length, L, and a particular load-bearing ability, P, but with the lowest mass or weight.

What attribute is being sought in this case? In other words, what is it that you want the most of — the bigger, the better?

The attribute you are seeking is the *reciprocal* of the mass, $1/m$. This will depend on the density, ρ, of the material and the dimensions of the bars (Figure 2.11). First write down an expression for $1/m$ in terms of ρ, L and A, the cross-sectional area required for a bar. Since the mass is given by

$$m = \text{density} \times \text{volume}$$

then

$$m = \rho A L$$

and therefore

$$\frac{1}{m} = \frac{1}{\rho A L}$$

Next, eliminate from the right-hand side

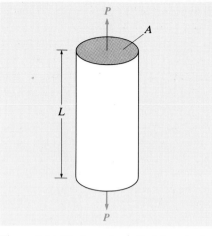

Figure 2.11 Tie bar of specified length, L

of the equation anything that is neither a *property* of the material (like ρ) nor a *prescribed quantity* (like L), replacing it by a combination of properties and/or fixed quantities. In this case you would want to eliminate A. Now A depends on strength, σ_t, a property of the material used and

$$\sigma_t = P/A$$

Thus

$$A = P/\sigma_t$$

You can now substitute P/σ_t for A in the equation for $1/m$. Doing this gives

$$\frac{1}{m} = \frac{\sigma_t}{\rho P L}$$

$$\frac{1}{m} = \frac{\sigma_t}{\rho} \times \frac{1}{PL} \qquad (2.4)$$

The combination of *properties* on the right-hand side, σ_t/ρ, is the *merit index* for this criterion. To get the lightest bar you should use the material with the largest value of σ_t/ρ. Note that if the density was

replaced by the cost of unit volume of material, c, you would have a merit index, σ_t/c, for the *cheapest* bar.

To summarize, the steps involved form a general method for deriving a merit index:

1 Write down any true expression for the attribute sought.

2 Eliminate any parameter that is not either a property or a fixed quantity.

3 The merit index is the combination of *properties* on the right-hand side of the equation.

SAQ 2.3 (Objective 2.4)
A suspension bar is required to be of length L and mass m. Derive a merit index for the maximum load, P, that the bar can carry if it is made of a material with a density ρ and tensile strength σ_t.

SAQ 2.4 (Objectives 2.4 and 2.7)
A long column of fixed length L and given load capacity P is required. The axial load that will just cause a column to buckle is $P_b = kEI/L^2$. Here k is a constant that depends on how the ends are constrained, E is Young's modulus and I is the second moment of area of the cross-section.

Derive a merit index that could be used to find the material for the lightest column in each of the following cases:

(a) When the cross-section is a square of side w.
(b) When the cross-section is rectangular, with a *specified* width w, which is greater than the thickness, t.

As long as it's aluminium

From this discussion you can conclude that for weight-saving a material with large values of the appropriate merit indices is required. Table 2.3 contains property values for a wide range of materials. At the top are the earliest airframe materials: two varieties of wood. These are notable for having low densities and anisotropic properties — properties that, because of the fibrous microstructure of wood, vary with direction. For a given mass of material, resistance to buckling is proportional to E/ρ^2 or E/ρ^3, depending on the conditions (see SAQ 2.4) and, although the

CHAPTER 2   NONFERROUS METALS

specific strengths of these materials are low and their specific moduli are not remarkable, their resistance to buckling is outstanding. In partnership with strong steel tie wires to carry tensile loads, wood was used widely on airframes to carry the compressive loads up until the 1940s. In the past twenty years, fibre-reinforced plastic 'composite' materials have been developed for aerospace use; they are man-made versions of wood, having anisotropy, low densities and good strength and stiffness. Of all the values in Table 2.3, the merit indices for carbon-fibre reinforced plastic are remarkable and it is therefore understandable that this material was chosen for *Voyager*, the ultimate aeronautical cargo carrier that circumnavigated the earth nonstop in 1986 without refuelling. In this case the cargo was the crew of two and the fuel to travel over 24 000 miles: it comprised 80% of the weight during take-off! It is clear that this type of material must have an assured future in aviation. Although still comparatively new, as experience accumulates of the design, building, inspection and repair of composite materials in structures, so their use in civil transport airframes will increase. This type of material was obviously not a contender for the airframe of the Comet.

The remaining materials in Table 2.3 are metals and alloys. Beryllium has by far the highest value of specific modulus in the table and a low density, comparable with that of carbon-fibre reinforced plastic. Unfortunately, this is not a typical nonferrous metal: ductile, tough, formable and chemically stable. It is brittle and very difficult to form to shape and, furthermore, it is toxic; the metallurgical industry appears largely to have given up its attempts to use this metal, in spite of its attractive indices. We shall not consider it further.

The alloys have virtually identical values of specific modulus. It follows that for resistance to buckling, the advantage increases as the density falls and, on this basis, the magnesium alloy is best, followed by the aluminium alloy. When the specific strength is considered, though, the magnesium alloy has by far the lowest value and this would appear to rule it out for a cabin skin. On all the indices, the aluminium alloy surpasses the stainless steel and it compares favourably with the titanium alloy, except in respect of specific strength.

At room temperature, these alloys thus appear to have no decisive advantage over one another, apart from cost, which strongly favours aluminium. Since the development of titanium alloys began only in the late 1940s, this type of material was not available for use on the Comet. Aluminium alloys were therefore the most favoured type of material for the cabin skin on the basis of their merit indices, cost and chemical stability.

By the 1940s, aluminium alloys had replaced varieties of wood for the construction of airframes for reasons other than those revealed by the merit indices in Table 2.3. With the advent of the stressed-skin,

49

cantilever-winged aeroplane in the 1930s, the aircraft industry had gone over to all-metal construction and, with the notable exception of the de Havilland Mosquito, had abandoned the distinct methods of design and construction appropriate for wood. The higher strengths available from alloys permitted thinner wings to be designed for higher speeds; alloy airframes had long lives and were generally more tolerant of exposure to moisture than 'biodegradable' timber.

The decision to use an aluminium alloy for the cabin skin of the Comet was an easy one in view of the industry's prior commitment to this type of material. It only remains to consider the choice of a specific alloy and the treatments that the alloy would require to develop its best properties.

2.4.2 The quest for strength in single-phase alloys

A material with a high specific strength is obviously required for airframe applications and pure nonferrous metals are so lacking in strength that they are rarely used for load-bearing duties. There are, however, three common ways of strengthening a single-phase metal:

(a) by work hardening;
(b) by reducing the average grain size;
(c) by alloying to make a solid solution.

Work hardening

Work hardening can be applied to an annealed, ductile metal by subjecting it to any form of plastic deformation at a temperature below about 0.3 of its melting point, T_m. The effects of plastic strain are to increase the strength and hardness and to decrease the ductility (e.g. the elongation-to-fracture). Figure 2.12 shows examples of the effects of work hardening on aluminium.

Useful increases in strength and hardness, up to about a factor of two, can often be obtained as a consequence of using cold-forming processes to shape the material into a product. For example, sheet of a given

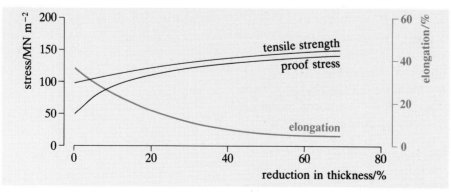

Figure 2.12 The effects of plastic strain on the properties of aluminium

composition can be bought in a variety of hardness levels, depending on how much strain it has received since last it was annealed. It can be obtained in the annealed condition or at one of several levels of work hardening: 'a quarter hard', 'half hard', 'three-quarters hard' or 'fully hard'. 'Fully hard' corresponds roughly to a cold reduction in thickness of 75% without annealing. As shown in ▼A dislocation model of work hardening▲, work hardening is a consequence of the accumulation of dislocations within the grains.

The loss of ductility that accompanies work hardening can be thought of as an 'exhaustion' of the finite capacity of the material to undergo plastic tensile strain. Cold working, by any process, prior to measuring the elongation-to-fracture in a tensile test, 'uses up' some of the material's ductility, leaving a reduced amount to be measured in the test. In the annealed state, the elongation-to-fracture is a maximum, equal to the **work-hardening exponent**, n, in Equation (2.7) — see ▼The stability of deformation during stretching▲. Cold working the annealed metal to a plastic strain ε reduces the elongation to about $(n - \varepsilon)$.

In cases where the product might be subjected to further plastic deformation by the purchaser, such as the use of drawn copper pipes for domestic hot water, it is usual to soften the product by heat treatment in order to avoid the risk of cracking when, for example, a plumber bends pipes to shape. This treatment involves heating the metal to a temperature of at least $0.3T_m$, to enable edge dislocations to undergo climb at a significant rate (see ▼The effect of heat on deformed metals: annealing▲).

Heat treatment at a sufficiently high temperature causes recrystallization and provides the means to change the grain size.

Grain-size strengthening

The second method of strengthening a single-phase material is to refine the grain size. It depends on the observed relationship that the proof strength, σ_p, depends inversely on the average grain diameter, d. This is given in the equation

$$\sigma_p = \sigma_0 + kd^{-\frac{1}{2}} \tag{2.14}$$

where σ_0 and k are constants for a given metal. This is generally referred to as the **Hall-Petch equation** and applies quite generally to polycrystalline metals. It implies that useful increases in strength can be obtained by decreasing the grain size: for example, doubling the $kd^{-\frac{1}{2}}$ component by reducing d from 0.3 mm to 0.075 mm. According to Equation (2.14) the grain size should be made as small as possible in order to promote strength. As explained in 'The effect of heat on deformed metals: annealing', the recrystallized grain size depends primarily on the amount of cold work the metal received prior to annealing—the larger the plastic strain received, the higher the dislocation density and the smaller the grain size produced by annealing.

▼A dislocation model of work hardening▲

In discussing the properties of bulk material in terms of dislocations, it is necessary to consider the aggregate effects of the dislocations present. The population of dislocations is represented by the **dislocation density**, ρ: that is, the number of dislocations that intersect an imaginary plane of unit area within the crystal. In polycrystalline material, ρ is taken to be independent of orientation of the plane. When a well-annealed metal crystal is examined in a transmission electron microscope the dislocations appear as dark lines (Figure 2.13) and can be counted. They reveal that ρ is of the order of 10^{12} m^{-2}. There are thus *about 1000 kilometres of dislocations present in a 1 cm cube of annealed crystal!*

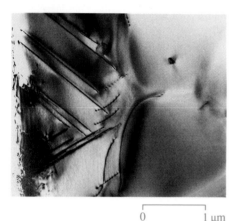

```
0              1 μm
```

Figure 2.13 Dislocations (dark lines) within a thin metal foil, about 0.1 μm thick

Studies have been made with electron microscopes of the way in which the dislocation density, ρ, changes with plastic strain, ε_p, and invariably it is found that, at temperatures below about $0.3T_m$, ρ increases with ε_p. The dependence is of the form

$$\rho = \rho_0 + C\varepsilon_p^{\beta} \qquad (2.5)$$

where ρ_0 is the dislocation density in the

annealed condition and C and β are empirical constants. The value of β lies in the range 0.5–2, most frequently being close to 1 at small strains.

The accumulation of dislocations that occurs when metals are cold worked (plastically strained at a temperature below $0.3T_m$) provides a ready explanation for work hardening.

Normally, two dislocations interact, that is they exert forces on one another. This happens because each dislocation has an elastic stress (or strain) field, such as that shown in Figure 2.14 for an edge dislocation. At any point in the dislocated crystal, the state of stress can be represented by six components, referred to arbitrary coordinate axes, $0x$, $0y$ and $0z$. With each stress component, σ, there is associated an elastic strain, ε, and an

```
0        25 mm
```

Figure 2.14 The stress field set up in a model of an edge dislocation in a plastic sheet. The model is made by inserting an oversize shim, representing an extra half plane of atoms in a crystal, into a slot. The dark lines are contours of equal maximum shear stress

elastic strain energy per unit volume of $\sigma\varepsilon/2$ or, since $\varepsilon = \sigma/E$, according to Hooke's law, of $\sigma^2/2E$. When the stress fields of two dislocations superimpose, the corresponding components of each, σ_1 and σ_2, can be added to give the stress component $(\sigma_1 + \sigma_2)$ at a given point and the strain energy per unit volume due to that component is $(\sigma_1 + \sigma_2)^2/2E$.

This is not usually the same as $\sigma_1^2/2E + \sigma_2^2/2E$, the sum of the strain energies of the dislocations when they are infinitely apart, so when the strain energies of all the points in the crystal are added, the total energy of the two coexistent dislocations usually differs from that of the two dislocations spaced far apart. In general, the total strain energy for the whole crystal *varies* with the distance apart of the dislocations. If the strain energy *decreases* as the dislocations approach one another there is a force of *attraction* between them. (Mathematically, force is equal to minus the energy gradient.) Conversely, dislocations *repel* one another if the strain energy *increases* when they approach. In all but a few special geometrical cases, dislocations interact mechanically.

Work hardening follows directly from this interaction. The higher the dislocation density, the smaller is the average spacing, x, of the dislocations. It can be shown that two parallel dislocations with the same Burgers vector, $\boldsymbol{b}$, exert on one another a force per unit length, F, where, in a material of shear modulus G,

$$F \simeq \frac{Gb^2}{2\pi x} \qquad (2.6)$$

If the dislocations are distributed uniformly, x is approximately equal to $1/\sqrt{\rho}$ and F is proportional to $\sqrt{\rho}$. In order to move a dislocation through the 'forest' of other dislocations requires an applied shear stress, τ, which is proportional to F, so the strength of the crystal increases with $\sqrt{\rho}$ and, hence, according to Equation (2.5), with $\varepsilon_p^{\beta/2}$, provided $\rho \gg \rho_0$.

This is work hardening and is consistent with the empirical relationship between tensile stress, σ, and tensile plastic strain, ε_p, measured in tensile tests on metals

$$\sigma = \sigma_0 \varepsilon_p^n \qquad (2.7)$$

The work-hardening exponent, n, is typically 0.1–0.6.

Finally, consider why dislocations accumulate during cold working. Imagine an edge dislocation gliding across a slip plane in a unit cube of crystal (Figure 2.15). When the dislocation glides completely across the plane it causes a shear displacement of b between the top and bottom faces of the cube. Since these faces are unit distance apart, the shear strain γ caused is simply $b/1$. It can be shown that if the dislocation glides a distance l across only part of the slip plane, so that $l < 1$, then $\gamma = bl$. It follows that if many dislocations, of density ρ, move in this manner then

$$\gamma = \rho bl \qquad (2.8)$$

Compare this with Equation (2.5) when $\beta = 1$. In both equations, strain is directly proportional to ρ. This implies that bl is

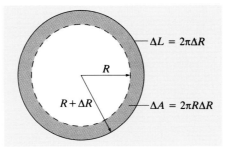

Figure 2.16 The expansion by glide of a dislocation loop lying in a slip plane

constant and therefore that the average distance l moved by a segment of dislocation is constant. In other words, during cold working, a segment of dislocation moves a distance l on average, causing an increment of strain before it is brought to rest by an obstacle — probably another dislocation. As it moves, it multiplies and the new segments of dislocation created move by l. These contribute to a subsequent increment of strain and so it goes on. Work hardening is a consequence of this accumulation of dislocations and the tendency of dislocations to 'get in one another's way'.

Finally, we shall consider briefly how dislocations 'multiply' as they move. The

simplest example is to consider a small loop of dislocation lying on its slip plane (Figure 2.16). If it is acted on by an appropriate shear stress, the segments of dislocation will all move, causing the loop to enlarge, thereby increasing the dislocation density. Various dislocation mechanisms have been identified for forming half a dislocation loop (with fixed ends) which then expands under stress; Figure 2.17 shows one example. Such events cause dislocations to multiply.

SAQ 2.5 (Objective 2.5)
Published measurements on polycrystalline iron indicate the following values for the parameters in Equation (2.5)

$$\rho_0 = 10^{12}\,\mathrm{m}^{-2}$$
$$C = 10^{15}\,\mathrm{m}^{-2}$$
$$\beta = 1 \quad (\varepsilon \text{ is the } fractional \text{ strain})$$

What do these numbers indicate about:
(a) the increase in dislocation density caused by a strain of 10%;
(b) the average distance moved by dislocations in this material?
Take $\varepsilon = \gamma/2$ and $b = 3 \times 10^{-10}\,\mathrm{m}$.

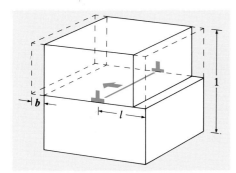

Figure 2.15 A unit cube of crystal containing an edge dislocation

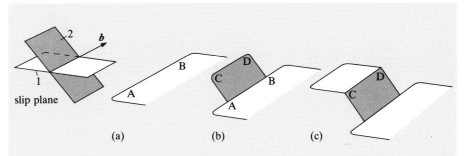

Figure 2.17 The formation of half a dislocation loop by the 'double cross-slip' of a screw dislocation from one slip plane to another. The half loop then expands

▼The stability of deformation during stretching▲

When metal test pieces are stretched in a tensile test at temperatures below $0.3T_m$, the plastic strain is uniform throughout the test piece up to the point of maximum load (e.g. Figure 1.13b). When the load reaches a maximum and begins to decrease, a small constriction forms in the test piece. From then on the plastic strain is nonuniform and is increasingly confined to the region of the constriction. A change from stable to unstable deformation therefore occurs at the maximum load.

If a small constriction accidentally forms when the load is increasing, the stress on it will be higher than elsewhere. It will then deform plastically and work harden until its strength becomes equal to the applied stress, whereupon it will stop deforming, allowing the strain in the other cross-sections to 'catch up'. Such deformation is stable.

In contrast, under a decreasing load, a constriction is not self-arresting because the stress on it due to the shrinking cross-section rises by *more* than the increase in strength that occurs by work hardening. Deformation is then unstable and a constriction develops preferentially. The strain at which deformation becomes nonuniform, ε_n, is determined by the way in which the material work hardens.

Recalling the general expression for tensile stress in terms of force, F, and cross-sectional area, A,

$$F = \sigma A$$

During a small increment of deformation, the change in F is given by differentiation

$$dF = \sigma dA + A d\sigma$$

At the onset of the instability $dF = 0$, so

$$A d\sigma = -\sigma dA$$

Rearranging this

$$d\sigma = -\sigma \frac{dA}{A} \qquad (2.9)$$

The volume is given by $V = Al$. Since plastic strain involves no significant change in volume, $dV = 0$ and hence

$$dV = A dl + l dA = 0$$

$$\frac{dl}{l} = -\frac{dA}{A}$$

The incremental change in strain is $d\varepsilon = dl/l$, so

$$d\varepsilon = -\frac{dA}{A}$$

Substituting this into Equation (2.9)

$$d\sigma = \sigma d\varepsilon$$

or

$$\frac{d\sigma}{d\varepsilon} = \sigma \qquad (2.10)$$

This is the condition for the onset of the instability.

This shows you that an instability is inevitable; σ increases steadily due to work hardening while the rate of hardening, $d\sigma/d\varepsilon$, is falling continually, so eventually the two must meet (Figure 2.18).

The relationship between the 'true' stress, σ, and the 'true' strain, ε, during the plastic deformation of metals is usually of the form

$$\sigma = \sigma_0 \varepsilon^n \qquad (2.11)$$

True stress is simply $\sigma = F/A$, the axial force, F, divided by A, the *actual*, rather than the original, cross-sectional area. True strain, ε, is defined in terms of the deformed gauge length, l, and the undeformed length, l_0, so that

$$\varepsilon = \ln(l/l_0).$$

Differentiating Equation (2.11) gives

$$\frac{d\sigma}{d\varepsilon} = n\sigma_0 \varepsilon^{n-1}$$

At the onset of the instability, $d\sigma/d\varepsilon$ must be equal to σ (Equation 2.10) and therefore using Equation (2.11)

$$\sigma = \sigma_0 \varepsilon^n = n\sigma_0 \varepsilon^{n-1}$$

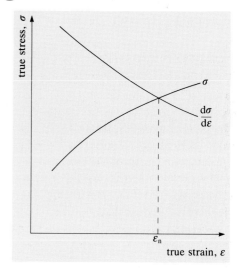

Figure 2.18 Changes in true stress, σ, and the hardening rate, $d\sigma/d\varepsilon$, as a function of the true strain, ε

Thus, if ε_n is the strain at which deformation becomes non-uniform,

$$\varepsilon_n = n \qquad (2.12)$$

You can now see the physical significance of n. It is equal to the tensile strain at which deformation becomes unstable during stretching, leading quickly to the formation of a 'neck' and fracture.

SAQ 2.6 (Objective 2.6)
In a tensile test, a test piece of annealed copper had a 1% proof strength of $50\,\text{MN}\,\text{m}^{-2}$ and an elongation-to-fracture of 50%. Assuming that the stress, σ, and the strain, ε, are related in the usual way ($\sigma = \sigma_0 \varepsilon^n$), what are the values of σ_0 and n?

Use these values to calculate the stress on the test piece when deformation becomes unstable.

Other factors also promote a small grain size: a fast heating rate, which causes nucleation of many new grains, and a short time at the annealing temperature in order to restrict grain growth.

Solution hardening

The third method of strengthening involves the dissolution of one or more alloying elements in the crystals of the host metal to form a 'solid solution': that is, the host and alloying elements are mixed together *on the atomic scale*. Solid solutions containing only metallic elements are almost invariably 'substitutional' in that each atom of solute added replaces an atom of solvent in the crystalline structure. Since an atom of the alloying element will differ from the host atom in both size and electronic structure, the crystal is distorted in the immediate vicinity of each solute atom. Interactions occur between a dislocation and these distortions and the effect is to increase the shear stress required to move the dislocation. The result is **solid-solution strengthening**.

The dislocation theory of solution hardening suggests that the proof strength, σ_p, of a solid solution should depend on the magnitude of the misfit between solute and host atoms $(1 - r_s/r_h)$, where r_s and r_h are the respective radii; and on c, the concentration of solute atoms in the alloy. They are related in the following way

$$\sigma_p \propto \left(1 - \frac{r_s}{r_h}\right)^{1.5} \sqrt{c} \qquad\qquad (2.15)$$

The dependence on $\sqrt{c}$ arises because the number of solute atoms present in unit area of a slip plane is proportional to $\sqrt{c}$ and when a dislocation glides across the slip plane it interacts with all these atoms. The magnitude of the interaction with each atom is proportional to

$(1 - r_s/r_h)^{1.5}$.

We mentioned the solution hardening of copper by nickel additions when we discussed the use of cupro-nickel for coinage. The copper–nickel system is unusual in that it shows a solid-solution phase that spans the entire composition range — it exhibits 'continuous solid solubility'. This phase shows solution hardening of copper by additions of nickel *and* solution hardening of nickel by additions of copper (see Figure 2.24). These trends cause the hardness, or strength, of the solid solution to attain a maximum near the 50% copper, 50% nickel composition.

Aluminium can be solution hardened by alloying additions of zinc, magnesium, copper, lithium or manganese, all of which can dissolve in aluminium in significant quantities. Provided that the alloy contains only a single phase, its resistance to corrosion should remain good, but this would restrict the choice to dilute, solid-solution alloys, which require no special heat treatment apart from annealing. The most common alloys of this type contain manganese (the '3000 series') or

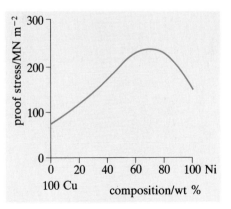

Figure 2.24 The variation of proof stress with composition for copper–nickel alloys

▼The effect of heat on deformed metals: annealing▲

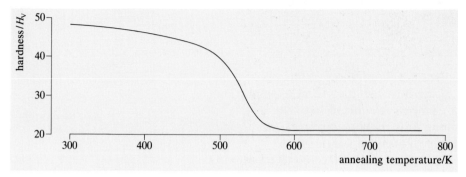

Figure 2.19 The hardness of aluminium, reduced 72% in thickness by cold rolling, after annealing for a fixed time at the temperature plotted on the horizontal axis

Figure 2.21 shows a specimen of aluminium that has been stretched and annealed in this way, and the grains are clearly visible.

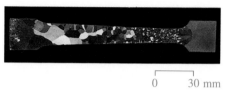

Figure 2.21 The electrolytically etched grain structure produced by recrystallization

When metals are cold worked they undergo changes in such properties as proof strength, elongation, hardness, density and resistivity. These changes can be reversed by annealing — heating at temperatures above about $0.3T_m$. The term **recovery** is used for the process whereby the properties change without any attendant change in the grain structure: that is, without changes in the shape and orientation of grains. If annealing produces an observable change in the grain structure, the process is called **recrystallization**. This is invariably accompanied by a major change in properties, such as shown in Figure 2.19 for annealing temperatures above 525 K.

Cold working creates a high density of dislocations within the grains, but these defects are not thermodynamically stable. The reduction in free energy — mainly elastic strain energy — accompanying the disappearance of these dislocations on annealing provides the driving force for the processes of recovery and recrystallization.

The size of recrystallized grains depends on the plastic strain that the metal has undergone during cold working. A simple experiment makes this point quite convincingly. Suppose that you have a tensile-test specimen with a *tapered* central

section and you stretch it plastically, with a force F, in a tensile-testing machine. Since the cross-sectional area of the specimen varies, so the stress set up by the force varies along the length of the sample. The stress–strain curve tells you that each value of stress is associated with a value of strain, so the amount of plastic strain must vary along the length of the specimen. You may find that Figure 2.20 helps to make this clear.

After being stretched, the sample is annealed and recrystallization occurs.

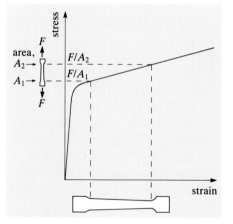

Figure 2.20 The variation of stress and strain along a tapered test piece

How does the recrystallized grain size depend on the plastic strain in the sample?

The smallest grains are to be found at the minimum area of cross-section, where the plastic strain is a maximum. Conversely, the largest grains have formed in those places having the smallest plastic strain. You should conclude that the grain size depends *inversely* on the plastic strain, as shown quantitatively in Figure 2.22.

This trend cannot continue to hold for ever smaller strains — with zero plastic strain the metal will not crystallize at all. The ends of the test specimen were

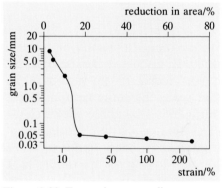

Figure 2.22 For a given annealing treatment the grain size depends on the prior plastic strain

clamped and therefore they were not strained and you can see that they have not recrystallized. There is a critical strain below which recrystallization does not occur; this strain varies with the metal, but is typically about 0.02.

Strains of this order are often involved in metal-forming operations, such as the deep drawing of a cup from a flat sheet, but they are undesirable when followed by annealing and further deformations. Annealing forms large grains which, when deformed, can produce unsightly 'orange peel' marks on the surface.

Recrystallization has been observed to begin with the formation of small recrystallized **nuclei** (grains), which grow until they meet one another. At this point recrystallization is complete and any further changes in the grain structure involve **grain growth**, which occurs by the growth of the larger grains at the expense of neighbouring smaller ones. The driving force for grain growth is the resulting reduction in the grain-boundary area and, hence, energy. Nucleation of a recrystallized grain involves the bulging out of a short segment of grain boundary between two fixed points, a distance L apart, leaving a volume of recrystallized metal with a low dislocation density (see Figure 2.23). For this to happen spontaneously, there must be a difference, U, in the stored energy per unit volume across the bulging boundary. Since the grain boundary has a surface energy, or 'surface tension', it resists bulging in the same way that a soap bubble does and the relationship between L, U, and the surface energy, Γ, is

$$U \geqslant 2\Gamma/L \qquad (2.13)$$

For a nucleus to grow, this condition must be satisfied. Nucleation is favoured by large values of U and L. A large value of U implies a nonuniform dislocation density and a large value of L would arise from having large cold-worked grains, where L is the length of a facet on a grain.

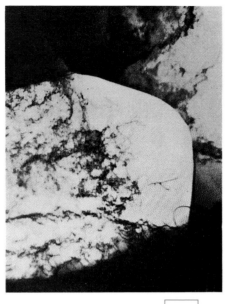

0 0.1 μm

Figure 2.23 A recrystallization nucleus. The bowing of a grain boundary facet leaves behind a crystal of low dislocation density

You can now interpret the dependence of grain size upon plastic strain (Figure 2.22). After small deformations, U is small and the nucleation of new grains is confined to a few widely spaced favourable sites. Grains grow from these sites to a large size before they meet other grains. The opposite is true after large deformations — the dislocation density is large and inhomogeneous so the value of U is large. Nucleation occurs abundantly at many sites and grains undergo little growth before meeting other grains. The resulting grain size is small. This is usually the preferred grain structure; not only is its strength higher than that of coarse grains, but it is free of the unsightly 'orange-peel' effect when the grains are deformed. The precise relationships between grain size, annealing time and temperature vary with the metal or alloy and have to be found by experiment.

magnesium (the '5000 series') and some alloys contain both elements. The numbering code for identifying these alloys is shown in Figure 2.25.

The 5000 series of alloys contain the largest range of solute concentrations, enabling high strength levels to be attained when the magnesium content is increased (see Table 2.4). According to the aluminium–magnesium phase diagram (Figure 2.26), the solubility of magnesium in aluminium is high at 723 K, but falls to about 1% at room temperature. This would appear to restrict solution-hardened alloys for use at ambient temperatures to those having a maximum magnesium content of 1%, with only a modest amount of solution

Table 2.4   Properties of a range of annealed 5000 series aluminium alloys

Alloy	Composition/weight %		0.2% proof stress/MN m^{-2}	Tensile strength/MN m^{-2}	Elongation/%
	Mg	others			
5005	0.5–1.1		40	125	30
5050	1.1–1.8		55	145	24
5052	2.2–2.8	0.15–0.35Cr	90	195	25
5454	2.4–3.0	0.5–1.0Mn 0.05–0.2Cr	120	250	22
5083	4.0–4.9	0.4–1.0Mn 0.05–0.25Cr	145	290	22
5456	4.7–5.5	0.5–1.0Mn 0.05–0.2Cr	160	310	24

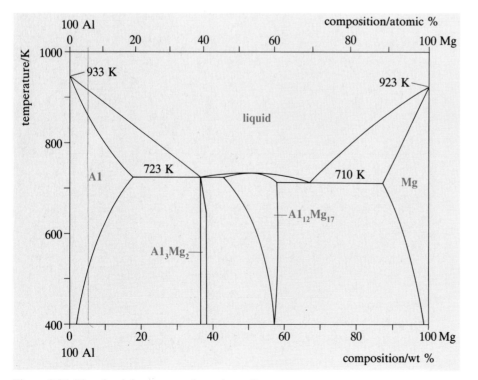

Figure 2.26 The aluminium–magnesium phase diagram

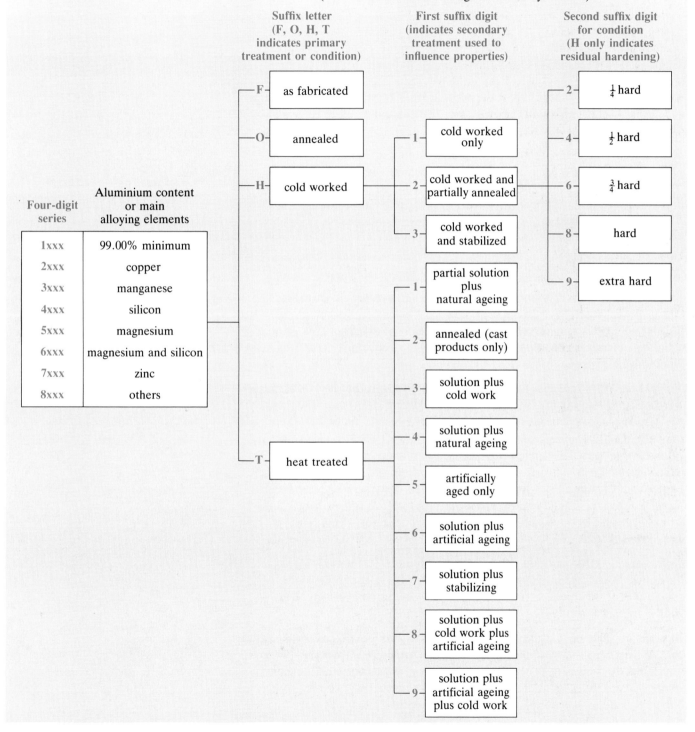

Figure 2.25 Aluminium and aluminium alloy temper and designation system

strengthening. In practice, contents up to about 5% can, fortunately, be retained in solid solution at room temperature if the alloy is cooled quickly enough from 723 K.

These solutions are said to be 'supersaturated' because they contain more than the equilibrium content of solute. Such alloys eventually begin to precipitate the Al_3Mg_2 phase, even at room temperature, particularly when they have been cold worked. To some extent, the microstructure — a solid solution containing a high dislocation density — can be stabilized by an 'H3 temper', heating for a short time at between 390 K and 420 K.

Because of their good appearance and resistance to corrosion, aluminium–magnesium alloys are used widely as the 'ring-pull' tops of drink cans, as trim on motor cars, as architectural panels, as boat hulls and for ships' superstructures. They are also used for welded structures such as truck bodies, storage tanks and pressure vessels.

2.4.3 Strengthening with a second phase

Finally, there is an alloying and heat treatment strategy for raising the strength of a metal by introducing small particles of another phase. The strategy is called **age hardening** or **precipitation hardening**. This requires an alloy containing a solid solution with a solid solubility that increases with temperature.

Actually, in dilute alloys the solubility almost invariably increases with temperature, so you might think that almost any dilute alloy would do for age hardening, but this is only qualitatively true. In practice, we are concerned with the *magnitude* of hardening and it is only in a minority of cases that this is significant. The reasons are not hard to find. If the *maximum* solubility is only small, then a correspondingly small solute content will change into precipitates and their spacing will tend to be large. Successful hardening requires the volume fraction of precipitates to be large and their spacing to be small and uniform. This tends to rule out very dilute alloys.

Even when a large volume fraction of precipitates is formed, the hardening may be meagre, since it also depends on the ability of a precipitate to obstruct a dislocation by one of the mechanisms that we shall consider later.

In practice, it is only aluminium alloys of the 2000, 6000 and 7000 series that are used in the heat-treated condition. All these alloys are strengthened by being taken through the treatment sequence of 'solution treat, quench, age'. The first two steps of the sequence produce a supersaturated solid solution, while the third step causes the precipitation of a second phase to occur. During the ageing treatment, the strength and hardness increase, as shown in Figure 2.27. When ageing is carried out 'artificially', that is, at above room temperature, a

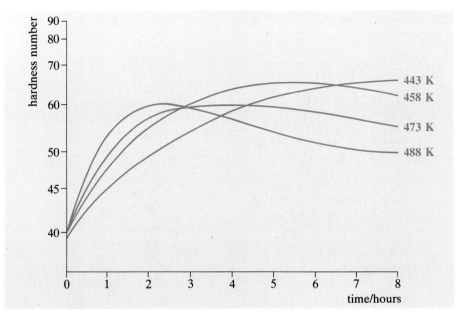

Figure 2.27 The change in hardness on ageing of a solution-treated, 6000-series alloy

maximum in strength is reached after a critical time and further ageing ('over ageing') causes softening. The most important point to grasp is that the highest strength is always associated with precipitates of very small size and spacing, and these are often not the equilibrium precipitate, but rather unstable precursors of it.

▼Mechanisms of deformation in age-hardened alloys▲ describes the competing mechanisms of hardening and softening that result in a maximum of hardness, or strength, on ageing. The properties of the most common alloys in the fully heat-treated state are presented in Table 2.5. The 7000 series is the strongest, the 6000 series the least strong; the 2000 series has intermediate strength.

Table 2.5   Composition and properties of some heat-treated aluminium alloys

Alloy	Composition/wt % (maximum or range)						Temper	0.2% proof stress/ MN m^{-2}	Tensile strength/ MN m^{-2}	Elongation/%
	Si	Cu	Mn	Mg	Zn	Others				
2014	0.5–1.2	3.9–5.0	0.4–1.2	0.2–0.8	0.25		T6	410	480	13
2618	0.1–0.25	1.9–2.7		1.3–1.8	0.10	0.9–1.3Fe 0.9–1.2Ni	T61	330	435	10
2024	0.5	3.8–4.9	0.3–0.9	1.2–1.8	0.25		T6	395	475	10
6063	0.2–0.6	0.10	0.10	0.45–0.9	0.10		T6	215	240	12
7010	0.10	1.5–2.0	0.30	2.2–2.7	5.7–6.7	0.11–0.17Zr	T6	485	545	12
7075	0.40	1.2–2.0	0.30	2.1–2.9	5.1–6.1	0.18–0.28Cr 0.25(Zr + Ti)	T6	500	570	11
7475	0.10	1.2–1.9	0.06	1.9–2.6	5.2–6.2	0.18–0.25Cr	T651	560	590	12

▼Mechanisms of deformation in age-hardened alloys▲

Consider the changes that occur in the microstructure of a solution-treated alloy when it is aged. First, some segregation of elements occurs and small precipitates are nucleated. As ageing continues, the precipitates increase in size and volume fraction. A point is soon reached during ageing at which the volume fraction stops increasing, but the average size of the precipitates and their distance apart continue to increase: larger-than-average precipitates grow while the smaller ones shrink and disappear. This is called **coarsening**, a process that is driven by the decrease in total surface area and, hence, surface energy of the precipitates.

Now consider the effects that these changes have on the modes of deformation. When the precipitates are very close together, a high enough applied stress causes a dislocation to cut through both the precipitates and the surrounding solid solution. The precipitates may resist movement of the dislocation for any of the following reasons:

(a) The precipitates may not fit perfectly within the 'holes' left for them by the surrounding matrix. This causes both the precipitates and the matrix to be strained elastically. In general, the strain, or stress, fields of the precipitates and an approaching dislocation will interact and the dislocation will be obstructed. This effect is usually most pronounced when the precipitate and matrix are coherent: that is, atomic planes are continuous across the interface between the two phases.

(b) The precipitate may have a high intrinsic strength, which means that the shear stress required to move a dislocation is high, especially if it is a hard intermetallic compound.

(c) Passage of a dislocation through a precipitate will cause a 'step' to form on the surface of the precipitate, thereby increasing its surface area and, hence, its interfacial energy. This 'energy cost' contributes to the obstacle effect, but its contribution is relatively small.

It can be shown using a model of a dislocation cutting through spherical precipitates that either mechanism (a) or mechanism (b) cause the shear strength, τ, of the alloy to depend on the microstructure according to the following equation

$$\tau \propto (rV_p)^{1/2} \qquad (2.16)$$

where r is the radius of a precipitate and V_p is the volume fraction. As ageing proceeds and r and V_p increase, the strength will rise (Figure 2.28), but this tendency does not continue indefinitely. A point is usually reached during coarsening when the gaps between adjacent precipitates become so large that the stress required to make a dislocation bend sufficiently to pass through the gap is less than the stress required to cause cutting of the particles. The onset of this 'bowing' or 'looping' marks the maximum in strength because, as the precipitates grow further, they become even further apart and the strength falls (Figure 2.29). This effect is called **over ageing** and it arises because the stress to cause bowing τ_b depends inversely on the size of the gap between precipitates, λ, as you will now see.

Consider a loop of mobile dislocation of radius r lying on a slip plane in a unit cube of crystal acted upon by shear stress τ. Suppose that the loop expands slightly to a radius of $(r + \Delta r)$. The circumference is now longer and a length of dislocation $2\pi\Delta r$ has been created. Since a unit length of a dislocation has an energy Γ associated with it, this requires an energy of $2\pi\Delta r\Gamma$ to be made available. Where does this energy come from? It can come only from the work done by the external agency (i.e. a machine) that is deforming the crystal. When the loop expands, the shear strain γ that occurs in the unit cube is, recalling Equation (2.8),

$$\begin{aligned} \gamma &= b\Delta A \\ &= b2\pi r\Delta r \qquad (2.17) \end{aligned}$$

where $\Delta A = 2\pi r\Delta r$ is the area swept out by the moving dislocation. The work done, $W = \tau\gamma V$, where V is the volume of the cube — unity in this case. Substituting for γ, from Equation (2.17) gives

$$W = \tau b 2\pi r\Delta r$$

The loop will expand only if the work done is equal to the energy required, $2\pi\Delta r\Gamma$: that is, if

$$\tau b 2\pi r\Delta r = 2\pi\Delta r\Gamma$$

Taking Γ as approximately equal to $Gb^2/2$ and rearranging gives

$$\tau \simeq \frac{Gb}{2r} \qquad (2.18)$$

This states that the stress required to bend a dislocation is *inversely* proportional to the radius of curvature. Now consider a dislocation approaching a pair of particles, between which there is a gap λ (Figure 2.30). We shall suppose that the particles are effective obstacles to the dislocation and therefore the dislocation will stop at or near the particle interface. What does the length of dislocation do *between* the particles? It bends into the gap with a radius dictated by Equation (2.18) and then stops. If the applied shear stress τ is steadily increased, the dislocation bows further into the gap with a decreasing radius of curvature. When the dislocation bends into a semicircle, $2r = \lambda$ and a critical point is reached. The dislocation can now pass through the gap without the need for any further reduction in radius, or increase in stress. Indeed, henceforth the radius increases and the dislocation 'bridgehead' breaks out in an unstable manner. Now you can see that the yield strength of the alloy is determined by Equation (2.18), with $2r = \lambda$, because this represents the maximum resistance to dislocation 'bowing' provided by the particles.

To summarize, in most age-hardened alloys, ageing at a given temperature causes initially an increase in yield strength as closely spaced precipitates form and grow, causing V_p and/or r in Equation (2.16) to increase. The strength is determined by the stress required to make dislocations *cut through* the particles. As ageing and coarsening proceed, however, the average gap between adjacent precipitates, λ, increases progressively. Eventually, the stress required to make dislocations bow through the gaps (Equation 2.18) becomes less than that required to make them cut through the precipitates and further ageing causes the strength to fall. Clearly, an optimum ageing time and attendant precipitate microstructure are required for maximum strength.

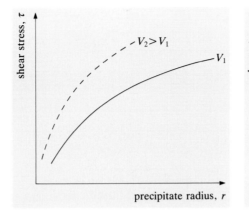

Figure 2.28 The initial growth of precipitates during ageing is accompanied by an increase in the volume fraction, V_p, and a corresponding increase in the shear stress required for dislocations to cut precipitates

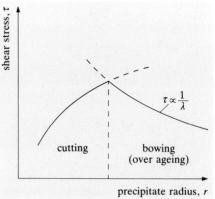

Figure 2.29 The maximum in strength corresponds to the precipitate size at which bowing becomes easier than cutting

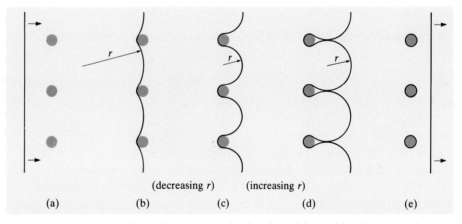

Figure 2.30 A moving dislocation encountering hard particles and bowing through the gaps between them

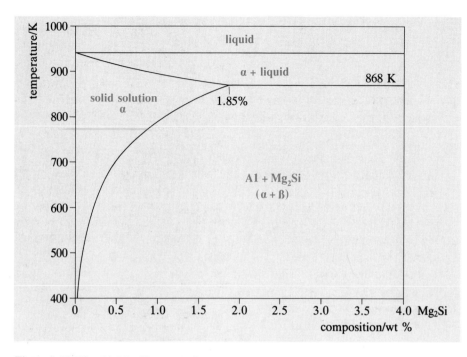

Figure 2.31 The Al–Mg$_2$Si 'pseudo-binary' phase diagram

As an example, we shall run through the sequence of events involved in the age hardening of a typical aluminium alloy — the alloy 6063. This is an aluminium–magnesium–silicon alloy of the 6000 series, in which the ratio of magnesium to silicon (atomic%) is two to one. Although such alloys contains three elements, their phase chemistry can be represented by a binary phase diagram (Figure 2.31), with the primary solid solution of aluminium, the α phase, and the equilibrium precipitate, Mg$_2$Si, the β phase. The 6063 alloy is solution treated for about one hour at 798 K before being quenched in water to room temperature. Ageing is subsequently carried out at 463 K and maximum strength is attained after about three hours.

You may wonder why the alloy is quenched to room temperature and then reheated to 463 K, instead of being quenched directly to 463 K. The answer is that it produces a greater hardening and the reason for this concerns the nucleation of the precipitate. Nucleation occurs when materials undergo changes such as recrystallization, solidification or other phase changes. The total energy required as a nucleus begins to grow *rises* at first; the energy required to provide its interface and the strain energy required when the nucleus is not a perfect fit within the host solid solution may actually *exceed* the free energy released by the phase change. This energy debt of small nuclei is reduced when they form on existent heterogeneities — defects in the crystal such as vacancies, dislocations and grain boundaries. Quenching aluminium alloy 6063 to room temperature rather than to 463 K introduces a larger supersaturation of vacancies which, being unstable, 'precipitate' to form vacancy clusters and dislocations (see Figure 2.32). These features in turn act as sites of heterogeneous nucleation of the precipitates of second phase that form on ageing.

Near grain boundaries, these features may be absent because a grain boundary has a powerful attraction for vacancies and sucks vacancies from the adjacent zone. This region then has no nuclei for precipitation on ageing and a 'precipitate-free zone' results.

As in most alloys, the precipitates that form first are not the equilibrium precipitates. The first precipitates are the so-called **Guinier-Preston zones**, or GP zones, which, in this case, are shaped like needles and lie in the direction of the edge of the unit cell of the face-centred cubic host crystal. These are 'coherent' with the host crystal: that is, all the crystal planes, although distorted, are continuous from the Guinier-Preston zones to the matrix crystal (Figure 2.33). The zones are not stable and, on ageing, they develop into rods of a β' phase, having a composition Mg_2Si and an hexagonal crystal structure. The atoms of this phase show long-range order: that is, the magnesium and silicon atoms are juxtaposed into a long-range pattern, like the atoms of sodium and chlorine in salt. Such precipitates form very effective obstacles to dislocations and the maximum hardness is reached when these precipitates are present.

Finally, as ageing continues, the β'-phase precipitates change in shape and crystal structure to precipitates of the equilibrium β phase — Mg_2Si platelets with a face-centred cubic crystal structure. By this stage the alloy is 'over aged', although it is much harder than it was before ageing began.

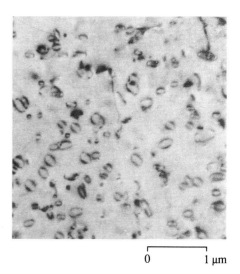

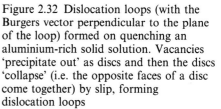

Figure 2.32 Dislocation loops (with the Burgers vector perpendicular to the plane of the loop) formed on quenching an aluminium-rich solid solution. Vacancies 'precipitate out' as discs and then the discs 'collapse' (i.e. the opposite faces of a disc come together) by slip, forming dislocation loops

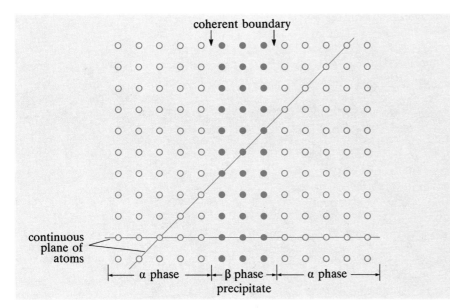

Figure 2.33 The continuity of crystal planes, indicated by the two marked lines, across the interface of a coherent precipitate

We must not fail to mention the roles of particles that do not participate in the precipitation-hardening process. There are two types of such particles. The first are coarse intermetallic compounds: that is, alloys such as $FeAl_3$ in which the metals are mixed in approximately simple proportions. They range in size from 0.5–10 μm and form, often from impurities, either during solidification of an ingot or during subsequent hot processing. The second type are smaller intermetallic compounds or 'dispersoids', 0.05–0.5 μm in size, which form during homogenization annealing of cast ingots.

Both types of particle are too widely spaced to cause significant increases in the proof or tensile strength, but they can affect other mechanical properties, both favourably and adversely. By interfering with the motion of grain boundaries and the attendant enlargement of grains during heat treatment, they can have a beneficial effect because, at low temperatures, 'small grains are better than big grains' for both strength and ductility. In some cases, the dispersoids may inhibit recrystallization altogether and the deformed grain structure is retained after annealing; after rolling the grains are pancake shaped, with fewer grain boundaries in the through-thickness direction than is the case with equiaxed grains (see Figure 2.34). This grain structure has superior corrosion resistance. On the other hand, the coarse, brittle particles reduce the work of fracture by forming cracks and voids during deformation.

Although the principle of age hardening was first discovered in aluminium alloys, it is not confined to them. Copper can be hardened this way by additions of beryllium or chromium to form strong, but expensive, alloys. (Since a precipitated microstructure does not scatter conduction electrons as effectively as a solid solution of the same composition, these alloys also have high electrical conductivities.) Titanium alloyed with 2.5% copper can also be age hardened, as you will see. Although the extent of strengthening is modest, some magnesium alloys are used in the age-hardened state. Nickel-based alloys can also be hardened this way, but more about this later. The properties obtained by heat treatment in a selection of these alloys appear in Table 2.6. You might like to compare the values of specific strength with those for the aluminium alloys.

In this section we have considered age hardening, or precipitation hardening, as the only new principle of strengthening metals to have been discovered in the past 2000 years. You saw how it depends on thwarting equilibrium and how it is widely applicable to nonferrous metals. In the next chapter, you will see that it can be applied to steels.

2.4.4 Modes of deformation

In the last section we said that products are usually designed so that, in service, they deform within their elastic range. The term 'strength' defines the effective upper limit to this range. The event occurring at this limit may vary from one material to another.

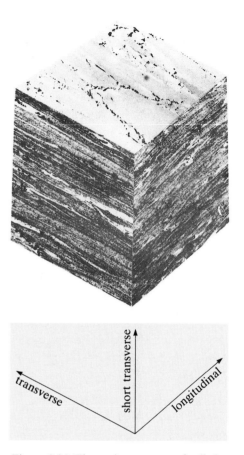

Figure 2.34 The grain structure of rolled and heat-treated alloy 7017

Table 2.6   Properties of a selection of age-hardened sheet alloys at room temperature

Alloy	0.2% proof stress, σ_p/MN m^{-2}	Tensile strength/ MN m^{-2}	Elongation/%	Young's modulus/GN m^{-2}	Fatigue strength at 10^8 cycles/% tensile strength	Density, ρ/Mg m^{-3}	Specific strength $\sigma_p \rho^{-1}$/N m g^{-1}
aluminium-based							
2014-T6	410	480	13	70	25	2.8	146
6082-T6	255	300	8	70	30	2.8	91
7075-T6	500	570	11	70	23	2.8	179
magnesium-based							
Mg–2Zn–1Mn	120	240	11	45	40	1.74	69
titanium-based							
Ti–2.5%Cu	550	690–920	10	105–120	60–65	4.56	121
nickel-based							
nimonic 263	600			200		8.9	67

Three possible events that could mark the end of wholly elastic deformation are: the onset of plastic strain; brittle fracture; the creation of cracks (e.g. by fatigue).

Each of these modes of deformation has its own loading criterion or strength.

Plastic deformation of metals can occur by a variety of atomic mechanisms. Under given conditions of temperature and applied stress, one mechanism is usually faster than the others and most of the plastic strain occurs by this mechanism. A change of conditions may result in a change in the dominant mechanism, accompanied sometimes by a radical alteration to the very nature of the deformation. For example, deformation may change from a strain-hardening to a non-hardening kind. In the latter case 'creep' occurs: that is, there is continuous deformation under a constant stress. This could be disastrous if it occurred in service!

The dependence of the mechanism of plastic deformation on the conditions of stress and temperature can best be represented by a 'deformation map'. As an example, the deformation map for aluminium is shown in Figure 2.35. Its axes give the deformation conditions and the space on the diagram is divided into areas within which a given mode of deformation is dominant (see ▼**Plastic deformation mechanisms and deformation maps**▲).

2.4.5 Modes of failure

You saw earlier that the loading on the cabin walls of an airliner arises primarily from pressurization and is greatest when an aircraft is at a high altitude. Under these conditions the outside temperature T is low (~ 215 K) and the homologous temperature, $T/T_m \simeq 0.24$. According to

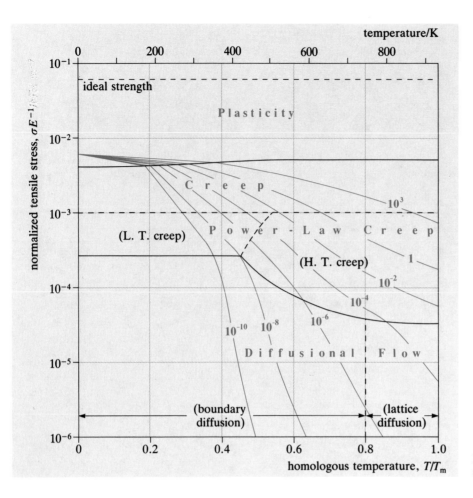

Figure 2.35 A deformation map for aluminium of small grain size

Figure 2.35, at this temperature the material is in the low-temperature creep regime and to produce a strain rate of $10^{-10}\,\text{s}^{-1}$ (a total strain of 5% over twenty years), a stress ratio, σ/E of 2×10^{-3} is required. Taking $E = 70\,\text{GN}\,\text{m}^{-2}$, this requires a tensile stress, σ, of $140\,\text{MN}\,\text{m}^{-2}$. This is about the same as the proof strength of pure aluminium, so it appears that creep is not a problem under these conditions. You would therefore not expect creep to occur at this temperature in aluminium and it is even less likely in aluminium alloys.

In designing an airliner, the designer must ensure that the material selected will not fail during the life of the aircraft. What do we mean by 'fail'? In general, it means 'fails to perform the intended function', which in this case is to contain the pressurized atmosphere. If it were to explode or even to develop a serious leak, it would have failed.

Three ways in which major leaks could develop in an aircraft cabin during service are:

• The wall could tear open on its first flight if the hoop stress was allowed to exceed the tensile strength of the material ('overload').

▼Plastic deformation mechanisms and deformation maps▲

Look at Figure 2.35 — begin at the top of the map where, at very high stresses, there is a line marked 'ideal strength'. When deformation is carried out under conditions that lie in the region above this stress, plastic flow is catastrophic and causes collapse. The stresses required to bring this about are very high, greater than $1600\,\mathrm{MN\,m^{-2}}$ for aluminium, and are equal to $0.06E$, where E is Young's modulus. Metals should not meet such stresses in normal service.

What are the 'carriers' of plastic deformation in crystalline solids such as metals?

They are dislocations — the mobile crystal defects that cause plastic strain when they move. When the ideal strength is exceeded, new dislocations are nucleated homogenously and move at high speeds within the crystal. At stresses below this, only pre-existing dislocations can move. They do so by gliding on their slip planes, as occurs in the region labelled 'plasticity'. As you come down the diagram to conditions of low applied stress, plasticity occurs at significant rates only when the crystal is sufficiently 'softened'.

Under what conditions does a crystal become softened?

When the temperature is raised sufficiently. For pure metals, this means raising the temperature to about $0.3T_{\mathrm{m}}$. Under these conditions the thermal activation, which is proportional to kT, is sufficiently vigorous to stimulate atomic events that soften the metal and thereby offset the effect of work hardening. The most important of these events is **dislocation climb**. At lower temperatures a dislocation is confined to move, or 'glide', only within its slip plane — the plane that contains both the Burgers vector *and* the dislocation line itself. Dislocation climb, however, occurs when the direction in which an edge dislocation moves has a component *normal* to the slip plane.

Climb gives edge dislocations a quite new degree of freedom. When they are held up by a strong obstacle in their slip plane, such as a particle, for instance, they can

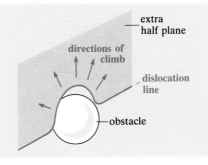

Figure 2.36 Climb past a precipitate

outflank the obstacle by climbing over it and then continue to glide in the slip plane, as shown in Figure 2.36. Note that pure climb causes no plastic strain, only glide does that, but climb is important because it can enable a dislocation that has been stopped to resume gliding.

There are two areas of Figure 2.35 where deformation occurs by creep under the control of dislocation climb. In both regions you have what is known as **power-law creep**, so called because the rate of steady-state creep depends on the applied shear stress raised to some power. One region is labelled 'low-temperature creep' and in this region the rate of climb is controlled by diffusion along fast paths, such as the cores of edge dislocations. Like grain boundaries, these slightly dilated regions provide paths for quicker diffusion than the perfect crystal does. In the region of 'high-temperature creep', diffusion within the crystals is sufficiently fast to control climb because it provides many more potential paths for diffusion than the dislocation cores and the rate of atomic jumps in the crystal is now significant, whereas at lower temperatures it is not.

Within these regions are drawn contours of equal strain rate, in units of fractional strain per second. At low temperatures the rate is too small to quantify.

The driving force for climb comes from the reduction in strain energy that accompanies the escape of a dislocation from a pile-up of dislocations at an obstacle. It is the arrest of dislocations

that causes work hardening and the rapid termination of any increase in strain under constant stress at low temperature. Climb frustrates this process and leads to steady-state plastic deformation occurring under a constant stress.

Finally, at the bottom of the diagram is a region of low stress labelled **diffusional flow**. This occurs in polycrystalline materials with a normal (i.e. rather small) grain size and does not involve dislocations because the applied stresses are too feeble to move them. After all we have told you about dislocations, you may well wonder how any plastic deformation is possible without them! As the name of the mechanism suggests, diffusion changes the shape of a crystal and thereby causes plastic strain.

Consider a small grain in Figure 2.37. If atoms are removed from the sides and attached to the top and bottom, the grain will become longer and the applied tensile stress will do work. This process can happen by diffusion of atoms under the driving force of the work done by the applied stress. The diffusion may occur through the volume of the crystal, in which case the atoms have a slower jumping frequency, but many potential paths, or along the grain boundaries, when the jumping frequency is higher, but there is a limited choice of paths. If diffusion through the lattice predominates, the creep is called **Nabarro-Herring creep** and if boundary diffusion prevails it is called **Coble creep** after the discoverers of these mechanisms.

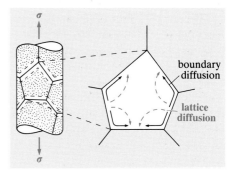

Figure 2.37 Deformation by diffusional flow

• The wall could become cracked during use due to 'fatigue'.
• The wall could become perforated by pitting corrosion (see Section 2.4.6). Alternatively, general corrosion could so reduce the thickness and raise the stress that the wall could tear open, as in the first example above.

Overload

To avoid the first mode of failure it simply requires the maximum stress to be kept at all times below the tensile strength. This danger is an obvious one and therefore such failures are rare.

Fatigue

Fatigue is more subtle. Although it has been recognized since the 1850s, as recently as the 1940s engineers did not have a realistic model of the phenomenon and they referred to it misleadingly as 'crystallization' (an example of this is to be found in Nevil Shute's popular novel *No Highway*). In the past forty years the consensus has emerged that fatigue is simply the nucleation of one or more small cracks, usually on the surface at a site where the tensile stress is large, and the subsequent slow growth of these cracks due to the load being repeatedly reversed. With continued growth, they eventually reduce the net cross-section to the point where the uncracked section is overloaded, whereupon sudden collapse occurs. Fatigue causes more service failures of machines, vehicles, bridges and similar structures than any other mode of fracture. The nucleation stage of fatigue is rendered unnecessary in cases where cracks already exist in the product: for example, within faulty welds. In crack-free components, nucleation is associated with plastic deformation, and cracks form and grow initially along slip bands: this is Stage I of growth shown in Figure 2.38. Such bands usually grow on planes carrying the largest shear stress. Later, at Stage II, these cracks change direction and grow on the plane of maximum *tensile* stress.

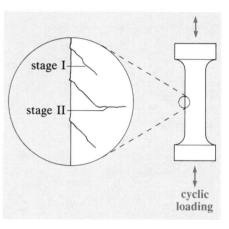

Figure 2.38 A sketch showing two stages of growth of a fatigue crack from the surface

In some cases, the crack faces are formed with parallel markings called **fatigue striations** (Figure 2.39). These lie perpendicular to the direction of crack growth and may be seen on the fracture surfaces with an optical or scanning electron microscope. They have been shown to be 'resting places' of the advancing crack front. A striation may correlate with one cycle of loading if all the stress cycles are identical, but in the case of a mixture of different stress amplitudes, some cycles may be too small to cause detectable crack growth and there may be no such correlation. Striations can provide valuable information on the history of a crack to an investigator trying to account for a service failure.

On a coarser scale, faces containing fatigue cracks usually have visible features called **shell marks** or **beach marks**. Like striations, these also reveal the shape of the crack front, but they do *not* show each cycle of growth. They are formed by changes that occur, throughout the life of the crack, in the loading conditions and/or in the extent of corrosion of

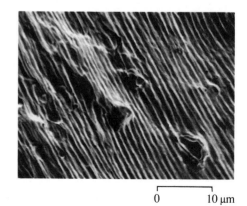

0   10 μm

Figure 2.39 Striation marks on the fatigue fracture surface of an aluminium alloy

the crack faces. Shell marks are what you can see with the naked eye on fatigued components (Figure 2.40).

In fatigue, the number of loading cycles that a small simply shaped test piece of material can withstand without breaking depends on the amplitude of the changing stress. This is given in the *S–N* **curve** for a material, of which Figure 2.41 is a particular example. To produce such a curve, a series of test pieces is subjected to different stress amplitudes and, in each case, the number of cycles before fracture is recorded. The stress amplitude that produces fracture is known as the **fatigue strength** for the corresponding number of stress cycles.

From Figure 2.41, the fatigue strength for aluminium alloy 2014-T6 corresponding to a life of 15 000 cycles — the number of pressurization cycles the cabin of the Comet was expected to go through in its designed lifetime — is 200 MN m^{-2}. To avoid fatigue in the walls of a cabin made from this alloy during the lifetime of an aircraft the stress changes occurring must not exceed this value.

How does this fatigue strength compare with the tensile strength of alloy 2014?

From Table 2.6, the tensile strength of 2014-T6 is 480 MN ,$^{-2}$ and, therefore, the fatigue strength for a life of 15 000 cycles is lower than the tensile strength. Thus, if a cabin wall is designed to avoid fatigue, it will automatically avoid failure due to overloading.

The proof stress for alloy 2014 given in Table 2.6 is 410 MN m^{-2}, so the fatigue strength is also less than the proof stress. This makes a general point that fatigue can occur under applied stresses that are well below the proof stress. It does not mean, though, that no plastic deformation is taking place during fatigue. It is, but only locally at regions of severe stress concentration at a crack tip (see ▼**Stress concentrations**▲).

Plastic deformation at the crack tip is accepted to be a necessary part of the mechanism by which fatigue cracks grow; since the precise mechanism is still a matter of dispute among experts, we shall not attempt to describe it here.

In some products, such as engine parts, for example, it is not unusual to require lives to be of the order of 10^8 cycles. Typical *S–N* curves for aluminium alloys, such as Figure 2.41, show that as the expected lifetime is increased to these high values the fatigue strength is reduced. Unlike ferrous materials, the fatigue strength of nonferrous metals continues to fall as the number of stress cycles before fatigue failure increases and they do not have a true fatigue limit: that is, a maximum stress amplitude below which no fatigue occurs. Therefore the fatigue *strength* at a given large number of cycles, commonly 10^8, is often quoted. It is frequently called the **fatigue limit**.

The fatigue strength, at 10^8 cycles, of a heat-treated aluminium alloy is a small proportion, less than 30%, of the tensile strength. As these alloys

Figure 2.40 Shell or beach marks on a broken steel crankshaft

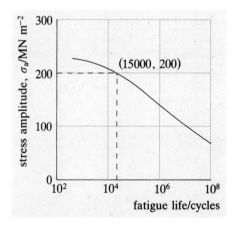

Figure 2.41 An *S–N* curve for the aluminium alloy 2014-T6, loaded cyclically between a stress of zero and the peak stress

▼Stress concentrations▲

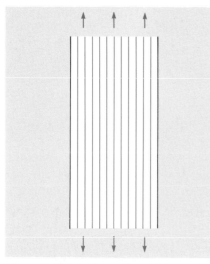

Figure 2.42 Imaginary lines of force within a bar of uniform section

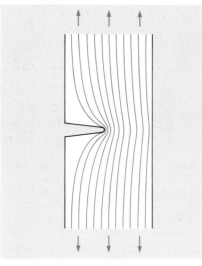

Figure 2.43 The redirection of lines of force caused by a change in section

Suppose that you have a parallel-sided bar of an elastic material carrying a total axial force F, applied uniformly across the ends of the bar. Within the bar, you might imagine this force to be carried by a series of equally-spaced filaments of material (Figure 2.42). Suppose we call the filaments 'lines of force'. What happens to these lines if we cut a piece out of the bar, creating a notch (Figure 2.43)? The lines will have to flow around the notch and, in so doing, they will intersect the smallest cross-section of the bar. Furthermore, within this section they will crowd together near the tip of the notch.

Since each line represents a given force, the *density* of lines (lines per unit area in the limit as the area considered approaches zero) represents stress (force per unit area). There is therefore a 'concentration' of stress. The maximum stress occurs at the tip of the notch and, provided the deformation is elastic, it is given by

$$\sigma_{max} = \sigma_{nom} \left(1 + 2 \sqrt{\frac{D}{r}} \right) \qquad (2.19)$$

where σ_{nom} is the 'nominal' stress, F/A, applied to the ends of the bar, D is the depth of the notch and r is the radius of curvature at the tip of the notch.

The term in the brackets is called the **stress concenration factor**, K_t, and it applies approximately to a variety of shapes (Figure 2.44) where the dimensions D and r are as shown.

A sharp crack may be regarded as a notch with a finite length D and a tip radius r that approaches the value of the interatomic spacing; essentially zero. According to Equation 2.19, as r approaches zero, the stress concentration factor approaches infinity, provided all the deformation is elastic. In practice, no real material can withstand an infinite stress and there will always be some nonelastic deformation, for example by slip, near the tip of a crack under stress.

SAQ 2.7 (Objective 2.7)
What is the stress concentration factor for (a) a semicircular notch and (b) a surface notch of length 10 mm and a tip radius of 0.5 mm?

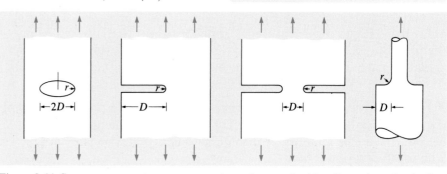

Figure 2.44 Some common stress concentrations characterized by dimensions depth, D, and radius, r

are progressively strengthened by alloying, this fatigue strength does not increase in proportion and it becomes a diminishing fraction of the tensile strength. This arises from the nature of the plastic deformation that occurs during fatigue of these alloys; it tends to be concentrated into a few narrow slip bands within which cyclic shear occurs. This localized shear chops up the strengthening zones and precipitates to the point where they become too small to be stable and, by a process called 'reversion', they dissolve, thereby causing softening. In this connection, the dispersoids, which are not cut by dislocations, perform the useful function of dispersing slip into a larger number of bands.

So far we have only considered the stress amplitude applied to a material as if the general stress level, or mean stress, did not matter. It does! An S–N curve should indicate the mean stress level used in the tests. To obtain Figure 2.41, for example, the load was varied cyclically between zero and the peak stress, giving a mean stress equal to the stress amplitude: a mean stress of zero, however, is a common choice for tests. Figure 2.45 shows a series of S–N curves for an aluminium alloy at different values of mean stress.

What happens to the fatigue strength as the mean stress is raised?

You can see that, at any particular value of N, as the mean stress is increased, the stress amplitude for failure is reduced. This trend is often described reasonably well by plotting the stress amplitude, σ_a, against the mean stress, σ_m (Figure 2.46). Suppose when the mean stress is zero, the fatigue strength is σ_{a0}. When the stress amplitude is zero, the mean stress is equal to the tensile strength, σ_t. On the basis of experimental results Goodman assumed a linear fall in the fatigue strength between these points, as shown. The equation of the Goodman line is

$$\sigma_a = \sigma_{a0}\left(1 - \frac{\sigma_m}{\sigma_t}\right) \qquad (2.20)$$

For the Comet, when the cabin is on the ground it is not pressurized and the stresses in the wall are zero. Thus, since the minimum stress is zero, the mean stress and the stress amplitude are equal. Therefore we can put $\sigma_m = \sigma_a$ in Equation (2.20) and, rearranging, this gives

$$\sigma_a = \frac{\sigma_{a0}}{1 + \sigma_{a0}/\sigma_t}$$

If σ_{a0} is the fatigue strength at 10^8 cycles and when the mean stress is zero, for a typical high strength alloy, $\sigma_{a0}/\sigma_t = 0.25$. Substituting for this in the above equation, $\sigma_a = 0.80\sigma_{a0} = 0.20\sigma_t$. Thus the effect of this mean stress is to reduce the safe working stress to 80% of the fatigue strength when the mean stress is zero.

To avoid fatigue, safe working stresses must be accurately assessed, otherwise 'safety factors' have to be used in the design. This is

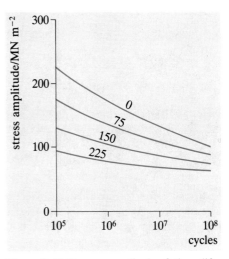

Figure 2.45 For any particular fatigue life, the stress amplitude to cause fracture — the fatigue strength — decreases as the mean stress, σ_m, is increased. The mean stress for each curve is given in MN m^{-2}

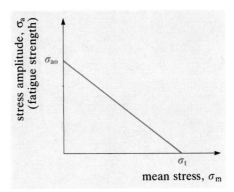

Figure 2.46 Goodman assumed that the relationship between fatigue strength and mean stress was linear

tantamount to assuming that the loads that have to be carried are the real loads inflated by a factor greater than one. Weight, however, is so important in aviation that factors of safety must be kept small and, in order to prevent fatigue, an accurate knowledge of stresses is required.

In many cases, the loading experienced by a component is not a single periodic cycle, like that of a watch spring, but is complex, like the loading of a car's suspension. ▼Miner's rule▲ provides a means of estimating the life of a component under these conditions.

Most of what is known about fatigue comes from observations of test pieces, but this knowledge must be applied to the design of engineering products, which are generally larger and more complex in shape. This is not easy. If a given stress amplitude applied for a given number of cycles causes a small test piece to break, applying the same stress for the

▼Miner's rule▲

Suppose that a component is subjected to a range of stress cycles: n_1 cycles of amplitude σ_1; n_2 cycles of amplitude σ_2; ... n_i cycles of amplitude σ_i. What will its fatigue life, N, be? Miner proposed a theory of 'linear damage' to answer this question.

He suggested that if the total life of a component when subjected to stress cycles of constant amplitude σ_1 is N_1, then subjecting a component to n_1 cycles would cause its life to be reduced by n_1/N_1. By extending this argument to cycles of other amplitudes, he concluded that fracture would occur when

$$\frac{n_1}{N_1} + \frac{n_2}{N_2} + \ldots + \frac{n_i}{N_i} = 1$$

That is, when

$$\sum_i \frac{n_i}{N_i} = 1 \qquad (2.21)$$

This is Miner's rule. It is used widely and, although it is known not to be very accurate, it has proved difficult to improve upon with simple procedures. Although the methods of fracture mechanics provide an alternative, they are complex and beyond the scope of this book.

It is a wonder that Miner's rule works as well as it does. The growth of a fatigue crack under stress cycles of constant amplitude is known to be nonlinear — the crack growth rate continuously increases. Also, it is now known that the order in which the stress cycles of different amplitude occur may affect the growth of a crack. Furthermore, the proportion of the fatigue life taken up in nucleating a crack varies with the stress amplitude and it can occupy a majority of the life of a product that is subjected to small stress amplitudes and therefore normally has a long life. In spite of all these complications, Miner's rule provides a useable method of estimating the life of a component subjected to stresses of different amplitudes.

SAQ 2.8 (Objective 2.8)
An aircraft component is subjected to 20 stress cycles of amplitude $210 \, \text{MN m}^{-2}$, 400 cycles of amplitude $140 \, \text{MN m}^{-2}$ and 1000 cycles of amplitude $70 \, \text{MN m}^{-2}$ in one month of operation. Use the S–N curve in Figure 2.41 to determine the number of months of operation before you would expect fatigue failure of the component if this pattern were continued.

same number of cycles would not necessarily break a large structure made of the same material, although it will usually cause a fatigue crack to appear. The reason for this is because fatigue causes *stable* cracks to develop and these cause total fracture only when they become *unstable*.

An important question about an airliner cabin is: to what length may a crack grow before it becomes unstable under service conditions? The answer affects how often aircraft that are in service need to be inspected. As an extra line of defence against fatigue, it is normal practice to examine the structure for cracks at specified intervals, both by eye and using portable instruments, such as those using eddy currents or the ultrasonic pulse-echo principle. If only small cracks are stable, this presents the inspectors with a needle-in-haystack problem.

The size of the largest stable crack that can be present under a given tensile stress depends on the toughness of the material (see ▼**Griffith and the stability of cracks**▲). It is given by Equation (2.22)

$$a_c = \frac{G_c E}{\pi \sigma^2}$$

where a_c is the size of the crack, known as the **critical crack length**; σ is the tensile stress and G_c is the toughness.

In choosing an alloy, some attention will have to be paid to the toughness of the materials being considered. Other things being equal, a large value will be an advantage because it permits the structure to be more 'crack tolerant'. Since we cannot yet guarantee to prevent the formation of any cracks during manufacture or service, this must be a consideration.

In aluminium alloys, the presence of coarse, brittle particles has an important influence on the toughness. As a crack grows during fracture and its tip approaches a particle, under the high stresses near the tip the particle will either crack or de-cohere from the surrounding aluminium. In this way a pattern of voids develops ahead of the crack and the crack extends by the growth and coalescence of neighbouring voids (Figure 2.49). The particles are beneficial in other ways so they should not be eliminated, but the toughness can be increased by refining the scale of the dispersoids and coarse particles.

2.4.6 The selection process

We derived earlier a merit index for lightness, which favoured the material with the highest specific strength. Inspection of Table 2.6 reveals that this is the aluminium alloy 7075. In the 1940s, however, this was a new alloy and de Havilland had no experience of its use. On the other hand, the alloy with the next highest specific strength, 2014, was well-known to the British aircraft industry and, accordingly, it was adopted for the Comet's cabin.

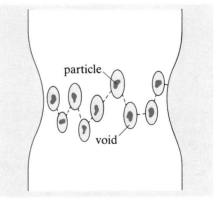

Figure 2.49 Fracture by the linking up (shown as broken lines) of elliptical voids containing particles

▼Griffith and the stability of cracks▲

In 1920 and 1924, A.A. Griffith, then working at RAE, Farnborough, published two papers that initiated a revolution, albeit a slow one, in the study of fracture. He was the first to recognize stress-concentrating flaws as fracture-initiation sites in brittle materials and showed that very high strengths were obtainable in flaw-free specimens. Of equal importance, he pioneered a quantitative approach to the mechanics of fracture by deriving a criterion for the stability of crack-like flaws. To do this he applied essentially thermodynamic reasoning to the energetics of a stressed body containing a crack.

Consider a plate of width w containing a central crack of length $2a$ through its thickness t and with $w \gg 2a$ (Figure 2.47). What happens if a uniform tensile stress, σ_a, is applied to the plate?

Three things need bearing in mind. The first is that by elastically deforming a body you introduce elastic strain energy, U, into it. This is a potential energy, available to do work. Secondly, a cracked body stores less strain energy than an identical, but uncracked one when deformed by the same amount. The crack faces separate

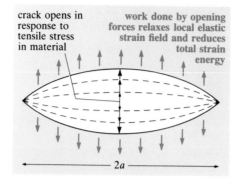

Figure 2.48 Reduction of strain energy due to the presence of a crack

and unload the material around it. The third thing is that surfaces have a specific surface energy, γ_0, per unit area associated with them — work is required to extend the area of any surface.

Griffith argued that the total potential energy U of an elastically deformed, cracked body like the plate in Figure 2.47 contained three components:

> U_0, the elastic strain energy of the equivalent *uncracked* body;
>
> U_s, the surface energy of the total area of cracked surface. This is positive since, if the crack healed up (which is allowed thermo-dynamically), U_s would *add* to the total elastic strain energy;
>
> $-W_r$, the work done by the forces moving the crack faces apart (Figure 2.48). This is *negative* because it reduces the elastic strain energy.

This he expressed as

$$U = U_0 + U_s - W_r$$

A crack can only grow if the total potential energy of the system *decreases*: that is, if

$$\frac{dU}{da} < 0$$

Since, by definition, U_0 is independent of crack length

$$\frac{dU}{da} = \frac{dU_s}{da} - \frac{dW_r}{da} < 0$$

expresses this condition and **Griffith's criterion** states that *for a crack to grow, the release of elastic strain energy due to*

that growth has to be greater than the surface energy of the extra cracked surfaces thus formed.

Now U_s = specific surface energy × total area of cracked surface

$$= \gamma_0 \times 2 \times 2ad$$

$$= 4ad\,\gamma_0$$

Therefore

$$\frac{dU_s}{da} = 4d\gamma_0$$

Evaluating W_r is more complex, but gives

$$W_r = \frac{\sigma_a^2 \pi a^2 d}{E}$$

Hence

$$\frac{dW_r}{da} = \frac{2\sigma_a^2 \pi a d}{E}$$

If we put

$$\frac{dU_s}{da} = \frac{dW_r}{da}$$

the equation can be used to define the stress at which a through crack of critical length $2a_c$ (a_c for an edge crack) will start to propagate: thus

$$\sigma_a = \sqrt{\frac{2E\gamma_0}{\pi a_c}}$$

Now Griffith was a bit naughty. In his first paper he had found good agreement between the results of fracture tests on precracked glass vessels and a somewhat different equation, which predicted values of σ_a more than double the above. What did he do with his Mark II equation? He varied the annealing treatments of his glass vessels until they produced results that agreed with it!

What Griffith did not realize was that, even in an archetypally brittle material such as glass, the work required to create new crack surfaces is more than just the thermodynamic surface energy. Other energy-absorbing processes, such as plastic deformation, need to be included. Nowadays these are taken into account by using the **toughness**, G_c, which replaces $2\gamma_0$ in his equation (and loses the thermo-dynamic link). Hence

$$\sigma_a = \sqrt{\frac{EG_c}{\pi a_c}} \quad \text{or} \quad a_c = \frac{EG_c}{\pi \sigma_a^2} \quad (2.22)$$

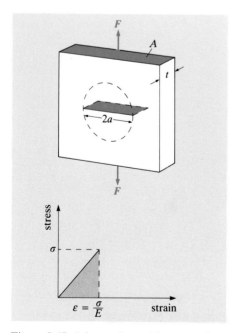

Figure 2.47 A large plate with a central crack

These are equivalent expressions of the critical condition for crack propagation in a material of toughness G_c. Both are referred to as the **Griffith equation**. Measured values of G_c for a variety of materials are given in Table 2.7, ranging from about $10\,\mathrm{J\,m^{-2}}$ for glass and concrete to more than 10^4 times this for some steels and titanium.

A related measure of toughness is the **fracture toughness, K_c** given by

$$K_c = \sqrt{EG_c}$$

From Equation (2.22) this may be written

$$K_c = \sigma_a \sqrt{\pi a_c}$$

The word 'toughness' is also used loosely to mean that quality of a material that is the opposite of brittleness and sometimes to describe the results of impact tests as 'impact toughness'. Both of these terms include the work to *initiate* a fracture as well as the subsequent work of propagation — the total **work of fracture**. G_c and K_c are both concerned solely with the *propagation* of pre-existing cracks.

> SAQ 2.9   (Objective 2.11)   What is the value of critical crack length for an edge crack in the alloy 2014 T 6 when stressed to half its proof strength? Take $G_c = 30\,\mathrm{kJ\,m^{-2}}$.

Table 2.7   Typical toughness values of some common materials

Material	Young's modulus $E/\mathrm{GN\,m^{-2}}$	Toughness $G_c/\mathrm{kJ\,m^{-2}}$
tool steel	210	10–110
mild steel	210	100
titanium alloy Ti-6Al-4V	120	20–110
aluminium alloy 2014	70	7–30
woods (across the grain)	0.2–1.7	8–20
GFRP (polyester glass SMC)	20	10
partially-crystalline thermoplastics	1–2.5	2–8
woods (along grain)	6–16	0.5–2
glassy thermoplastic (PMMA)	3	0.3–0.4
thermosetting plastic (epoxy)	4	0.1–0.3
alumina (Al_2O_3, a typical polycrystalline ceramic)	390	0.06
concrete	16	0.03
window glass	72	0.01

Although alloy 2014 was suitable on mechanical grounds, it posed a potential problem. When metals react chemically with their environment it is called **corrosion**. The Comet's design called for a skin material that would be stable for at least twenty years, so the alloy chosen had to have adequate resistance to corrosion.

'Wet' corrosion — corrosion in the presence of water — is an electrochemical process involving an anode and a cathode joined electrically by an aqueous electrolyte and it can be a serious problem when two different metals are joined together.

Corrosion cells do not always require two *separate* electrodes of dissimilar metals, however. The electrodes may be two different phases within the surface of the same piece of alloy. Provided the two phases have different electrode potentials and are bridged by an aqueous electrolyte, they can form a microscopic cell on the surface and the anodic phase is attacked. If the anodic phase is a small particle embedded in a cathodic matrix, it corrodes rapidly, but is soon removed, whereupon the reaction stops. More dangerous is the reverse situation, where a cathodic phase is embedded in an anodic matrix, such as copper in aluminium. Corrosion of the matrix surrounding the particles occurs and continues unabated because the particles are *not* removed. Both types of reaction can lead to **pitting corrosion**. Thus, for alloys based on a given metal, aluminium, say, single-phase alloys generally resist corrosion better than alloys containing a phase mixture.

Pitting corrosion, however, does sometimes occur in single-phase alloys. Variations in microstructure can exist within a single phase due, for example, to chemical segregation such as 'coring', or non-uniform states of plastic strain (i.e. of dislocation density). Another variation in the microstructure of a single-phased alloy occurs because a phase usually contains grains. The attendant differences in electrode potential because of these microstructural variations may be sufficient to create local corrosion cells. Furthermore, differences in the chemical environment can develop during corrosion; for example, concentration gradients of oxygen or chlorine ions may build up in the electrolyte and this may stimulate further local attack. Although corrosion is generally seen as a problem, metallography routinely exploits 'controlled pitting corrosion' in order to reveal microstructural features to the eye by etching with the appropriate reagent.

Alloy 2014 contains copper and copper reduces the corrosion resistance of aluminium more than any other element. Some of the copper resides in intermetallic compounds such as $CuAl_2$, which has an electrode potential 0.32 V higher than that of aluminium and therefore causes pitting corrosion. This was well known in the 1940s and a remedy had been devised. It was to enclose the alloy between two thin layers of 99.7% pure aluminium, which then acted as 'cladding'. These composite sheets can be made by hot rolling the three sheets of material together, a process known as roll bonding.

Although the aluminium cladding protects the alloy from corrosion well, it causes a significant reduction in life because fatigue cracks form earlier in the soft surface layers. This must be offset by reducing the design stress in the clad sheet.

The choice of alloy 2014, suitably clad with aluminium, was the cautious one of staying with a tried and trusted material. The designers had a radically new design specification with which to cope and did not want the additional challenge of a new material. Furthermore, Comet was ordered for production 'straight off the drawing board' and was therefore to be built without going through the prototype stage.

Over the period since Comet was designed, favourable experience has been gained with the stronger alloy 7075 and, today, it would be an alternative.

2.4.7 The outcome

The Comet entered airline service with BOAC in 1952 to become the world's first operational turbo-jet airliner by some seven years. Over the next two years, three aircraft were lost in service and the fleet was grounded in 1954 while the problem was investigated. It was found to be a cabin-skin problem, but *not* one arising from the choice of material. In two cases, after many hundreds of flights, the cabin exploded just as the aircraft reached its cruising altitude and the mode of failure was found to be the unstable growth of a fatigue crack. The skin was found to be over-stressed, particularly in the vicinity of cut-outs such as windows. The investigation into the accident revealed that, when the cabin was pressurized, the stress near the corner of a window was about $280 \, \mathrm{MN \, m^{-2}}$.

If the toughness of the alloy 2014 is about $30 \, \mathrm{kJ \, m^{-2}}$, what would the critical crack length have been under these conditions?

Equation (2.22) in the section on 'Griffith and the stability of cracks' gave the expression for critical crack length, a_c, as

$$a_c = \frac{G_c E}{\pi \sigma^2}$$

$$= \frac{30\,000(70 \times 10^9)}{\pi(280 \times 10^6)^2} \, \mathrm{m}$$

$$= 8.5 \, \mathrm{mm}$$

This is a very rough calculation because the real geometry was much more complicated than that for which the equation was derived. It was found by testing, however, that the critical crack length was indeed small — of the order of $20 \, \mathrm{mm}$.

The estimates of stress concentrations had not been accurate enough and this had not been picked up on subsequent testing of full-sized cabins. These tests suggested that the cabin should have been capable of withstanding pressurization for an adequate lifetime, but later the test cabins were found to be unrepresentative of the cabins used in service. A later test carried out on a grounded aircraft confirmed that the fatigue life was only some 2000 flights. Subsequent aircraft were modified with thicker skins and redesigned cut-outs.

Comets were put back into service with satisfactory results. In 1958, a Comet Mark 4 started the first jet-propelled trans-Atlantic service and the aircraft still flies today as the Nimrod. The Comet has already entered the history books as a milestone in aviation development.

Although this case history may appear to be rather old, it is still relevant from a materials point of view. Airframes are still constructed today from heat-treated aluminium alloys such as 2014 and the use of the higher-strength alloys such as 7075 has been very limited, due to their susceptibility to stress corrosion cracking and their low fatigue strength ratios (Table 2.6). However, the use of advanced composite materials for both military and civil aircraft is increasing steadily.

2.5 Skin material for a Mach 3 airliner

The Mach number attributed to an aircraft, in this case Mach 3, is the speed of the aircraft relative to the surrounding air, divided by the speed of sound in that air. The speed of sound in air depends on the square root of the air temperature, but it is independent of the air density. Above a certain altitude the air temperature becomes a constant 215 K and in this region, called the stratosphere, the speed of sound is about $290 \, \text{m s}^{-1}$. We now want to consider the needs of a new aircraft designed to cruise at just over $3000 \, \text{km h}^{-1}$, the speed chosen in the 1960s for an American supersonic transport (SST) aircraft. It was designed, but not built.

What are the main problems that affect the choice of material for the skin of such an aircraft?

The cruising altitude, about 20 km, is higher than that of the Comet in order to take even further advantage of reduced air density. At this height the air density is only about a tenth of that at sea level and the pressurization of the cabin will be about $90 \, \text{kN m}^{-2}$, compared with $57 \, \text{kN m}^{-2}$ for the Comet. This requirement can be met by an appropriate choice of wall thickness and does not present a materials problem.

Kinetic heating of the airframe, however, does present a problem. During high speed flight, huge amounts of work are done by the engines against the drag of the air and this work is transformed into heat, some of which ends up in the airframe.

The maximum temperature attainable at a given speed can be estimated as follows. Consider Figure 2.50, showing a model of a fuselage in a

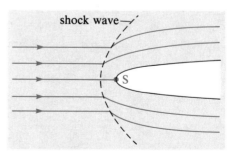

Figure 2.50 The flow of air past an aircraft

wind tunnel. Air is deflected either side of the nose, but at the apex, point S in the diagram, some of the air gives up its kinetic energy and is brought to rest. This kinetic energy is transformed into heat, raising the temperature of the stopped air by, say, ΔT. Thus

$$\frac{mv^2}{2} = mc_{\mathrm{p}}\Delta T$$

where m is the mass of air considered, c_{p} is its specific heat at constant pressure and v is the speed. Rearranging this you get

$$\Delta T = \frac{v^2}{2c_{\mathrm{p}}}$$

The rise in temperature therefore depends on the square of the speed. Table 2.8 gives the cruising speed for different aircraft and the corresponding temperature rise.

Since the ambient temperature at the cruising altitude of the proposed supersonic transport aircraft is 215 K, the skin must be capable of withstanding 630 K throughout its life. At points away from the nose of the cabin, temperatures are less than this by up to about 50 K

2.5.1 Follow the Concorde?

The only supersonic aircraft in service at present is the Anglo-French Concorde and from Table 2.8 the peak temperature endured by Concorde is 400 K. This is just within the scope of the best creep-resistant aluminium alloys and alloy 2618 is used. It is precipitation hardened by ageing at 463 K (Figure 2.51). Even this alloy is not truly stable at the service temperature (see Figures 2.52 and 2.53) and it undergoes creep deformation over the 'hot' design life of 20 000 hours (see Figure 2.54). Concorde's designers required the total creep strain during service to be kept below 0.1%.

As you can see from Figure 2.54, the creep strain in alloy 2618 when it is stressed to 176 MN m^{-2} for 20 000 hours is just within the design

Table 2.8 Cruising speed and kinetic heating temperature for various aircraft

Aircraft	Speed/Mach	ΔT/K
Comet	0.7	23
B747	0.85	33
Concorde	2.0	185
SST	3.0	415

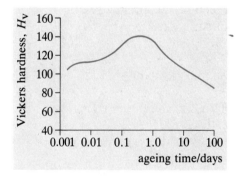

Figure 2.51 The effect of ageing at 463 K on the hardness of alloy 2618

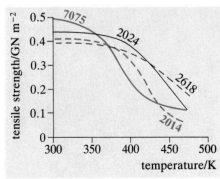

Figure 2.52 The tensile strength of four alloys at the indicated temperature, measured after 20 000 hours at this temperature

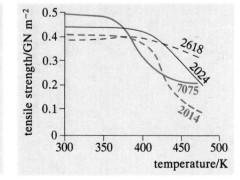

Figure 2.53 The tensile strength of four alloys at room temperature, measured after 20 000 hours soaking at the indicated temperature

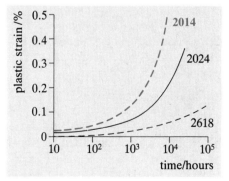

Figure 2.54 Creep curves for three alloys at 393 K under a stress of 176 MN m^{-2}

81

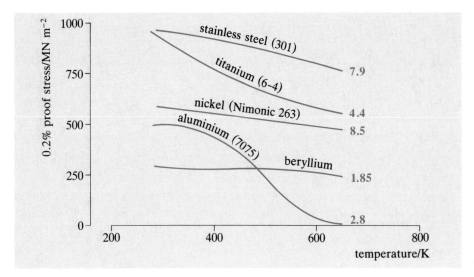

Figure 2.55 The proof stress of some aerospace alloys as a function of temperature. The density, ρ, for each alloy is given in $\mathrm{Mg\,m^{-3}}$

requirement of less than 0.1%, whereas the alloy used for the Comet, alloy 2014, creeps excessively. Both these alloys contain aluminium, copper and magnesium, but they differ in the ratio of copper to magnesium. Alloy 2618 also contains iron and nickel, which are deliberately added to form intermetallic particles to inhibit grain growth, and it resists creep better because the grain structure and the precipitates are more stable at 400 K.

Would this alloy be suitable for the skin of a Mach 3 aircraft?

The maximum temperature due to kinetic heating is 630 K and this is well above the alloy's ageing temperature of 463 K, at which temperature over ageing occurs after only one day (see Figure 2.51).

This temperature of 630 K is about 70% of the melting temperature of aluminium alloys, so we must abandon them and seek other alloys that have significant strength at 630 K. Figure 2.55 shows how the proof stress varies with temperature for the main families of structural metals. If you compare their specific strengths at 630 K, you find that the order (measured in $\mathrm{N\,m\,g^{-1}}$) is: titanium. 141; stainless steel, 108; nickel, 62; aluminium, 45. Titanium looks promising and we shall therefore examine its metallurgy.

Titanium alloys

This metal differs from all the other nonferrous metals you have met so far in that it shows **allotropy**: that is, on heating it undergoes a change of crystal structure. This fact dominates the metallurgy of titanium. At 1155 K it changes from the α (HCP) structure to the β (BCC) structure and at 1951 K the β phase melts. The addition of an alloying element can have one of three effects:

(a) Aluminium and oxygen, whose atoms form less than four bonds

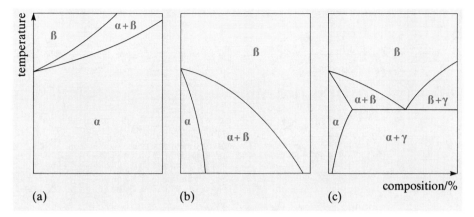

Figure 2.56 Types of phase diagram found among titanium alloys

with other atoms, stabilize the α phase: that is, raise the temperature range over which the α phase is stable, as shown in Figure 2.56(a).

(b) Elements that can form more than four bonds per atom, stabilize the β phase in two ways. With additions of elements such as molybdenum, niobium and vanadium, this happens continuously, as shown in Figure 2.56(b). Additions of copper, for example, stabilize the β phase up to the formation of a eutectoid, as shown in Figure 2.56(c).

(c) Elements that are similar to titanium in that they form four bonds per atom (e.g. zirconium, tin and silicon) are neutral in their effect on the α and β phases.

Allotropy offers the prospect of alloys with a choice of three structures —α, β and α plus β. It also offers the possibility of producing metastable microstructures by heat treatment, as is done in steels (see Chapter 3).

Two titanium alloys are manufactured in the appropriate form (sheets) for the cabin of a Mach 3 aircraft: one with 6% aluminium and 4% vanadium, an alloy of the first type mentioned in (b) above, and the other with 2.5% copper, an alloy of the second type described in (b). The latter is the only commercial titanium alloy to be precipitation hardened.

The creep performance of alloys is often described by a graph of stress plotted on a logarithmic scale against the **Larson–Miller parameter**, as shown in Figure 2.57. This empirical parameter, ϕ, contains the temperature, T, measured in kelvin, and the time, t, in hours, for which the material has been under stress and is defined by the equation

$$\phi = T(A + \log_{10} t)$$

where A is an empirical constant. A graph of this type refers to the conditions required to cause either fracture or a given amount of plastic strain by creep, 0.2% in this case.

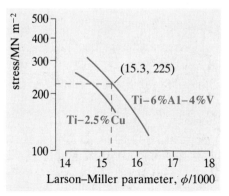

Figure 2.57 The Larson–Miller parameter, a function of time and temperature for a plastic strain of 0.2%, plotted against the applied stress

The graph may be used to compare the specific strengths of alloys in the following way. Suppose you would like to know the stress that can be endured for 20 000 h at 630 K without the strain exceeding 0.2%. First you would use the values $T = 630$ K and $t = 20\,000$ h to find ϕ.

$$\phi = 630(20 + \log_{10} 20\,000)$$
$$= 15\,300$$

Drawing a vertical line corresponding to this value of ϕ, you find that it cuts the graph for titanium alloy with 6% aluminium and 4% vanadium at a stress of 225 MN m^{-2}. By the same procedure, the required stress for the alloy with 2.5% copper is 160 MN m^{-2}.

Under the stated conditions the former alloy has the higher specific strength and is therefore preferred. For sheet material, it is used in the annealed condition. There are two alternatives: anneal at a temperature above 1278 K where the alloy consists of the β phase, or anneal at a lower temperature where there is a mixture of $\alpha + \beta$ phase. In both cases, when the annealing time has elapsed the alloy is cooled slowly to room temperature. In the former case, the alloy consists entirely of grains of the β phase when it begins to cool, but as it cools through the $\alpha + \beta$ phases field (Figure 2.56b) the β phase transforms progressively to the α phase by diffusion of atoms. The new grains of α phase have a distinctive flattened shape called **Widmanstätten** which produces a basket-weave pattern (Figure 2.58). The reason this is so regular is that the flat α grains form parallel to certain crystal planes of the β grains.

The alternative annealing is usually carried out at about 973 K, where the structure consists of intermingled equiaxed grains of α and β. On cooling slowly, the grains of α phase remain, while those of the β phase transform to Widmanstätten α grains, giving the structure shown in Figure 2.59.

In both cases, the titanium is strengthened by the solute atoms dissolved in each phase and by the refinement of the grains brought about by the transformation of β grains to α grains.

Although it is high in the electrochemical series, titanium forms a very stable protective film consisting of titanium dioxide, less than a nanometer thick. This good resistance to oxidation is maintained to 820 K, the highest temperature at which titanium alloys retain useful creep strength. The oxide film is self-healing, reforming almost instantanously if it is mechanically damaged in an oxidizing atmosphere.

The metal is unusual in the large extent to which it will dissolve various other elements: oxygen (14.5 weight %), nitrogen (20 weight %), carbon (0.5 weight %), hydrogen (0.01 weight %). By comparison, the solubilities of these elements in aluminium are all less than 10^{-4} weight %. During hot working, which is normally done in air, oxygen is absorbed to a depth of about 0.2 mm, forming a brittle surface layer which must be removed by machining or pickling. Because of this high

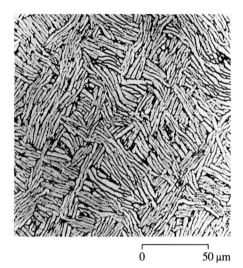

0   50 μm

Figure 2.58 The basket-weave grain structure in Ti–6%Al–4%V

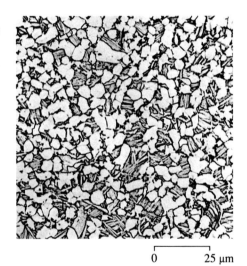

0   25 μm

Figure 2.59 Equiaxed α- and transformed β-phase grains in Ti–6%Al–4%V

affinity for oxygen, titanium cannot be melted in oxide refractory moulds and the production of titanium billets has to be carried out by electric arc melting onto a copper hearth cooled by liquid alkali metals (potassium and sodium). This contributes to the high cost of titanium.

The titanium alloy, Ti–6%Al–4%V, should be satisfactory for the cabin skin of a Mach 3 supersonic transport aircraft, for both creep and oxidation resistance. It has a reasonably high toughness, 2.4–2.7 kJ m^{-2}, depending on the heat treatment it receives, and also meets the requirement for fatigue, with a fatigue strength at 10^8 cycles as high as that of the tensile strength at room temperature. Its use in such a situation, however, is a nettle that has yet to be grasped.

2.6 Material for a gas-turbine combustion chamber

Sir George Edwards, the designer of the Viscount airliner, once said that a new aircraft is designed around the engine that is available. This underlines the fundamental importance of the engine to the whole performance of an airliner. For example, the operating costs are dominated by the specific fuel consumption and the specific thrust of the engine. These parameters are strongly influenced by the temperature of the gas entering the turbine — the 'turbine entry temperature'. As Figure 2.60 shows, for a given specific fuel consumption, the specific thrust can be increased by increasing the turbine entry temperature (TET), thereby allowing the required thrust to be obtained from a smaller, lighter engine without using more fuel.

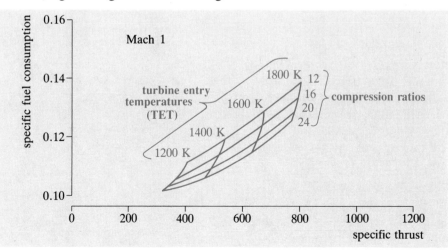

Figure 2.60 Performance graphs for a jet engine

The turbine entry temperature is determined by the temperature of the gas leaving the combustion chamber, where it has been heated by burning fuel (Figure 2.61). The combustion chamber is the 'hot spot' of

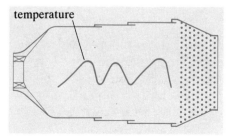

Figure 2.61 The temperature distribution along a pipe combustion chamber

the whole engine and therefore requires materials with high temperature resistance: in existing chambers, temperatures may exceed 1320 K locally. Severe temperature gradients are set up and these change with the different phases of flight. Such gradients are accompanied by thermal stresses and since these vary with time there is the risk of **thermal fatigue**. This is the same in principle as normal fatigue and differs only in the way in which the stress cycles are generated. Of course, at the high temperatures concerned, creep occurs, too, and **creep fatigue** contributes to, and interacts with, the overall fatigue. The chamber is also exposed to an oxidizing environment which it must withstand for the whole of its designed life without unacceptable material loss by oxidation and corrosion. Under these complex conditions, it is difficult to find simple performance criteria based on standard properties and so materials for a gas-turbine combustion chamber are assessed by empirical tests that simulate service conditions.

Like airliner cabins, combustion chambers are constructed from sheet material, but they have complicated shapes which must be assembled by welding, so the materials must be capable of being welded.

We now want to look at the kind of material that is used in combustion chambers. You have seen that one important factor in deciding how a material will respond to stress is the homologous temperature, T/T_m. Most metals creep at temperatures above 0.4 of their melting point (although certain alloys can be used to much higher fractions of T_m). Thus to avoid creep at 1320 K would require a melting temperature above 3300 K. As Table 2.9 shows, there are few candidates. Graphite sublimes at 3800 K, but it burns in oxygen; the metal tungsten melts at 3650 K, but it oxidizes rapidly in oxygen and has a very high density of $19.3 \, \mathrm{Mg \, m^{-3}}$. We therefore have to settle for a material with a lower melting point and a tendency to creep in service.

Nickel has an acceptable property profile — melting point is not everything — and virtually all the materials in a gas turbine working at temperatures above 800 K are made of nickel-based alloys. They are descended from nichrome, the material used in the element of an electric fire. Since they work at higher fractions of the melting temperature than any other alloys, they have come to be known as **superalloys** to distinguish them from lower-temperature nickel-based alloys such as monel.

2.6.1 The metallurgy of nickel-based superalloys

The nickel–chromium phase diagram is shown in Figure 2.62. It is a eutectic diagram and it indicates that solid nichrome (Ni–20%Cr) is all γ phase, a nickel-rich solid solution with a face-centred cubic structure. The dissolved chromium forms a layer of Cr_2O_3 on the surface, which protects the alloy from the air. This gives the alloy good resistance to oxidation and for this reason chromium is present in every nickel superalloy. Chromium also causes solution strengthening of the nickel, but this is less important; as you will see, there are much more effective ways of hardening.

Table 2.9 Melting points of some refractory metals and compounds

Material	T_m/K
Al	933
NaCl	1074
Cu	1356
Be	1550
Ni	1726
Co	1765
Fe	1808
SiO_2	1880
Ti	1950
Pt	2042
Zr	2125
Cr	2160
V	2160
Al_2O_3	2290
Hf	2420
B	2600
Nb	2740
Mo	2880
SiC	3000
MgO	3100
Os	3300
W	3650
C(sublimes)	3800

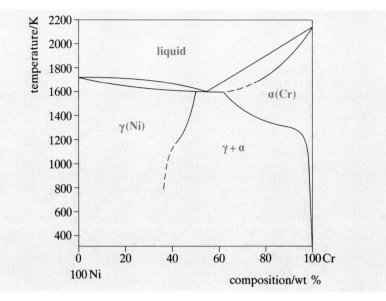

Figure 2.62 Nickel–chromium phase diagram

The first commercial superalloy was Nimonic 75, so called because it could be used at up to 750 °C (1023 K). It consists of nichrome with small additions of titanium (0.2–0.6 wt %) and carbon (0.08–0.15 wt %). The carbon is virtually insoluble at temperatures below about 1300 K and takes the form of two carbides: large particles of TiC in the grain interiors and, at the grain boundaries, metal carbides of the type $M_{23}C_6$, where the metal M is rich in chromium. The metal carbides are important for strengthening the grain boundaries. At temperatures above about $0.6T_m$, grains are able to shear or 'slide' along the grain boundaries, causing voids and cracks to nucleate and develop into fractures (see Figure 2.63). Grain-boundary particles inhibit sliding by 'pegging' neighbouring grains together and carbides are used for this purpose in most of today's superalloys. Nimonic 75 is easy to shape and weld so it is still used for the cooler parts of combustion chambers, but it lacks the higher-temperature strength and resistance to oxidation required for the hottest parts.

The next advance in the development of alloys was to employ precipitation hardening. It had been known since the work of the famous French metallurgist Chevenard in 1929 that nickel could be strengthened in this way by additions of aluminium or titanium. The effect of such additions is illustrated by the nickel–aluminium phase diagram (Figure 2.64). It takes the form of a eutectic on the left and a peritectic on the right, with their reaction temperatures close together.

In the middle of the diagram a remarkable phase, called γ', is stable. It is an intermetallic compound with compositions close to that of Ni_3Al and shows long-range order in that the nickel and aluminium atoms are juxtaposed rather than mixed randomly. The crystal's structure is

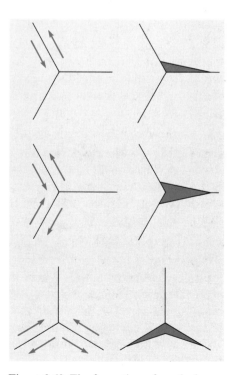

Figure 2.63 The formation of cracks by grain-boundary sliding

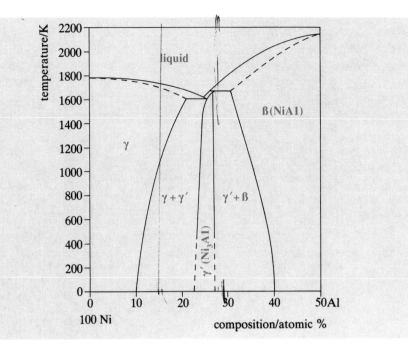

Figure 2.64 Nickel–aluminium phase diagram

similar to that of a face-centred cubic crystal (Figure 2.65a), but with nickel atoms located in the corners of the cube and aluminium atoms at the centres of the faces (Figure 2.65b). Although there are eight corners to a cube, each corner atom is shared among eight adjacent cubes, so there is, on average, only one aluminium atom per cube. Similarly, each of the six face-centred atoms is shared by two neighbouring cubes, so there are three nickel atoms per cube. If the composition of the phase is not exactly Ni_3Al, some sites of the deficient element are left empty. In contrast, the γ phase is a nickel-rich, face-centred solid solution in which nickel and aluminium atoms are mixed almost randomly.

The crystal structure and the planar spacing of the γ and γ' phases are very similar, so when γ' phase forms within supersaturated γ phase, during heat treatment, the crystals of the γ and γ' phases align with the planes of their cubes parallel. Thus there is continuity between their crystal planes, they are said to be **coherent**, and the energy of the interface between γ and γ' phases is small.

Consider the effect on the arrangement of atoms caused by a dislocation gliding from the γ phase into the γ' phase. Passage of a dislocation with a Burgers vector b caused a shear displacement b of atoms on one side of the slip plane relative to those on the other side. The magnitude of the Burgers vector, b, is usually the distance of closest atomic spacing, so in the face-centred γ phase it is $a/\sqrt{2}$, the distance between a corner atom and a face-centred atom, where a is the length of an edge of the cubic unit cell (see Figure 2.65a). Since the aluminium and nickel atoms are mixed randomly in γ phase, or nearly so, there is no loss of order as

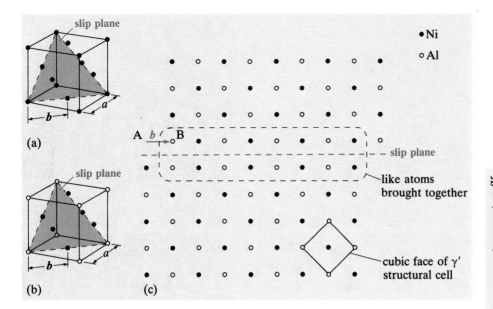

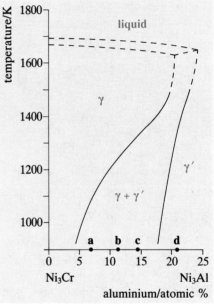

Figure 2.65 The structural cells of (a) γ phase and (b) γ′ phase. In heat-treated nickel superalloys these are aligned parallel to one another. (c) Dislocations of Burgers vector *b* glide across both phases, but in the γ′ phase, similar atoms become aligned across the slip plane

Figure 2.66 The 'pseudo-binary' phase diagram for certain nickel–aluminium–chromium alloys

a dislocation passes through it. This is *not* so in the γ′ phase. When the dislocation passes from the γ into the γ′ phase, its Burgers vector and slip plane remain unchanged. Its effect in the γ′ phase is to displace nickel atoms to sites previously occupied by aluminium atoms (i.e. from A to B in Figure 2.65c) and vice versa.

Across the slip plane this disturbs the natural long-range order between aluminium and nickel atoms and thereby increases the free energy of the crystal. The increase in energy manifests itself as a resistance of the γ′ phase to the passage of a dislocation (see 'Mechanisms of deformation in age-hardened alloys'). Thus **order hardening** makes the precipitates of γ′ phase very effective obstacles to slip. The cutting of γ′ precipitates does occur, but it requires the application of a large stress.

The nickel–titanium phase diagram is similar in possessing γ and γ′ phases and if both aluminium and titanium are present they will mix freely with one another as the M constituent of Ni_3M.

As we showed earlier with aluminium alloy 6063, you can sometimes use 'pseudo-binary' phase diagrams to represent the phases in an alloy with more than two elements. For example, Figure 2.66 shows a phase diagram for chromium–aluminium alloys containing 75 atomic % nickel. As you move along the composition axis from left to right, the aluminium content increases as that for chromium diminishes.

How do the strengths of the compositions marked **a** to **d** in the phase diagram vary with temperature (see Figure 2.67)?

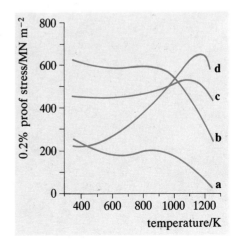

Figure 2.67 The temperature dependence of strength of alloys **a**–**d** in Figure 2.66

Alloys **a** and **b** show the usual trend of a proof stress that falls as the temperature rises. Alloy **d**, the γ' phase, is unusual in showing the reverse trend. Complex dislocation models have been proposed to explain this, but they are beyond the scope of this book. Such an unusual, but favourable, effect is found in the γ' phase of many compositions, including pure Ni_3Al and those titanium and niobium alloys that have a γ' phase.

At temperatures up to 900 K, alloys **b** and **c** are stronger than either **a** or **d**, and yet they contain a mixture of phases very similar to the phases of compositions **a** (γ) and **d** (γ'). This shows that there is a strengthening *interaction* between the phases.

At high temperatures γ' is the stronger phase and the strength of the phase mixture increases with the proportion of γ' (Figure 2.68). For high-temperature strength, an alloy should have a high proportion of γ', but it should not be 100% γ' because then the interactive strengthening is lost at lower temperatures. In the strongest alloys, γ' forms about 70% of the volume.

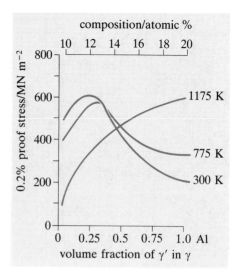

Figure 2.68 The effect on strength of the volume fraction of γ' phase

What type of heat treatment might be used to control the formation of γ' phase in a superalloy?

Precipitation hardening: solution treat; quench; reheat; age.

The nickel–aluminium phase diagram in Figure 2.64 has a $\gamma/\gamma + \gamma'$ boundary with a positive slope, so precipitation hardening is feasible. The result of this for the alloy Nimonic 90 (see Table 2.10) is shown in Figure 2.69, where the dark areas are γ' particles. These are so closely spaced that they obstruct dislocations very effectively. The stress required to make a dislocation bow out between the particles is greater than that to make the dislocation cut through the particles, so cutting occurs (Figure 2.70). Exposure to high temperatures causes **coarsening** — the growth of large particles at the expense of smaller ones.

Table 2.10   Compositions of three nickel-based superalloys in weight %

Alloy	C	Si	Cu	Fe	Mn	Cr	Ti
Nimonic 75	0.08–0.15	1.0 max	0.5 max	5.0 max	1.0 max	18.0–21.0	0.2–0.6
Nimonic 90	0.13 max	1.0 max	0.2 max	1.5 max	1.0 max	18.0–21.0	2.0–3.0
Nimonic 263	0.04–0.08	0.1–0.4	0.2 max	0.7 max	0.2–0.6	19.0–21.0	1.9–2.1

Alloy	Al	Co	Mo	B	Zr	Pb	S
Nimonic 90	1.0–2.0	15.0–21.0		0.020 max	0.15 max	0.002 max	0.015 max
Nimonic 263	0.3–0.6	19.0–21.0	5.6–6.1	0.0001 max		0.002 max	0.007 max

Eventually the particle spacing would be expected to increase until the bowing of dislocations between particles becomes the easier deformation process and the strength would then fall due to over ageing (see Figure 2.29). This occurs very slowly in these alloys because the driving force for coarsening — the interfacial energy between γ and γ' — is small, as a consequence of the similarity between the phases. This is an important factor in the success of the γ/γ' superalloys.

In some of the most advanced superalloys, the volume fraction of γ' phase is very high ($\sim 70\%$) and in such cases dislocation bowing does not occur, even after prolonged coarsening, because the interparticle spacing, λ, is always too small. In such cases, plastic deformation involves cutting of the precipitates.

Although Nimonic 90 is not used for combustion chambers, it resembles an alloy that is — Nimonic 263 (see Table 2.10). Both alloys, like nichrome, contain about 20% chromium, but they also have about 20% cobalt, which solution hardens the nickel. The main difference between them is the presence in 263 of about 6% molybdenum, another strong carbide former. Both alloys are age hardened by γ' precipitates.

You may well wonder how an alloy such as Nimonic 263 can be welded — surely melting and solidifying would destroy the precipitates? This is true, but fortunately they can be reformed in the weld. The precipitation process is sluggish in this alloy and it does not occur during cooling of the weld. The welding process acts as a solution treatment and quench. After welding a simple ageing treatment is all that is required to complete the precipitation hardening of the weld by γ' phase.

Nimonic 75 is also used to make combustion chambers: it is a development of nichrome and utilizes solution hardening and pinning of grain boundaries by carbides.

These and other nickel-based superalloys are used to carry stress at higher homologous temperature than any other materials ($\sim 0.8 T_{\mathrm{m}}$) and they stand out as exceptional cases of successful materials development. Both use chromium to confer protection from the atmosphere.

2.7 Summing up

This completes the chapter on nonferrous metals. Since there are over eighty of them, we have had to take representative examples and the emphasis has been placed on aluminium, titanium, nickel and their alloys. We have looked at the metallurgy of these metals and have seen the significance of allotropy in titanium: it creates the possibility of having α–β changes in dilute alloys and therby the creation of a fine grain structure. Aluminium alloys are notable for the occurrence of a wide range of precipitation-hardening reactions which, together with their low density, allow high specific strengths to be achieved. The nickel-based superalloys are characterized by the γ/γ' precipitation-

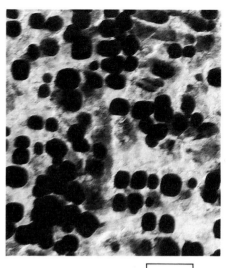

0   0.25 μm

Figure 2.69 A transmission electron micrograph of heat-treated Nimonic 90 showing dark γ' phase in a matrix of light γ phase

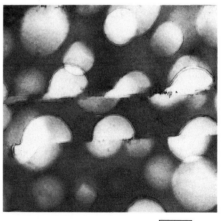

0   0.1 μm

Figure 2.70 Particles of γ' phase that have been sheared by the passage of dislocations

hardening reaction, which permits stable, high-temperature alloys to be made. In this and other applications we have seen the value of protection from oxidation by passive surface films and that problems may occur when these films start to break down. In choosing materials for service, we have explored how a merit index may be used, provided there is a clear criterion for choice.

SAQ 2.10 (Objective 2.5)
Starting with an annealed plate of aluminium with a large grain size (~ 1 mm), explain how you would reduce the grain size by at least an order of magnitude.

Estimate the minimum difference in dislocation density across a grain boundary facet, $20\,\mu$m wide, that would be required to cause this facet to form a recrystallized nucleus. Take the grain boundary energy to be $Gb/25$, the energy of unit length of dislocation to be $Gb^2/2$ and b to be 3.6×10^{-10} m.

SAQ 2.11 (Objective 2.5)
Explain in terms of the motion of dislocations why age-hardened alloys usually show a maximum in strength as a function of time at a given ageing temperature.

On ageing, an aluminium alloy attains a maximum proof stress of $400\,\mathrm{MN\,m^{-2}}$. Estimate the average spacing of the precipitates when the alloy is at its peak strength. Take $G = 27\,\mathrm{GN\,m^{-2}}$, $b = 0.36$ nm and the shear stress to be half the tensile stress.

SAQ 2.12 (Objective 2.9)
It is planned to use fine-grained aluminium trays in an industrial oven for two months at 473 K. To avoid becoming unacceptably distorted in service, the material in the trays should strain by no more than 5%. Use a deformation map (Figure 2.35) to identify the mode of deformation and to estimate the maximum stress that the trays should be allowed to carry.

Objectives for Chapter 2

After studying this chapter you should be able to:

2.1 Describe qualitatively the 'property profile' of a typical nonferrous metal and compare it with those of other classes of materials (SAQ 2.2).

2.2 Name six nonferrous metals and give a typical application of each.

2.3 Describe four processes of wear due to contact between surfaces — adhesive, abrasive, sub-surface fatigue and chemical wear (SAQ 2.1).

2.4 Derive merit indices for performance under specified loading conditions (SAQs 2.3 and 2.4).

2.5 Describe and explain, in terms of crystal defects, four treatments that are used for strengthening alloys — work hardening, grain-size hardening, solution hardening and precipitation hardening. Indicate those that involve achieving nonequilibrium microstructures (SAQs 2.5, 2.10 and 2.11).

2.6 State the conditions for the formation of a neck in a stretched test piece (SAQ 2.6).

2.7 Recognize and be able to manipulate formulae for:

(a) the stresses in the thin walls of a pressurized cylinder;
(b) the stress concentration factor of a simple notch (SAQ 2.7);
(c) the onset of buckling in a strut of uniform section (SAQ 2.4).

2.8 Describe the empirical correlation between stress amplitude and fatigue life, in cycles, and show how Miner's hypothesis can be used to estimate the life of a cyclically loaded component (SAQ 2.8).

2.9 Describe the representation of the stress and the temperature ranges over which creep may be observed in crystalline materials (e.g. nonferrous alloys) and describe, in terms of crystal defects, the mechanism of deformation (SAQ 2.12).

2.10 Describe the stages of fatigue: crack nucleation; stages 1 and 2 of crack growth, critical crack propagation.

2.11 Determine the stability of a crack in terms of the toughness, the stress and the crack size (SAQ 2.10).

2.12 Define and use the following terms and concepts:

    age hardening or precipitation hardening
    annealing
    beach marks or shell marks
    Burgers vector
    coherency
    critical crack length
    dislocation climb
    fatigue
    fatigue limit
    Guinier–Preston (GP) zones
    grain growth
    grain-size hardening (the Hall–Petch equation)
    Larson–Miller parameter
    merit index
    Miner's rule
    order hardening
    over ageing
    pitting corrosion
    recovery
    recrystallization
    solution hardening
    specific modulus
    specific strength
    strain hardening or work hardening
    stress concentration factor
    striations
    toughness

Answers to self-assessment questions

SAQ 2.1 The actual area of contact is given by Equation (2.1), where N in this case is the weight of the coin and P is $1\,\mathrm{GN\,m^{-2}} = 10^3\,\mathrm{N\,mm^{-2}}$. Taking the coin's mass, m, to be $10^{-2}\,\mathrm{kg}$

$$N = mg$$
$$\simeq 10^{-2} \times 10\,\mathrm{N}$$
$$= 0.1\,\mathrm{N}$$

From Equation (2.1)

$$A_c = \frac{N}{P}$$
$$= \frac{0.1}{10^3}\,\mathrm{mm^2}$$
$$= 10^{-4}\,\mathrm{mm^2}$$

This is rather small compared with the area of the face of the coin, say $500\,\mathrm{mm^2}$.

You would expect little welding between the glass, whose surface may be greasy, and the coin alloy, with its passive oxide surface, so f and therefore μ would be small.

SAQ 2.2
Mild steel It is ferromagnetic, so it will not pass through the electromagnet in the vending machines; it has poor resistance to oxidation (it rusts); its hardness, about $140\,H_V$, is rather too high.
Stainless steel Its hardness, about $180\,H_V$, is too high.
Titanium Its density is less than half that of the silver alloy it would have to replace, so it would feel different; it has too high an intrinsic value, about £20 per kg.

SAQ 2.3
Step 1 Write down any true expression for the attribute sought, P.

$$P = \sigma A$$

where A is the cross-sectional area.

Step 2 Eliminate any parameter on the right which is *not* a property or a fixed quantity. In this case you wish to eliminate A. Since the mass, m, is given by $m = \rho L A$, then

$$A = \frac{m}{\rho L}$$

This can now be used to eliminate A from the expression for P, which becomes

$$P = \frac{\sigma m}{\rho L}$$
$$= \frac{\sigma}{\rho} \times \frac{m}{L}$$

Step 3 The property term, σ/ρ, is the merit index for the maximum load.

SAQ 2.4
a) *Step 1* The property sought is lightness or inverse weight.

$$\frac{1}{m} = \frac{1}{\rho L A}$$

where A is the sectional area.

Step 2 You now need to eliminate A. The column is described as long, so it is expected to fail in compression by elastic buckling. The maximum load, P, must equal the buckling load, $P_b = kEI/L^2$. From Table 1.1, $I = w^4/12$ so

$$P = \frac{kEw^4}{12L^2}$$

Since

$$A = w^2$$
$$P = \frac{kEA^2}{12L^2}$$

This gives

$$\frac{1}{m} = \left(\frac{kE}{12L^2 P}\right)^{1/2} \times \frac{1}{\rho L}$$
$$= \frac{\sqrt{E}}{\rho} \times \left(\frac{k}{12L^4 P}\right)^{1/2}$$

Step 3 The merit index is $\sqrt{E}/\rho$.

(b) *Step 1* The property sought is again lightness.

$$\frac{1}{m} = \frac{1}{\rho L A}$$

Step 2 Again, A needs to be eliminated. This time $I = wt^3/12$ (see

Table 1.1) and w is *fixed*. Therefore, since $A = wt$,

$$P = \frac{kEA^3}{12w^2 L^2}$$
$$\frac{1}{A^3} = \frac{kE}{12w^2 L^2 P}$$
$$\frac{1}{A} = \left(\frac{kE}{12w^2 L^2 P}\right)^{1/3}$$

Substituting this into the equation from Step 1

$$\frac{1}{m} = \frac{1}{\rho L}\left(\frac{kE}{12w^2 L^2 P}\right)^{1/3}$$
$$= \frac{E^{1/3}}{\rho} \times \left(\frac{k}{12w^2 L^5 P}\right)^{1/3}$$

Step 3 The merit index is $E^{1/3}/\rho$.

SAQ 2.5
(a) $\rho = \rho_0 + C\varepsilon_p^\beta$
If the strain is 10%, $\varepsilon_p = 0.1$ and $\beta = 1$

$$\rho = 10^{12} + 10^{15} \times 0.1$$
$$\simeq 10^{14}\,\mathrm{m^{-2}}$$

(b) From Equation (2.5), the rate of increase of ρ with respect to ε_p when $\beta = 1$ is

$$\frac{d\rho}{d\varepsilon_p} = C$$
$$= 10^{15}\,\mathrm{m^{-2}}$$

From Equation (2.8), when $\varepsilon_p = \gamma/2$,

$$\varepsilon_p = \tfrac{1}{2} b \rho l$$
$$\rho = \frac{2\varepsilon_p}{bl}$$
$$\frac{d\rho}{d\varepsilon_p} = \frac{2}{bl}$$

Therefore

$$\frac{2}{bl} = 10^{15}\,\mathrm{m^{-2}}$$
$$l = \frac{2}{10^{15}(3 \times 10^{-10})}\,\mathrm{m}$$
$$= 6.7\,\mathrm{\mu m}$$

SAQ 2.6 Rearranging Equation (2.11) gives

$$\sigma_0 = \sigma \varepsilon^{-n}$$

Since the proof stress, $\sigma_p = 50\,\mathrm{MN\,m^{-2}}$ when $\varepsilon = 0.01$, substituting these values into this equation gives

$$\sigma_0 = 50 \times 0.01^{-n}\,\mathrm{MN\,m^{-2}}$$

If n is the *true strain*, ε_n, at which deformation becomes unstable

$$n = \varepsilon_n = \ln\left(\frac{l}{l_0}\right)$$

where l is the deformed length and l_0 the undeformed length. Thus

$$n = \ln\left(\frac{l_0 + \Delta l}{l_0}\right)$$

$$= \ln\left(1 + \frac{\Delta l}{l_0}\right)$$

Now, $\Delta l/l_0$ is the elongation, which is given as 0.50, so

$$n = \ln 1.5$$

$$\approx 0.4$$

Thus

$$\sigma_0 = 50 \times 0.01^{-0.4}$$

$$= 315\,\mathrm{MN\,m^{-2}}$$

When the deformation becomes unstable, $\varepsilon = \varepsilon_n = n = 0.4$ so

$$\sigma_p = 315 \times 0.40^{0.4}$$

$$= 218\,\mathrm{MN\,m^{-2}}$$

SAQ 2.7 The stress concentration factor is

$$K_t = [1 + 2\sqrt{(D/r)}]$$

(a) For a semicircular notch, $D = r$ and the stress concentration factor is three.

(b) For a notch at the surface, $D = 10\,\mathrm{mm}$ and $r = 0.5\,\mathrm{mm}$. Therefore the stress concentration factor is $[1 + 2\sqrt{(10/0.5)}] = 9.9$.

SAQ 2.8 Equation (2.21) gives the condition for failure

$$\sum_i \frac{n_i}{N_i} = 1$$

According to Miner, the fraction of the life used up in a month's operation is

$$\sum_{i=1}^{3} \frac{n_i}{N_i} = \frac{n_1}{N_1} + \frac{n_2}{N_2} + \frac{n_3}{N_3}$$

Here N_1, N_2 and N_3 are the fatigue lives at the stress levels of 210, 140 and $70\,\mathrm{MN\,m^{-2}}$, respectively. From Figure 2.41

$$N_1 = 10^4 \qquad N_2 = 10^6 \qquad N_3 = 10^8$$

Therefore

$$\sum_{i=1}^{3} \frac{n_i}{N_i} = \frac{20}{10^4} + \frac{400}{10^6} + \frac{1000}{10^8}$$

$$= 24.1 \times 10^{-4}$$

Thus the total fatigue life is expected to be $1/(24.1 \times 10^{-4})$ months or 34.6 years.

SAQ 2.9 The critical crack length is given by Equation (2.22)

$$a_c = \frac{G_c E}{\pi \sigma_a^2}$$

From Table 2.6, the proof stress of alloy 2014 is $410\,\mathrm{MN\,m^{-2}}$, so the tensile stress is $205\,\mathrm{MN\,m^{-2}}$. Also from Table 2.6 the Young's modulus for the alloy is $70\,\mathrm{GN\,m^{-2}}$. Therefore

$$a_c = \frac{(30 \times 10^3)(70 \times 10^9)}{\pi(205 \times 10^6)^2}\,\mathrm{m}$$

$$= 16\,\mathrm{mm}$$

SAQ 2.10 The plate must be recrystallized by annealing. To make this happen it must be cold worked and then heated to a temperature of at least $0.4T_m$. The grain size obtained depends inversely on the amount of cold work the plate has received (Figure 2.22) and a reduction in thickness of at least 20% is required in this case. This treatment introduces a large density of dislocations which on heating causes many recrystallized nuclei to form, which grow until they meet to give a small grain size ($\leqslant 0.1\,\mathrm{mm}$).

A nucleus is formed when a grain boundary facet moves and bows out under the influence of the difference, U, in the strain energy per unit volume between the

two sides of the boundary. This happens when

$$U \geqslant \frac{2\Gamma}{L}$$

where Γ is the grain boundary energy and L is the size of the facet. If the difference in dislocation densities across the boundary is $\Delta\rho$ and the dislocation energy per metre is $Gb^2/2$ then

$$U = \Delta\rho Gb^2/2$$

Recalling that $\Gamma = Gb/25$ and combining these equations you get

$$\Delta\rho \geqslant \frac{2Gb/25}{Gb^2 L/2}$$

$$\Delta\rho \geqslant \frac{4}{25bL}$$

$$\Delta\rho_{min} = \frac{4}{25(3.6 \times 10^{-10})(20 \times 10^{-6})}\,\mathrm{m^{-2}}$$

$$= 2.2 \times 10^{13}\,\mathrm{m^{-2}}$$

SAQ 2.11 A peak in strength is caused by a change from a mode of deformation that causes an increase in strength on ageing, for example precipitate cutting, to a mode, such as dislocation bowing, that causes the strength to drop as ageing proceeds. Throughout ageing, the average size of the precipitates and, hence, their average spacing, λ, increases. The peak strength corresponds to the precipitate size and spacing at which the stresses required for the two modes of deformation are the same.

The shear stress, τ, required for the bowing mechanism is found from Equation (2.18) by putting $2r = \lambda$. Thus

$$\tau = \frac{Gb}{\lambda}$$

The maximum tensile proof stress in the alloy is $400\,\mathrm{MN\,m^{-2}}$ and this corresponds to a maximum shear stress of $200\,\mathrm{MN\,m^{-2}}$. Rearranging the equation to give the spacing

$$\lambda = \frac{Gb}{\tau}$$

$$= \frac{(27 \times 10^9)(0.36 \times 10^{-9})}{200 \times 10^6}\,\mathrm{m}$$

$$= 0.05\,\mu\mathrm{m}$$

This illustrates just how close together the precipitates must be for maximum strength.

SAQ 2.12 The maximum permitted strain of 5%, or 0.05, develops over a life of two months, about 5×10^6 s, so the average strain rate is $0.05/5 \times 10^6 = 10^{-8}\,\mathrm{s}^{-1}$. The service temperature is 473 K ($\sim 0.5T_\mathrm{m}$), so the service conditions are described by the point of intersection between the vertical line on Figure 2.35 corresponding to $T = 473$ K and the $10^{-8}\,\mathrm{s}^{-1}$ strain-rate contour. This point lies approximately on the horizontal line corresponding to $\sigma/E = 10^{-4}$ so

$$\begin{aligned}
\sigma &= 10^{-4}E \\
&= 10^{-4}(70 \times 10^9) \\
&= 7\,\mathrm{MN\,m}^{-2}
\end{aligned}$$

This is a modest level of stress!

The point describing the deformation conditions lies within the area of the map labelled 'diffusional flow'; plastic deformation occurs mainly by the directed motion of vacancies.

Chapter 3
Iron and Steel

by Ken Reynolds and George Weidmann

Chapter 3 Iron and steel

3.1 Introduction

3.1.1 The versatility of iron and steel

Our engineering achievements have depended upon iron and steel for well over a millenium. Industries as diverse as agriculture, minerals, timber, oil and chemicals, as well as the most modern semiconductor plants, all rely on ferrous alloys for their processing machinery and equipment. An engineer might describe these alloys as a family of materials based on the iron–carbon system possessing a wide range of useful properties, which can be formed into complex shapes, joined by welding, and heat treated to impart useful combinations of toughness and strength, but with an unfortunate tendency to corrode.

Dictionary definitions state that steels are alloys of iron and carbon. They always contain both, with iron being the major constituent. Other elements may be present, too, sometimes as undesirable impurities or, more often, as deliberate alloying additions for specific purposes. The family based on the iron–carbon system includes not only one of the strongest materials in absolute terms (tensile strength of $3000\,MN\,m^{-2}$), but also materials so versatile that within the same component (like a chisel) a soft and malleable shaft can adjoin a hard cutting edge. Figure 3.1 shows how steels compare with other materials in terms of strength and elastic modulus.

You can assess this versatility quite easily in and around the house with just a pin and small magnet. Iron is one of the four ferromagnetic elements and there are not likely to be many things made from the other three (nickel, gadolinium and cobalt) around the home. Articles that are attracted to the magnet can therefore be assumed to be made of iron or steel. Actually, there are a few steels — principally within the class known as 'stainless' — that are not attracted to a magnet, but household articles made of these are often so marked. The point of a needle or pin is used to see whether the object can be scratched, or whether the pin point simply glides over the surface. Different places on the same object, for example the tip and the shaft of a cold chisel, can behave differently. A typical set of results of such a scratch test is shown in Table 3.1.

In effect this is a scratch *hardness* test. A bit rough and ready maybe, but it nevertheless gives quite a good indication of the relative hardness of different objects. (A workshop technician may occasionally do the same thing by running a file across an object to see whether it's too hard to machine.) ▼Hardness measurements▲ discusses what is involved in quantitative hardness testing. The range from $80\,H_V$ for the softest iron to $1000\,H_V$ for the hardest steel is particularly large. How is it

Table 3.1 Scratch-test results

Scratched by the pin	Not scratched by the pin
Cutlery	
handle of stainless steel knife	stainless steel knife blade
prongs of fork	old carving knife blade
teaspoon	knife 'steel' sharpener
Tools	
side of hammer head	hammer face
top of tenon-saw	teeth of saw
handle of pliers	jaws of pliers
top of screwdriver shaft	blade of screwdriver
sides of vice	jaws of vice
nails and screws	masonry nail
Garden tools	
top of fork	tines of garden fork
top of spade	blade of expensive hoe
handles of pruning shears	blade of axe
sides of wheelbarrow	jaws of pruning shears
blade of cheap hoe	
piece of wire	

possible to obtain such widely differing hardnesses in materials whose compositional differences are measured in fractions of a percent? How can adjacent parts of the same object differ so greatly in hardness?

The main themes of this chapter are the ways in which the microstructures of ferrous metals can be manipulated — by alloying, by hot and cold working, and by heat treatment — to achieve such a variation in properties and the mechanisms involved.

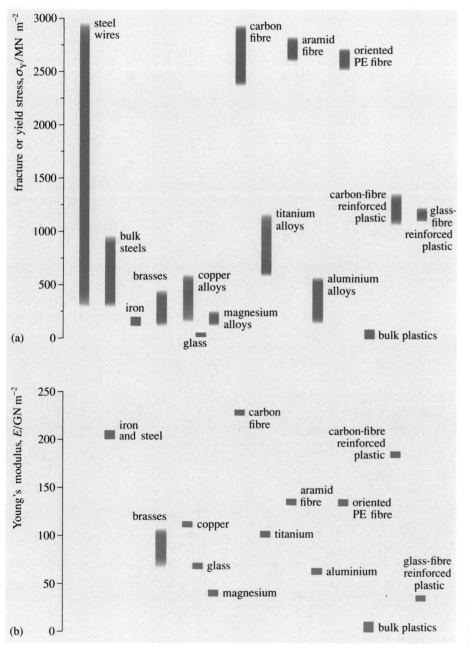

Figure 3.1 Comparison of yield or fracture stress and Young's modulus of steels with those of other materials

▼Hardness measurements▲

There is a bewildering variety of tests purporting to give some measure of the hardness of materials. They range from a scale of what scratches what (Mohs), through measuring the size of the impression left by an indenter of prescribed geometry under a known load (Vickers, Knoop and Brinell hardness) or measuring the depth to which an indenter penetrates under specified conditions (Rockwell B and C, Shore A), to the height of rebound of a ball or hammer dropped from a given distance (Shore scleroscope). Not surprisingly, perhaps, each test produces a different number (some on arbitrary scales) for the hardness of a given material. The approximate correlation between different scales of hardness is shown in Figure 3.2.

The chief attractions of the various hardness tests are that they are relatively simple and quick to perform and that they are virtually nondestructive. Thus they are very well suited for quality control purposes. However, there is not a well-defined materials property called hardness,

and what all these tests measure is differing combinations of the elastic, plastic and sometimes fracture behaviour of materials. Relating the results to properties such as yield stress and Young's modulus is not straightforward.

At one extreme, Shore A (also known as the International Rubber Hardness test, IRHD) measures solely the elastic response of elastomers. It is widely used as a check on the degree of cure or crosslinking of an elastomer and there is an approximate correlation between the hardness value and the shear modulus.

At the other extreme, the size of the plastically deformed impression produced by indenters such as the 136° diamond pyramid in the Vickers hardness test (Figure 3.3) must obviously bear some relation to the yield stress, σ_Y, but the response of materials depends on their ratio of yield stress to Young's modulus. For the softer metals, with low σ_Y/E, when they are either annealed or fully work hardened, it is found that

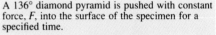

A 136° diamond pyramid is pushed with constant force, F, into the surface of the specimen for a specified time.

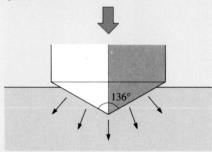

Material flows, plastically, away from the indenter.

At the end of the loading period the pyramid is removed and the diagonal lengths, d of the indentation are measured.

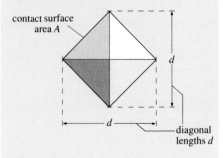

The hardness is the force divided by the contact surface area of the indentation

$$H_V = \frac{F}{A} \text{ kg force mm}^{-2} = \frac{2F \sin (136°/2)}{d^2}$$

Most machines have a set of tables for each loading force. The user measures the average diagonal length and reads the hardness from the tables. Loads vary from 30 kg down to 5 g (microhardness).

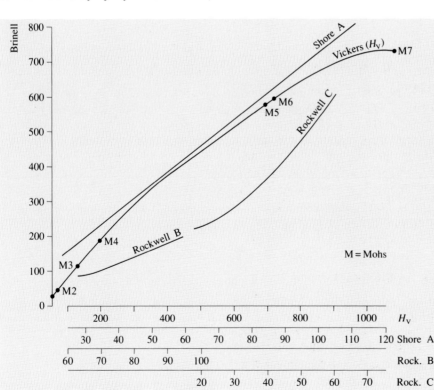

Figure 3.2 Approximate correlation of different hardness scales

Figure 3.3 The Vickers hardness test

Table 3.2   Typical H_V values of materials

Material	H_V/kgf mm^{-2}
tin	5
aluminium	25
gold	35
copper	40
iron	80
mild steel	140
fully hardened steel	900
limestone	250
MgO	500
window glass	550
fused silica	720
granite	850
quartz	1200
Al$_2$O$_3$	2500
tungsten carbide	2500
polypropylene	7
polycarbonate	14
PVC	16
polyacetal	18
PMMA	20
polystyrene	21
urea formaldehyde	41
epoxy	45

$$H_V \approx 3\sigma_Y$$

where H_V is the **Vickers hardness number**. (Note that H_V is conventionally cited in units of kgf mm^{-2}, so σ_Y must have the same units in the above equation.)

For materials with higher σ_Y/E, which includes a wide range of metals, glasses and plastics, the mode of deformation changes, and the relationship between H_V and σ_Y becomes more complex. However, for steels, there is a useful empirical relationship between the tensile strength (in MN m^{-2}) and H_V (in kgf mm^{-2}): namely,

$$\sigma_{TS} \approx 3.2 H_V$$

The values of H_V for a range of materials are shown in Table 3.2.

Finally, it should be borne in mind that, in those materials which exhibit time-dependence of either σ_Y (for example metals above $0.5T_m$) or both σ_Y and E (for example most plastics), the size of the indentation will increase with time, and thus their H_V value will depend on how long the load is applied.

3.1.2 Practical considerations and terminology

One reason for iron and steel being so widespread and cheap is that iron is the fourth most abundant element in the earth's crust and is fairly easily extracted from iron ores. Reserves are presently estimated to be 7×10^{10} tonnes, which are enough to keep us all going at present rates of consumption for at least another 200 years or so. Iron requires considerably less energy to extract it from its ore than do most of the other engineering metals, such as those in Chapter 2. For example, 1 kg of iron requires about 54 MJ compared with 280 MJ for aluminium, 108 MJ for copper, 415 MJ for magnesium and 550 MJ for titanium. Not surprisingly, iron and steel are thus considerably cheaper than the nonferrous metals.

Before going any further, let's distinguish between the terms iron and steel. The metal extracted from the ore is called **iron**, but, because of the way it is produced (usually in a blast furnace), it's far from pure. Typically it contains something over 4 wt % carbon, about 1 wt % each of manganese and silicon and smaller amounts of highly undesirable sulphur and phosphorus. This has to be refined before it is used, either to make **cast irons**, with more than 2% carbon, or to make **steels**, with a carbon content between 0.05 and 2%. Steels can be used 'as-cast' or wrought into various products like forgings, sheet and wire, but cast irons can only be made as castings — with few exceptions they cannot be mechanically worked. Finally, materials in which the carbon has been reduced to below 0.05% are usually referred to as **irons**. ▼Iron and steel nomenclature▲ summarizes the different classes of iron and steel.

Although it's the carbon content that has the major influence on the properties and response to heat treatment, additional elements are always present in commercial steel — even when it is described as **unalloyed**, or **plain carbon**. Undesirable impurities, which may be dug out of the ground with the iron ore or picked up during extraction, refining and melting, need to be eliminated or otherwise rendered innocuous within the metals. A standard specification for the chemical analysis of any 'unalloyed' carbon steel will always involve determination of the 'big five': carbon, silicon, manganese, sulphur and phosphorus, though none of them is likely to be present in an amount exceeding 1.5% by weight. Silicon and manganese are beneficial as deoxidizers and as solid-solution strengtheners. Sulphur and phosphorus have an embrittling effect and are usually kept below 0.05% unless deliberately increased to impart specific properties such as machinability.

All commercial steels contain entrapped nonmetallic **inclusions** and we talk about **clean** steels or **dirty** steels according to the quantity present. Nonmetallic inclusions become aligned as the steel is subsequently worked and give rise to a so-called **fibre structure**. It's principally the inclusions that enable the directions of metal flow to be seen, as illustrated in the etched section of how not to make a gear wheel in Figure 3.4.

> EXERCISE 3.1   What, if anything, is 'wrong' with the way the teeth have been produced in the gear section shown in Figure 3.4? (Hint: what if it were wood rather than steel?)

▼Segregation and its implications▲ discusses other ways in which non-homogeneous structures can arise. **Wrought iron**, which is not much used nowadays, can be thought of as a dirty, very low-carbon steel. Its significant slag content is dispersed by forging or hot rolling and ends up as extended, orientated filaments.

3.1.3 Iron and carbon on the atomic scale

The minimum atomic spacing in pure BCC iron at 293 K is 0.248 nm, and for carbon in graphite it's 0.142 nm. This difference in size, with the volume of a carbon atom being only

$$\left[\left(\frac{142}{248} \right)^3 \times 100 \right] \approx 19\%$$

of the volume of an iron atom, is responsible for much of the behaviour of steels.

Figure 3.4 Etched section through a forged gear stamped out of a flat bar

▼Iron and steel nomenclature▲

Iron and steels are classed below according to their carbon content.

Iron. Less than 0.05 wt % C, but can contain other elements (e.g. Si for transformer cores).

Low-carbon steels (mild steels). Up to about 0.2 wt % C; insufficient for useful quenching and/or tempering treatments.

Medium-carbon steels. Usually 0.2–0.6 wt % C; engineering steels whose properties can be tailored by quenching and tempering.

Hypoeutectoid steels. Less than the eutectoid composition of 0.8 wt % C.

Hypereutectoid steels. More than 0.8 wt % C (remember 'hyper-', as in hyperspace, hypertension and so on, means 'more than').

High-carbon steels. 0.6 to 2 wt % C (it's rare for commercial steels to exceed 1.4 wt % C).

Cast irons. More than 2 wt % and up to about 4.5 wt % C.

Wrought irons. Less than 0.03 wt % carbon, but containing quantities of slag from the iron-making which are dispersed by hot working.

Apart from 0.8% and 2%, the dividing lines between the classes are only approximate. In addition, there are alloy steels which, as well as carbon, contain deliberate additions of alloying elements (e.g. Mn, Si, P, S, Ni, Cr, Ti, W) to modify behaviour.

▼Segregation and its implications▲

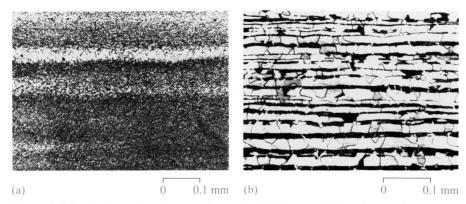

(a)   0   0.1 mm   (b)   0   0.1 mm

Figure 3.5 Banded structures in wrought steels. (a) Alloy steel (b) carbon steel

Etchants reveal flow lines not just because there are inclusions in a specimen. The etchant also attacks individual areas within the microstructure at different rates, because they have different grain orientations and different chemical compositions. When steel solidifies, composition gradients exist between the liquid and the growing solid crystals at most practical cooling rates (which are too fast for equilibrium to be maintained by diffusion). Across an individual crystal this effect gives rise to the phenomenon of **coring**. When a large mass solidifies, composition gradients also develop through its cross-section, as well as across the individual grains. This lack of chemical homogeneity due to macro-segregation is on a larger scale than the dendrite arm spacing. It persists into wrought products and is difficult to eliminate. Because the effect becomes visible as bands in sections prepared for metallography, it is referred to as **banding**.

As might be expected, banding can become troublesome in alloyed steels (Figure 3.5a) where there is more solute to become segregated, and where the alloying element is required to bring about a specific response to heat-treatment. However, even in unalloyed carbon steels (Figure 3.5b), macrostructural segregation of phosphorus, silicon, manganese and carbon may give rise to directional properties: for example, dramatic variations of impact energy with the direction of fracture (this is especially likely if the steel also happens to be 'dirty').

Sometimes the structural effects of segregation become so pronounced that the material behaves as if it were made up from several layers (see Figure 3.5b, which is a commercial low-carbon steel). When this occurs, particularly in thick sections, it is referred to as **lamination** and can cause serious problems in welded structures, where the laminations may provide paths of easy crack propagation. The possibility of laminated steels developing internal ruptures is of serious concern to any structural engineer.

Banding in alloy steels also poses problems in obtaining uniformity of properties across a section during heat treatment. Such banding is extremely difficult to eliminate since long heat-treatment times are required, and frequently banding persists down into extremely thin sections.

You should not forget these factors during your study of the science underlying the properties and heat treatment of steels. Some directionality is always with us, and large objects in particular seldom exhibit uniform mechanical properties in every direction!

SAQ 3.1 (Objective 3.2)
A male thread may be formed on steel bar in one of two ways, by machining with a cutting tool or by forming the thread by deformation under shaped rollers. Which of these two is likely to give the strongest and toughest thread given that both have exactly the same hardness level initially?

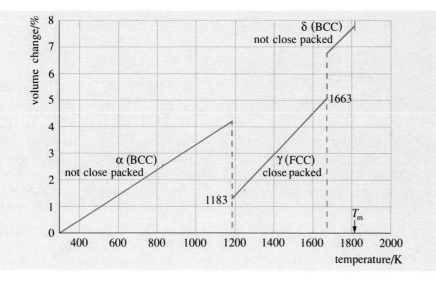

Figure 3.6 Volume change versus temperature for iron. Such sharp transitions only occur in very pure iron. Solutes spread the transitions over a range of temperatures

Iron is somewhat unusual among metallic elements in that it undergoes two solid-state phase changes with temperature (allotropy, Section 2.5.1). The volume change of a sample of pure iron against temperature (Figure 3.6) clearly shows these transitions. Up to 1183 K the structure of iron is body-centred cubic, α phase. Between 1183 K and 1663 K it is the face-centred cubic, γ phase, and from that temperature until it melts at 1811 K, the δ phase is body-centred cubic again. (The 'missing' β phase is the paramagnetic form of α, which is still BCC and for most metallurgical purposes can be regarded as the same as α.)

EXERCISE 3.2
(a) Which has the higher coefficient of expansion, BCC or FCC?
(b) Why are there abrupt volume changes at 1183 K and 1663 K?

SAQ 3.2   (Revision)
Draw sketches to show:
(a)   The tetrahedral interstitial holes in a pair of BCC unit cells and in a single FCC unit cell.
(b)   The octahedral interstitial holes in a pair of BCC unit cells and a quartet of FCC unit cells.

Despite the greater unoccupied volume in BCC (packing factor 68%) than FCC (74%), the largest interstitial holes in BCC iron (diameter 0.072 nm) are smaller than those in FCC (0.104 nm). Thus, there is a significantly larger difference between the diameters of the BCC interstice and the carbon atom, than between those of the FCC interstice and carbon. It is not surprising, therefore, that carbon is almost insoluble in α iron at room temperature (less than 0.005 wt %)

since a much larger lattice distortion is required to accommodate it in BCC than in FCC. In other words, there is practically no room for it, except at defects, grain boundaries and dislocations. The internal energy required is greater than the free energy gained by the effect on the entropy of introducing foreign atoms. However, with increasing temperature the hole sizes increase and the energetic penalty decreases so that more carbon can be accommodated. At 990 K, its solubility in α iron has risen to 0.02 wt %. This may still not seem much, but nevertheless the difference in solid solubility can be exploited to allow strengthening by precipitation hardening in low-alloy steels, though sometimes there can be undesirable embrittlement effects.

EXERCISE 3.3   If an α-iron crystal is strained uniaxially by external mechanical forces, where would you expect the carbon to locate itself?

3.1.4 The Fe–Fe$_3$C system

The equilibrium phase diagram for the Fe–Fe$_3$C system is shown in Figure 3.7. Its essential features are:

• The horizontal axis is in wt % carbon with Fe$_3$C at 6.67 wt %. This represents the useful limit of the diagram.

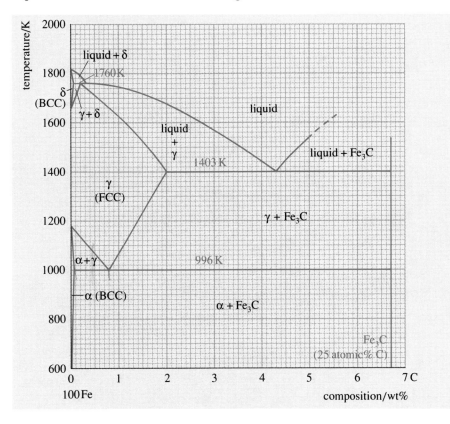

Figure 3.7 Fe–Fe$_3$C phase diagram

• There are five phases indicated namely, liquid, α, γ, δ and Fe₃C.
▼Microconstituents of iron and steel▲ gives details of these and other
phase structures found in ferrous metals.

• There are three horizontal lines which refer to the following reactions:
(a) at the peritectic at 1760 K, (liquid + δ) → γ on cooling;
(b) at the eutectic at 1403 K, liquid → (γ + Fe₃C) on cooling;
(c) at the eutectoid reaction at 996 K, γ → (α + Fe₃C).

• At cooling rates which are slow enough to allow the eutectoid
reaction to take place under near-equilibrium conditions, the γ solid
solution of austenite with 0.8% C breaks down into α ferrite containing
0.02% C and cementite, Fe₃C, containing 6.67% C.

• The 2% C composition separates those alloys which undergo the
eutectic reaction on slow cooling from those below 2% that don't.

EXERCISE 3.4   From the above compositions, estimate the amounts of
ferrite and cementite resulting from the eutectoid decomposition of a
steel containing 0.8 wt % C.

As the eutectoid reaction takes place, layers of cementite grow in a
branching manner, with ferrite filling in the spaces and forming the
matrix. This intimate mixture in the microstructure of steels has a
characteristic appearance and was called pearlite by the nineteenth-
century microscopists on account of its lustre in etched sections, said to
resemble mother of pearl. The name has been retained because it is a
very useful way of describing the near-equilibrium product arising from
austenite decomposition.

The 2% composition is chosen to differentiate between cast irons and
steels because cast irons undergo the eutectic reaction during
solidification, whereas steels do not. Notice that both undergo the
eutectoid decomposition of γ at 996 K. However, although this diagram
is used extensively, it is *not* the true equilibrium diagram for the system!
(See ▼True equilibrium in the iron–carbon system▲.)

SAQ 3.3   (Revision — see Figure 3.7)
(a) Describe the sequence of phase changes when liquid steel
containing 0.3 wt % carbon is slowly cooled from 1800 K to room
temperature.
(b) Describe the sequence when liquid steel containing 2.3 wt %
carbon is cooled from 1800 K to room temperature.

▼Microconstituents of iron and steel▲

Ferrite (α and δ). The BCC form of iron
and of solid solutions based on it. In pure
iron, α ferrite is stable up to 1183 K,
whereas δ ferrite occurs between 1663 K
and the melting temperature (1811 K).

Austenite (γ). The higher density, FCC
form of iron and of solid solutions based
on it. In pure iron, it is stable between
1183 K and 1663 K.

Cementite (Fe₃C). The compound formed
with 6.67 wt % carbon (25 atomic %). It
has a complex hexagonal crystal structure
with carbon in interstitial positions. It is
very hard (harder than martensite) and
brittle.

Graphite. The most stable form of carbon
in the Fe–C system, but usually found
only in the cast irons. Its form and
distribution (flakes, aggregates, nodules or
spheroids) control the strength and
ductility of cast irons.

Pearlite. A microstructure formed by the
breakdown of austenite at the eutectoid
(0.8 wt % C), it consists of an intimate
lamellar mixture of α ferrite and
cementite.

Martensite. Nonequilibrium
microstructure formed by cooling
austenite too rapidly for carbon to diffuse
out of solid solution to form Fe₃C. The
entrapped carbon distorts the lattice and
retards the shear transformation from
FCC to BCC, causing the product of the
shear transformation to be a tetragonal
lattice. Its hardness increases and ductility
decreases with increasing carbon content.

Bainite. Nonequilibrium microstructure
consisting of supersaturated ferrite and
cementite, formed when austenite breaks
down at large undercooling by a
combination of shear and diffusive
processes.

▼True equilibrium in the iron–carbon system▲

Figure 3.8 Cast-iron fire grate distorted by thermal cycling causing 'growth' by decomposition of Fe_3C to graphite

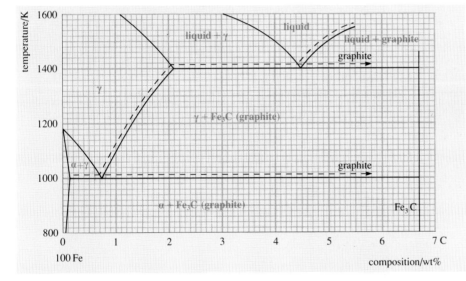

Figure 3.9 Relationships of phase boundaries in the iron–graphite diagram (broken lines) compared with the $Fe–Fe_3C$ diagram (full lines)

For most purposes in considering the structure and heat treatment of steels, the phase diagram is generally accepted as the portion between pure iron and Fe_3C. However, Fe_3C is *not* the true equilibrium state. Free energy considerations show that the stable phases are those based on iron and graphite. There's also some strong practical evidence for this.

• Medium-carbon steels held for very long times at temperatures in the region of 800 K (for example under creep test conditions) sometimes show a breakdown of Fe_3C to graphite.

• Eutectic Fe_3C formed when cast irons are rapidly cooled, readily breaks down into graphite and iron after a few hours annealing at temperatures around 1250 K. The reverse reaction never occurs.

• Fe_3C present in pearlite in cast irons breaks down to graphite and ferrite during thermal cycling through the eutectoid transformation. This is the cause of warping of objects such as fire baskets (Figure 3.8). The graphite has a higher specific volume than Fe_3C, thus parts exposed most frequently to temperatures above the eutectoid actually swell over a period of use.

The equilibrium diagram for iron–graphite is very similar to that normally shown for $Fe–Fe_3C$, the main differences being that the eutectic and eutectoid temperatures are a few degrees higher and that the phase boundary between γ and $\gamma + Fe_3C$ and the eutectic and eutectoid compositions are all shifted to slightly lower carbon percentages. The two diagrams are shown superimposed in Figure 3.9 for comparison.

The 'graphite' diagram is more important when one is considering cast irons and, even then, in the solid state, it is usual to consider the eutectoid as being between ferrite and Fe_3C.

3.2 Near-equilibrium structures in plain carbon steels

3.2.1 The properties of annealed steels

Plain carbon steels are those alloys with compositions up to 2% by weight of carbon. Some of their basic mechanical properties as a function of carbon content up to 1.2 wt % carbon are shown in Figure 3.10. These data are for steels in the annealed condition. The steels were slowly cooled from the austenite phase, giving a microstructure near to that produced by true equilibrium cooling. As a number of properties depend on grain size, the data refer to steels with similar grain size. These property changes may be accounted for by the increase in the amount of cementite in the steel as the carbon content rises. Cementite is classed as an **intermediate compound**; very similar to the intermetallic compounds (e.g. $CuAl_2$). It is both hard and brittle — increasing amounts of it increase the strength and decrease the ductility of steel.

The quantity of carbon present relative to the eutectoid composition (0.8 wt %) critically affects the microstructures resulting from slow cooling. Figure 3.11(a) shows the microstructural development in a steel of eutectoid composition. The developments of annealed hypoeutectoid and hypereutectoid steels are shown in Figures 3.11(b) and 3.11(c) and the corresponding microstructures in Figures 3.12(a) and 3.12(b).

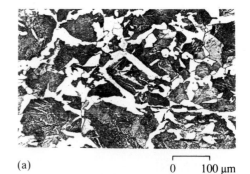

(a)   0   100 μm

(b)   0   40 μm

Figure 3.12 Examples of hypoeutectoid and hypereutectoid steels. (a) 0.4 wt % C, (b) 1.1 wt % C

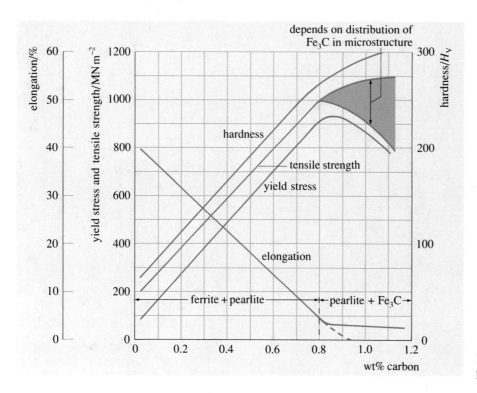

Figure 3.10 Typical properties of annealed plain carbon steels

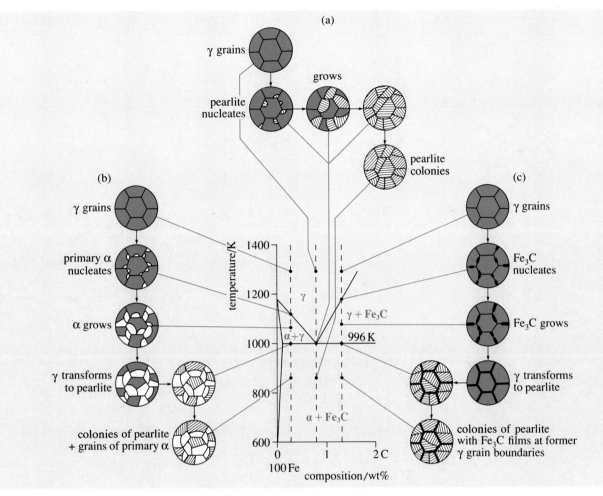

Figure 3.11 Microstructures during slow cooling from the austenite region of (a) a eutectoid steel, (b) a hypoeutectoid steel, (c) a hypereutectoid steel

EXERCISE 3.5   The dark areas in Figure 3.12(a) are pearlite. How does this microstructure confirm that the sample contains about 0.6 wt % carbon and is a hypoeutectoid steel?

In annealed hypoeutectoid steels, the cementite is present only as a constituent in the lamellar pearlite, but its behaviour, coupled with the intimate structure, confer on the pearlite quite different properties from those of the soft and ductile ferrite. So, in hypoeutectoid steels the pearlite may be regarded as behaving like a strong second phase. A rule-of-mixtures argument (see Chapter 7), suggests that, as the amount of ferrite decreases and the amount of pearlite increases, there should be an increase in hardness and strength, but a fall in ductility. Figure 3.10 shows that this is indeed the case.

The data in Figure 3.10 are room-temperature values. Figure 3.13 shows how yield stress and tensile strength vary with temperature in a 0.15 wt % carbon steel. Both parameters drop sharply above 700 K. This

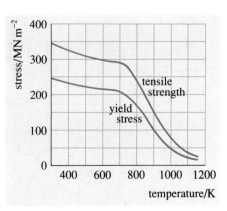

Figure 3.13 Temperature dependence of tensile strength and yield stress for a 0.15 wt % carbon steel

phenomenon is utilized in order to work steels to shape. Forging and extrusion are usually performed on steels in their softer, austenitic state. From the phase diagram (Figure 3.7), for 0.15% C, the minimum temperature for austenite is 1140 K, or $0.63T_m$. This temperature is just about at the point where the steep drops in yield stress and tensile strength level out in Figure 3.13. In practice, hot working is performed at higher temperatures (around 1550 K) since lower forces are required.

▼Plain carbon steels▲ summarizes their properties and applications.

▼Plain carbon steels▲

Plain carbon steels have one great advantage — they are cheap. Alloy steels are not only more expensive because the alloying ingredients usually cost more than iron, but the steels into which they are introduced are more difficult to manufacture and hence more costly. Wherever their properties will suffice, plain carbon steels are therefore used. Table 3.3 is a summary of their principal classifications with indications of typical engineering applications. Of course, even these 'plain' carbon steels have controlled amounts of manganese and silicon (which are present in the iron when it is extracted) and minimal amounts of the impurities sulphur and phosphorus.

Low-carbon steels are readily weldable and this makes them extremely useful for large structures such as ships and bridges, as well as car-body panels. However, we can't have it all ways — if we choose a steel with enough carbon to respond to heat treatment then this very virtue makes it difficult, or impossible, to make good reliable welds.

Table 3.3 Properties and applications of plain carbon steels

Carbon content wt %	General properties	Typical applications
0.01–0.1	Soft, ductile, no useful hardening by heat treatment except by normalizing, but can be work hardened. Weldable.	Pressings where high formability necessary
0.1–0.25	Strong, ductile, no useful hardening by heat treatment except by normalizing. Weldable. Can be work hardened. Ductile behaviour can become brittle at temperatures just below room temperature.	General engineering uses as 'mild steel' e.g. sheet plate sections, pressings etc. Can be made free machining by controlled additions of MnS (0.5 wt % S, 1.5 wt % Mn).
0.25–0.6	Very strong, heat treatable to produce wide range of properties in quenched and tempered conditions. Difficult to weld. Can become brittle at subambient temperatures.	Bars and forgings for a wide range of engineering components, e.g. connecting rods, springs, hammers, axle shafts, requiring strength and toughness.
0.6–0.9	Strong, whether heat treated or not, but ductility lower than when less carbon present. Not weldable.	Where strength is more important than maximum toughness, e.g. tools, wear-resistant parts, etc. The strongest metallic materials (e.g. piano wire, yield stress $> 2000\,\mathrm{MN\,m^{-2}}$) and the so-called silver steel are in this group.
0.9–2.0	Wear resistant and can be made very hard at expense of toughness and ductility. Cannot be welded. Tend to be brittle if structure not carefully controlled.	Cutting tools like wood chisels, files, saw blades. Press and blanking tools. Wear-resistant applications where free Fe_3C particles in structure are useful, e.g. rolling element bearings.

3.2.2 Annealing below the eutectoid

Low-carbon steels, with less than about 0.2 wt % carbon, account for the largest proportion of steel production. Their high ductility is important because of the consequent good formability (hence 'mild' steels) and they are extensively used in the construction industry (beams, concrete reinforcement bars and so on) and for many applications where plate and sheet are used ('tin' cans, cabinets, car bodies and so on). The initial shaping from the cast state requires hot working to keep the forces involved manageably low. This is also effective in breaking down the cast microstructures, but hot working produces oxide scale, preventing the achievement of good surfaces and accurate dimensions, so sooner or later cold working will be involved. Cold working results in work hardening, which can lead to cracking if taken too far without some intermediate softening process.

Two distinctly different types of treatment for softening are possible: heating below the eutectoid and heating above it, into the austenite region. Heating to a temperature below the eutectoid (996 K) causes recrystallization of the ferrite (α) without modifying the pearlite. This process is known as **subcritical annealing** or **process annealing**.

For low-carbon steels with small amounts of pearlite, subcritical annealing has four distinct advantages over annealing in the austenite region. First, the lower temperature means that the cost of equipment and energy is less. Second, the subsequent cooling rate is of little consequence, since no austenite transformation is involved. Third, there is less risk of sagging or distortion because the metal is more rigid at the lower temperature. Fourth, there is less scaling when the treatment is done in air. The metal is to all intents and purposes as soft and as ductile as it would be after softening in the austenite region.

The main disadvantage of subcritical annealing is the longer time required to complete the recrystallization. Moreover, it is only possible to recrystallize solid solutions that have been plastically deformed, where the driving force for recrystallization comes from the elastic strain energy associated with dislocations. The high dislocation density in cold-worked steel gives an adequate rate of crystallization of ferrite at temperatures just below 0.6 of the phase-transformation temperature (Figure 3.14).

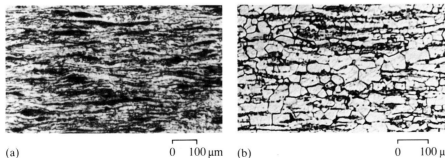

(a)   0   100 µm   (b)   0   100 µm

Figure 3.14 Microstructures of fully annealed 0.4 wt % plain carbon steel. (a) Cold worked. (b) After subcritical annealing

3.2.3 Annealing and normalizing above the eutectoid

Heating to temperatures above the eutectoid completely recrystallizes steel, whatever its initial condition. This is invaluable for recrystallizing materials containing no stored strain energy, such as castings and hot-worked products which have been processed at a high temperature.

Heating large-grained structures of steel into the γ-phase field results in the formation of a completely new set of smaller γ grains which will not grow much larger unless the temperature is allowed to rise too far above the (α + γ)/γ boundary. On cooling, heterogeneous nucleation occurs around the austenite grains, which are then replaced after transformation by smaller grains of α phase plus some pearlite colonies (Figure 3.15). Thus steels are rare among the common engineering metals in that they can be converted from a coarse grain structure to a fine one by thermal treatment alone, without any need for intervening mechanical deformation (titanium is another example). Grain growth in the ferrite and in the austenite conditions occurs in the usual way, being time and temperature dependent. Recrystallization as a result of transformation introduces a refining mechanism not possible with single-phase systems. The importance of this is shown by the effect of grain size on yield stress in Figure 3.16 and is described by the Hall–Petch relation (Chapter 2).

Two slightly different procedures are used for heat treatment in the austenite region:

Full annealing The steel is heated into the austenite region, held long enough to form austenite grains, and cooled slowly (usually in the furnace) to give the equilibrium products. Sometimes this is called a **furnace anneal**. A typical structure is shown in Figure 3.17(a).

Normalizing The steel is heated to about 50 K above the austenite phase boundary, held for a short time to allow formation of new γ grains, but not for long enough to allow them to grow significantly, and then cooled in air — that is, at a faster rate than after full annealing. The advantages are obvious: the furnace can be kept running constantly at the appropriate temperature and does not have to be repeatedly heated to temperature and then cooled down. The normalized product has roughly similar properties to the fully annealed one, but the faster cooling rate results in a smaller grain size (Figure 3.17b).

As well as the softening processes discussed above, components such as welded assemblies may be given a low-temperature treatment called **stress-relief annealing**. The primary purpose here is to allow long-range residual stresses to be relaxed, rather than to cause a change in the visible microstructure of the steel. Welds invariably contain residual stresses caused by the differential contractions of weld metal and adjacent material that has not been heated to the welding temperature.

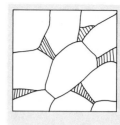

Original structure of equiaxed ferrite crystals and pearlite.

Heating into γ region causes nucleation of γ crystals at pearlite colonies and at triple points, leading to

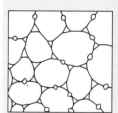

equiaxed crystals of γ. Provided the temperature is not too high, these new crystals will not grow.

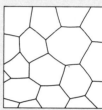

Cooling back into ferrite regions nucleates new crystals at triple points and along boundaries; sometimes also near centre of γ crystals.

Final cooling through eutectoid completes transformation of γ crystals, but structure of ferrite and pearlite is now much finer than original

Figure 3.15 The refinement of a ferrite + pearlite structure

In practice, steel fabrications and weldments are usually stress relieved at temperatures in the region of 900 K, where some microstructural change may take place. This relaxes the stresses to a safe level in a reasonably short period of time: that is, in hours rather than days. Cooling after stress relieving must not be too fast, otherwise there is a risk of sections of different thicknesses reaching room temperature at different times and so creating a new pattern of residual stress.

In addition to these normalizing and annealing procedures, a further heat treatment may be employed for medium and high-carbon steels where maximum ductility is required.

The lamellae of ferrite and cementite in pearlite possess a large interfacial area and hence a large surface energy per unit volume. If a pearlite structure is held for a period of hours just below the eutectoid temperature, the cementite plates have sufficient time to adopt a spherical shape in order to reduce this surface energy (Figure 3.17c). Although there is some loss of strength, this structure is much tougher and more ductile than even that associated with full annealing. Such a **spheroidizing anneal** is often carried out to improve cold workability of medium and high-carbon steels.

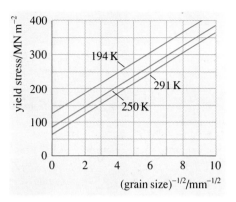

Figure 3.16 Dependence of yield stress on grain size for a mild steel tested at various temperatures

EXERCISE 3.6   A 0.1 wt % carbon steel has a grain size of 0.1 mm. How could its yield stress at 291 K be increased? (Hint: Figure 3.16 may help with ideas, but notice the horizontal scale is (grain size)$^{-1/2}$.)

SAQ 3.4   (Objective 3.4)

(a) Why does the spheroidizing anneal of a hypoeutectoid plain carbon steel have to be performed below the eutectoid temperature?

(b) What is the energetic driving force behind the restructuring that occurs during this annealing?

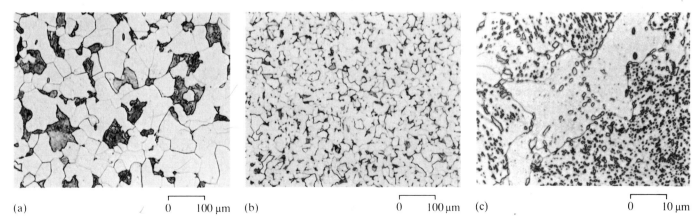

(a)   0   100 μm   (b)   0   100 μm   (c)   0   10 μm

Figure 3.17 Microstructures of (a) fully annealed 0.2 wt % carbon steel and (b) the same steel in normalized condition. (c) Microstructure of a spheroidized 0.4 wt % carbon steel

3.2.4 High-tensile steel wire

In wire form, steel is one of the strongest materials available, being rivalled only by some of the fibres and by single-crystal whisker materials (see ▼Other high-strength materials▲). It is widely used in ropes and cables for shipping, cranes, suspension bridges and lifts, and in applications ranging from musical instruments to valve springs in car engines.

Technically it is a very straightforward material, nothing more than a medium- to high-carbon (0.8–0.9%) steel with about 0.8% manganese for solid-solution strengthening of ferrite. (Manganese also combines with, and hence neutralizes, any sulphur present.) The secret of the high tensile strength lies in the way in which austenite-transformation products are caused to form in the microstructure, coupled with subsequent work hardening by carefully controlled, but heavy, degrees of cold work. Tensile strengths of $2000\,MN\,m^{-2}$ are easily attained on 'wires' of up to 15 mm diameter, and considerably higher — up to 2800–$3000\,MN\,m^{-2}$ — on the finer ($\approx 25\,\mu m$) diameters.

A range of wire strengths is used in rope manufacture. It is not always essential to use steel wire having the highest tensile strength and, in any case, the highest tensile strengths are only attainable with finer wires. For larger diameters, above about 2 mm, the tensile strengths obtained by cold working can be surpassed by heat treatment of martensitic structures, but their fatigue strength and fracture toughness are not as good.

EXERCISE 3.8   Why do you suppose these properties are inferior in a quenched and tempered wire compared with cold drawn?

A major advantage of steel wire is that it still has ductility at these extremely high strength levels, even if recorded as only 1 or 2% elongation in the tensile test. This is rather misleading as the reduction of cross-sectional area on fracture is about 50%, and the low overall elongation reflects the rapid onset of necking in the tensile test. Because of this, steel may be loaded safely to a higher fraction of its tensile strength than other 'strong' materials, where catastrophic failure may occur without significant prior deformation. The common form of failure of springs made from steel wire is by fatigue originating at some small imperfection like an inclusion, after several millions of load cycles. There are very many ways of twisting wires together to form a rope, and rope construction is critical in determining the flexibility and extensibility of a rope (see Chapter 6). This, in turn, affects its breaking load and extension. In practice 90% of the tensile breaking load of the individual wires can be achieved. Whatever the construction, individual wires are, however, subject to bending and torsion in operation and the strength of the rope is dependent in a complex manner on the strength and elastic moduli of the material from which it is made. If we want a

▼Other high-strength materials▲

Materials in competition with high-tensile steel wire are also only available in fibrous form. They fall into two groups: those based on polymers (or polymer precursors) and those based on single-crystal whiskers of metals and ceramics. (Inorganic glass fibres can be prepared with strengths in excess of $3\,GN\,m^{-2}$, but these strengths can rarely be utilized because of the fibres' extreme susceptibility to surface damage.) Their principal mechanical properties are summarized in Table 3.4. Fibre materials are considered in more detail in Chapter 6.

Table 3.4   Properties of high-performance fibres

Material	Tensile strength σ_t	Young's modulus E	Density ρ
	$GN\,m^{-2}$	$GN\,m^{-2}$	$kg\,m^{-3}$
steel wire	3.0	210	7860
Polymer			
carbon fibre (UHS)	5.2	270	1750
aramid fibre	3.1	124	1440
oriented polyethylene	2.6	120	970
Inorganic			
asbestos	2.1	160	2500
alumina	1.0	100	2800
silicon carbide	4.0	410	2500

EXERCISE 3.7   Which material has the highest (a) specific strength, and (b) specific stiffness, and how do these compare with those of steel wire?

rope with a high breaking load, we therefore need to start with an individual wire of high elastic limit — which usually means a high tensile strength.

The strength of high-tensile steel wire is achieved through work hardening arising from the cold deformation of fine pearlitic structures. Free ferrite at the prior austenite grain boundaries detracts from the strengthening, so a eutectoid steel is often preferred. For very high strengths, hypereutectoid steels can be used which increase the amount of carbide particles dispersed through the microstructure.

EXERCISE 3.9   How could a very fine pearlitic microstructure be obtained in a eutectoid steel?

Fast cooling, yet avoiding martensite formation, is not easy when wire is made on a production scale. Figure 3.18 is the time–temperature–transformation (TTT) diagram for a eutectoid steel. In practice, the wire is austenitized at 1125–1175 K and then quenched into a bath of lead or molten salt at a temperature of around 850 K. It is left in the bath for a sufficient time for all the austenite to transform, and is then removed and cooled in air. It is a traditional process that was found to be successful long before TTT diagrams came on the scene to explain why the desired structure was produced. It is called **patenting**. Nowadays it is possible to achieve a similar structure by cooling in a controlled air blast.

Patented wire is found to possess a higher ductility than all other possible structures offering a similar hardness level. Greater ductility is possible with spheroidized structures obtained by tempering, but their hardness is less; harder structures may be obtained by tempering at

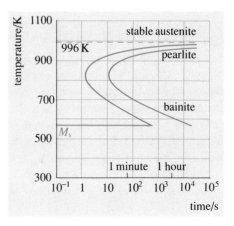

Figure   3.18 Time–temperature–transformation (TTT) diagram for 0.8 wt % plain carbon steel

115

lower temperatures, but their ductility (or rather, reduction of cross-sectional area in the tensile test) is lower than fine pearlite resulting from patenting. Cold working of patented wire develops a strong, fibrous structure. The high degree of cold work raises the strength, but the fibrous structure prevents significant loss of toughness. A typical microstructure of patented, cold-drawn wire is shown in Figure 3.19. Although the detail of the pearlite is too fine to be resolved under the optical microscope, the pronounced texture can easily be seen, whilst the higher magnification of the electron microscope allows the carbide to be resolved.

There is, however, one problem with very strong wire. This is a phenomenon known as **strain-age embrittlement**, which occurs after a period of loading at ambient or slightly higher temperatures. To avoid this highly undesirable effect, it is usual to use a wire of lower tensile strength, which is much less susceptible to strain-age embrittlement.

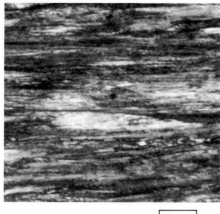

0   40 µm

Figure 3.19 Longitudinal section of piano wire, showing microstructural banding

3.3 Steel for car bodies

3.3.1 Structural requirements

Both the structure of a motor car and the materials chosen to build it should satisfy the demands made on them.

What are these demands?

The practical ones are mechanical. The structure must be rigid and strong enough to support the weight of the passengers and their goods, and the mechanical components (such as engine, gearbox) of the car. In an increasingly safety conscious age, it also must protect the passengers against impacts. These functions must be performed at temperatures ranging from $-40\,°C$ to $+50\,°C$, say, and in relative humidities between practically zero and 100%. In addition, the structure should shield the passengers from the weather, be durable in the face of exposure to agents such as sunlight and salt, be suitable for mass production and be as cheap and as lightweight as possible.

Let's examine the forces acting on the structure. At rest, apart from the direct loads at the points of support or attachment of the different components, the main forces are flexural, due to the bending moments induced between the downward acting weights of engine, passengers, body and so on, and the upward reactions at the four support points. Figure 3.20 shows a simplified representation of this. With the car in motion, forces due to various accelerations (longitudinal due to forward motion or braking, sideways due to cornering, vertical due to road unevenness or braking) are superimposed on the static loading. When the vertical forces are unequal, the structure is also subjected to torsional loading. The structure needs to respond elastically to all these forces, and the deflections need to remain small (otherwise doors might fly open, or windscreens fall out).

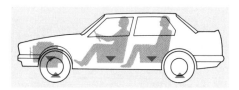

Figure 3.20 Main forces acting on a static car body

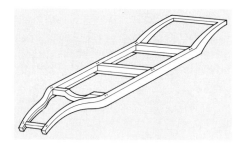

Figure 3.21 A typical steel-channel chassis

Until the 1930s, the structure of cars was based on the chassis, a frame with two steel members running the length of the car that were joined by a number of cross members (Figure 3.21), and to which the body and all the other parts of the car were attached. Initially the body panels were of wood fastened to wooden frames, then steel panels were introduced (sometimes aluminium was used instead on higher priced cars), then the steel panels were welded together and the wooden frames eliminated. This produced a stiffer overall structure. Finally, in 1934, the first chassisless body was produced (Figure 3.22) in which all the stiffness resided in the **monocoque** body shell. This has remained the dominant structure for mass-produced cars ever since. ▼Chassis versus monocoque▲ highlights its advantages in structural terms over the earlier designs.

Figure 3.22 The monocoque body of the 1930s Citroen

▼Chassis versus monocoque▲

To explore the differences between these two methods of automobile construction, we're going to take simple models of each and compare their flexural rigidities.

From Chapter 1, the bending equation is

$$\frac{M}{I} = \frac{E}{R}$$

For two different cross-sectional geometries, subjected to the same bending moment, M, and made of material with the same Young's modulus, E, a measure of their relative rigidities is the ratio of the radii R to which they are bent. This is given by

$$\frac{R_1}{R_2} = \frac{I_1}{I_2}$$

Let's consider the chassis to comprise two steel U-channel sections running the length of the car, facing each other and connected by cross members (Figure 3.23a). The second moment of area I_U of one of these U-channels is given in Figure 3.23(b). Taking representative values of $d_1 = 100$ mm, $w_1 = 30$ mm and $t_1 = 8$ mm, gives

$$I_U = 1.4 \times 10^{-6}\,\text{m}^4$$

For the model chassis, assuming that the cross members do not contribute to the bending stiffness, I_C is just twice this, or $I_C = 2.8 \times 10^{-6}\,\text{m}^4$.

The model for the monocoque structure is a hollow rectangular box (Figure 3.24). Taking values reasonably close to those of a car, we have $w_2 = 1.5$ m, $d_2 = 1.2$ m and $t_2 = 0.4$ mm. Putting these into the

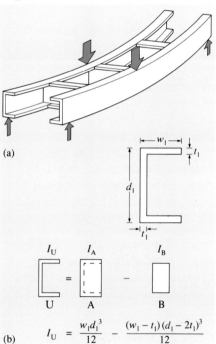

(a)

(b)

$$I_U = \frac{w_1 d_1^3}{12} - \frac{(w_1 - t_1)(d_1 - 2t_1)^3}{12}$$

Figure 3.23 (a) Model of car chassis and bending geometry (b) U-channel cross-section and second moment of area

expression given in Figure 3.24 yields $I_M = 5.5 \times 10^{-4}\,\text{m}^4$, thus

$$\frac{R_C}{R_M} = \frac{2.8 \times 10^{-6}}{5.5 \times 10^{-4}} = 5 \times 10^{-3}$$

This is the ratio of the stiffness in bending of the two model structures. The monocoque is about 200 times as rigid as the U-channel.

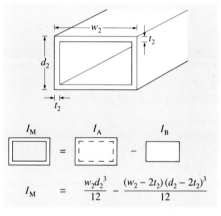

$$I_M = \frac{w_2 d_2^3}{12} - \frac{(w_2 - 2t_2)(d_2 - 2t_2)^3}{12}$$

Figure 3.24 Model of monocoque structure and second moment of area

EXERCISE 3.10   Compare the stiffness to weight ratios of the two models.

A similar relationship is found in the case of torsion, although the modelling is less straightforward.

This comparison of flexural rigidity is a bit unfair on the chassis-based body, since it ignores any contribution to the stiffness from the body mounted on it. However, despite both this and the simplicity of the models, it does illustrate the benefits of changing to monocoque construction. In practice, monocoque body shells have their stiffness further enhanced by designing in the equivalent of corrugations (Chapter 1), frequently disguised as styling features.

3.3.2 Which steel?

A series of events took place during the period from 1923 to 1933 which created the mass-production industry we know today. A major turning point was the introduction of a sheet material that could be handled and formed by machines rather than by skilled workers. This material was thin sheet steel.

In 1923 the American Rolling Mill Company opened the first continuous hot-strip mill for rolling thin steel. A continuous hot-strip mill consists of a series of rolling mills through which the hot sheet passes consecutively, each mill producing successively thinner sheet. The mills of the 1920s would roll sheets nearly one metre wide down to a thickness of 1.3 mm. The important point about this sheet was that it was considerably wider than the steel that had previously been available. It was this development, together with progress in welding technology, that started the series of changes in the design of the car-body structure summarized above.

EXERCISE 3.11 From the preceding section, and elsewhere, summarize the requirements for a steel for car-body panels.

The most important of these, apart from availability in sheet form, are weldability, formability and deep-drawing ability, a ductile–brittle transition temperature below the minimum service temperature, a good surface finish (people concerned with cars seem obsessed by this requirement), and cost. So what sort of steel fills the bill? On cost grounds we can eliminate medium and high-alloy steels, leaving unalloyed ones as prime contenders. 'Plain carbon steels' in Section 3.2 surveyed the general properties and typical applications of the various steels.

Although these properties are qualitative, it is apparent that only steels below 0.25 wt % C meet the formability and weldability criteria (see ▼Welding the body panels▲). However, in steels with carbon contents over 0.1 wt % there might be problems with the transition from ductile to brittle behaviour at low temperatures.

In practice, steels with less than 0.1 wt % C are used for car-body panels. Unfortunately they are liable to an effect known as the **yield-point phenomenon**, which can seriously affect the surface appearance of the finished body panel. Let's examine why, and what's involved.

▼Welding the body panels▲

A motor car body is too complex to be made in one piece, largely because of re-entrant angles and the need for stronger sections around apertures. Consequently, whatever material is used it must be capable of being joined to other panels in the car body, and some of these must be capable of transferring loads, as they are not simply a decorative assembly. Joins must consequently have a high integrity. One of the great advantages of mild steel over other materials is that it can readily be joined by welding. However, although fusion welding was already an established process when the first steel from hot-strip mills became available, the process was unsuitable for welding thin sheet material. In the 1930s a new welding process, electric-resistance welding, was introduced, and this has become the mainstay of car-body assembly techniques. Several thousand electric-resistance spot welds are used in assembling a typical car body.

Welds are made by clamping two sheets together between two electrodes and passing a large current between them. The sheets are thus locally heated. Clamping forces are so high, the current so large and the time for which it passes so short that manual control is impractible. The electrodes are made of copper–chromium alloy, shaped as truncated cones, and have internal passages for cooling water. The clamping force is supplied pneumatically or hydraulically. The pressure used for clamping steel components is usually about $70 \, N \, m^{-2}$. This is below the yield stress for mild steel at room temperature, but serves to squeeze the two sheets together.

Figure 3.13 showed how the yield stress of the steel varies with temperature. At some stage during the heating by the electric current, the yield stress falls to below the stress produced by the clamp, and the steel deforms plastically: it is hot forged.

The temperature continues to rise until the melting temperature is reached. When the interface between the two sheets is destroyed, the current is automatically switched off.

The liquid region, known as the 'nugget', solidifies very rapidly because it's surrounded by a large amount of cold metal and the water-cooled electrodes. Figure 3.25 shows the microstructure of a weld nugget. The growth direction of the grains reveals the direction in which heat was abstracted from the solidifying liquid.

Here the grains have grown perpendicular to the surface, indicating that most of the heat was conducted through the surface into the water-cooled electrodes.

The changes in microstructure around the spot weld have a considerable effect on the behaviour of the joined panels. The grains nearest the weld grow considerably. As a result, these grains have a lower yield stress than the fine grains in the remainder of the sheet (remember the Hall–Petch relation). For this reason it should not be totally unexpected that spot welds usually fail around the weld: that is, in the zone in which grain growth has occurred. The larger grain size also leads to higher ductile–brittle transition temperatures, increasing the risk of failure.

Figure 3.25 Micrograph of a spot-weld nugget joining pieces of thin steel sheet

3.3.3 The yield-point phenomenon

What is the yield-point phenomenon, what effect does it have and how can it be overcome?

To start with, consider Figure 3.26, which shows the stress–strain curves for three steels with different concentrations of carbon. Note that only (a), the low-carbon steel, shows a yield point (the 'blip' at the end of the elastic portion of the curve).

Which of the three steels would be most suitable for cold working?

Clearly the one with 0.1 wt % carbon, since it has the lowest yield stress and greatest elongation.

In practice, the steel used in car bodies is hot rolled from a large block down to fairly thin sheet and subsequently cold rolled to produce the desired surface quality and thickness.

What are the disadvantages of hot working?

The dimensional accuracy attainable is not as great as in cold working, and the surface finish is not as good because in air at high temperature

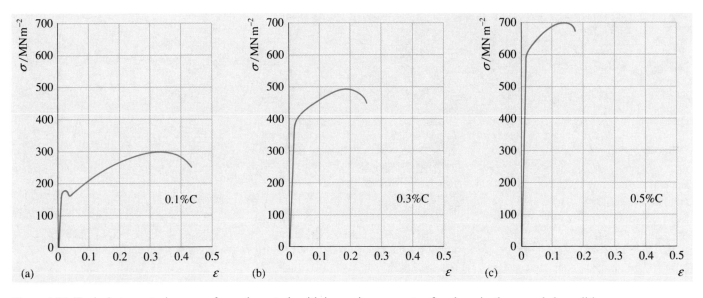

Figure 3.26 Typical stress–strain curves for carbon steels with increasing amounts of carbon, in the annealed condition

an oxide scale forms which leaves a roughened, pitted surface on the sheet. Cold rolling can produce material to very close tolerances with almost a mirror finish.

Why does the stress required to produce further deformation normally increase above the yield stress?

Work hardening occurs, that is the interactions between the increasing number of dislocations require an increasing force to drive the dislocations through the crystal.

Although cold rolling raises the strength of the steel by work hardening, it also reduces the ductility. If the sheet has to be extensively deformed to make the shaped body panel, this is bad news, for three reasons:
- because ductility is necessary to make the pressing,
- because the deformation forces are higher,
- the 'spring back' of the pressed shape is greater and more difficult to control.

Hence the cold-rolled sheet has to be annealed before it is suitable for forming a body panel. This is usually done in an oxygen-free atmosphere to prevent oxidation of the highly finished surface.

During pressing, the sheet is strained by different amounts in different directions. If the strain in any one direction is too great, failure will occur by necking. The material will become unstable in this region and become thinner and thinner until a crack appears. Hence it is useful to have a high rate of work hardening and a large plastic extension before necking occurs.

Figure 3.26 shows that low-carbon steels do not simply reach their elastic limit and then start to work harden. They suffer an effect whereby, at the elastic limit, the stress suddenly drops and a strain of 2–3% is needed before the normal work hardening begins. *This only occurs when the steel is first taken to its elastic limit after annealing or has been allowed to stand for a few days after previous deformation* as illustrated in Figure 3.27.

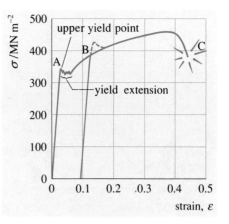

Figure 3.27 A stress–strain curve for a mild-steel specimen

When an annealed low-carbon (less than 0.2%) steel is tested in tension it exhibits a yield point (A in Figure 3.27) and the yield extension appears as a serrated, flattish portion of the curve. If testing is continued to the region where work hardening becomes apparent, say to point B, and the load is then removed, the yield drop does not occur if the load is re-applied. At least, not if the test is continued within a few days. Immediate re-application of the loading produces a smooth work hardening curve which continues through necking and on to tensile fracture C in the normal way.

However, if the test piece taken to B is rested for a few days before the test is continued, a new yield drop appears as indicated by the dotted curve above B. The effect first becomes apparent after a few days at room temperature and the new yield point becomes fully established after about 2–3 weeks.

To explain this phenomenon, we need to analyse what happens when a metal yields. At yield, large numbers of dislocations are nucleated and start to move, producing plastic deformation, although the stress is still too low to move the pre-existing pinned dislocations.

In a car-body steel, the carbon concentration is less than 0.1 wt % and is usually around 0.04 wt %. The iron–carbon phase diagram predicts that at room temperature such a composition should contain a small amount of iron carbide (Fe_3C) as a precipitate in ferrite (α iron). Ferrite is the interstitial solid solution and can hold about 0.005 wt % carbon in solution. But carbon atoms are just a bit too large to fit into the spaces between the iron atoms, which are consequently forced apart, producing localized strains. Now, all dislocations also have localized strain fields associated with them.

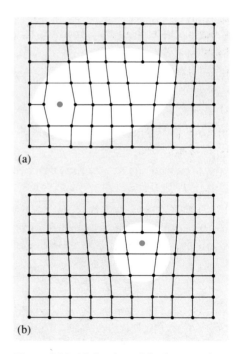

Thermodynamics being what they are, the system should try to minimize its total internal strain energy. How might it do so from the state just described?

If the carbon atoms could move to the dislocations, the overall strain energy of the lattice would be reduced. The net result of the above effects is that, given time, the carbon atoms in a ferritic structure will diffuse to dislocations (Figure 3.28). A dislocation is then said to possess an **atmosphere** of carbon atoms.

Figure 3.28 (a) Section of ferrite crystal containing an edge dislocation and a carbon atom at a different site. (b) A section of a ferrite crystal in which the carbon is located at a dislocation. The strained areas are highlighted. Note the smaller strained area in (b) compared with (a)

Why is there no yield point if the specimen is immediately retested, as shown in Figure 3.27 at **B**, whereas the yield point does return after a few days?

At **B**, the newly generated dislocations have not developed atmospheres. It takes time for the carbon atoms to diffuse to the new positions and form the atmospheres.

When the new atmospheres have formed, why is the elastic limit above **B** greater than if the test had been continued immediately after unloading?

The atmospheres have pinned the dislocations, and a greater force is now required to nucleate fresh dislocations and get them moving.

It is not the increase of yield stress on its own that affects a car-body panel. It is the accompanying, and sudden, yield extension which is the greatest nuisance. What happens is that deformed bands appear on the surface of the sheet. These rumple the surface and move over it as deformation proceeds until the whole area has yielded and begins to work harden. Such bands are known as **Lüders bands** and are shown in a sheet specimen in Figure 3.29. Lüders bands will show through paintwork. In a complex pressing, different portions of the same sheet may have been strained from zero right up to where necking begins. Somewhere there is bound to be a region where the deformation lies just in the strain range where Lüders bands rumple the surface. (In pressings, the bands are often referred to as **stretcher strains**.)

3.3.4 Overcoming the undesirable effects

What can be done to prevent the appearance of Lüders bands?

The carbon could be eliminated altogether, which is impractical, or it could be tied up by combining it with a strong carbide former. A more cunning method exploits the time lag between a light deformation and the formation of dislocation atmospheres.

Removal of carbon from interstitial solution can be achieved by adding small amounts of elements like titanium, niobium or vanadium which have a stronger affinity for carbon than has iron. Such methods put up the cost though they are used to advantage in HSLA steels (see Section 3.3.5) where the fine carbides can be exploited for precipitation hardening. The more cunning method is the one most widely used by pressing manufacturers and involves deforming the steel a little (about 2–5%) before it is pressed.

Consider Figure 3.27 again. When the specimen is reloaded immediately after unloading the yield drop does not appear. There is insufficient time for the new dislocations to acquire carbon atmospheres so they remain as free dislocations. In practice, the steel is passed through a set of rollers which slightly deform the surface in such a way as to cause plastic deformation and to exceed the deformation that would leave Lüders bands anywhere on the surface.

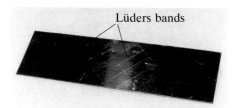

Figure 3.29 The criss-cross lines are Lüders bands in a stainless-steel sample

But this cannot be a permanent solution. If the steel is allowed to stand for a matter of weeks at room temperature, carbon atoms will diffuse to the mobile dislocations and pin most of them. Steel kept in storage for longer than this period is likely to develop Lüders markings in pressings.

How might such old stock be made suitable for pressings?

By repeating the light mechanical deformation carried out at the steel mill before using it in any pressing operation.

In view of these problems associated with the pressing of low-carbon steel, why is it used rather than, say, a metal like aluminium which does not normally exhibit a yield drop? To answer this we have to look at the behaviour of the materials under stress. In a tensile test, the slope of the stress–strain curve when steel undergoes plastic deformation is much steeper than for aluminium.

What does this indicate?

That the rate of work hardening is higher in steel than in aluminium.

A material which work hardens rapidly is less likely to form a neck and tear into a hole than one which work hardens slowly. As one region becomes strained its strength increases, so further straining has to take place in adjacent regions. This is very important in press-forming. Materials with low rates of work hardening are notoriously difficult to use because they neck down in one place and this quickly develops into a split. The consequence is that the depth of form is severely limited. Apart from this, steel is cheaper and stiffer than aluminium and easier to weld.

3.3.5 High-yield, weldable steels

The pressure from car designers to produce even lighter components, yet without sacrificing rigidity or strength, has led to the use of **high-strength low-alloy (HSLA)** or **micro-alloyed** steels for critical body panels. Figure 3.30 shows one example of this. Use of such materials is becoming widespread. Although they are marginally more expensive than conventional mild steels, the advantage of their greater yield stress is more than ample compensation.

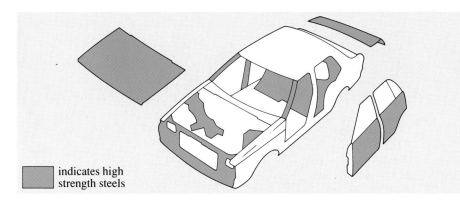

indicates high strength steels

Figure 3.30 High-strength low-alloy and micro-alloyed steels used in critical car-body panels (Datsun *Sunny*)

SAQ 3.5 (Objective 3.3)
A manufacturer has discovered a coil of deep-drawing-quality steel in store, which has been there for about 12 months. Explain what would be likely to happen if this coil were used for making pressings in its present condition and indicate how any difficulty might be overcome.

Figure 3.31 compares the strengths of various metals used in car bodies. It includes a group of commercially available steels, based on BS4360 'Weldable structural steels'. These have ductilities and properties like formability and weldability that are virtually the same as traditional low-carbon steel. However, they all have a yield stress higher than that of the traditional steel.

These higher yield stresses are obtained by judicious alloying with very small amounts of various elements. For example, controlled amounts of boron (0.005%), niobium and vanadium (up to 0.1% each), coupled with controlled hot rolling to produce a fine grain size, allow the higher yield stresses to be obtained without losing the desirable properties. These effects are further enhanced by precipitation hardening from nitrides, carbides and borides as the steel cools following hot working (Figure 3.32).

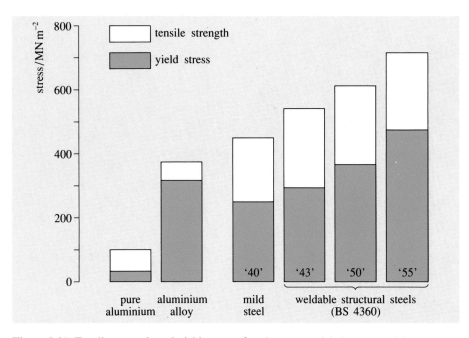

Figure 3.31 Tensile strength and yield stress of various potential sheet materials for use in automobile body panels

Other micro-alloying methods include solid-solution strengthening of ferrite in, for example, rephosphorized steels. Phosphorus is usually regarded as an undesirable impurity since it migrates readily to grain boundaries and decreases ductility. Figure 3.33 shows phosphorus is nonetheless a very powerful ferrite strengthener compared with other solutes. It is also 'cheap' in the sense that it is present in the steels anyway from the iron-making processes. Although their production demands great care, rephosphorized steels offer the potential for improved yield stress and enhanced work-hardening ratios.

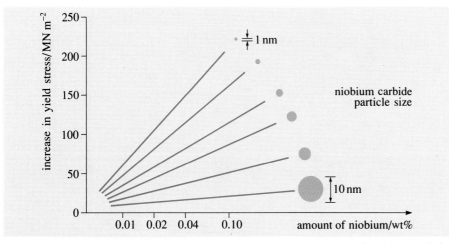

Figure 3.32 The increase in yield stress depends on both the amount of niobium and the size of the precipitated particles of niobium carbide

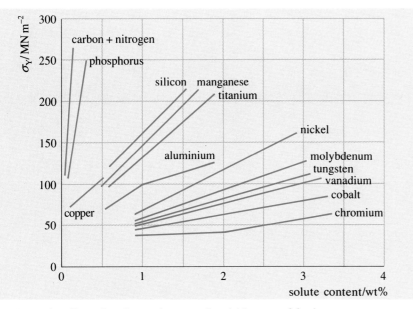

Figure 3.33 The effect of various solutes on the yield stress of ferrite

Micro-alloying also enables dual-phase structures to be produced, consisting of ferrite and low-carbon martensite (see next section). Islands of higher-strength martensite in a ductile ferrite matrix increase the steel's yield stress and work-hardening rate, while retaining its high ductility.

There is no doubt that car makers will continue to make more extensive use of micro-alloyed steels. They do, however, present some problems:

• Their higher yield stress means greater spring back in press forming, which makes accurate and reproducible pressings harder to achieve. Different batches of steel respond in slightly different ways. The formability of the strongest types is less than that of conventional mild-steel sheet.

• Repairs to body panels made from these steels are more difficult. Gas welding is unsuitable because the spread of heat tends to soften too much of the surrounding bodywork, so that they have to be brazed or spot welded. Also, where body distortion has occurred, a greater force is needed to restore the correct alignment.

What of alternative, nonferrous materials? It seems surprising that, given the developments in materials technology since the 1920s, no viable alternative to steel has emerged for mass-produced car bodies. One reason is that steel technology has been developing, too, hence the HSLA steels discussed above. Another is the huge investment, not just in capital equipment, but also in design, manufacturing and servicing expertise and experience, committed to steel bodies. However, changes are afoot — although so far, these are evolutionary rather than revolutionary. ▼Competing materials▲ considers some of these.

Remember, though, that apart from rare examples like the ill-fated, stainless steel De Lorean, steel body shells already come with a polymeric coating: the paint.

3.4 Nonequilibrium structures in plain carbon steels

3.4.1 The martensite transformation

Some of the objects tested with the needle point in Section 3.1.1 were harder than, or as hard as, the needle point itself. How was the hardness of the needle point attained? (Or that of the hammer face or the saw teeth?). The answer lies in the phenomena resulting from the γ (FCC) to α (BCC) phase transformation in iron and the way carbon interferes with it to give structures such as martensite under nonequilibrium cooling conditions. It is this, above all else, that gives steels their special place among engineering materials. (The martensite transformation is reviewed in ▼Martensite in steel▲.)

Is the levelling off in hardness at higher carbon contents in Figure 3.37(c) due to incomplete transformation of austenite?

Retained austenite does occur in high-carbon steels. In the microstructure (Figure 3.41) of a 1% C steel, the white areas between martensite needles are pools of retained austenite. However, this is not the complete explanation for the levelling off of hardness at about 900 H_V, even though it probably does account for the spread observed with different steels. Electron microscopy of the quenched structures reveals that there are two different formations: one a bundle of rods and the other a series of plates. The former is based on low-carbon martensite and is soft; the latter forms in 0.7% C and above and is hard — about 900 H_V. Steels containing between 0.1 and 0.7% carbon form mixtures of these two when quenched and the hardness is directly related to the amount of each type present.

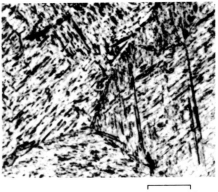

Figure 3.41 Martensite structure as seen under an optical microscope

▼Competing materials▲

The oil crisis of the 1970s, together with predictions of the limited lifetimes of the world's energy resources, concentrated the minds of car manufacturers on ways of reducing the total energy cost of manufacturing, maintaining and running cars. Part of this involved, and still involves, looking at materials other than steel for the bodies of mass-produced cars. The goal is to produce lighter, more damage- and corrosion-resistant bodies than steel bodies, but with similar strength, stiffness, protection, processability and cost. Aluminium, as already mentioned, was an early rival on some of the more up-market cars. Nowadays, aluminium alloys can provide structures of similar stiffness and strength, yet weighing only half as much as comparable steel ones, with the bonus of better corrosion resistance. However, they are more expensive than steel, are more susceptible to damage both in production and service, due to their lower hardness, and are more difficult to weld (although adhesive bonding could provide an alternative).

Of the other, more traditional, materials, wood has been used as the chassis material in, for example, the Marcos. More recently, the Africar (Figure 3.34), developed to cope with the unmetalled roads and the climate and terrain of Africa, had both its chassis and body made from epoxy-bonded plywood.

Various specialist car makers have long made polyester/glass-fibre bodies, the panels of which are mounted on a tubular steel space frame, whilst practically all Formula One racing cars today have carbon-fibre/epoxy, monocoque body shells. However, in nearly every case these are made either by hand lay-up (contact moulding) or by closed mould techniques with relatively long cycle times. Thus they are unsuited to mass-production.

Increasingly, though, the volume car manufacturers are incorporating plastics or plastics composite components for the exteriors of their cars. The most striking example of this is the front and rear bumpers, where a variety of materials has been used. This includes classes of material dealt with in Chapters 5 and 7, such as sheet moulding compound, or SMC (a polyester/chopped glass fibre composite); RIM polyurethane with glass fibres (RIM stands for reaction injection moulding); injection moulded polycarbonate blended with poly(butylene terephthalate); and elastomer-modified nylon. Among other developments were the introduction of an SMC bonnet and injection moulded polyester/glass fibre tailgate on the Citroen BX in the early 1980s; and in 1983 the Pontiac *Fiero* (Figure 3.35) was the first volume production car (120 000 units p.a.) to have its entire body shell made from polymer based materials — a combination of SMC and RRIM (reinforced RIM) panels.

Apart from their superior corrosion-resistance and their potential for making lighter body shells, materials based on polymers have one further advantage over steel. Their mechanical damping is much higher (see Chapter 5), so that vibrations and noise are much less of a problem.

The fact that the Africar and Pontiac *Fiero* have not been commercial successes is no reflection on their body materials.

Figure 3.34 The Africar

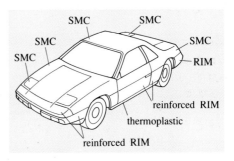

Figure 3.35 Materials used in the 1983 Pontiac *Fiero*

▼Martensite in steel▲

Martensite transformations occur in many materials. Their chief characteristic is the transition from one crystal structure to another by *shear* (that is, displacement) rather than diffusion, nucleation and growth. This means that the transition is very much more rapid than a diffusion-controlled transition.

In pure iron the $\gamma \rightarrow \alpha$ transition is martensitic if the rate of cooling is fast enough. The martensitic α ferrite is BCC, just like ordinary diffusion-controlled ferrite, but it is much finer grained and is soft and ductile. (That it is martensitic can be seen from the characteristic rumpling effect in Figure 3.36 produced by displacements normal to the surface.)

In steel, the martensitic transformation is unusual because the transformed steel is *harder* than the untransformed steel. The hardening is due to the role of carbon in the transformation. Dissolved carbon in the γ FCC lattice impedes the transformation to BCC on rapid cooling. It is much less soluble in BCC than in

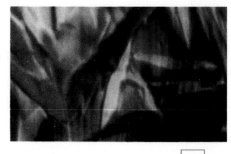

0   10 µm

Figure 3.36 Martensitic iron

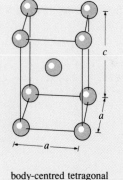

body-centred tetragonal
unit cell of martensite

(a)

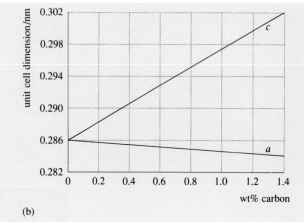

(b)

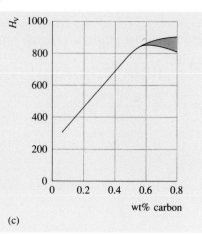

(c)

Figure 3.37 Martensite lattice constants and Vickers hardness as functions of carbon content

FCC (Section 3.1.3), but with insufficient time to diffuse to Fe_3C nucleation sites, it remains in solution and distorts the lattice away from BCC to body-centred tetragonal (BCT) (Figure 3.37a). The amount of the distortion depends on the carbon content, as Figure 3.37(b) shows, and this relates directly to the hardness of the resulting martensite (Figure 3.37c).

The process of martensite formation is characterized by two temperatures (Figure 3.38): the temperature at which it starts to form, M_s, and the temperature at which it finishes, M_f. Both M_s and M_f vary with carbon content (Figure 3.39).

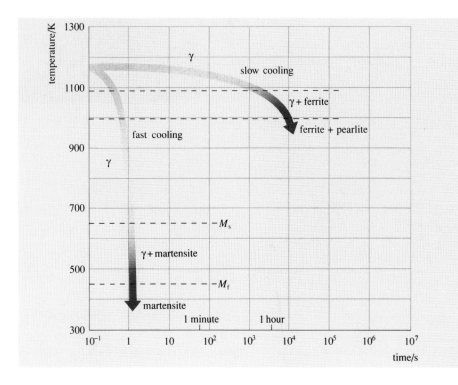

Figure 3.38 Cooling curves for a hypoeutectoid steel showing transformations and products

Why is martensite in steel so hard and brittle, when other martensites are soft? Essentially there are two reasons. Firstly, the BCT lattice doesn't have the five independent slip systems necessary for ductility and, secondly, since the shear transformation involves the nucleation and movement of many dislocations, the martensite has a very high dislocation density (see 'A dislocation model of work hardening' in Chapter 2). In fact, if the yield stress of steel is plotted as a function of (dislocation density)$^{1/2}$, it's found that the values for martensite fall on the same line as those for cold-worked non-martensitic steels (Figure 3.40).

Finally, it's worth noting that there are materials in which the martensitic transformation does not compete with a diffusion-controlled one (for example in some Ni–Ti alloys, and in quartz). In some of these the *reversible* change in volume associated with the transformation can be exploited (for example self-clamping pipe couplings, thermostatic switches in electric kettles and window openers in green-houses). These are the so-called **shape-memory alloys**.

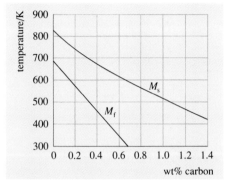

Figure 3.39 M_s and M_f as a function of carbon content for plain carbon steels

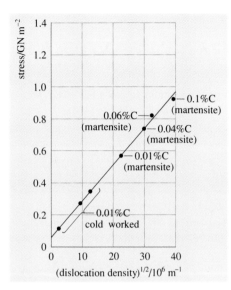

Figure 3.40 0.2% proof stress versus (dislocation density)$^{1/2}$ for cold-worked ferritic steels and for martensitic steels

SAQ 3.6   (Objective 3.5)
Compare the transformation times of the following processes.
(a) An austenite grain 0.5 mm across transforms to martensite. Assume that the speed of transformation v_m is given by $v_m \approx (E/\rho)^{1/2}$.
(b) A pearlite structure with lamellar spacing of 10 μm is formed. Take the diffusion coefficient of carbon in austenite at 996 K to be $5 \times 10^{-13} \, \mathrm{m^2 \, s^{-1}}$.

When a component is quenched to induce martensitic transformation, there is a risk of distortion or even of cracking occurring. This arises from the volume changes on quenching shown in Figure 3.37. Consider what would happen in a fairly thick cross section. The outer skin of the component reaches the transformation range before the underlying core, so it is the first to undergo the transformation. Consequently, the outside of the component transforms to the hard and extremely brittle martensite, whose expansion subjects the austenitic core to plastic strain.

Subsequent expansion as the core transforms can easily crack the brittle martensitic skin. If it does not cause cracks (called 'quench cracks'), then distortion may occur (straight bars come out banana shaped) unless the way the article is plunged into the quenching bath is not carefully directed to even out the cooling. Where such differential volume changes are involved, residual stresses can become very high and can sometimes result in delayed cracking a few hours after the quenching operation has been carried out. (See ▼Quenching media and cooling rates▲.)

Fortunately, there are ways of minimizing these problems by alloying the steel, and this is considered in Section 3.6.

▼Quenching media and cooling rates▲

The medium used to bring about the cooling of the steel is important not only for its effect on structural changes within the metal, but also for controlling the distortion and residual stress likely to be produced in a component. An order of merit may be assigned to various quenchants, indicating which gives the highest and which the lowest overall cooling rate (Table 3.5). The critical region is from 600 K down to 350 K, where the martensite transformation takes place in most engineering steels. The temperature and rate of circulation of quenchant are also important factors.

By using different quenchants a range of cooling rates can be achieved. The way the article is introduced (for example, end-on or lengthways), the speed at which it moves through the quenchant or the circulation rate of the bath, and the way gas films which form at the interface are dispersed, are all important factors in industrial operations. Beware of thinking that what happens within the metal is all important; in commercial heat treatment the correct quenchant and mode of quenching have a major influence on whether the steel will crack, distort or retain high residual stress.

Table 3.5 Ranking order of quenchants

FAST	brine	
	water	
	water-soluble polymers	more effective if stirred to break up the vapour blanket and bring fresh quenchant into contact with objects
	oil	
	helium gas	more efficient if blown at work piece
	air	
SLOW	vacuum	

3.4.2 Case hardening with martensite

Civil engineering machines such as bulldozers, cranes and excavators, to say nothing of tanks and military vehicles, are fitted with tracks rather than rubber-tyred wheels. Such vehicles are heavy and have to be able to work on uneven ground, which can range from soft and wet to dry and highly abrasive. What are the requirements for a track plate (Figure 3.42) in such vehicles.?

The most important requirements are for abrasion resistance coupled with high strength and toughness. Unfortunately, these are conflicting. For dry abrasion resistance, high hardness is necessary (usually 550 H_v or above) and this calls for either martensite or large amounts of iron (or alloy) carbides in the microstructure.

EXERCISE 3.12   Why should this conflict with strength and toughness?

The conflicting requirement could be resolved by arranging for the component to have a wear-resistant skin on the outside of a tough, strong core, and this has great attractions. In fact, it is a widely used solution to problems of combining conflicting requirements in a finished component. How is this achieved? If we wanted a finished component

Figure 3.42 Details of track components

like a track shoe to have a hard skin surrounding a tough core there are two possible ways of achieving this:

1 Start with the component made in a low-carbon steel and diffuse extra carbon into the surface layers wherever high hardness is required. Then austenitize the whole article and quench. Only the high-carbon (carburized) skin will produce hard martensite. The underlying low-carbon regions will merely transform to BCC and remain soft and tough. It is usual practice to heat afterwards to about 550 K to relieve gross quenching stresses without lowering the hardness.

2 Start with a medium- to high-carbon steel in a strong, tough condition throughout, obtained by a conventional heat-treatment process. Then heat to austenitizing temperature only those surface regions where high hardness is required. After sufficient time for heat to diffuse and austenitize the desired depth of skin, quench the surface so that the transformation product is martensite. Because of its high carbon content, this will be hard and brittle, but the underlying core, which did not reach the austenitizing temperature, will retain its original properties.

Both of these techniques are referred to as **case hardening**. The first, where carbon is diffused, is called **carburizing** (or sometimes, when nitrogen is introduced as well, it is known as cyaniding, carbo-nitriding or by some other proprietary process names). The second is known as **flame** or **induction hardening**, depending on the heating method used to raise rapidly the surface into the austenitizing temperature range.

EXERCISE 3.13   For our track component, we need a strong, tough material on which to develop a hard, abrasion-resistant skin. Which of the two methods of case hardening is most suitable?

So we must opt for a composition with sufficient carbon to form a hard martensite when suitably quenched: that is, with more than 0.5% carbon. Before treatment, the steel must have a pearlite or tempered martensite structure to impart toughness and strength in the core material. The surface layer to be hardened, which may be quite localized, is then rapidly heated by electrical induction or by a gas flame to a temperature above the transformation point, which converts the surface layer into an austenitic solid solution. When this austenitic zone reaches the required depth, the heated area is quenched, causing the martensitic transformation. Transformation will occur only in the surface layers and not in the core because the latter did not reach the austenitization temperature.

Hence, the final structure is hard martensite covering the original strong pearlite or tempered martensite.

The depth of hardening of the track plate is not too critical, and hardened zones of 8–12 mm are readily produced. This contrasts with carburizing, where case depths seldom exceed 1 mm, or with precision

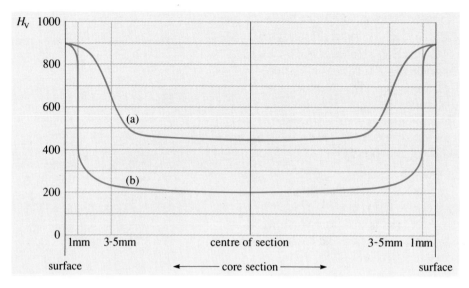

Figure 3.43 Hardness profiles across sections of case-hardened components (a) flame-hardened medium-carbon steel, (b) carburized low-carbon steel

components like axles or crankshaft journals, which may be induction hardened to 1–2 mm only. A typical hardness profile across a flame-hardened section (curve (a) in Figure 3.43) reflects the higher strength of the core material. A carburized component would have a softer centre and shallower case, as indicated by curve (b) in Figure 3.43.

SAQ 3.7   (Objectives 3.4, 3.5 and 3.6)

Two steel shafts, 20 mm diameter, have been case hardened. One (a) is a 0.15% plain carbon steel which has been carburized to a depth of 0.8 mm and a surface hardness of 850 H_V. The other (b) is a 0.4% carbon steel which has been induction hardened to 2.5 mm depth and a hardness of 700 H_V. Sketch a hardness profile across a sectional diameter of each shaft and identify the significant microstructures. Assume that prior to case hardening the low-carbon shaft was in a normalized condition (hardness 160 H_V) and the medium-carbon shaft was quenched and tempered to a hardness of 250 H_V.

One other advantage of case-hardening is the volume expansion that takes place when martensite forms from austenite. When only the surface layers are subjected to the transformation, they develop high residual compressive stresses when quenched, balanced by tensile stresses in the underlying material. The net effect is that the compressive stress remains in the surface and any externally applied loads must overcome this residual stress before the surface layers themselves are subjected to tensile forces. Since fatigue cracking usually initiates at the surface under the action of tensile stress, fatigue life in particular is greatly improved by such treatment, and this is why numerous components such as axle shafts and journals are induction-hardened to depths of a few millimetres or so. Not only do the martensitic surface layers improve wear resistance, but the residual compressive stress greatly extends life under fluctuating service loads (i.e. fatigue).

SAQ 3.8 (Objectives 3.5 and 3.6)

A manufacturer has for several years been making 16 mm diameter toggle pins, which in service are subjected to impact loadings, by machining a mild-steel bar and then case hardening the surface by carburizing. By mistake, the manufacturer was supplied with 0.5% carbon-steel bar which went through the same process.

How would the properties of the pins be affected and would you recommend that they be allowed to go into service?

3.4.3 Tempering of martensite

Owing to the inherent brittleness of the martensite formed on quenching, to say nothing of residual stresses induced by the volume changes associated with the transformation, a wholly martensitic component would be of little use for service. (In addition, it poses certain ▼Welding problems▲.) Hence it is frequently necessary to relieve

▼Welding problems▲

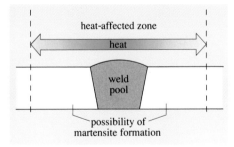

Figure 3.44 The heat-affected zone

Martensite enables a wide range of strength and toughness combinations to be obtained in steels which could not be achieved by any other method. However, the martensitic transformation can cause serious problems if it appears after welding. Why should this be?

Consider what takes place during the welding of two plates such as indicated in Figure 3.44. Molten steel is produced during welding and forms the weld pool between the two plates. The molten steel will be at a temperature in excess of 1800 K and the plates initially at room temperature. During solidification of the weld pool, heat is conducted along the plates and some of each plate adjacent to the pool will be heated into the austenite phase. As heat is dissipated rapidly from these regions, they can be cooled quickly enough to form martensite. The area that is affected in this way is called the **heat-affected zone**. It is not a serious problem in mild steels because of their low carbon levels. If the plate contains more than about 0.25% carbon it can become brittle in these areas and cracks can often be observed a few millimetres away from the weld pool. Should there be any risk of this occurring the whole area should be preheated prior to welding. This reduces

the cooling rate immediately after welding and thus avoids the problems.

Very-low-carbon steels can be welded easily, because the austenite transformation on cooling occurs at fairly high temperatures and the transformation products are soft. Between 0.2 and 0.4% carbon, components can be welded, but care has to be taken with the cooling and post-weld treatment because the transformation occurs at lower temperatures, because the products tend to be harder, and because the volume changes are more likely to cause residual stresses and cracking. Above 0.4% carbon it is virtually impossible to make sound welds because of the transformation behaviour and the hardness (that is, high strength) of the transformation structures.

Similar considerations apply to alloy steels, where the carbon level still dominates welding characteristics.

the quenching strains and to allow the martensite itself to transform towards an equilibrium structure which, though somewhat less hard, will be much tougher than the as-quenched material.

In plain carbon martensites, all the carbon atoms are in solid solution. The carbon atoms are frozen into interstitial holes which are too small to accommodate them easily (as we saw in 'Martensite in steel', Section 3.4.1). This causes large numbers of dislocations, and a hard steel with low ductility and poor toughness is the result.

How might we begin to restore some ductility to the material without, of course, sacrificing too much of the hardness which has been gained?

Martensite is a nonequilibrium structure and is therefore metastable. Diffusion of carbon is the process by which a structure more akin to an equilibrium one may be produced. Let's see how fast this might occur at room temperature.

> EXERCISE 3.14   How long will it take a carbon atom to diffuse 1 nm and 10 nm in a ferrite matrix at room temperature. Take the diffusion coefficient D at 300 K to be $5 \times 10^{-21}\,\mathrm{m^2\,s^{-1}}$.

So diffusion of carbon in ferrite is quite a rapid process on an atomic scale, even at room temperature. Since the martensitic structure, like ferrite, is body centred and full of dislocations, we might expect that carbon in this would have a similar diffusivity and that an equilibrium structure would be approached at ambient temperature. In fact, this does not happen because there are other factors to consider.

For the formation of the equilibrium Fe_3C phase there are two sets of conditions that must be fulfilled. The free-energy change must be sufficient to supply enough surface and strain energy for the nucleation of the crystals and their growth against the resistance of the surrounding material. The rates of diffusion must also be sufficient to permit formation of the precipitates in a sensible time. We have seen that diffusion can take place rapidly. The microstructural evidence tells us that there are close parallels with the age hardening of aluminium alloys. The first precipitates to form are not the equilibrium phase (Fe_3C), but carbide with the formula $Fe_{2.4}C$. Raising the temperature enables the equilibrium Fe_3C phase to form and, of course, diffusion is much faster, too (Figure 3.45).

As the precipitates of equilibrium carbide grow larger, and more and more carbon comes out of the solution, the distortion of the crystal lattice is reduced and slip becomes possible. The ductility of the martensite begins to increase. The new structure is called **tempered martensite**. The overall martensitic plate morphology does not change because the diffusion coefficient of iron atoms in their own matrix is still relatively slow.

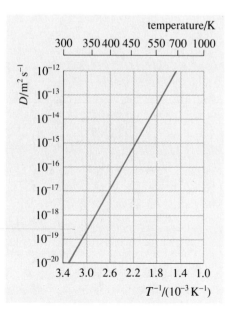

Figure 3.45 Diffusion coefficient of carbon in ferrite as a function of temperature

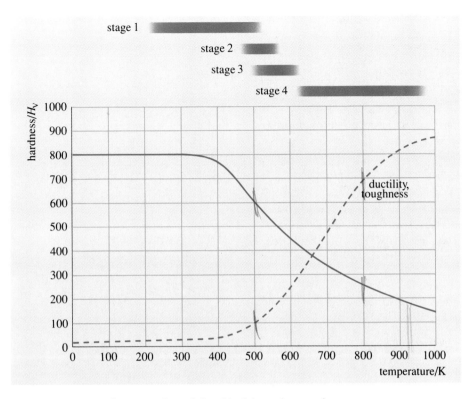

Figure 3.46 Tempering curve for a 0.4 wt % plain carbon steel

The hardness falls with increasing tempering temperature as shown in Figure 3.46, whilst the ductility rises. Four distinct stages of microstructural change can be identified:

Stage 1 (up to 520 K). This is just an extension of the slow relaxation that occurs at room temperature. The carbide which forms with the nonstoichiometric ratio $Fe_{2.4}C$ is generally known as **epsilon (ε) carbide**. This is a very fine carbide which can add to the hardness of low and medium-carbon martensites. This process is equivalent to the precipitation hardening reaction discussed in Chapter 2 where we considered the formation of intermediate precipitates in aluminium alloys.

Stage 2 (470–570 K). Diffusion of carbon atoms to the ε carbide reduces the overall carbon content of the martensitic matrix. Accordingly the temperature of the matrix at which martensite finishes, M_f, will rise (Figure 3.39) and any retained austenite will now become unstable. However, at these tempering temperatures bainite (see next section) will form rather than martensite.

Stage 3 (500–620 K). Fe_3C can now form in preference to ε carbide, and all excess carbon comes out of solution. Since all the lattice strain in martensite was caused by these carbon atoms, the tetragonal martensite matrix will contract in the c direction (see Figure 3.37a) to become BCC. There is now no distinction between this BCC structure and normal α ferrite except that the plate-like morphology is retained. Thus the steel now contains $\alpha + Fe_3C$, which is the equilibrium structure. However, there is one very important difference on the microstructural scale. The Fe_3C is present in the form of short rods and *not* as branching lamellae as in pearlite. (Remember, pearlite is only formed by the eutectoid decomposition of austenite.)

Stage 4 (above 620 K). Figure 3.17(c) showed a spheroidized structure produced after lamellar pearlite had been heated to just below the eutectoid temperature. In order to reduce the surface energy of this structure, Fe_3C forms spheres which minimize the surface area to volume ratio. The Fe_3C in tempered steels is rod-like (stage 3), but at high enough temperatures this agglomerates to form a spheroidized structure very similar to that in Figure 3.17(c).

The exact choice of tempering temperature for a given steel will be governed by its intended application. Strength is closely related to hardness and if a high hardness is required, for example the edge of a wood chisel, then a low tempering temperature will be used. However, if ductility and toughness are needed, as in a shaft for example, then a higher temperature is required. Thus tempering is a critical finishing process that allows a desired combination of strength, ductility and toughness, within the limits offered by the particular carbon content. The structure characteristic of a quenched and tempered 0.4% C steel offering an optimum combination of strength and toughness is shown in Figure 3.47.

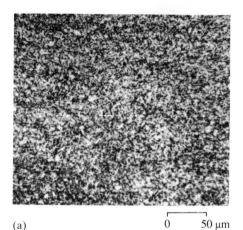

(a) 0 50 μm

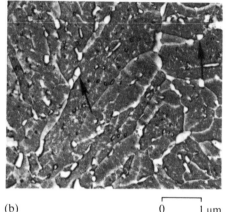

(b) 0 1 μm

Figure 3.47 0.4 wt % C steel tempered at 600 K. (a) Optical micrograph. (b) Electron micrograph showing particles of Fe_3C. Yield stress 450 MN m^{-2}; tensile strength 750 MN m^{-2}; elongation 18%; Charpy impact 40 J

SAQ 3.9 (Objective 3.5)

In the annealed condition a particular 0.3% carbon steel has a tensile strength of 480 MN m^{-2} and an elongation of 20%. Outline two ways in which this strength might be increased to 600 MN m^{-2} whilst still maintaining reasonable toughness and ductility. Give *one* limitation for each method you suggest.

SAQ 3.10 (Objectives 3.4 and 3.5)

Sketch the microstructure you would expect in each of the following and describe in general terms the hardness, tensile strength, ductility and toughness you would expect:

(a) Mild-steel strip (0.1% carbon) cold worked 50% after subcritical annealing.

(b) 0.6% carbon steel hammer head, normalized from 1130 K.

(c) Face of the hammer head in (b) after heating surface to 1100 K, quenching, finally tempered at 500 K.

(d) 1.2% carbon steel chisel, heated to 1100 K, quenched, and tempered at 500 K.

▼Bainite▲

Bainite is produced at cooling rates intermediate between those for martensite and for ferrite plus pearlite (Figure 3.48). Its structure is the result of a mixture of a diffusion-controlled reaction and a shear transformation. In optical micrographs (Figure 3.49) it appears similar to martensite, although the plates are not as pronounced. In bainite, not all the carbon is in solid solution, and small Fe_3C carbides can be seen. In bainite formed at higher temperatures (upper bainite) they appear at the plate boundaries; at lower temperatures (lower bainite) they are nucleated within the plates (Figure 3.50).

The hardness of bainite is intermediate between those of pearlite and martensite, and there is a range of hardness between upper and lower bainite, reflecting the increasingly fine microstructure as the transformation temperature is reduced. For example, in a medium-carbon steel, where the hardness of pearlite is about 300 H_V and martensite 700 H_V, the bainite would range from about 400 H_V if formed at 700 K, to up to 600 H_V when formed just above the M_s temperature (≈ 600 K).

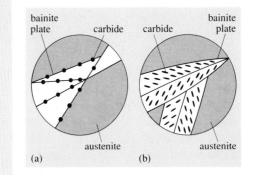

Figure 3.49 Microstructure of 0.4 wt % carbon steel, quenched, showing ferrite at original austenite grain boundaries (light etching) in a matrix of pearlite and bainite

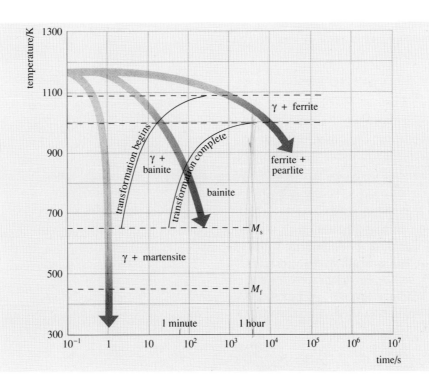

Figure 3.48 Cooling curves for a hypoeutectoid steel showing bainite formation

Figure 3.50 Schematic representation of bainite: (a) upper bainite, (b) lower bainite. The plates can be thought of as forming in a similar way to those in martensite

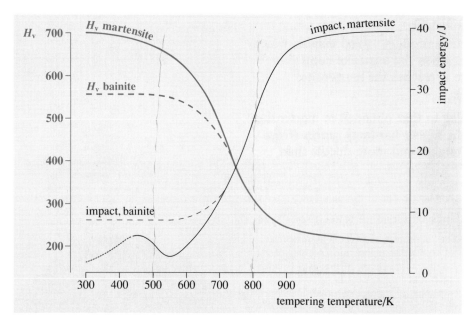

Figure 3.51 The properties of martensite and bainite structures in a medium-carbon steel as a function of the tempering temperature

martensites that have been only lightly tempered. If bainite is heated above its formation temperature it undergoes structural changes not unlike those of martensite during tempering, where the ferrite loses its supersaturated carbon and the Fe_3C rods grow larger. When this happens the properties of tempered martensite and bainite heated to the same temperature are practically identical (Figure 3.51).

How is the greater toughness of bainite at the higher hardness levels exploited and in what kind of products?

The commonest application is one which may surprise you; the ordinary masonry nail for hammering into brickwork. Toe caps of protective footwear and some high-quality golf-club shafts are other examples. Let's look at the first of these.

What properties are required of an artefact that has to be capable of penetrating masonry for some kind of fixing? What kind of materials profile is needed? It's not simply a question of being harder than the masonry. Many substances would qualify for that. The material must also be capable of being shaped to a point and a head, and it must be tough and rigid enough not to fracture or flex under hammer blows. Ordinary nails made of low-carbon steel wire are too soft to penetrate brick and would buckle and bend over when hammered. Something harder is needed.

EXERCISE 3.15   Could we use a soft nail, but with the surface of the point case hardened, as outlined in the previous section?

EXERCISE 3.16   Suppose a heat-treatment process like that just described were to go wrong. What would be the effect on the hardness and impact resistance of the nails of the following?

(a) The austempering temperature falls 20 K *below* the M_s of the steel instead of being 20 K above it.

(b) The quench into the salt bath is too slow, taking say 1 minute to reach the correct austempering temperature?

It's time to come clean! Although we've been considering what are nominally plain carbon steels, it is extremely doubtful whether the bainite transformation could ever be achieved in alloys of iron and carbon only. The pearlite 'nose' of the TTT curve occurs after only such a short time that it's practically impossible to determine the shape of the TTT curve below the pearlite transformation. Arguments about whether bainite can be formed on continuous cooling of a plain carbon steel have long raged between practical metallurgists and theoreticians. In principle, yes; in practice, no. The reason that we are able to draw TTT diagrams like those in Figure 3.52 is that commercial steels always contain manganese (among the other elements essential to make acceptable quality steel) and it is *this* which delays the transformation kinetics so that we can determine such TTT diagrams.

So, the 'plain carbon' steels we have been discussing so far are really not that at all — they already carry small amounts of elements other than iron and carbon that affect the transformation behaviour. In fact, the typical steel used for masonry nails would contain about 0.6% carbon and 1% manganese, along with small quantities of silicon (0.3%) and sulphur and phosphorus impurities, plus residual or trace amounts of nickel, chromium and so on. One thing that these small amounts do, over and above their expected solid-solution hardening or precipitation effects, is to dramatically alter the shape of the TTT diagrams, as we shall see in Section 3.6.

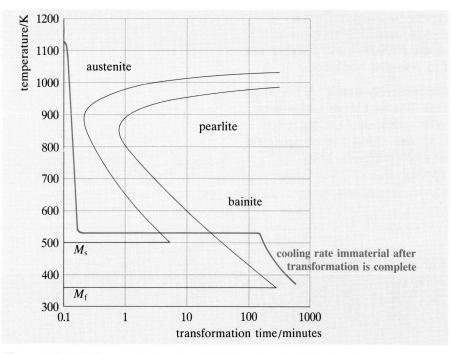

Figure 3.52 TTT diagram for 'unalloyed' 0.8% carbon steel showing the heat-treatment cycle for austempering

SAQ 3.11   (Objective 3.7)

Refer to Figures 3.52 and 3.56 (Section 3.6.1).

(a) What structure would you expect to find in a bar of 0.8% plain carbon steel heated to 1200 K, cooled rapidly to 900 K, held for 1 hour and then quenched into water?

(b) What is the quickest way of producing a structure of 50% bainite with 50% martensite in a small sample of the steel used for Figure 3.56? (Ignore different cooling rates between surface and centre.)

3.5 Cast irons

3.5.1 The solidification process

Let's go now to the other side of the Fe–Fe$_3$C phase diagram (Figure 3.7) and look at the materials with carbon contents between 2 and 5 wt % — the cast irons. Traditionally these have been regarded as brittle materials, but cheap and easy to cast. Why is this, and can anything be done to overcome their lack of tensile strength?

One feature *not* revealed by the phase diagram is the effect of carbon on the fluidity of the liquid phase (remember that 'eutectic' means 'easy flowing'). This is over and above its direct effect of lowering the liquidus. Compositions containing between 2 and 4.3% carbon have a high fluidity and are thus widely used for making intricate castings. These compositions have a second useful property, namely that when the liquid solidifies to austenite and graphite there is no significant volume contraction.

EXERCISE 3.17   Why should this be useful in the manufacture of castings?

The reason for the lack of shrinkage when graphite separates from liquids near the eutectic composition is that graphite has a lower density ($2300 \, \text{kg m}^{-3}$) than the liquid from which it is formed ($\approx 7600 \, \text{kg m}^{-3}$). Hence the normal liquid-to-solid contraction is counterbalanced by the volume increase due to graphite formation.

It is very important to recognize that this compensation does not occur if the liquid solidifies to $\gamma + Fe_3C$. Cementite has a density similar to that of iron, so if the liquid transforms to the $\gamma + Fe_3C$ eutectic, the shrinkage reverts to normal. Hence it is more difficult to make sound castings from steel than from cast iron.

There is yet another advantage of cast irons. When the liquid is poured into the mould, some of the carbon starts to oxidise and forms a blanket of mainly CO gas. This helps the metal to run freely into all the crevices of the mould by forming a thermally insulating layer between the liquid and the mould surface, delaying solidification and maintaining the fluidity. At the same time, the gas tends to stop the metal reacting with the mould surface. The result is that castings strip easily from the mould, with practically no adhesion of mould material. Small wonder, then, that graphitic cast iron is the dominant material for the manufacture of castings.

Cast irons contain several elements other than carbon, and they all have an effect on the phase boundaries. For practical purposes cast irons can be related to the iron–carbon system by what is referred to as their **carbon-equivalent** (or **CE**) **value**. The CE value frequently includes the concentrations of phosphorus and silicon, since these are the commonest impurities. For silicon and phosphorus,

$$\% \, \text{CE} = \% \, \text{C} + \tfrac{1}{3}(\% \, \text{Si} + \% \, \text{P})$$

Hence the CE value is used instead of %C in Figure 3.7.

EXERCISE 3.18   Describe the solidification sequence of liquid cast iron containing 4% carbon and sketch the expected microstructure at room temperature, given that the carbon separation is graphite at the eutectic temperature and Fe_3C at the eutectoid.

The 'real' microstructure of such a cast iron is shown in Figure 3.53. It contains:

(a) primary austenite dendrites, which subsequently decomposed to pearlite after passing through the eutectoid;

(b) graphite flakes formed in the eutectic reaction;

(c) white ferrite (α) regions, where rejected carbon has joined onto graphite already present;

(d) pearlite colonies formed in the primary austenite as well as the eutectic austenite.

It's apparent that the microstructure of cast iron is very complicated (and it can get much worse!). Fortunately, it's not our purpose to delve into the metallography of cast irons. However, there are certain other features which are needed in order to understand the wide range of properties and applications of cast irons.

For over a century, cast irons have been classified by the appearance of a freshly produced fracture. If the eutectic reaction has formed graphite, the fracture path follows the graphite flakes and so the exposed surface appears dark grey in colour and quite dull. Hence the name **grey cast iron**, indicative of a graphitic structure. Further subdivisions are based on how coarse the structure appears. These are useful indicators of quality, but we won't consider them further.

If the casting is cooled fairly quickly, the constituent which forms in the eutectic is not graphite, but cementite, Fe_3C, which is a hard and brittle constituent. When this material is fractured, the fracture path runs through the cementite and the surface exposed has a silvery-white colour. Hence, these are called **white cast irons**.

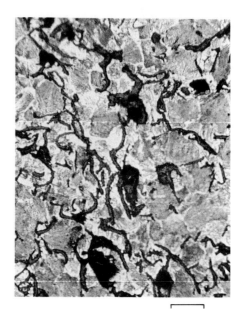

$\overline{}$
0   100 μm

Figure 3.53 Etched section of flaked-graphite cast iron

EXERCISE 3.19   How would a fracture appear if a thick section of casting had cooled at a rate where the outer skin had undergone the eutectic reaction, liquid $\rightarrow \gamma + Fe_3C$, while the slower-cooled core material had undergone, liquid $\rightarrow \gamma +$ graphite?

3.5.2 Controlling graphite form and distribution

Both the graphite and cementite structures are brittle, and cast irons have long had a reputation for lack of both shock resistance and ductility. Tensile strength varies with the amount of pearlite and the distribution of graphite, but the elongation in a tensile test is virtually zero. However, this only applies when the eutectic structures form interconnecting flakes and needles, such as in Figure 3.53.

EXERCISE 3.20   Would the structures be brittle if the graphite were not in the form of interconnected flakes?

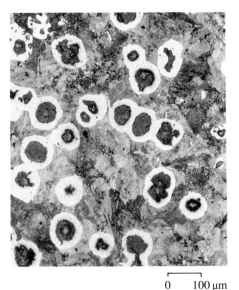

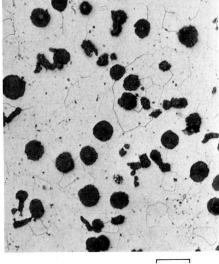

0   100 µm   0   100 µm

Figure 3.54 Pearlitic SG cast iron   Figure 3.55 Ferritic SG cast iron

It's possible to modify the form of eutectic graphite by making minute additions of elements which change the interfacial energy between the graphite nucleus and the surrounding liquid. Magnesium and cerium are two elements which do this (in amounts of the order of 0.05%) and both exert a dramatic influence on the form of the eutectic graphite as can be seen from Figure 3.54 and 3.55. In the Figure 3.54 the graphite has formed in isolated spheroidal shapes, instead of interconnecting flakes as in Figure 3.53. Apart from the presence of the magnesium, both structures could well be of similar composition and follow an identical solidification sequence: namely, primary γ ; followed by eutectic transformation to γ + graphite; carbon rejected on further cooling to the eutectoid temperature; and all the austenite finally breaking down to pearlite by the eutectoid reaction.

EXERCISE 3.21   Why is there a rim of ferrite (white) around the nodules of graphite?

One of the limitations of phase diagrams emerges from the above phenomenon — they can only depict the phases and amounts likely to be present under near equilibrium conditions. They cannot predict the form and distribution of those phases, which are largely dependent on nucleation and growth characteristics in the particular system. Another example of this is what happens when a structure like that in Figure 3.54 is annealed or cooled slowly through the eutectoid transformation. There is no nucleation of Fe_3C and, consequently, no pearlite. Instead, the austenite transforms to ferrite and graphite, with all the graphite growing onto the nodules already present. Such an annealed nodular iron is shown in Figure 3.55.

Would a casting with this structure be brittle?

No, on the contrary it would be quite ductile.

The **nodular** or **spheroidal graphite (SG)** cast irons have, therefore, very different properties compared with flaked-graphite irons, yet still possess the attribute of excellent castability. Some typical properties are shown in Table 3.6. (Note that flaked-graphite irons depart from linearity in the early part of the stress–strain curve, though they exhibit very little overall ductility, since one phase, the ferrite, starts to deform plastically at low stress levels.)

Another tune may be played on the microstructures obtainable from the eutectic part of the iron–carbon system. If the liquid is solidified quickly, the eutectic transformation is wholly to γ + Fe$_3$C and yields a white iron. This is extremely hard, but also extremely brittle, much more so even that the flaked-graphite irons. However, if such a structure is heated to around 1300 K (that is, below the eutectic temperature), the Fe$_3$C decomposes into graphite and austenite. This time the graphite

Table 3.6   Properties of cast irons

Type	Microstructure	0.2% proof stress MNm^{-2}	Tensile strength MNm^{-2}	Elongation %	Typical applications
low duty, grey iron	ferrite + pearlite + coarse flake graphite	–	150	0.5	General low duty engineering castings. Beds of large machines. Automobile engine block.
high duty, grey iron	pearlite + fine flake graphite	–	400	0.2–2	Load-bearing parts, vehicle brake drums, camshafts, oil pumps, gears.
ferritic SG iron	ferrite + graphite nodules	250	375	20	Maximum shock resistance. Brackets, bearing housings, valve casings.
pearlite SG iron	pearlite + graphite nodules	450	750	5	Tough, load bearing applications e.g. engine crankshafts.
annealed mild steel	ferrite	250	400	25	

(Compressive strengths of cast irons are approximately 4 × tensile strengths.)

does not form flakes but, instead, very ragged agglomerates of a roughly spheroidal shape. Depending on the subsequent rate of cooling, the austenite surrounding these may break down at the eutectoid temperature to either $\alpha + Fe_3C$, which produces pearlite, or to $\alpha +$ graphite, which grows onto the existing agglomerates. These materials are known as **malleable** irons (since they have a useful degree of ductility and toughness). Yet more variations on the theme are possible, since the annealing may be conducted in either an inert or an oxidizing atmosphere. This affects the amount of carbon which can oxidize in the outer skin and gives rise to **blackheart** or **whiteheart** malleable castings according to how the microstructures appear in the core and the outside of the casting when fractured.

> SAQ 3.12   (Objective 3.1)
>
> Which of the following would be classed as cast irons?
> (a) Microstructure exhibits large pearlite colonies in matrix of pearlite and graphite flakes.
> (b) Composition: 1.0% C, 13% Cr, 1.1% Mn, 0.5% Si.
> (c) Composition: 4.0% Si, 0.03% C.
> (d) Composition: 3.4% C, 1.8 Si, 1.3% Mn.
> (e) Brittle, extremely hard outer skin with shiny, silvery fracture surrounding coarse-grained, dull-grey core.
> (f) Microstructure: large colonies of pearlite surrounded by feathery networks of ferrite.

Cast irons, then, are complicated materials — and we haven't even started to discuss the heat treatments that may be applied to austenite in the solid state. However, all that needs bearing in mind is that almost every hardening and tempering treatment and transformation that can be used for steels may be applied equally well to the austenite in a cast iron. The main difference is that the cast iron will always contain Fe_3C or graphite resulting from the eutectic, and the form of this may override the effects of heat treatment on the properties of the surrounding austenite products.

> EXERCISE 3.22   Why is it undesirable for cast iron to contain the equivalent of more than 4.3 wt % carbon?

3.6 Alloy steels

3.6.1 Alloy steels and hardenability
We've seen that, although it is carbon which dominates the hardness and strength of steel, nevertheless, a large number of the steels used in engineering contain small to moderate amounts of additional elements

(Section 3.4.4). Some of these are very expensive compared with iron, so there must obviously be good reason or reasons for incorporating them. Let's look first at their effect on the hardening behaviour of steels.

EXERCISE 3.23 Why is it that carbon, rather than other metallic elements, dominates the hardness and heat treatment of steel?

Consider two shafts, one of 6 mm diameter, the other 60 mm, whose mechanical property profile can be met by a medium-carbon steel in a quenched and tempered condition. Say,

- yield stress 1000 MN m^{-2}
- tensile strength 1200MN m^{-2}
- ductility 15% minimum elongation in tensile test
- impact energy 20 J

The hardness corresponding to the strength properties is 375 H_V (see 'Hardness measurements') which, from Figures 3.37(c) and 3.46 suggests that around 0.4 wt % C is required in order to be able to achieve this level after quenching and tempering at 600 K.

Referring back to Figure 3.52, which is the TTT diagram for a plain 0.8 wt % carbon steel, can you foresee problems getting 100% martensite in each of the two shafts?

The small-diameter shaft should respond satisfactorily since the whole cross section will avoid the pearlite nose of the transformation during the quench. Of course, one can't be certain of this just by looking at the TTT diagram but, with a radius of only 3 mm, and something like 5–6 seconds before pearlite transformation starts around 900 K, it seems quite reasonable to suppose that the centre of the shaft will miss the nose of the pearlite curve and thus form martensite as it is quenched. So, provided we choose the appropriate quenching medium and take care to avoid cracking and distortion, the quenched martensite will have a hardness of approximately 600 H_V. This can be tempered back to the required level of 375 H_V and thereby produce the property profile required.

What about the 60 mm diameter shaft? It seems clear that the cooling rate at the centre will *not* miss the pearlite nose of Figure 3.52. In fact, the cooling rate would be so slow, despite the fact that the outside of the bar quenched out within a matter of seconds, that near-equilibrium quantities of ferrite and pearlite would be formed in the centre. The reason is that the cooling rate at the centre is determined by the thermal conduction properties of the steel. In this example, only the outermost 3 mm or so of the shaft would form martensite and the inner region would vary from mixtures of martensite and pearlite to ferrite and pearlite at the very centre. Moreover, the outside of the shaft could well crack owing to the volume expansion as the central regions transform later.

So what can be done? Clearly with the larger diameter shaft it would be desirable not to quench at all in view of the problems, but it would not meet the mechanical property specification if it were cooled slowly in air.

Figure 3.56 indicates a possible way out. It is the TTT diagram for one of the common low-alloy engineering steels. Its shape is not all that dissimilar to the TTT diagram for plain carbon steel, *but notice the time scale*.

EXERCISE 3.24   These questions relate to the low-alloy steel in Figure 3.56.
(a) What is the least delay before transformation to pearlite begins at any temperature?
(b) How long does it take to complete transformation at 870 K?
(c) For what period is austenite stable if the steel is cooled to 820 K and held constant at this temperature?

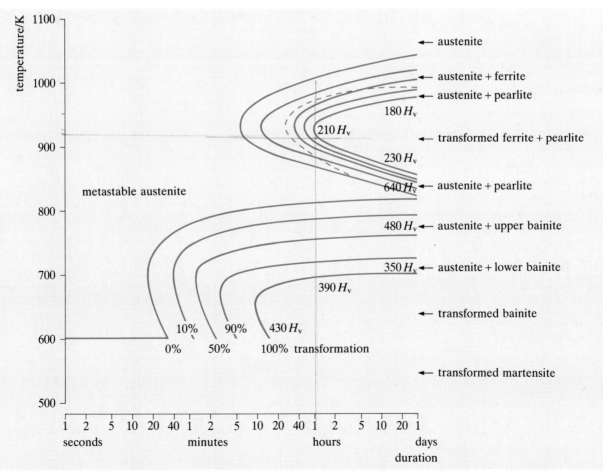

Figure 3.56 TTT diagram for low-alloy engineering steel. Composition: 0.36% C; 0.52% Mn; 1.52% Ni; 1.17% Cr; 0.27% Mo. Austenitized at 1130 K. H_v values are for transformed steel, either fully transformed at the temperature shown or quenched after 24 hours

Notice also how the alloying elements have opened up a gap between the pearlite transformation and the bainite, to give what is called an austenite 'bay'. Various alloying elements exert different effects on each of these transformations, some retarding one, but not the other quite so much, and others retarding both to roughly the same extent. Molybdenum, for example, greatly retards the pearlite and thus exposes the bainite, whereas nickel pushes both back by about the same amount.

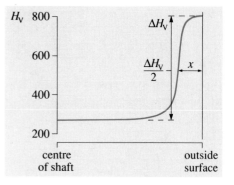

Figure 3.57 Definition of hardenability

Now we have a solution to our problem with the shafts. By all means let's make the 6 mm shaft with a plain carbon steel because it is cheap and will do the job, but let's make the 60 mm shaft from the low-alloy steel so that it can be cooled *in air* and still produce martensite right to the centre. It's worth the extra cost of the alloying elements since they allow the shaft to be cooled more slowly. This minimizes the differential shrinkage and distortion, yet the section produced is 100% martensite throughout.

We can also exploit the delay of the austenite bay. If there are likely to be distortion problems with a complex shape, the cooling can be discontinued in the region of 820 K until the thermal gradients have evened out. It is then continued to just above M_s and the temperature held until the gradients have evened out again. Finally the whole mass is allowed to cool slowly through the M_s–M_f range. This type of treatment is called **martempering**.

This shifting of the transformation diagram to longer times gives rise to a property known as **hardenability**. It can be expressed as the depth at which the hardness has dropped to its mean value (Figure 3.57). Controlling hardenability is a very important reason for adding alloying ingredients to engineering steels — perhaps the most important. There is a synergistic effect in that combinations are more potent than separate individual additions, which is why nickel, chromium, molybdenum and vanadium are often added in pairs or threes (remember that manganese is invariably present as well). It is essential to recognize that it is the carbon which gives rise to the potential for *hardening*, but it is the substitutional alloying additions which produce the *hardenability* of the steel.

SAQ 3.13 (Objective 3.8)

Distinguish between hardness and hardenability as these terms apply to engineering steels and sketch the form of TTT diagram for a steel having high hardenability.

3.6.2 Effects of alloy additions on steels

A quick glance at a book of steel specifications shows a bewildering array of special compositions which have been developed for specific purposes. They tend to fall into three classifications, namely:

Low alloy, less than 2 wt % of the major alloying addition.

Medium alloy, between 2 and 10 wt % of the major alloying addition.

High alloy, more than 10 wt % of the major alloying addition.

The alloying elements in question are, with the exceptions of phosphorus, nitrogen and silicon, other transition metals that can be easily dissolved in iron. Probably the best known alloy steels are stainless ones, developed to overcome one of the biggest disadvantages of the ferrous family, their susceptibility to ▼Rusting▲

Table 3.7 lists the common alloying elements used in steels and notes some of their principal effects.

▼Some typical engineering alloy steels▲ gives the compositions and properties of a small selection of the common steels. Let's look at low-medium-and high-alloy steels in turn, to get a flavour of what alloying is about.

Low-alloy steels

Low-alloy steels have been developed in recent years in response to the need for high-strength materials which may be easily welded. They are used for such structural applications as I-beams, box sections and pipelines, as well as numerous smaller components and the high-strength, car-body panels in Section 3.35.

The carbon content has to be low in order to prevent brittle martensite forming in the heat-affected zone around a weld.

Given this restriction, how can the yield stress be increased in these steels?

• By adding alloying elements which form solid solutions to harden the ferrite. Note that, with the exception of very small atoms (e.g. nitrogen), these will form substitutional solid solutions.

• By reducing the grain size. This is achieved by introducing particles that pin grain boundaries, and by grain refining heat treatments.

• By redistributing the harder carbide phase as a fine precipitate. Although distributing the pearlite colonies more uniformly in the ferrite matrix and reducing the interlamellar spacing (both controlled by heat treatment) will have a strengthening effect, a uniform dispersion of fine carbide precipitates throughout the ferrite grains is much more satisfactory.

Let's look at each of these methods in more detail.

Figure 3.33 showed how various alloying elements contribute to an increase in yield stress by solid-solution strengthening. However, there is a problem with making additions like this.

What might this problem be, bearing in mind the ductility and toughness expected of a low-carbon steel?

▼Rusting▲

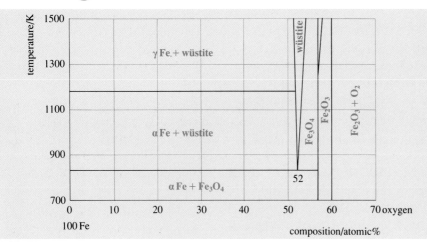

Figure 3.58 Part of the Fe–O equilibrium phase diagram

Unlike the other transition metals in their group, iron and steel corrode readily in the presence of oxygen and water. The process is enhanced by the presence of ions such as sulphides and chlorides, which are common in industrial and marine atmospheres.

Why do iron and steel continue to rust in oxygenated, aqueous environments, whereas metals such as chromium and aluminium cease reacting once oxide films have formed on their surfaces? The three desirable characteristics of a protective oxide film are:

● It must adhere to the metal (it helps if metal and oxide have coherent structures).

● The volume changes when oxide forms should be small.

● Diffusion through the oxide should be slow (minimal crystal defects).

Chromia (Cr_2O_3) fulfils two of these (there is a large increase in volume), while alumina (Al_2O_3) fulfils all three. In contrast, rust satisfies none of them. Thus Cr_2O_3 and Al_2O_3 are said to **passivate** the surfaces of the metal.

In dry, unpolluted atmospheres, a thin ($\approx 4\,nm$), protective oxide film *can* form on iron surfaces. This is anhydrous FeO, or **wüstite**. However, as the iron–oxygen equilibrium phase diagram shows (Figure 3.58), its composition is nonstoichiometric (i.e. there isn't an equal number of Fe^{2+} and O^{2-} ions). Its formula is more accurately written as $Fe_{0.953}O$, with the charge balance made up by Fe^{3+} ions.

> **EXERCISE 3.25** What proportion of the Fe ions need to be Fe^{3+} in order to give $Fe_{0.953}O$. In perfectly stoichiometric FeO, all Fe ions would be Fe^{2+}. What proportion of these must be replaced by Fe^{3+} to make $Fe_{0.953}O$?

The deficiency of iron in the oxide leads to a defect structure, which means that the FeO film is not as protective as Cr_2O_3 or Al_2O_3, although it is still better than rust. So what is rust, and how does it form?

Rust is the name given to mixtures of iron oxides formed on iron and steel surfaces in oxygenated aqueous environments. These oxides (Figure 3.59) are:

● green rust: hydrated ferrous oxide, FeO.OH (**goethite** or **lepidocrocite**);

● brown rust: hydrated ferric oxide, Fe_2O_3 with $Fe(OH)_3$ (**haematite**);

● black rust: mixed ferrous and ferric oxides, Fe_3O_4 ($= FeO.Fe_2O_3$) (**magnetite**);

Above a relative-humidity threshold of about 60% (less in the presence of, for instance, hygroscopic dust), there is sufficient water around for some droplets to condense on the metal surface. These provide the electrolyte for electrochemical corrosion to start. At the anode

$$Fe \rightarrow Fe^{2+} + 2e^-$$

At the cathode

$$\tfrac{1}{2}O_2 + H_2O + 2e^- \rightarrow 2OH^-$$

In combination these give

$$Fe + \tfrac{1}{2}O_2 + H_2O \rightarrow Fe(OH)_2$$

and the ferrous hydroxide further oxidizes

$$Fe(OH)_2 + \tfrac{1}{4}O_2 + \tfrac{1}{2}H_2O \rightarrow Fe(OH)_3$$

This ferric hydroxide changes rapidly to hydrated ferrous oxide

$$Fe(OH)_3 \rightarrow FeO.OH + H_2O$$

which can oxidize further

$$2FeO + \tfrac{1}{2}O_2 \rightarrow Fe_2O_3$$

and combine to give

$$FeO + Fe_2O_3 \rightarrow Fe_3O_4$$

So we have our mixture of nominally $FeO.OH$, Fe_2O_3 and Fe_3O_4. Each of these oxides tends to be nonstoichiometric, because of the close structural relationship between them. They are all based on a cubic, close-packed array of oxygen ions, with iron as Fe^{2+} and Fe^{3+} occupying the tetrahedral and octahedral interstices. If all the octahedral sites are occupied by Fe^{2+} ions we have the ideal stoichiometric FeO. If 14.1% of these are replaced by Fe^{3+} (Exercise 3.25) we get $Fe_{0.953}O$. With 66.7% replacement by Fe^{3+} ions, half of which migrate to tetrahedral sites, we get Fe_3O_4; and 100% replacement gives Fe_2O_3. Thus interconversion of the oxides is readily achieved merely by redistribution of ions between interstitial sites, and with no major structural changes.

Their lack of stoichiometry leads to defect structures which provide diffusion paths through their lattices. In Fe_2O_3 and Fe_3O_4, vacancy defects appear in the oxygen anion sublattice, so that oxygen ions can diffuse through the film towards the metal–oxide interface. At the same time, and also due to the nonstoichiometry, electrons can diffuse in the other direction (all three oxides are semiconductors), thus completing the electrochemical circuit. The net result is that further oxidation takes place at the interface between the metal and the oxide. At first sight, the reaction should slow down as the oxide layer thickens and the diffusion paths lengthen. However, there is an approximate doubling in volume accompanying the reaction, and it is this expansion against the existing iron-oxide layer that leads to spalling, exposing fresh metal to further direct attack.

0   5 μm

Figure 3.59 (a) Crystals of green rust (FeO.OH). (b) Columnar crystals of black rust (Fe_3O_4) and rounded crystals of brown rust (Fe_2O_3)

Many of the solutes which harden the ferrite solid solution also reduce ductility and raise the ductile–brittle transition temperatures. Unfortunately, all the common alloying elements except nickel actually increase the transition temperature and are, therefore, not immediately desirable. In particular, silicon and phosphorus, which have the greatest effect on the yield stress after carbon and nitrogen, also cause a marked rise in the transition temperature.

The next most effective element in solid-solution strengthening is copper. Unfortunately, copper in steels can give rise to processing problems because it produces phases with low melting temperatures. These phases reduce the steel's hot ductility and, hence, workability. This effect is known as **hot shortness**. Copper is also difficult to remove from steel scrap.

This leaves manganese, vanadium, titanium and nickel as the main contenders. Titanium and vanadium are strong carbide formers and, of the other two, manganese is the cheaper. In Section 3.3 we also saw the potent strengthening effect of niobium carbides (Figure 3.32).

To summarize then, the yield stress of a normalized low-carbon steel can be raised by 300 MN m^{-2} from around 250 MN m^{-2} by the addition of very small amounts of alloying element (2 wt % in total). The ductility and toughness at operating temperatures are preserved in such a steel, which is also weldable. These steels are an important group of materials offering much higher yield stresses than the traditional 'mild' steel. They are the micro-alloyed or HSLA steels mentioned in Section 3.3.5.

Medium-alloy steels

In contrast to the low-alloy steels, medium alloys usually have a medium-carbon content and are heat treated, either by transforming to bainite or by quenching and tempering. The hardenability and properties produced by tempering are therefore important considerations.

Whatever other functions they may have when added to a steel, alloying elements, with one exception, all increase the hardenability; the exception is cobalt. The mechanism by which alloying increases hardenability is complex but the effects may be seen on the TTT diagram in Figure 3.56.

By delaying both the $\gamma \rightarrow \alpha + Fe_3C$ and $\gamma \rightarrow$ bainite reactions, more time becomes available for a steel to cool to form martensite. Even the interior of thick sections will form martensite as the hardenability of the steel is increased, to the extent that some will do so on cooling in air. Avoiding the need to quench into oil or water removes the risk of cracking in thick sections. This also greatly minimizes dimensional distortion associated with the transformation.

Nickel and chromium are the two elements that have the greatest effect in delaying the formation of equilibrium products. Steels containing up

151

Table 3.7   Specific influence of common alloying elements in steels

Element	Solid solubility/%		Influence on ferrite	Influence on hardenability	Tendency to form hard carbides	Principal functions
	In ferrite	In austenite				
Chromium	Unlimited	12.8 (increases with carbon)	Strengthens slightly. Great increase in corrosion resistance	Moderately increases	Strong	1 Corrosion resistance 2 Increases hardenability 3 Abrasion and wear resistance with high carbon contents 4 Strength and oxidation resistance at high temperatures
Cobalt	75	Unlimited	Strengthens by solid solution hardening	Decreases slightly	Similar to iron	1 Contributes 'red' hardness, i.e. strength at moderately elevated temperatures
Manganese	3	Unlimited	Very powerful solid solution strengthener	Moderately increases	Greater than iron, less than chromium	1 Increases hardenability cheaply 2 Takes care of sulphur as MnS and thereby improves machinability
Molybdenum	37.5	3 (increases with carbon)	Age hardening possible	Strongly increases	Strong, greater than chromium	1 Increases hardenability 2 Prevents a form of embrittlement encountered during tempering of Ni/Cr steels 3 Raises hot strength and hardness 4 Restricts austenitic grain growth 5 Improves corrosion resistance of stainless steels 6 Forms carbides with high abrasion resistance
Nickel	10	Unlimited	Strengthens and toughens	Mild improvement Stabilizes austenite		1 Improves strength and toughness especially at subzero temperatures 2 With high chromium contents renders steels austenitic
Phosphorus (usually regarded as an impurity)	2.8	0.5	Hardens strongly but tends to embrittle	Increases	None	1 Powerful solid solution strengthener of low carbon steels 2 Improves corrosion resistance mildly 3 In small amounts, improves machinability of free cutting steels
Silicon (residual element)	18.5	2 (increases with carbon)	Hardens but reduces ductility appreciably	Moderately increases		1 Deoxidation of liquid steel 2 Increases electrical resistivity (used in transformer steels) 3 Improves oxidation resistance 4 Strengthens low alloy steels
Titanium	6	0.75	Age hardening possible	Very strongly increases	Extremely strong	1 Forms hard carbides 2 Locks up carbon in stainless steels to prevent local depletion by formation of chromium carbides
Tungsten	35	6 (increases with carbon	Age hardening possible	Strong, especially in small amounts	Strong	1 Hard, abrasion resistant carbides in tool steels 2 Hot hardness and strength 3 Delays tempering to higher temperatures
Vanadium	Unlimited	1 (increases with carbon)	Moderate solid solution hardening	Very strongly increases	Very strong	1 Restricts grain coarsening of austenite 2 Increases hardenability 3 Delays softening during tempering

▼Some typical engineering alloy steels▲

A selection of commonly used engineering steels with their properties is given in Table 3.8. These are, however, only a tiny selection from the wide range covered by International Standards. The first few show the increasing influence of carbon on what are essentially plain carbon steels. The later ones contain deliberate alloying ingredients for specific effects. Notice the last column which indicates the maximum diameter of section that can be heat treated to give the properties quoted.

Table 3.8   Commonly used engineering steels and properties

Type of steel		Composition/%					Properties at centre of bar of diameter D				Condition	Limiting diameter → max.
Designation	En number	C	Mn	Ni	Cr	Mo	Yield strength	Tensile strength	Elongation (minimum)	Impact (minimum)		mm
							$\frac{MN}{m^{-2}}$	$\frac{MN}{m^{-2}}$	%	J		
015A 03	En 1	0.03	0.15				160	320	30	90	normalized	
							380	460	10	–	cold drawn	
220M 07 (free machining)	En 1A	0.07	1.0	(+ 0.2% S)			210	360	22	60	hot rolled	
							380	470	7	–	cold drawn	
070M 20	En 3A	0.20	0.70				210	420	21	50	normalized	150
							390	520	12	–	cold drawn	
							350	600	20	30	heat treated	18
080M 40	En 8	0.40	0.80				500	630	8	–	cold drawn	
							270	540	16	25	normalized	150
							380	670	16	35	heat treated	63
							450	760	16	35	heat treated	18
080M 50	En9	0.50	0.80				570	640	8	–	cold drawn	
							300	600	14	–	normalized	150
							420	750	14	–	heat treated	63
							550	900	12	–	heat treated	18
150M 36	En 15	0.36	1.50				540	700	7	–	cold drawn	
							380	600	14	45	normalized	150
							390	670	18	45	heat treated	150
							460	750	16	40	heat treated	63
							600	900	12	32	heat treated	18
817M 40	En 24	0.40	0.55	1.0	1.25	0.3	660	900	13	50	heat treated	150
							830	1050	12	45	heat treated	63
							1200	1600	5	12	heat treated	28
410S 21 (stainless)	En 56 (ferritic)	0.08	1.0	1.0	13.00		270	410	20	–	normalized	150
							360	600	20	35	heat treated	150
302S 25 (stainless)	En 58 (austenitic)	0.08	1.5	9.0	18.0		260	500	40	–	annealed	
							580	840	12	–	cold drawn	

to a total of about 5 wt % of these elements are often referred to as **air-hardening steels**, since even air cooling results in a fully martensitic structure. A typical composition might be 0.55 wt % Mn, 4.25 wt % Ni, 1.25 wt % Cr, 0.3 wt % Mo, 0.3 wt % C.

Unfortunately, nickel/chromium steels, such as the above suffer from a condition known as **temper brittleness**, which is a loss of toughness after tempering in the region of 950 K. This occurs because certain trace impurities (e.g. P, Sn, As and Sb) migrate to the grain boundaries during tempering, reducing the cohesive strength of the boundaries and leading to embrittlement. The addition of small amounts of molybdenum overcomes this problem by anchoring the atoms and preventing their migration.

How can we account for the other elements in the typical steel above? The manganese is there for the reason mentioned in Section 3.2.4., to mop up any sulphur present. The presence of molybdenum is mainly to overcome the temper brittleness, but it also forms a hard carbide (as do chromium, tungsten and vanadium). Furthermore, the carbides of all four elements produce a precipitation-hardening effect when the steel is tempered, which enhances the strength and hardness imparted by the carbon alone.

Hardness and strength are retained by the precipitated carbides while toughness is increased by the removal of carbon from the martensite.

Molybdenum, in common with tungsten, cobalt and to a lesser extent chromium, has the effect of slowing down the effect of tempering, as indicated in Figure 3.60. This means that higher temperatures or longer times are required to soften a steel, allowing greater control of tempering. It is a very useful characteristic for steels used at moderately elevated temperatures, for instance in steam plant.

An example of a medium-alloy steel would be one used for high quality spanners. You may well have a set stamped 'chrome vanadium steel'. There are several other alloying elements, too, a typical composition being: 0.3 wt % C, 0.75 wt % Mn, 0.3 wt % Si, 1.8 wt % Ni, 0.87 wt % Cr, 0.4 wt % Mo, 0.1 wt % V.

High-alloy steels

Many important engineering steels are high alloy: for example, those used for cutting tools, for dies used in forgings or extrusion, for furnace parts and other demanding applications. ▼**Combining wear resistance with toughness**▲ looks at one example. However, we'll concentrate on those steels known as 'stainless'. The term 'stainless' just describes the corrosion properties of the steels and tells us little about their composition, treatment or other properties. Chromium confers this stainless property when at least 12 wt % is present. Usually stainless steels contain between about 12 and 18 wt % Cr. The chromium forms a protective oxide surface which can only be broken down under certain conditions. ▼**The Statue of Liberty**▲ describes one application of stainless steel to overcome a corrosion problem.

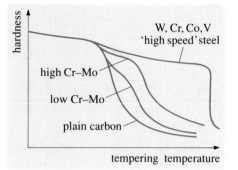

Figure 3.60 Strong carbide formers delay the tempering of medium-carbon steels

Stainless steels can be ferritic, austenitic, martensitic, or precipitation hardening, depending on carbon content and heat treatment. A common example of the first is the steel used for such articles as forks and spoons, where a high hardness is not particularly required and ease of shaping is more important. A typical composition is 0.04 wt % C, 0.45 wt % Mn, 14 wt % Cr, which is a ductile, formable, ferritic material fairly easily shaped from thin sheet, though not as ductile as some of the FCC stainless types.

Where hardness is important, for example in the blade of a carving knife or a pair of scissors, a martensitic grade might be used with a composition of 0.3 wt % C, 13 wt % Cr, 0.4 wt % Mn. After hardening the structure is tempered to give hardnesses in the range 500–700 H_V. Sometimes titanium, niobium or vanadium are added to form carbides in preference to chromium carbide; the removal of chromium as carbide would reduce corrosion resistance. Molybdenum may also be present to improve resistance to attack by chloride ions in, for example, marine environments.

The well known '18-8' stainless steel is nominally 18 wt % Cr, 8.5 wt % Ni, 0.8 wt % Mn, 0.05 wt % C. The presence of nickel and manganese stabilize the γ down to room temperature, so this steel is neither ferritic nor martensitic, but austenitic. The steel is also non-magnetic. In common with other FCC metals, austenitic steels are ductile and can be formed into shapes by cold working. Their work hardening rates are high, making them very suitable for such applications as sink units and mixing bowls manufactured by drawing or pressing. However, their high ductility makes them difficult to machine.

One final point to be considered is the cost of alloying. Some elements, such as manganese, affect the cost but little. Even when the alloying element is much dearer than iron, provided small amounts are used the overall effect on the cost is minimal. Low and medium alloy steels are

▼Combining wear resistance with toughness▲

It is not normally possible to combine maximum hardness with maximum strength (or toughness) in a homogeneous material. However, there is one steel that comes close to it. This is **Hadfield's manganese steel**, an austenitic steel containing 13 wt % manganese and 1.2 wt % carbon. For many years it's been used for the most demanding applications. Its structural condition is retained austenite which, being FCC and holding all the manganese and carbon in solid solution, is tough. However, as soon as it is deformed, the austenite undergoes a martensitic transformation nucleated by the plastic deformation. It hardens to levels in the region of 500–550 H_V, but only in the deformed skin. Thus it resists abrasion at the surface while still maintaining the toughness of the underlying material. For certain types of arduous applications requiring wear resistance, such as at railway intersections (the 'frogs' of points and crossovers) Hadfield's manganese steel is unsurpassed. However, its corrosion resistance is no better than ordinary steel.

SAQ 3.14 (Objective 3.9)

Alloy steels are more expensive than plain carbon ones, though mechanical properties and response to heat treatment are largely dominated by the carbon content. Explain whether an alloy steel might be justified for each of the following and, where appropriate, indicate the function of the alloying additions.

(a) A lathe tool for machining steel, where tool tip temperatures are likely to reach 970 K.

(b) A small-diameter wire for the spring in the locking mechanism of a car door.

(c) Thin knife blades for a domestic food processing machine.

(d) 50 mm diameter × 300 mm long high tensile bolt for chain testing machine.

therefore only marginally more expensive than their plain carbon counterparts, and often the cost difference is outweighed by the fact that less material is needed to perform the same function.

With high alloy steels the situation is different, especially if appreciable amounts of some of the rarer alloying elements such as tungsten, molybdenum or cobalt are used. The high alloy 'high speed' tool steels are relatively expensive compared with their plain carbon counterparts. However, their properties are far superior: they will retain their hardness even when they become almost red hot during high speed cutting operations.

▼The Statue of Liberty▲

(a)

(b)

Figure 3.61 (a) The 'crinoline', or armature, of the Statue of Liberty. (b) Initial construction in Paris

consists of sheets of copper (2.4 mm thick) attached with copper rivets to the bars of the crinoline by copper saddles (see Figure 3.62). The arrangement of crinoline and saddles puts two dissimilar metals (copper and iron) into close proximity, with the risk that in the presence of water a galvanic cell might be formed. Iron, the metal lower in the electrochemical series, would form the anode in such a cell and it would corrode at a rate that depends on the current flowing in the cell.

This risk was foreseen before construction and wads of asbestos felt, impregnated with shellac, were used within the saddles to separate the two metals. Unfortunately,

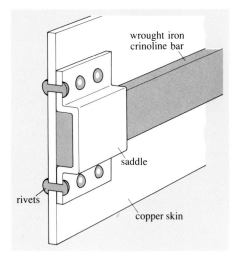

Figure 3.62 Method of attaching skin to 'crinoline'

On 4 July 1986 the centenary was celebrated of 'Miss Liberty', a gift from France to the USA, designed by the sculptor August Bartholdi assisted by engineer Gustave Eiffel. In the two years preceding the centenary, $75 million had been spent on restoration, mainly required on account of wet corrosion. Although many people think the statue is made of stone, it actually consists of a copper skin over a steel skeleton. From a central steel pylon there extends a secondary structure of horizontal steel bars which supports a 'crinoline' of wrought iron. This defines the external shape of the statue, as shown in Figure 3.61. The surface of the statue

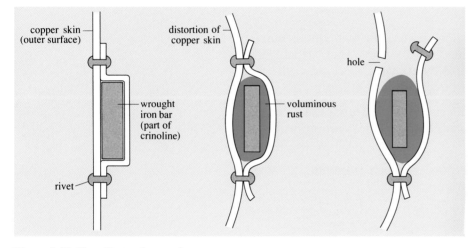

copper skin
(outer surface)

distortion of
copper skin

hole

wrought
iron bar
(part of
crinoline)

voluminous
rust

rivet

Figure 3.63 The effects of corrosion

this device did not prevent electrolytic cells from being set up when water accumulated inside the statue from condensation and from leaks of rainwater, particularly via the torch at the top of the statue. Over the 100 years, about one third of the 1500 saddles were affected by corrosion of the iron within. The anodic reaction is the same as that described in 'Rusting'

$$Fe \rightarrow Fe^{2+} + 2e^-$$

The cathodic reaction might be expected to be the deposition of copper ions from solution onto the copper cathode

$$Cu^{2+} + 2e^- \rightarrow Cu$$

However, there are insufficient copper ions in solution for this to be significant and the dominant cathodic reaction is

$$\tfrac{1}{2}O_2 + H_2O + 2e^- \rightarrow 2OH^-$$

This leads to the same mixture of products as described in 'Rusting'.

We have already seen that when a volume of iron is converted to rust it will occupy a much larger volume. When this swelling beset the statue's crinoline, the saddles became distorted and their rivets pulled out of the copper skin (see Figure 3.63). The skin became deformed and disfigured, and there were now additional holes for rainwater to enter. In some places, over half the section of the wrought iron had been lost by corrosion, so the ability of the crinoline to carry the load of the skin

was increasingly in doubt.

The remedy was two-fold: to replace the wrought-iron crinoline by one of stainless steel with an electrode potential almost the same as that of copper. There would then be little driving force for galvanic corrosion. As a belt to these braces, a layer of the 'noble plastic', polytetrafluoroethylene (PTFE) carried on a self-adhesive tape of woven glass fibre was used to cover the steel and thereby separate it from the copper. Together with the repair of the leaking torch and air-conditioning of the interior, the restoration is considered to have cured the problem. Just to be on the safe side, accessible saddles will be inspected periodically!

Miss Liberty is an example of copper used as a 'roofing' material. It has been remarkably successful, as shown by the decision of the restorers to retain almost the whole of the original skin. Its average rate of corrosion over a century has been only 1 µm per year, so at this rate it would last over 1000 years. The reason for this apparent chemical stability is that the copper became passive. Within the first ten years of exposure, the copper skin acquired a green patina consisting mainly of basic copper sulphate $CuSO_4.3Cu(OH)_2$. This is insoluble in water (but not in acid rain) and it effectively separates the corrosion reagents, copper and water containing dissolved oxygen and sulphur dioxide, and makes further reaction very slow. It is this conspicuous green patina which appeals to architects because it is both attractive and protective.

Objectives for Chapter 3

After studying this chapter, you should be able to:

3.1 Distinguish between steels and cast irons in terms of composition, microstructure and mechanical properties. (SAQ 3.12)

3.2 Outline the reasons why commercially produced steels may exhibit directionality and variation of properties. (SAQ 3.1)

3.3 Explain the yield point phenomenon in low carbon steels and how it may be avoided. (SAQ 3.5)

3.4 Sketch the optical microstructure and property profile for the following types of steel in the annealed and heat treated conditions:
low carbon (mild) steel
medium carbon (0.2–0.6 wt %) steel
high carbon steel
(SAQs 3.4, 3.7 and 3.10)

3.5 Describe the properties and structures of steels in non-equilibrium structural conditions and specify the methods by which such structures may be achieved in practice, for example by alloying, working, heat treatment, and by combinations of all three. (SAQs 3.6–3.9)

3.6 Define and explain how carburizing or flame hardening (also called induction hardening) may be used to case harden a steel component. (SAQs 3.7 and 3.8)

3.7 Identify the products of austenite transformation from an isothermal (TTT) transformation diagram. (SAQ 3.11)

3.8 Define the term 'hardenability' and relate it to the compositions and TTT diagrams for low and medium alloy steels. (SAQ 3.13)

3.9 Explain the reasons for making alloying additions to steels. (SAQ 3.14)

3.10 Define and use the following terms:

austenite
bainite
carburizing
cementite
eutectoid
ferrite
full anneal
hardenability
induction hardening
Lüders bands (stretcher strains)
martensite

micro-alloying
normalizing
patenting
pearlite
quenching
solute atmosphere
spheroidizing anneal
stress-relief anneal
subcritical annealing
tempering
yield point

Answers to exercises

EXERCISE 3.1 The flow lines run *along* the teeth near the 3 and 9 o'clock position, but *across* those at 12 and 6 o'clock. The latter would be much less tough than the former, rather as you would expect if the gear were made of wood. The flow lines reveal how the gear had been formed by punching a hole in a length of bar and expanding it into a circle. It should have been formed so that all the flow lines run radially in the teeth (for example by upset forging).

EXERCISE 3.2

(a) FCC, because the steeper slope is exhibited by the close-packed FCC state.

(b) The BCC structure is loose packed (the packing factor is only 68%) whereas the FCC is close packed (74%). Hence, when the transformation $\alpha \rightarrow \gamma$ occurs at 1183 K, there is a reduction in specific volume and when the $\gamma \rightarrow \delta$ occurs at 1663 K, an increase.

Notice also that if it weren't for the insertion of the FCC γ region, the α and δ points of the curve would be continuous.

The existence of these phase changes has significant implications for the solubility of carbon in iron. In turn, the introduction of carbon modifies the specific volume relationships shown in Figure 3.6, as well as altering the temperatures at which they occur.

EXERCISE 3.3 Interstitial carbon atoms occupy the octahedral sites in both FCC and BCC lattices, because that is where the most space exists.

In BCC iron, of the six iron atoms surrounding these sites, two are closer than the rest. Tensile strain in one direction thus separates these closer atoms on that particular axis and makes the octahedral space larger. The carbon atoms favour these strained sites since, by moving to them, the overall strain energy is lowered. This phenomenon is readily observed in a piece of wire twisted clockwise and anticlockwise about its axis as a torsional pendulum (see Chapter 5). Measurements of internal friction reveal that the time for the carbon atom to jump from one axis to another in the same unit cell at ambient temperatures is of the order of 1 second.

EXERCISE 3.4 Apply the lever rule. The amount of Fe_3C will be

$$\frac{0.8 - 0.02}{6.67 - 0.02} = 0.117$$

or roughly one tenth of the area observed under the microscope will be cementite, Fe_3C, and the other nine tenths ferrite.

EXERCISE 3.5 The area occupied by pearlite is about 75%. Since the transformation occurred under near-equilibrium conditions, the areas which are now pearlite represent the proportion of austenite that was present at 996 K. As the austenite at 996 K can only contain 0.8% carbon and the ferrite contains negligible carbon, if 70% was present the overall composition of the sample must have been $0.75 \times 0.8\% \simeq 0.6\%$ carbon. Thus it is a hypoeutectoid steel. It's really only a matter of applying the lever rule at the eutectoid temperature.

EXERCISE 3.6

(a) It could be cold worked to increase the strength by work hardening.
(b) It could be cold worked and annealed for a given time and at a given temperature so that the grains would be refined to a much smaller size.
(c) It could be austenitized and cooled quickly so that a new set of ferrite grains of smaller grain size would be produced.

EXERCISE 3.7

(a) Ultrahigh-strength (UHS) carbon fibre has the highest specific strength:

$$\frac{\sigma_f}{\rho} = \frac{5.2 \times 10^9}{1750}\,N\,m\,kg^{-1}$$

$$= 3.00\,MN\,m\,kg^{-1}$$

which is eight times that of steel wire.
(b) Silicon carbide has the highest specific stiffness:

$$\frac{E}{\rho} = \frac{410 \times 10^9}{2500}\,N\,m\,kg^{-1}$$

$$= 0.16\,GN\,m\,kg^{-1}$$

which is six times that of steel wire.

EXERCISE 3.8 There is no fibrous structure produced by the heat treatment, so when a crack initiates it can more easily spread through the cross section. Also, wires which have been heat treated often lose carbon from their outer surface and this lowers the tensile strength exactly at the place where it is most needed to resist fatigue stresses.

EXERCISE 3.9 Reference to the phase diagram (Figure 3.7) shows that a 0.8% carbon steel would be wholly pearlitic on slow cooling, but the pearlite would be quite coarse. To obtain a *fine* pearlitic structure it is necessary to cool rather quickly, but not so fast as to produce martensite, which would be totally unsuitable for any subsequent cold-working operation.

EXERCISE 3.10 The weight of each model is given by $W = AL\rho g$, where

$$A = \text{cross-sectional area}$$
$$L = \text{length}$$
$$\rho = \text{density}$$
$$g = \text{gravitational acceleration}$$

For the same length and same material, the ratio of their stiffness-to-weight ratios is:

$$\frac{R_C}{A_C}\frac{A_M}{R_M} = 5 \times 10^{-3}\frac{A_M}{A_C}$$

$$= \frac{5 \times 10^{-3}[w_2 d_2 - (w_2 - 2t_2)(d_2 - 2t_2)]}{2[w_1 d_1 - (w_1 - t_1)(d_1 - 2t_1)]}$$

$$= \frac{2.5 \times 10^{-3}[1.8 - (1.4992 \times 1.1992)]}{0.003 - (0.022 \times 0.084)}$$

$$= \frac{2.5 \times 2.16 \times 10^{-6}}{1.15 \times 10^{-3}}$$

$$= 4.7 \times 10^{-3}$$

So with the dimensions chosen, the stiffness to weight ratios of the two models are in a similar proportion to their stiffnesses.

EXERCISE 3.11 Any list should include the following:

(a) ability to be produced as thin sheet which can easily be...
(b) formable to the complex curvatures of the body shape;

(c) capable of being deep drawn (that is, pressed into deeply recessed shapes without tearing);

(d) good surface finish which will not deteriorate in atmospheric conditions ranging from arctic conditions, through humid and temperate climates, to hot desert with strong ultraviolet radiation;

(e) high strength-to-weight ratio;

(f) high elastic modulus (you don't want the body to quiver like a block of jelly!) for dimensional integrity under dynamic conditions;

(g) cheap to manufacture — not only the panel itself, but also building into the whole body;

(h) toughness and/or ductility to withstand accidental damage;

(i) easily repairable after damage.

EXERCISE 3.12 Martensite is very brittle unless it is tempered, but if it is tempered its hardness falls below $550 H_v$ and its abrasion resistance is lost. Carbides, too, are brittle compounds and have poor shock resistance; hence, although they are very hard, if they are present in large amounts, especially a extensive grain boundary agglomerations, the material will have low toughness. Additionally, because of their complex shape, track shoes and plates often have to be made by casting, and this tends to produce interlinking eutectic structures containing the brittle carbides. This is not good for a product certain to be subjected to all kinds of knocks and impacts.

EXERCISE 3.13 We must go for flame or induction hardening, not carburizing. Carburization is only suitable when we start off with a very low-carbon steel which will *not* form a hard martensite on quenching. Although very tough and ductile, such a steel can never be as strong as a medium-carbon steel (roughy 0.2 to 0.6% C) with a pearlite or tempered martensite structure. Hence, as our tract component requires strength as well as toughness, it has to start off with enough carbon to respond to the localized heat treatment and retain its core properties after the surface has been quenched.

EXERCISE 3.14 We can use distance $s = \sqrt{Dt}$, or $t = s^2/D$. For $s = 1$ nm,

$$t = \frac{(1 \times 10^{-9})^2}{5 \times 10^{-21}} \text{ s} = 200 \text{ s}$$

For $s = 10$ nm

$$t = \frac{(10 \times 10^{-9})^2}{5 \times 10^{-21}} \text{ s} = 20\,000 \text{ s}$$

This is 5.6 hours.

EXERCISE 3.15 Not much point (if you'll excuse the pun). Although the tip would be hard enough to penetrate the brick, the rest of the nail would still be mechanically weak. We could start off with a stronger heat-treatable steel nail and locally harden the point, but imagine what a fiddle such individual flame-hardening treatments would be for thousands upon thousands of nails coming through a manufacturing process. Even if this were done, the martensite points would be dangerously brittle when the first hammer blow was struck.

EXERCISE 3.16

(a) At 20 K below M_s, some martensite would form, which would render the nail dangerously brittle. It would hardly matter that the structure was still predominantly bainite, the damage would have been done because the first martensite needles would be large, extending right across the prior austenite grains, from boundary to boundary.

(b) As the pearlite nose of the TTT diagram (Figure 3.52) occurs after only a short time a slow cool through this region would allow soft products to form. Thus the final structure would not be wholly bainitic, so the nail would be soft and less suitable for penetrating masonry. It would however be safe to hit with a hammer since it would be tougher than it needed to be, whereas the martensite in (a) would make the nails highly dangerous.

EXERCISE 3.17 When most liquids (metals or nonmetals) solidify, there is a volume contraction usually of between 3 and 10%. This can lead to porosity within the casting and elaborate feeding arrangements have to be made to minimize it. In complex castings it is practically impossible to ensure that feeder liquid can gain access to every solidifying pool. What an advantage, then, if no shrinkage takes place — because no feeding is necessary and no shrinkage voids are likely to form anywhere within the casting.

EXERCISE 3.18 On cooling to 1420 K the liquidus is reached and primary γ starts to form. At 1403 K the liquid has reached the eutectic composition. The lever rule gives the quantity of γ phase as

$$\left(\frac{4.3 - 4}{4.3 - 2.0}\right) = 13\%$$

The remaining 87% liquid then breaks down to γ + graphite. Applying the lever rule between 2% and 100% carbon indicates that the eutectic reaction yields

$$\frac{4.3 - 2.0}{100 - 4.3} = 2.4\% \text{ graphite}$$

So the 87% eutectic liquid produces a further $0.87 \times (100 - 2.4) = 85\% \gamma$ phase and $0.87 \times 2.4 = 2.1\%$ graphite. Thus just below the eutectic temperature the structure consists of primary austenite with secondary austenite and graphite flakes formed in the eutectic breakdown.

On further cooling, all this γ phase has to reject carbon as it moves down the solvus to the eutectoid at 996 K. The rejected carbon would be expected to grow onto the graphite already present, rather than nucleate afresh. Remember, because of its high specific volume, the rejected carbon would need to find a lot of space between the iron atoms and would thus introduce a high strain energy. By joining onto the graphite already present it merely extends an existing grain boundary, although it still produces a significant increase in the overall volume.

At 996 K, the γ has to break down to the α + Fe_3C eutectoid. Diffusion distances are generally too great for the rejected eutectoid carbon to form onto the existing graphite (though this *can* happen with very slow cooling of thick sections), and the strain energy factor militates against further graphite nucleation. As Fe_3C is almost as stable as graphite and has a much lower specific volume, this is the favoured phase in the eutectoid breakdown, exactly as in the 'steel' compositions you considered earlier. So the austenite (γ) forms pearlite in the normal way (see Figure 3.53).

EXERCISE 3.19 The outer skin would exhibit a 'white' fracture while the graphite core would be 'grey'. An intermediate zone might be evident where the two were mixed and this would be described as 'mottled'. A fracture test is a very good way of discovering the 'depth of chill' of a cast section, for example, when a hard skin is deliberately sought for a wear-resistant application, or to discover what depth of hard skin may be present on a casting that has to be machined.

EXERCISE 3.20 Quite obviously not. If they are not interconnected, a crack could run through a block or particle of graphite, but would then have to traverse a region of pearlite or ferrite before meeting another piece of graphite. As both pearlite and ferrite are tough, this would increase the overall toughness of the cast iron.

EXERCISE 3.21 The austenite adjacent to the graphite decomposed to α + graphite during the eutectoid breakdown, since there was a convenient graphite interface within the diffusion range of the carbon atoms. This was energetically more favourable than nucleation of Fe_3C which had to take place farther away from the nodule.

EXERCISE 3.22 From the phase diagram (Figure 3.7), this is the eutectic composition of the liquid. Hence if the carbon exceeds 4.3%, primary graphite or primary Fe_3C will form. These will be so coarse within the structure that they render the metal virtually useless for any load-bearing application. A small knock and it may literally fall apart. This is why the cast iron tapped from a blast furnace is unsuitable for engineering applications and has to be remelted under controlled conditions to make acceptable castings.

EXERCISE 3.23 Carbon is an interstitial solute, whereas all the metallic ones form substitutional solutions. Martensites in substitutional alloy systems are not hard. It is the way interstitial carbon interferes with the $\gamma \to \alpha$ transformation in iron that gives rise to the lattice strain and all the ensuing hardening, tempering and transformation products. Even when substitutional elements form precipitates, the resulting lattice strains are less dramatic than carbon in BCC iron.

EXERCISE 3.24

(a) 5 to 6 minutes.
(b) 10 hours.
(c) Over 1 day!

EXERCISE 3.25 One mole of $Fe_{0.953}O$ contains 0.953 of a mole of Fe ions, some of which is Fe^{2+} and some Fe^{3+}. Suppose there are x moles of Fe^{2+} ions and y moles of Fe^{3+} ions. Then

$$x + y = 0.953$$

The positive charges must balance the negative charge on one mole of O^{2-}.

$$2x^{(+)} + 3y^{(+)} = 2^{(-)}$$

So we have a pair of simultaneous equations

$$2x + 2y = 2 \times 0.953$$
$$2x + 3y = 2$$

Subtraction gives $y = 0.094$. So, in the 0.953 mole of Fe ions there is 0.094 mole of Fe^{3+} ions. The proportion of Fe ions which are Fe^{3+} is therefore

$$\frac{0.094}{0.953} = 0.099 = 9.9\%$$

One mole of Fe^{2+} would combine stoichiometrically with one mole of O^{2-}. However, one mole of $Fe_{0.953}O$ has $(0.953 - 0.094) = 0.859$ mole of Fe^{2+}. Thus $(1 - 0.859)$ mole = 0.141 mole or 14.1% of the Fe^{2+} has been replaced by a lesser number of more highly charged Fe^{3+} ions.

Answers to self-assessment questions

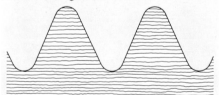

(a) machined thread

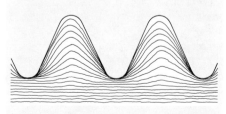

(b) reoriented inclusions

Figure 3.64 Sections of (a) machined thread and (b) rolled thread

SAQ 3.1 The bar starts off with a fibre structure running parallel to the axis, arising from the way it was worked down from the original cast structure. Segregates and banding will thus lie along the length of the bar.

Machining a thread cuts across the banding, so each thread profile will have lines of inclusions running across it, as illustrated in Figure 3.64(a).

Thread forming by deformation re-orients the flow lines to follow the profile of each turn, as illustrated in Figure 3.64(b). As well as having a better surface finish and being stronger by cold work, this structure is much more resistant to service loadings (such as fatigue) than the machined thread. Additionally, rolling produces a compressive residual stress which also improves the fatigue life.

SAQ 3.2 Figure 3.65 overleaf shows the tetrahedral and octahedral interstitial holes in all four cases.

SAQ 3.3

(a) At 1780 K the δ liquidus is reached and some BCC δ starts to form, but at 1760 K this is consumed by the peritectic reaction liquid + $\delta \to \gamma$, to leave γ and residual liquid. Cooling continues until all the liquid solidifies as γ at about 1740 K. Then this cools to 1110 K, where the $\gamma/(\alpha + \gamma)$ boundary is reached, so α starts to separate. The remaining γ gets richer in carbon as cooling continues and more α forms until it reaches the eutectoid at 996 K. The amount of ferrite (α) is then

$$\frac{0.8 - 0.3}{0.8 - 0.005} = 63\%$$

During the eutectoid reaction, the residual 37% γ forms pearlite. This structure of 63% ferrite and 37% pearlite will be the room temperature condition since the tiny changes in carbon solubility in the ferrite will have negligible effect.

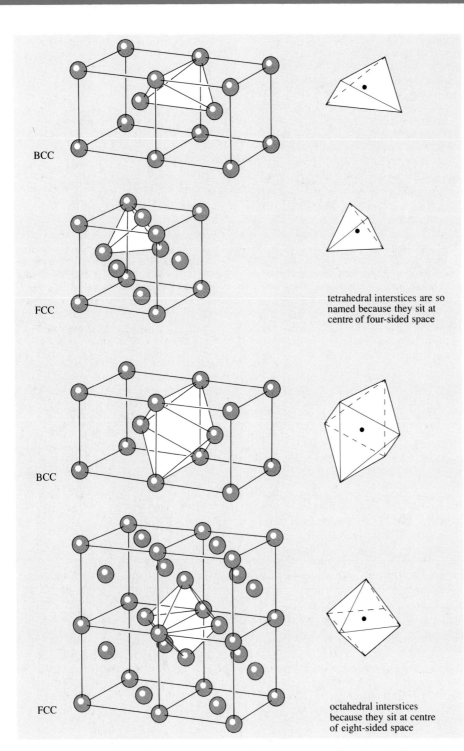

BCC

FCC

tetrahedral interstices are so
named because they sit at
centre of four-sided space

BCC

FCC

octahedral interstices
because they sit at centre
of eight-sided space

Figure 3.65 Tetrahedral and octahedral interstices. Some atoms have been left out of the FCC lattice for clarity

(b) This misses the peritectic reaction by a wide margin, so the first solid will begin to appear at the γ liquidus at 1640 K. Then γ continues to separate down to the eutectic reaction at 1403 K, by which stage the amount of γ will be

$$\frac{4.3 - 2.3}{4.3 - 2.0} = 87\%$$

According to the Fe–Fe$_3$C diagram (but remember this is not the true equilibrium diagram) the remaining liquid then breaks down by the eutectic reaction to form γ + Fe$_3$C, so the quantity of the eutectic will be 13%. Applying the lever rule just below 1403 K, the amount of Fe$_3$C will be

$$\frac{2.3 - 2.0}{6.7 - 2.0} = 6.4\%$$

and the *total* γ will be 93.4% of which, remember, 87% was primary.

Nothing happens to the eutectic Fe$_3$C on further cooling, but the γ will reject carbon as the temperature falls and its composition moves down the solvus line, until it reaches 0.8% carbon at the eutectoid temperature of 996 K. The quantity of γ will then be

$$\frac{6.7 - 2.3}{6.7 - 0.8} = 74.6\%$$

and the extra Fe$_3$C will have joined onto that already present. All the γ now undergoes the eutectoid breakdown to form pearlite. We finish up with dendrite-shaped colonies of pearlite formed from primary γ plus eutectic colonies of pearlite, interlaced with Fe$_3$C.

SAQ 3.4

(a) The object is to form spheroids of Fe$_3$C within a ferrite matrix. If austenite is produced during the treatment, as it must be at any temperature above the eutectoid, then on cooling it will transform to pearlite. The lamellar distribution of Fe$_3$C is harder and less machinable than the spheroidal distribution.
(b) The total surface energy of dispersed particles is high if they have a large surface area (i.e. when the particles are small and finely dispersed). By agglomerating into fewer, larger particles the total surface energy is reduced because the ratio of surface area to volume is less. Hence, although the volume proportion may remain the same, large particles will

grow at the expense of smaller ones, by diffusion across the matrix, in order to reduce the total energy of the system.

SAQ 3.5 When delivered this material would have been deformed lightly so as to avoid the yield-point phenomenon. After 12 months in store, solute atmospheres will have formed around most of the mobile dislocations, so the yield extension will occur and Lüders bands will spoil the surface of the pressings.

The material could be restored by plastically deforming it by 2% or so prior to pressing. The easiest way to do this would be by roller levelling which would minimally affect the thickness, though any type of deformation would achieve the desired result. It would not be possible to eliminate the yield extension by heat treatment — this would simply assist the formation of the carbon atmospheres.

SAQ 3.6

(a) The transformation time t is

$$t = \frac{x}{v_m}$$

where x is the transformation length (that is, the grain size). Thus

$$t = x\left(\frac{\rho}{E}\right)^{1/2}$$

and, assuming room temperature values for E and ρ,

$$t = 5 \times 10^{-4}\left(\frac{7860}{210 \times 10^9}\right)^{1/2} \text{s}$$

$$= 5 \times 10^{-4} \times 1.9 \times 10^{-4} \text{s}$$

$$\approx 10^{-7}\text{s} \quad \text{or} \quad 0.1\,\mu\text{s}$$

(b) $\quad t = \frac{x^2}{D}$

where the diffusion length x is *half* the lamellar spacing (the maximum average distance the carbon has to diffuse to form Fe_3C). Then

$$t = \frac{(5 \times 10^{-6})^2}{5 \times 10^{-13}} \text{s}$$

$$= 50\,\text{s}$$

Thus, with the data used, the martensite transformation is 500 000 000 times faster than the diffusive one!

SAQ 3.7 See Figure 3.66. The hardness will show a steep change from the surface to the core in each section. Note that the core of the low-carbon steel will be martensitic, since the whole component is quenched in the process. The core hardness of the 0.4% C shaft will be unchanged because it was not heated when the surface layers were austenitized, except that, where the induction-heated zone begins, the core will have softened slightly due to being tempered at a higher temperature than that which produced the original hardness of $250\,H_V$.

SAQ 3.8 Disaster! The idea of case hardening the low-carbon steel is to produce a wear-resistant, martensitic skin over a tough core; and, as the final heat treatment is quenching from the austenite region, the carbon content of the core has to be low enough not to produce a brittle transformation product.

When the starting material contains 0.5% carbon the core *will* transform to become hard and brittle in the final quench and will thus lose its toughness. Carbon will diffuse into the surface layers to form a skin containing about 1% carbon, which will of course give the desired wear-resistant surface. However, not only will the core be brittle, it will also almost certainly quench crack during transformation. It would be most unwise to let these pins go into service, especially as they are to be subjected to impact loading.

If the manufacturer knows that the wrong steel has been supplied, but still wants to use it, the pins could be case hardened by flame (or induction) hardening. This heats only the outer surface. Before this is done it would be essential to develop maximum toughness in the bar by normalizing or, preferably, quenching and tempering to a spheroidal carbide structure. (This

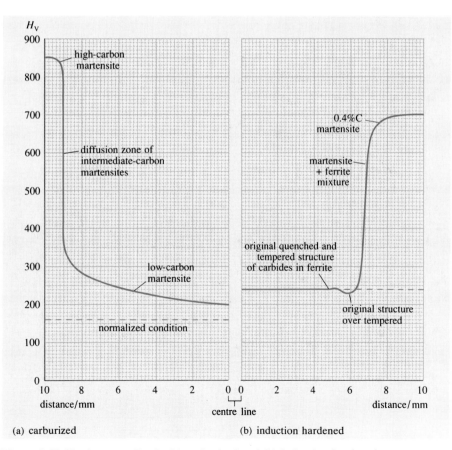

(a) carburized

(b) induction hardened

Figure 3.66 Hardness profiles in (a) carburized and (b) induction-hardened component

assumes that the customer agrees to the change in specification of the final product.)

SAQ 3.9 The two methods most likely are:

(a) Work hardening. Elongation will fall but there will still be 'residual' ductility of the order of 10% at the strength level quoted. The limitation is that the shape will be altered by the plastic deformation.
(b) Achieve the strength by quenching to martensite and then tempering it to the desired strength level. (Elongation of tempered martensite in this steel will be 18–20%). The limitation is that the steel has to be quenched to form the martensite and this may lead to distortion and, possibly, cracking. Thus, although a shaped component may be strengthened in this way and, in contrast to work hardening, no mechanical working is involved, some correction of dimensions may be necessary. There is also a risk of decarburization of the surfaces of the component if the austenitizing is carried out in oxidizing conditions. This could lead to serious loss in strength at the surfaces — where it is most needed.

SAQ 3.10 Figure 3.67 shows the microstructures.

(a) This will consist of elongated grains of ferrite, roughly twice as long as they are thick in the longitudinal section, and some pearlite in bands. Hardness about 200 H_V, tensile strength $\approx 600\,MN\,m^{-2}$, elongation about 15–20%, and tough, especially in directions at 90° to 'fibre'.
(b) 0.6% carbon will consist of 75% pearlite and 25% ferrite in the normalized condition. The pearlite colonies will be resolvable as lamellae of cementite and ferrite. Hardness $\approx 260\,H_V$, tensile strength $800\,MN\,m^{-2}$, elongation $\approx 18\%$ and tough.
(c) The heated face will form 100% martensite, with possibly a few isolated undissolved carbide particles, hardness $650\,H_V$, tensile strength indeterminable, ductility and toughness low.
(d) Austenitizing at 1100 K will not dissolve all the Fe_3C carbide (see Figure 3.7) so on quenching the structure will be martensite with carbides. Tempering at 500 K will reduce the stresses in martensite without significantly altering its structure. Hence the final structure will be globules of Fe_3C in a martensite matrix. Typically

hardness will be above $750\,H_V$ and tensile strength will be indeterminable. Ductility and toughness will be low, but an excellent cutting edge will be possible.

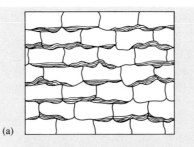

Figure 3.67

SAQ 3.11

(a) The steel would transform completely to pearlite in the time stated. Hence, no matter how rapidly it is cooled afterwards, no martensite can form because all the austenite had transformed during the isothermal holding period.

(b) Beginning with austenite, we need to cool rapidly to a temperature opposite the 'nose' of the bainite transformation region, about 700 K. Hold until transformation is 50% complete (second line) which would be in the region of 1¾ minutes. Then quench to room temperature. The 'nose' of the bainite transformation would give the shortest time, though anywhere between 670 K and 700 K would produce 50% bainite within 2 minutes.

SAQ 3.12 (a), (d) and (e) are cast irons.

(a) This is cast iron because graphite flakes derive from the eutectic reaction, liquid $\rightarrow \gamma$ + graphite.
(d) This is cast iron because 3.4% carbon will undergo the eutectic reaction, liquid $\rightarrow \gamma$ + graphite.
(e) These are the characteristics of a chill cast iron, where the outer skin has solidified as Fe_3C plus austenite, while the slower-cooling core has produced graphite plus austenite. The austenite in both regions subsequently transforms to pearlite.

(b), (c) and (f) are steels.

(b) This is steel because the carbon percentage is well below the austenite solubility limit, so the eutectic reaction will not occur on cooling.
(c) This would be classed as a silicon 'iron', since the carbon level is too low for any response to the kinds of heat treatment applied to engineering steels.
(f) The absence of free graphite or cementite, and the structure, indicate a composition below 0.8% carbon.

SAQ 3.13 Hardness is a measure of the resistance to deformation of a steel and is related to its strength. It depends principally upon the carbon content and structural condition of the steel.

Hardenability is a measure of the degree to which martensite formation will occur in steel sections on cooling. It is not related to the absolute hardness, but is dictated by the delaying effect of alloying additions on the transformation of austenite to pearlitic or bainitic products. A steel of high hardenability will form martensite at the centre of thick sections even when cooled slowly, in air, whereas a low hardenability steel would transform to softer products under the same cooling condition.

Hardenability is conferred by substitutional alloying elements, often in combination, although the percentage carbon also has an effect.

Comparison of Figures 3.52 and 3.56 shows the form of TTT diagram for 'low' and 'high' hardenability, respectively. Your diagram should resemble Figure 3.56, with clear spaces between the vertical axis and the noses of the pearlite and bainite transformation 'start' curves.

SAQ 3.14

(a) The temperature generated would quickly soften an unalloyed steel. Hence additions would be required to delay the tempering process and to induce secondary hardening. In practice, tungsten, chromium, molybdenum, cobalt and vanadium would be used in a 0.8–1.0 wt % carbon steel.

(b) There's no special requirement for this apart from strength and toughness which can both be imparted by the carbon content. A small section can easily be heat treated to produce tempered martensite, or a plain carbon steel could be cold drawn. There is no need for alloying additions, though in practice a manganese content in the region of 0.8–1.0 wt % with 0.6–0.8 wt % carbon would be suitable. But this would not be classed as an alloy steel.

(c) Sharpness, of course, is required but this could be easily obtained in a thin section without alloying. However, for food processing a high degree of corrosion resistance is essential, so a stainless heat-treatable alloy is required. Chromium content would need to be at least 12 wt %, with about 0.3–0.4% carbon. This would be classed as an alloy steel.

(d) A high tensile strength can best be achieved by quenching and tempering a medium carbon steel. However if the bolt is 50 mm diameter a plain carbon steel would have nowhere near high enough hardenability, so a combination of chromium, nickel and molybdenum would be needed. The steel described in Figure 3.56 would be suitable, though cheaper combinations would be possible as the diameter of the bolt is not especially big. Definitely a low-alloy steel of high hardenability is required for this application.

Chapter 4
Ceramics and Glasses

by Nick Reid (consultants: John Briggs and Christina Doyle)

Chapter 4 Ceramics and glasses

4.1 Types of ceramic

In the preceding chapters, you have seen something of the properties of metals and their alloys. You should now understand why these materials are used so widely and also have a clearer view of their limitations. It is quite clear that technology could not be based on metals alone and, even where they have been used previously, other materials are increasingly offering alternative choices because they have certain advantages, such as temperature resistance, chemical stability or ease of processing.

In this chapter we will look at the typical property profiles of ceramics and glasses, beginning with a covalently bonded ceramic. We shall next consider ionic bonding in ceramics and their progress to glasses and glass ceramics. Finally, in order to illustrate the 'battle of the property profiles', we shall discuss three applications of glasses and ceramics in competition with other materials. These materials not only have a long and distinguished history, but also a future full of the promise of advanced applications in the next century.

4.1.1 An archetype?

Diamond, a natural form of carbon, is probably the most coveted of gem stones and we shall start by considering this familiar material because it typifies much of the 'property profile' of a ceramic. 'Diamonds are for ever', so the saying goes, although it certainly needs qualifying by a phrase such as 'in normal use'. Diamonds have lives measured on a geological timescale and, provided they are not subjected to very high stresses or temperatures, they are apparently stable — mechanically, chemically and thermally. This is a feature of most ceramics.

If you wear a diamond ring, you will know to be careful when cleaning windows — if you accidentally drag it across the pane you may scratch the glass. Indeed, a glazier has a diamond-tipped tool for marking and cutting glass. This well illustrates the high hardness of diamond, which is why diamond crystals are used for the indenter of many hardness-testing machines and for the tips of rock-drilling bits. A high hardness at ambient temperature is another feature of ceramics.

Gem diamonds invariably have a faceted surface. Before final polishing, each facet is formed by cleaving a larger crystal: that is, by causing it to fracture in a brittle manner along certain planes. In other words, diamond has low toughness and this, too, is a feature of ceramics at ambient temperatures. Indeed, the main thrust of current research in engineering ceramics is to increase toughness, because this shortcoming is holding back many potential applications.

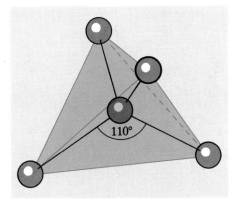

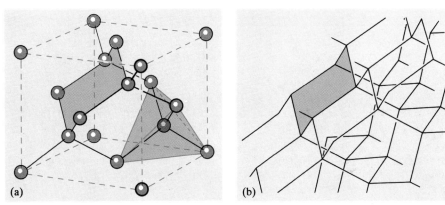

Figure 4.1 The arrangement of bonds around a carbon atom in diamond

Figure 4.2 The diamond crystal structure. (a) The relationship between tetrahedral and FCC structures; and the formation of six-membered rings leading to (b) extended layer structures.

The geometrical symmetry of the facets makes another general point. Most ceramics are crystalline and when the crystals cleave, they do so only on certain planes.

4.1.2 Cause and effect

These properties of diamond all have a common origin — the nature of the chemical bonding. In this case the bonding is covalent, which means that each pair of nearest neighbour atoms shares a pair of electrons, thereby forming a chemical bond. Each carbon atom has four outermost electrons which it shares with its four nearest neighbours. Atoms are arranged symmetrically so that the bonds to every pair of neighbour atoms are at the same angle, 110° (see Figure 4.1). When this arrangement is repeated extensively in all three dimensions the crystal structure of diamond is produced (Figure 4.2). It is 'cubic' in the sense that half the atoms are located on the sites of a face-centred cubic (FCC) structure, and the other half are located on an identical FCC structure shifted by a quarter of a body diagonal from the first (Figure 4.3). Silicon, also from Group 4a of the Periodic Table, adopts the same crystal structure and so does carborundum, SiC, the compound formed from these two elements. Carborundum is familiar as a black grit bonded to paper or cloth and sold as an abrasive. In this case the carbon atoms are located at the sites of one of the two interpenetrating FCC structures, while the silicon atoms are associated with the other FCC structure. Crystallographers call this the 'zinc blende' structure. Cubic boron nitride (BN) also has this structure and is used as a cutting-tool material on account of its high hardness.

The stability of diamond stems from the high energy associated with the covalent bonding. Chemical changes or thermal changes, such as a phase change, would require bonds to be broken. Actually, at ambient

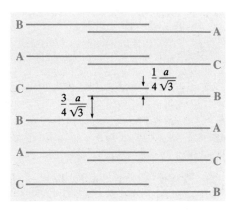

Figure 4.3 Diamond can be regarded as two superimposed, but displaced, face-centred cubic structures. This gives two alternating separations of the close-packed planes, shown here 'edge-on' as ABC

▼The width of a dislocation▲

As you saw in Chapter 2, a dislocation may be regarded as forming the boundary in the slip plane between the area in which slip has occurred and that in which it has not. Since all solids deform elastically, this boundary must have a finite width — there must be a *gradual* change in the slip displacement from zero to **b**, the Burgers vector. (If there were *no* elastic distortion of a dislocated material, the material would be cracked beyond the 'extra' half plane of atoms — see Figure 4.4). The

dislocation width, *w*, is defined as the distance in the slip plane over which the displacement of atoms on one side of the plane relative to those on the opposite side of the plane changes from $+b/4$ to $-b/4$ (see Figure 4.5 for an edge dislocation in a simple cubic crystal). This quantity is important because it determines the shear stress required to move the dislocation and it depends on the nature of the bonding in the material. In crystals in which the bonding is directional (for

example, diamond) the width is small ($w/b = 1$ to 2), in order to minimize the size of the region in which large changes of shape occur. (Note how the square pattern of atoms in Figure 4.5 is distorted near the centre of the dislocation.) On the other hand, in materials with nondirectional bonding (e.g. most metals) this distortion is associated with a smaller energy, so it spreads out more in the slip plane and the width is larger ($w/b = 5$ or more).

The famous physicist Sir Rudolph Peierls considered how the width, *w*, would affect the shear stress, τ, required to move a dislocation and he used a simple theoretical model to conclude that

$$\tau = \frac{2Gw}{b} \exp\left(-2\pi w/b\right) \qquad (4.1)$$

where *G* is the shear modulus.

The stress depends on the relative width, *w/b*, which appears twice in the equation: as a factor and as a power of e. The latter has the greater influence so an *increase* in *w/b* causes a *decrease* in the exponential term, because of the minus sign, and therefore a *decrease* in τ.

The *wider* the dislocation, therefore, the *lower* the stress required to move it.

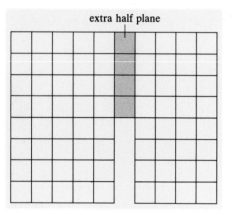

Figure 4.4 An edge dislocation associated with a crack in a 'rigid' material

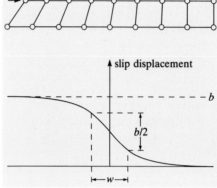

Figure 4.5 The change in atomic displacement relative to the centre of an edge dislocation

pressure the thermal stability is limited and above about 1500 K diamond begins to transform into graphite, but this can be prevented by the application of sufficient hydrostatic pressure. In the presence of oxygen, its stability is even more limited and diamonds cannot be used in air at temperatures above about 1300 K without oxidizing.

Diamond's bonding has a profound effect on the mobility of dislocations (see ▼The width of a dislocation▲). Because of the directional nature of its covalent bonds, dislocations in diamond have a narrow core and require a large shear stress to make them move. This accounts for the high hardness of diamond.

The brittleness of a diamond crystal can be thought of as the outcome of a competition between plastic deformation and cleavage (see ▼The race between cleavage and flow▲). When a cracked crystal of diamond is loaded in tension, the maximum tensile stress at the crack tip exceeds the fracture strength *before* the maximum shear stress at the crack tip attains the shear strength. Therefore the crystal is brittle.

4.1.3 Other properties

Diamond has at least two claims to an entry in *The Guinness Book of Records*: it has the largest known values of Young's modulus, $1200\,GN\,m^{-2}$, and thermal conductivity, up to $3000\,W\,m^{-1}\,K^{-1}$. Its thermal conductivity is over twice that of silver or copper and explains why diamonds feel cold when first touched. There is a connection between these two properties because the propagation of elastic waves makes a major contribution to the thermal conductivity and the velocity of these waves depends on Young's modulus. With its fairly low density, $3.5\,Mg\,m^{-3}$, the velocity of compression waves in diamond is very high, $18\,km\,s^{-1}$, and this fact, too, deserves an entry.

Diamond does provide a good *qualitative* property profile for a ceramic: thermally and chemically stable, hard with a tendency to brittleness. It would be wrong, however, to claim that diamond is a *typical* ceramic. First, it is an element, whereas all other ceramics are compounds; and some of its properties have quite exceptional values.

What, then, is a ceramic?

The name is derived from the Greek for 'burnt earth' and suggests that the formation or processing of these materials normally involves the application of heat, for example, to fire or sinter them. This is true, but so does the extraction or processing of metals. We shall therefore take ceramics to be materials that have a diamond-like property profile and contain at least some nonmetallic elements. They are held together by covalent and/or ionic bonds.

SAQ 4.1   (Objective 4.1)
What is the coordination number of an atom in the diamond cubic structure?

Contact between two diamonds can produce small amounts of plastic deformation if the force of contact is large enough. The slip direction is the direction of minimum atomic spacing and the slip plane is the plane of maximum interplanar spacing. Draw a structure cell and indicate the slip plane and direction.

4.1.4 An ionic ceramic

Traditional ceramics can be regarded as mixed oxides. They are made from clays, which are mainly blends of oxides produced by both chemical and physical interactions. We shall now look at the oxides of the elements belonging to the third row of the Periodic Table (sodium, magnesium, aluminium, silicon, phosphorus, sulphur and chlorine) and see how the melting points vary (Table 4.1). Only the first four can be regarded as having 'ceramic' melting points; the first three oxides are held together mainly by ionic bonding because these elements have electronegativities which differ significantly from that of oxygen (see

▼The race between cleavage and flow▲

Suppose that a material contains a sharp crack and that a tensile stress is applied to the material at right angles to the plane of the crack. With the stress slowly increasing, if the first event to occur at the crack tip is the generation of dislocations, then the crack will become blunt and will not propagate; the material is then tough. On the other hand, if the first event is an advance of the crack tip, the material is brittle.

You can estimate for a number of materials the ultimate shear strength, τ_{max} — the shear stress to nucleate a dislocation at the surface. It depends on the nature of the chemical bonding and is related to the value of the shear modulus, G. You can also estimate the ultimate cleavage strength, σ_{max} — the tensile stress required to cause a sharp crack to advance. Again this depends on the bonding and is difficult to estimate accurately.

It is, however, the *ratio* of these quantities, $Q_{max}(=\sigma_{max}/\tau_{max})$, compared with the ratio of the largest tensile stress to the largest shear stress at the crack tip, Q_c, that is important. If $Q_{max} < Q_c$, then the material will be brittle. For diamond it is estimated that $Q_{max} = 1.16$ and $Q_c = 3.66$ which is consistent with its known brittleness. On the other hand, for the soft metals copper and silver, Q_{max} is about 30 and Q_c is 13–14, which explains why cleavage is unknown in these pure metals.

While this approach is successful at distinguishing between these two extremes, it is not precise enough to account for the many materials, such as iron and steel, that undergo a ductile-to-brittle transition over a range of temperatures. The problem of explaining this phenomenon quantitatively is still the subject of active research.

Table 4.1 Oxides of elements in the third row of the Periodic Table

Element	Electronegativity	Oxide	Melting point/K	State at 290 K
sodium	0.9	Na_2O	1548 (sublimes)	crystalline
magnesium	1.2	MgO	3073	crystalline
aluminium	1.5	Al_2O_3	2320	crystalline
silicon	1.9	SiO_2	1880	crystalline/glassy
phosphorus	2.2	P_2O_5	855	crystalline/glassy
sulphur	2.6	SO_2	200	molecular gas
chlorine	3.2	Cl_2O_7	unstable	molecular gas
oxygen	3.5			

▼Electronegativity▲). The oxide of this group with the highest melting point is magnesia, MgO, a material used to make refractory bricks for furnaces. Because it is already oxidized, magnesia has excellent chemical stability in air.

Since magnesium and oxygen have the same valency, they can combine in equal atomic proportions. Magnesia crystallizes into the structure illustrated in Figure 4.6. The oxygen ions form a face-centred cubic lattice, with the magnesium ions in the octahedral interstitial sites. There are four normal atom sites in the face-centred structural cell and four octahedral spaces. When these are all occupied by oxygen and metal ions, there is a balance between the electrical charges so that the net charge is zero. With equal populations of the ionic species we say that there is a **stoichiometric compound** — one that obeys the formula MgO. By having some of the lattice or interstitial sites empty, it is possible to have small departures from the stoichiometric composition. It is

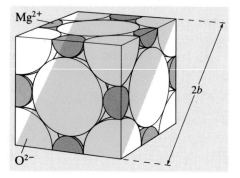

Figure 4.6 The crystal structure of magnesia, showing a slip plane and slip vector, **b**

▼Electronegativity▲

The concept of electronegativity provides a simple means of quantifying the proportion of the apparent ionic character in the bond between two given elements.

When two atoms A and B are covalently bonded, they share bonding electrons and the bond can be represented as [A•B], where electron • comes from atom A and ○ comes from atom B. When both atoms are of the same element (e.g. carbon atoms in diamond) the sharing is equal and the bond is purely covalent. If, however, the atoms come from two different elements and atom B, say, has a stronger attraction for electrons than atom A, then the electrons may not be shared equally. Any such polarization of electrons

can be thought of as giving the bond some ionic character because purely ionic bonding involves the transfer of one or more electrons from one atom to another. If, for example, an electron were transferred from atom A to atom B, the ionic bonding may be represented as $[A]^+[•B]^-$.

The extent of the ionic character will depend on the relative 'affinities' of atoms A and B for electrons and an element with a strong affinity is said to have a high electronegativity. More than one scale has been devised for expressing this. On one scale, the electronegativity is proportional to the arithmetic mean of the element's ionization energy (the energy in $kJ\,mole^{-1}$

required to remove an electron from an atom) and the electron affinity (the energy in $kJ\,mole^{-1}$ given out when an atom receives an extra electron). The greater the difference $(x_A - x_B)$ between the electronegativities of two atoms A and B, the greater will be the ionic character of the bond between them. Examples are given in Table 4.2.

SAQ 4.2 (Objective 4.11)
Based on the electronegativities given in Tables 4.1 and 4.2, what is the minimum proportion of ionic bonding in magnesia, MgO?

Table 4.2 Electronegativities, x_i, for a range of chemical bonds

Bond	C—H	N—H	O—H	F—H	F—C	Cl—C	Br—C	I—C
$x_B - x_A$	2.6 − 2.1	3.0 − 2.1	3.5 − 2.1	4.0 − 2.1	4.0 − 2.6	3.2 − 2.6	3.0 − 2.6	2.7 − 2.6
% ionic	7	16	29	44	29	9	5	1

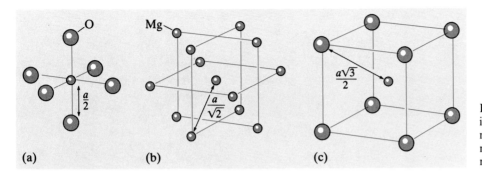

Figure 4.7 The spacing of a magnesium ion from its neighbours in a crystal of magnesia: (a) nearest neighbours; (b) next-nearest neighbours; (c) third-nearest neighbours

understandable that the magnesium ions should prefer the octahedral to the tetrahedral spaces in view of the ratio of the ionic radii of magnesium to oxygen (0.45); that of the tetrahedral site (0.225) is much too small, whereas that of the octahedral space (0.414) requires only a modest dilation. The magnesium ions push the oxygen ions apart, thereby reducing their electrostatic energy.

You need to look closer, inside the cell, to see the local environment of each ion. The chemical bonding in the crystal is the resultant of all the electrostatic attractions of ions of opposite signs *minus* all the repulsions of ions of the same sign. The fragment of crystal depicted in Figure 4.6 has a magnesium ion at its centre. The nearest, next-nearest and third-nearest neighbours, as shown in Figure 4.7, are of opposite, the same and opposite electrical sign, respectively, and form a pattern which has a minimum electrostatic potential energy.

The resulting crystal structure has important consequences for the way in which a material deforms plastically. In a crystal consisting of one type of atom, such as diamond (see SAQ 4.1), slip occurs by the glide of dislocations (Burgers vector *b*) on planes spaced *d* apart, such that b/d is a minimum for that crystal structure. This criterion does *not* apply to ionic crystals because modes of slip are ruled out if slip would cause ions of like charge to become nearest neighbours, even transiently. Consequently, the preferred mode of slip in magnesia is that indicated in Figure 4.6. Of course, there are six such 'edge-to-opposite-edge' slip planes in a crystal, but even if slip occurs on all these planes, there are certain changes of shape that cannot be achieved: for example, a change in length parallel to the diagonal of a cube. This limitation causes polycrystalline magnesia to be brittle for reasons other than that given for diamond (narrow dislocations of low mobility) and is explained in ▼The von Mises criterion▲. At temperatures above 2000 K, slip occurs additionally on face planes of the cube and this permits a grain to achieve *any* change of shape. Under these conditions, polycrystalline magnesia is ductile.

> SAQ 4.3 (Objective 4.1)
> When slip occurs on cube face planes in magnesia, what is the slip direction? (Hint: ions of like charge must not become nearest neighbours during slip.)

tensile stress

⇧

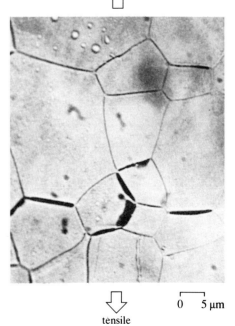

⇩

tensile stress

0   5 μm

Figure 4.8 Cracks at the grain boundaries of deformed magnesia

173

▼The von Mises criterion▲

When slip occurs, a crystal shears like a pack of cards, thereby changing its shape, *but not its volume*. In a polycrystalline aggregate, neighbouring grains differ in crystalline orientation and their slip planes are not parallel to one another. If the grains are to slip without cracking, a grain must deform such that it remains in contact with its neighbours and this requires a grain to undergo a different change of shape from the one it would undergo if it had no neighbours. To avoid cracking, a grain must therefore be capable of undergoing *any* change of shape. Put more pedantically, it must be capable of 'an arbitrary strain with no change in volume'.

How many pieces of information (or 'components') are required in order to describe an arbitrary strain? In Chapter 1 we described six general components of stress or strain — a change in length parallel to each of three coordinate axes and a shear of each of three pairs of faces on a cube. If there is no change of volume, then you only need to know the change in length parallel to *two* co-ordinate axes, which reduces the components of strain to five.

von Mises first showed that such a strain can be achieved by the simultaneous operation of five slip systems — provided the systems are 'independent': that is, the change in shape caused by any one of the five systems cannot be achieved by any combination of the other systems. In specific cases, there is a mathematical method for checking whether given slip systems are independent or not, but it is beyond the scope of this book. When the method is applied to the slip systems shown in Figure 4.6 it reveals that only two of the six are independent. When, however, slip can occur additionally on cube planes (at temperatures above 2000 K), five independent systems exist and polycrystalline magnesia then becomes ductile.

All the metals with cubic crystal structures (FCC and BCC) have five independent slip systems. Any brittleness occurring in these metals cannot be ascribed to a deficiency of slip systems.

▼Roof tiles▲

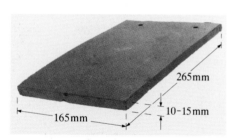

Figure 4.9 British Standard plain tile

Clay tiles, in common with building bricks and terra cotta, are made from types of clay known as loams or marls, in which the clay is mixed with sand or chalk, and their characteristic reddish colour is derived from haematite (Fe_2O_3). Since they are fired at relatively low temperatures, the end product consists of a crude glass surrounding isolated crystals of unmelted quartz, together with some transformed mullite ($3Al_2O_3 \cdot 2SiO_2$).

The use of clay tiles for cladding pitched roofs dates back to the seventh century BC and perhaps even earlier, although the oldest known example in Britain is Roman from the first century AD. By 1212, their use instead of wooden shingles and thatch was mandatory in the City of London to reduce the fire hazard. There were evidently problems with their quality, since a Statute of the Realm was passed in 1477 to regulate their quality and punish makers and sellers of substandard tiles.

What is interesting, however, is how little the specification has changed in the five-hundred years up to British Standard BS402:1979 *Clay Plain Roofing Tiles and Fittings* which states that

> The tiles shall be either hand made or machine made, as may be specified and shall be manufactured from well weathered or well prepared clay or marl.

> The tiles shall be well burnt throughout and free from fire cracks, true in shape, dense, tough and, when broken, show a clean fracture.

The specified dimensions are shown in Figure 4.9 and it is remarkable how close these are to those in Edward IV's Statute '. . . And that every such plain Tile so to be made, shall contain in length ten inches and a half, and in breadth six inches and a quarter of an inch, and in thickness half an inch and half a quarter at the least . . .'

Unfortunately, terms such as 'true', 'fire crack', 'dense' and 'tough' are not further defined, but the Standard does specify a strength test for the tiles. This is a three-point bending test along the long axis of the tile in which the tile is required to withstand a minimum load of 778 N (= 150 lbf) after immersion in water for twenty-four hours.

One of the reasons for 'fire cracking' in tiles is that quartz — like iron, see Chapter 3 — undergoes a phase change at 846 K, with an associated sharp volume change. On cooling a tile from its firing temperature of around 1100 K, the sudden contraction of the quartz crystals at 846 K induces localized cracking around them. The presence of crack stoppers (see Chapter 7), such as needle-shaped mullite crystals usually prevents these cracks from growing to macroscopic size.

4.2 From ceramic to glass

The two most abundant elements in the Earth's crust are oxygen (47%) and silicon (27%), so it is not surprising to find that these elements are found in one of the most common types of naturally occurring material — the silicates. Sand and clays contain silicates and are the raw materials for the production of the traditional ceramics, such as bricks and pottery, which have contributed so much to meet basic human needs (see ▼Roof tiles▲). Consider some of the useful materials that can be made from oxygen and silicon.

4.2.1 Crystalline silica

Like carbon, an atom of silicon has four outermost electrons, while oxygen needs two electrons to complete its outer shell. Clearly, a composition SiO_2 would allow either ionic or covalent bonds to be formed. Since these elements differ significantly in electronegativity (Si 1.9, O 3.5), there is some ionic character to the bonding, but it is predominantly covalent. Four oxygen atoms surround a silicon atom to share (most of) the silicon's four outer electrons. In the resulting structure, the smaller silicon atom is located in an interstice between the four larger oxygen atoms. The larger oxygen atoms can be thought of as four mutually touching cannon balls (Figure 4.10); the interstitial space within can accomodate a sphere with a radius up to 22.5% of the balls' radius. In practice, the silicon atom manages to fit into a smaller space because of the overlap of the electron shells in the covalent bonding. The result is a tetrahedron of oxygen atoms with a silicon atom at its centre (Figure 4.11).

This is the basic building block of silica and the silicates.

Bulk material is made up by repeating these blocks in three dimensions, with each oxygen atom being shared by two neighbouring tetrahedra (Figure 4.12). Although the composition of a single tetrahedron is SiO_4,

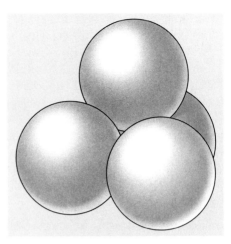

Figure 4.10 Four neighbouring oxygen atoms in silica are arranged like a pile of cannon balls

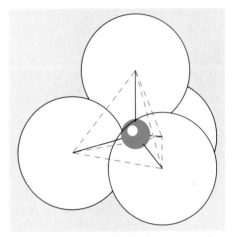

Figure 4.11 The tetrahedral structure of a SiO_4^{4-} ion

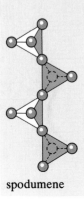
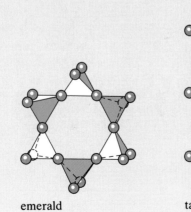
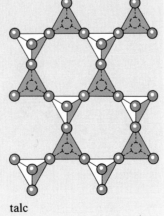

spodumene   emerald   talc

Figure 4.12 Different silicate crystals are formed by joining tetrahedra in different ways

the composition of the bulk is SiO_2, due to the fact that each oxygen atom is shared by *two* tetrahedra. One form of silica is called high cristobalite ('high' because it is stable only at temperatures above 1750 K) and consists of tetrahedra arranged so that the silicon atoms form a diamond cubic structure (Figure 4.13). The oxygen atoms are located mid-way between each nearest pair of silicon atoms. The geometry of this structure suggests that the bonding is largely covalent, as in diamond, and in fact the bonding is about 65% covalent, 35% ionic. Quartz, the low-temperature form of silica that regulates your watch, can also be visualized in terms of tetrahedra. In this case there are two chains of tetrahedra spiralling around opposite sides of a hexagonal cell (Figure 4.14).

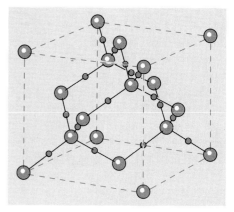

Figure 4.13 The crystal structure of the high-temperature form of silica

4.2.2 Amorphous silica

Consider what happens when crystalline silica is heated to the liquid state and then cooled. If it is heated slowly, when the melting temperature of 1983 K is reached, it transforms into a liquid, causing an abrupt change in specific volume (see Figure 4.15). If liquid silica is now cooled at a moderate rate, its volume contracts, but the melting temperature is passed without any crystals being formed. This is shown by the absence of a step-change in the cooling curve in Figure 4.15. The liquid is *supercooled*. As cooling continues a temperature T_g is reached at which the slope of the graph decreases fairly abruptly and this marks the formation of a glass — amorphous silica. It has the mechanical properties of a solid and the molecular structure of a liquid. Both the positions of the tetrahedra and the connections between them lack the long-range order present in high cristobalite and it is widely accepted that the tetrahedra form a random network (Figure 4.16).

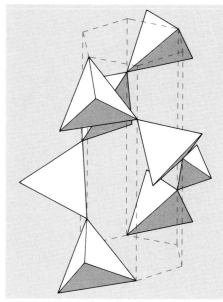

Figure 4.14 Quartz, the low-temperature form of silica

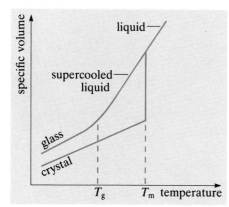

Figure 4.15 A sketch of the variation in specific volume, the volume of unit mass, with temperature, as crystal is first heated and then supercooled to glass

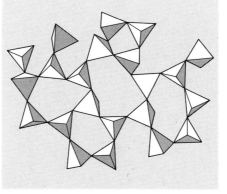

Figure 4.16 Silica tetrahedra connected in a random network to form a glass

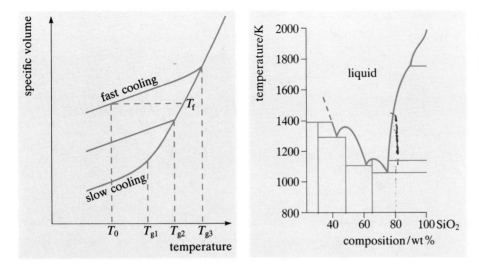

Figure 4.17 The effect of cooling rate on the variation of specific volume with temperature for a glass, showing different glass-transition temperatures, T_g, and a fictive temperature, T_f

Figure 4.18 Part of the equilibrium phase diagram for Na_2O–SiO_2

Why does a glass form so easily in this kind of material? The answer involves the interplay between the cooling rate, molecular mobility in the liquid and the nucleation of a crystal of complex form. Near the melting temperature, glass formers like silica have large values of viscosity, of the order of $100\,N\,m^{-2}\,s$, compared with $0.1\,N\,m^{-2}\,s$ for a molten metal. This implies a low mobility of the tetrahedra. It therefore takes a comparatively long time for the tetrahedra to get into correct alignment so as to form crystal nuclei. Even a modest cooling rate takes the liquid through the melting temperature without nucleation occurring. As cooling proceeds, the mobility decreases steadily and the **glass-transition temperature**, T_g, is reached, at which point the mobility changes from large-scale cooperative motion to localized vibrational motion. Such motion is inadequate to allow the molecular structure of the liquid to keep changing fast enough during cooling to that which is in thermal equilibrium. From this point onwards the molecular structure characteristic of the liquid at a certain temperature (the **fictive temperature**) is 'frozen in'. For a given melt, the values of both the fictive temperature and T_g depend on the cooling rate used (Figure 4.17).

Most glasses actually consist of silica mixed with metal oxides. These 'alloying' additions greatly reduce the melting temperature (Figure 4.18), thereby making the glass easier and cheaper to make. For the high-volume products, such as windows, bottles and jars, soda-lime silica glass is used consisting of 70–75% SiO_2, 13–17% Na_2O, 5–10% CaO and small amounts of other oxides (see ▼ Glass making ▲).

Consider what happens to soda (Na_2O) when it is added to silica glass. Since the electronegativity of sodium is very low (0.93) it tends to give up its outermost electron to the oxygen, the oxide becomes ionized and adds oxygen to the random network. The result is that some oxygen atoms become nonbridging: that is, they become corners of tetrahedra which are not attached directly to another tetrahedron and at these

▼Glass making▲

The manufacture of commercial glasses is based on the readily available ingredients, silica and sodium carbonate. Glasses made from these alone are vulnerable to attack by water, so to reduce this effect, some of the sodium carbonate is replaced by calcium carbonate to make **soda-lime silica glass**, which is used for such products as windows, bottles and jars, light bulbs and spectacle lenses. Since the carbonates decompose during glass making to form oxides and carbon dioxide, the glass, as its name suggests, is really a mixture of oxides.

The first stage in making the glass is to melt the pulverized raw materials in a furnace. One of these (SiO_2) has a very high melting point (~ 2000 K). To avoid the need and expense of heating to this temperature, a quantity of powdered scrap glass, known as 'cullet', is added to the raw materials. This makes up 15–30% of the total and by melting before any of the raw constituents, it conducts heat and dissolves the other ingredients together. This significantly reduces the time required to melt the mixture and hence it saves energy. Figure 4.19 shows a type of continuous glass-melting furnace, with a sketch of the temperature profile along its length. The processes taking place are complex and interdependent. They fall into the three categories of melting, refining (or fining) and homogenization.

The furnace is either gas-fired or oil-fired and is usually arranged so that the maximum temperature, about 1800 K for soda-lime silica glass, is reached approximately one third of the way along the melting chamber, where the refining process takes place.

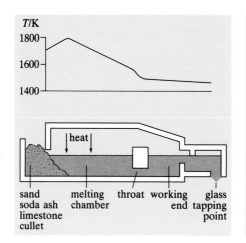

Figure 4.19 A tank furnace for continuous glass melting with a typical temperature profile

Associated with the melting is the reaction of the silica with the sodium and calcium carbonates, which leads to the evolution of carbon dioxide gas. The process of refining is the clearing away of these gas bubbles. To aid this process, fining agents such as sodium sulphate are sometimes used to sweep out the bubbles and this movement, together with convection and the flow of the melt through the tank, aids the homogenization of the melt.

Beyond the throat of the furnace is the working end where the fining and homogenization occur, as the temperature is reduced to the working range of the glass. Since there are no abrupt changes in the viscosity of glass as the temperature changes, glasses are frequently characterized by a set of reference temperatures at which their viscosities have defined values. These are shown in Figure 4.20 for soda-lime silica glass. In

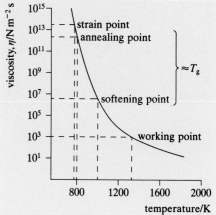

Figure 4.20 The viscosity of a soda-lime silica glass as a function of temperature

the melting chamber the viscosity, η, is of the order of 10 N m^{-2} s, similar to that of thick treacle. When the glass is being worked (e.g. moulded or drawn) the viscosity is between 10^2 and 10^5 N m^{-2} s, so the temperature at which $\eta = 10^3$ N m^{-2} s is arbitrarily called the **working point**. The **softening point** is defined to be the temperature at which $\eta = 10^{6.6}$ N m^{-2} s, which is a viscosity associated with the creep at a prescribed rate of a given length of a glass rod or fibre under its own weight. The **annealing point** ($\eta = 10^{12.4}$ N m^{-2} s) is the temperature at which any internal stresses are relieved after several minutes. The glass transition temperature, T_g, lies somewhere between the softening and annealing points. Finally, the **strain point** ($\eta = 10^{13.5}$ N m^{-2} s) is the highest temperature from which the glass can be cooled rapidly without causing significant levels of internal stress to be developed.

points gaps are formed in the network. The sodium ions go into these gaps, thereby providing ionic bonds between unbridged tetrahedra (Figure 4.21). Thus, when soda is added to silica, some of the covalent bonds are replaced by ionic bonds of lower energy which, being nondirectional, reduce the viscosity of the melt, enabling the glass to be worked at lower temperatures.

One of the most important aspects of the property profile of such glass is the extent to which it facilitates economic processing into products. First, the material can be melted at readily achievable temperatures and, secondly, its melt is resistant to the formation of necks when it is stretched (it behaves like chewing gum) and this property enables glass melts to be 'spun' into thin fibres or blow-moulded into bottles and jars. Many polymer melts show this property (see Chapters 5 and 6).

Of course, since the processing temperatures have been deliberately lowered by alloying, the service temperature ceiling is also lowered and for high-temperature applications, such as the envelopes of quartz halogen lamps, silica glass must be used. Indeed, fibres of silica glass are used to form the 'ceramic tiles' that provide the thermal barrier coating that enables the space shuttle, despite being made largely from aluminium alloys, to re-enter the Earth's atmosphere. Since glasses are made from oxides, they do not react chemically with oxygen in the air, being oxidized already!

The maximum service temperature of a glass is usually set by the onset of unacceptable softening. Glasses may also be sensitive to *changes* of temperature, which can induce cracking and failure due to thermal stress (see ▼Thermal shock▲). Soda-lime silica glass has poor resistance to thermal shock because its coefficient of thermal expansion, α, is large ($\sim 9 \times 10^{-6}\,\mathrm{K}^{-1}$). To overcome this problem the borosilicate (e.g. Pyrex) glasses were developed for use in the kitchen and the laboratory, with the value of α reduced threefold. They have compositions in the range 60–80% SiO_2, 10–25% B_2O_3, 2–10% Na_2O and 1–4% Al_2O_3.

network modifier

Figure 4.21 'Fluxing' ions, such as sodium, provide ionic bonds between unbridged silica tetrahedra

Table 4.3   Some materials properties

Property	Units	Materials									
		silica glass	borosilicate glass	annealed soda-lime silica glass	thermally strengthened soda-lime silica glass	glass ceramic	silicon nitride (RBSN)	magnesia (MgO dense fine-grain)	alumina 99.9% (fine-grain)	silicon carbide (RBSC)	stainless steel (18/8)
density, ρ	$\mathrm{Mg\,m}^{-3}$	2.2	2.4	2.5	2.5	2.6	2.4–2.6	3.55	3.9	3.2	7.9
specific heat at 470 K	$\mathrm{J\,g}^{-1}\mathrm{K}^{-1}$	0.91	0.94	0.98	0.98	0.83	0.86	1.05	0.99	0.88	0.51
strength (in bending)	$\mathrm{MN\,m}^{-2}$	70	70	70	250	> 110	300–350	170	380	450	600
Young's modulus, E	$\mathrm{GN\,m}^{-2}$	70	70	72	72	92	150–180	276	400	400	200
expansion coefficient, α	$10^{-6}\mathrm{K}^{-1}$	0.62	3.2	7.8	7.8	1.0	2.6	11.6	7.7	4.5	16
thermal conductivity at 470 K	$\mathrm{W\,m}^{-1}\mathrm{K}^{-1}$	1.8	1.5	1.8	1.8	2.4	12.5	25	25	100	150
hardness	H_v	490	420	460		600	750	600	1900	2500	180
temperature limit in air (no creep)	K	1220	720	720	720	970	1400	1500	1700	1200	970
toughness, G_c	$\mathrm{J\,m}^{-2}$	~ 1	~ 1	~ 1	~ 1	~ 10	10	1	25	25	2×10^5
Poisson's ratio, v	–	←————— 0.25 assumed —————→					0.24	0.23	←——— 0.25 assumed ———→		0.35
transparent	–	yes	yes	yes	yes	yes	no	no†	no†	no	no

†Depends on grain size, but even the smallest levels of porosity (> 0.2%) renders oxides opaque

SAQ 4.4 (Objectives 4.2 and 4.11)
What is the fictive temperature of a glass? Describe qualitatively how the fictive temperature would vary with the rate of cooling during the formation of a glass.

SAQ 4.5 (Objectives 4.3 and 4.4)
Compare quantitatively silica, borosilicate and annealed soda-lime silica glasses for their resistance to thermal shock. Consult Table 4.3 for values of the properties you require.

▼Thermal shock▲

Figure 4.22 A cylindrical rod with constrained ends

In general, an unconstrained material expands or contracts when its temperature changes. When it is constrained, so that its natural expansion or contraction is prevented, a thermal stress is set up in the material. If this is tensile and exceeds the fracture strength, the material may become cracked by **thermal shock**.

Consider a cylindrical rod, the ends of which are fixed (Figure 4.22). Suppose the rod is heated from temperature T_0 to T. If it was unconstrained it would undergo a thermal strain of $\alpha(T - T_0)$, but this is thwarted by the fixed ends and the net strain is zero. This is explained by saying that in addition to the thermal strain, there is a deformation strain of $-\alpha(T - T_0)$ giving a total strain of zero. If the deformation is purely elastic, according to Hooke's law it is associated with a stress given by

$$\sigma = -\alpha(T - T_0)E \qquad (4.2)$$

This is a negative or compressive stress, but if the rod was *cooled* from T_0 to T, the change in temperature $(T - T_0)$ would be negative and the stress would be tensile. If the stress, σ, is less than the fracture

strength, σ_f, fracture will not occur provided

$$\sigma_f > \alpha(T_0 - T)E \qquad (4.3)$$

or

$$(T_0 - T) < \sigma_f/\alpha E$$

Constraint can come from causes other than fixed ends. For example, if a thin surface layer is cooled very quickly while the underlying layers of the same material change little in temperature, the contraction of the layer will be constrained by contact with the underlying bulk material. This is what happens if you plunge a hot glass plate into cold water — the surface layer immediately attains the water temperature, T_c, before there has been time for heat to flow from the underlying glass at temperature T_h. The contraction of the layer is constrained, creating parallel to the surface a biaxial tensile stress given by

$$\sigma = \frac{\alpha E(T_h - T_c)}{1 - v} \qquad (4.4)$$

where v is Poisson's ratio. This is similar to Equation (4.2), but with a factor of $1/(1 - v)$, which arises simply because the plate's surface is constrained in two dimensions whereas the rod was constrained in one. For a soda-lime silica glass $\alpha \simeq 10^{-5}\,\text{K}^{-1}$, $E = 70\,\text{GN}\,\text{m}^{-2}$ and $v = 0.2$. If it is quenched from 373 K to 273 K, the maximum surface stress is about 90 MN m^{-2} and exceeds the mean fracture strength. This explains why glass jars and milk bottles are so prone to thermal shock.

Equation (4.4) can be rearranged to give the temperature difference ΔT that is required to create a thermal stress σ equal to the fracture stress σ_f.

$$\Delta T = \frac{\sigma_f(1 - v)}{\alpha E} = R \qquad (4.5)$$

The quantity R can be used as a merit index of a material's resistance to thermal shock, but it applies only if the rate of heat transfer is infinite. Since all real materials have finite thermal properties, this is never truly the case.

To take account of finite rates of heat transfer, the thermal properties such as the conductivity λ, measured in units of $\text{W}\,\text{m}^{-1}\,\text{K}^{-1}$, must be included. For example, in the case where heat is transferred across the surface at a constant rate, h, the merit index is

$$R' = R\lambda \qquad (4.6)$$

The biaxial surface stress set up is then equal to that in Equation (4.4) multiplied by the factor rh/λ where r is a characteristic dimension — for a plate, it would be half the thickness.

Under conditions where the surface *temperature* changes at a constant rate, the appropriate merit index of thermal shock resistance for material of a given size and shape is

$$R'' = \frac{R\lambda}{\rho c} \qquad (4.7)$$

Where ρ is the density of the material, measured in kg m^{-3}, and c is its specific heat, measured in J kg^{-1} K^{-1}.

4.2.3 Crystals in a glass

Mass production of articles from a glass melt involves a range of highly efficient processes. Following extensive development work in the 1950s and 1960s by Corning, Schott and other companies, a number of glass compositions that could be crystallized after the shaping process were identified. The heat treatment required for both controlled nucleation and crystal growth is shown schematically in Figure 4.23. The important step is the provision of large numbers of crystalline nuclei, which allow a fine uniform crystal structure to develop. Réaumur had successfully crystallized ordinary glass bottles back in the eighteenth century by heating them in sand in furnaces at about 1070 K. Unfortunately, the incidence of nucleation was totally inadequate so that large needle-shaped crystals developed to give very weak products.

These crystallized or 'cerammed' glasses, better known as **glass ceramics**, have mechanical, electrical and thermal properties that are quite different from those of the parent glasses from which they are derived. Since a fully dense glass is the intermediate product, pore-free microstructures can be expected after the crystallization process.

The vast majority of commercial glass ceramics are based on two aluminosilicate systems: $MgO-Al_2O_3-SiO_2$ and $Li_2O-Al_2O_3-SiO_2$. Various other oxides, such as P_2O_5, TiO_2 or ZrO_2, are added and are essential to ensure large numbers of crystal nuclei.

Besides their electrical properties or aesthetic qualities, which can often be quite important, the crystalline phases produced in glass ceramics, such as cordierite ($2MgO.2Al_2O_3.5SiO_2$); eucryptite ($Li_2O.Al_2O_3.2SiO_2$) and spodumene ($Li_2O.Al_2O_3.4SiO_2$) have low thermal-expansion coefficients which ensure that they have good resistance to fracture due

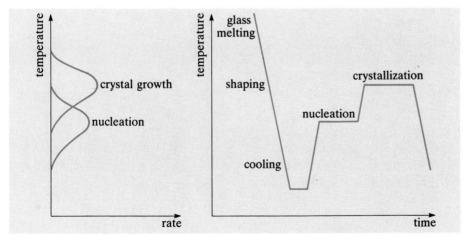

Figure 4.23 A schematic view of the time–temperature schedule required for glass-ceramic production

to thermal stresses or thermal shocks. (The names of these glass ceramics are mineralogical rather than chemical, since the phases were first identified by geologists in natural rock specimens.)

The main advantage of the glass-ceramic route for production is the prospect of being able to use automated mass production based on established glass-manufacturing techniques.

Having now completed a brief survey of the main types of ceramics and glasses, we shall now look in depth at three cases of applications taken from disparate environments: the home, the hospital and the garage.

4.3 The glass-ceramic hob

4.3.1 Introduction

In the industrialized countries, most domestic cooking is now done using either electricity or gas. This is understandable since they are clean and smokeless, and the 'fuel' can be 'piped' to the cooker. Furthermore, heat is almost instantly available on demand — nobody has to get the fire going first. Electric cookers were first used in about 1895 and three types were employed:

1 Cast-iron hot plates with resistance wires cemented under the plates, the heat being transferred by conduction through the plate.

2 Bare wire spirals which simulated the flames of gas cookers and transferred heat by radiation. This system was decidedly dangerous when mains voltage was used.

3 Flat spiral elements operating on a low-voltage, high-current supply (12 V/100 A) from a transformer. This involved heat transfer by both conduction and radiation.

Starting in the 1950s, the preferred design became the insulated 'radiant' flat spiral, where the nickel–chromium resistance wire was sheathed inside an 'Inconel' tube from which it was insulated by a refractory powder, such as magnesia packed to a high density. This system could operate at mains voltage and showed reasonably rapid temperature response to changes of power setting. The only major problem with this system is that of cleaning up after spillages. The iron hot plate is easier from this point of view, but its thermal response is slower.

In the 1950s, Corning Glass Company in the USA developed a new family of materials based on the controlled crystallization of glasses. The properties of these glass ceramics were largely dictated by those of the silicate crystalline phases they contained. Some of these phases had extremely low thermal-expansion coefficients and were rapidly exploited for their thermal shock resistance. One of the markets which rapidly opened up was cooking ware, in particular casseroles, jugs and crockery, with trade names such as 'Pyrosil', 'Pryoceram' and so on. Other glass companies around the world lost no time in developing and patenting their own glass-ceramic materials.

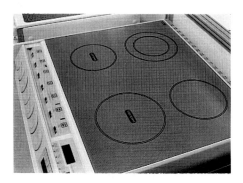

Figure 4.24 A typical glass-ceramic cooker hob

In 1957 Corning introduced the first glass-ceramic cooking surface and over the years other suppliers have further developed this product. The research, development and patent activity in this area was most intense in the period 1968–72, but the product became commercially significant only in the 1980s.

Essentially, the glass-ceramic sheet separates and insulates the heating coil from the cooking utensils and at the same time provides a flat, easy to clean and attractive surface — 'the counter that cooks' according to early advertising copy! Although most of the glass-ceramic hobs incorporate resistance heaters, a further type exists based on the use of quartz halogen lamps. They transmit infrared radiation quite efficiently through the glass ceramic to give a better rate of heating and energy efficiency than the hobs heated by resistance heaters.

4.3.2 The properties required

Consider further the goal of making a cooker with a continuous flat working surface and the properties required of the material used for it. Could a metal or alloy be used? Could a glass be used?

The advantage of a continuous surface is the ease of cleaning. The surface has to be able to let heat pass from the heater to a pan by conduction or, if the surface is transparent, by conduction and radiation. A high thermal conductivity, however, is undesirable because heat will then be conducted from the heaters to the whole surface. Hot regions that are not supporting a pan will waste heat and will be a potential hazard for the cook. This rules out the use of materials with thermal conductivities that are high compared with those of glass and glass ceramics ($1–4\,W\,s^{-1}\,K^{-1}$). If low-conductivity materials are used, they must be transparent in the infrared spectrum so that most of the heat can be conveyed from the heater by radiation.

If the surface is to be cleaned easily, it must be smooth so that it does not retain dirt. Furthermore, it must stay smooth throughout its lifetime, so it must resist scratching. This requires a material with a higher hardness than that of the materials that make contact with it: for example, a glass pan with a hardness of about $550\,H_V$ (recall 'Hardness Measurements' in Chapter 3). The surface must have sufficient robustness not to become cracked by thermal shock due to rapid heating or cooling; a hob at $900\,K$ should be able to resist having cold water spilled on it without cracking. You have already seen that the merit indices for resistance to thermal shock (R, R' and R'') favour a large fracture strength, σ_f, and small values of thermal-expansion coefficient, α, and Young's modulus, E. Methods of increasing the strength of glasses are described in ▼Strengthening glass▲

The surface must also have a good resistance to damage caused by impact. Such damage is usually caused by contact stresses (see ▼Hertzian stress▲) rather than by bending stresses. The typical

▼Strengthening glass▲

When glass breaks, a crack propagates rapidly from a pre-existing stress-raising defect. This process almost invariably begins at the surface where the material has suffered damage due to contact with its environment. The internal structure of a glass is usually quite featureless down to about 10 nm — it displays no voids or porosity, no phases and associated boundaries. Fracture initiation at the surface is also favoured in those loading situations where the applied stress is largest at or near the surface: for example, when a sheet of material is loaded in bending or a rod is loaded in torsion. Indeed, since loads are usually transferred to a component by surface contact, the locations of maximum stress usually occur on or near the contact area and are associated with the existence of contact points between microscopically rough surfaces. This is a major problem in the tensile testing of brittle materials because the gripping of the test piece creates severe local stress concentrations. In a ductile material such as a soft metal, these would become relaxed by plastic deformation and the contact stresses would become more uniformly distributed over the area of contact. For glasses and ceramics, bending tests are preferred because there is no need for gripping (see 'Flexural tests' in Chapter 8) and bending strength is frequently cited instead of tensile strength.

Since the surface is the material's weakest link, the strategy involved in strengthening glasses is to treat the surfaces in order to increase their strength. The treatments used put the surfaces into compression, which, for equilibrium, means that the interior is put into tension. The material is then said to contain **self-stresses** (also called **internal** or **residual stresses**), because, unlike applied stresses, these stresses exist without any external forces being applied to the material. A surface under compressive self-stress, σ_s, is stronger because it can withstand the application of a larger applied tensile stress, σ_a, before fracture occurs. The resulting net tensile stress is $\sigma_a - \sigma_s$ and the surface will fracture when this becomes equal to the fracture strength, σ_f. That is, when

$$\sigma_a - \sigma_s = \sigma_f \qquad (4.8)$$

Thus the applied stress must exceed the

fracture strength by σ_s before the glass will break. Therefore the glass is strengthened, compared to untreated annealed glass, for which $\sigma_a = \sigma_f$.

There are two approaches used to introduce self-stresses. One involves the use of heat treatment — **thermal toughening** — while the other involves diffusing appropriate ions into the surface — **chemical toughening**. It is unfortunate that these have come to be known as types of toughening because the toughness G_c of the material is *not* increased significantly by treatment and that is why we took σ_f above to be about the same in treated and untreated glasses — see Equation (4.8).

Thermal toughening

The first step of the thermal toughening treatment is to heat, for example, a car window, to a uniform temperature close to the softening point (see Figure 4.20). At this temperature, the glass is viscous enough to retain its shape for the treatment time required, while fluid enough to relax any internal stresses by viscous flow. The surfaces are then cooled rapidly and uniformly by an array of cold air jets, taking the surface temperature below the strain point of the glass (see Figure 4.25).

The thermal contraction of the surfaces occurs freely because the interior, which is still hot, is soft and therefore contracts by viscous flow in order to remain compatible with the surfaces. This state of affairs ceases when the interior falls below the strain point and becomes too viscous to flow. Since the interior is hotter than the surfaces, if free to do so, it would contract more than the surfaces on cooling to ambient temperature. It is at this stage that the self-stress is created. The greater thermal contraction of the interior causes a matching elastic contraction of the

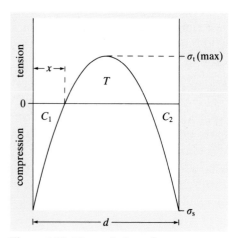

Figure 4.26 The variation of stress across the thickness of thermally toughened glass

surface layers, thereby putting a compressive stress into the surfaces and a tensile stress into the interior.

A typical distribution of stress across the thickness of a glass plate is shown in Figure 4.26. Typically, for a thickness d the stress reaches zero at a distance of about 0.2–$0.25d$ from the surfaces. The areas enclosed between the curve and the horizontal axis represent force distributions, tensile above the axis and compressive below it. If the glass is to be in equilibrium these areas must be equal.

The magnitude of the surface stress, σ_s, depends on a number of parameters, such as the coefficient of thermal expansion and the maximum temperature difference that can be established between surface and core during cooling. This temperature difference depends on the rate at which heat is removed and the thickness and thermal properties of the plate. These are arranged to give σ_s a value of about $-100\,\mathrm{MN\,m^{-2}}$. This approximately doubles the strength of the glass.

There is another benefit of the treatment. If fracture *does* occur (Figure 4.27) the glass breaks into very small, almost cubic, fragments, which are much less dangerous than the sharp daggers of glass produced when annealed glass breaks. The treated glass produces more cracks and therefore smaller fragments, because, at the point of fracture, it contains more elastic strain energy to 'feed' the cracks than annealed glass.

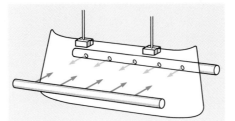

Figure 4.25 Air jets cooling glass windscreens

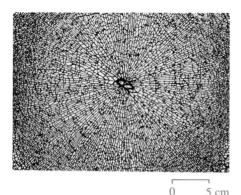

Figure 4.27 A pattern of cracks in thermally toughened glass

Chemical toughening

Chemical toughening induces a compressive stress in the surface by exchanging some of the original ions of alkali metal for ions of a larger diameter. For example, sodium ions, with a radius of 0.098 nm, are replaced by potassium ions of radius 0.133 nm. The treated surface layers would then occupy a greater volume, if free to do so, but expansion is resisted by contact with the untreated core material, which therefore exerts a compressive stress on the surface layers. This process of 'ion stuffing' is carried out by immersing the glass in a hot bath of a molten salt containing the larger ions. For example, a soda-lime silica glass can be treated by immersion in potassium nitrate, KNO_3. The temperature must be kept below the annealing point, otherwise the self-stresses induced would relax by viscous flow, but it must be high enough to allow the ions to diffuse to a reasonable depth, about 0.1 mm, in a cost-effective time. Much higher compressive stresses, up to $400\,MN\,m^{-2}$, can be induced by chemical toughening, but they exist over a much shallower depth, about $100\,\mu m$ below the surface. This limited depth of the compressive surface stress is a drawback of this type of strengthening as only a small surface defect can penetrate into the tensile zone, making fracture *easier* than in the untreated glass. Another drawback is the relatively long treatment times required for ion exchange.

SAQ 4.6 (Objectives 4.2 and 4.3) Glasses with small coefficients of thermal expansion have good resistance to thermal shock. What potential do such materials have for thermal toughening?

▼Hertzian stress▲

Pressing an indenter into the surface of a silicate glass, or a typical engineering ceramic of low toughness, invariably results in a crack being formed (Figure 4.28). This may be accompanied by some plastic deformation. Ceramics and brittle materials in general contain numerous crack-like defects, which are usually subcritical in size: that is, too small to propagate under the normal surface stress. If the stress is raised, even locally, so that the Griffith energy condition (Equation 2.22) is fulfilled, cracks will propagate. The propagation, however, may peter out after a short distance if the stress field is not sufficiently far reaching.

The stresses in an elastic medium subjected to contact by a spherical object were first analysed by Hertz. Figure 4.29 shows that where surfaces are actually touching there is no tensile stress parallel to the surface. Immediately outside the circular area of contact, however, the surface stress in the radial directions is tensile and has a maximum value at the boundary of the contact area. Its value is given by the expression

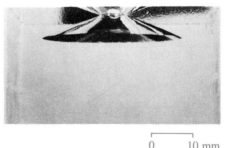

Figure 4.28 A cone crack produced by a ball indenter

$$\sigma_{max} = \frac{(1-2v)F}{2\pi a^2} \quad (4.9)$$

where F is the force on the ball, v is Poisson's ratio for the material under the ball and a is the radius of the area of contact.

It is these Hertzian stresses which cause the characteristic conical ring cracks to propagate (Figure 4.28). Contact by hard materials with high curvature and hence a small radius is especially dangerous. Since the radius, a, is small, for a given force the Hertzian stress generated is high.

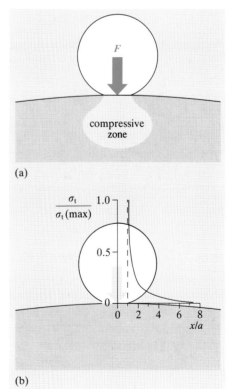

Figure 4.29 Stresses under a spherical indenter

185

situation visualized is that of a small salt cruet falling from a shelf. The more flexible the hob under impact, the less likely it is that Hertzian damage will occur. Therefore, a larger hob is more difficult to fracture than a smaller one of the same thickness. Paradoxically, dropping large heavy pans on the cooker's surface is less likely to cause damage than a sharp pointed object because the load is carried by a much larger area of contact.

Since the resistance heaters are in contact with the underside of the hob, the hob material must be capable of resisting long-term exposure to a temperature of about 900 K. The variation in viscosity of glass-based materials is shown in Figure 4.30 as a function of temperature.

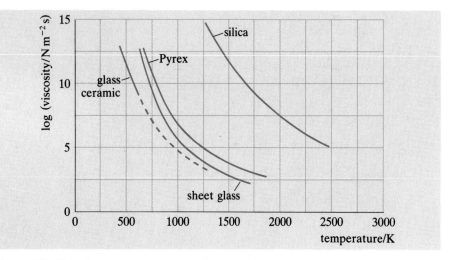

Figure 4.30 Viscosity–temperature curves for some glasses

SAQ 4.7 (Objective 4.4)
Based on the preceding discussion of the properties required, select from Table 4.3 the material whose properties best meet the requirements of a continuous flat hob.

4.3.3 Manufacture of hobs

The outline of the glass-ceramic process was given in Section 4.2.3 and a schematic representation of the production route for the glass-ceramic hobs is shown in Figure 4.31. The choice of raw materials is governed by many factors, including cost, ease of fabrication and the properties of the end product.

A mixture of up to fifteen oxides is melted to form a homogeneous glass at about 1920 K. Some of the main ingredients of typical commercial compositions are listed in Table 4.4. The major ingredients are SiO_2, Al_2O_3 and Li_2O.

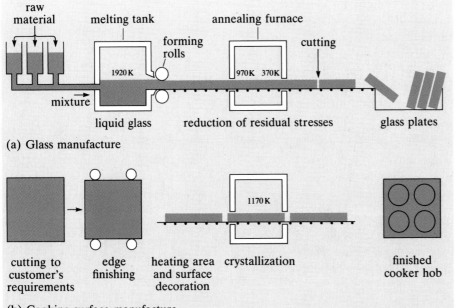

Figure 4.31 Production of glass ceramic hobs

Table 4.4 Compositions of a typical glass-ceramic cooker top expressed as molar ratios relative to alumina (= 1)

Oxide	Proportions
SiO_2	6.1
Al_2O_3	1.0
Li_2O	0.68
MgO	0.30
ZnO	0.13
$Na_2O + K_2O$	0.04
TiO_2	0.11
ZrO_2	0.07

Magnesia, MgO, improves the working properties of the glass melt and acts as a cheap partial replacement for Li_2O. The oxides TiO_2 and ZrO_2 are the nucleating agents for crystallization.

The glass is formed into continuous sheets by a system of rollers at about 1650 K before being annealed and cooled to room temperature. The sheet is then cut to the desired shape including the trimming of the edges. The transparent glass is next crystallized by submitting it to a controlled temperature/time schedule (see Figure 4.23) within the range 950–1200 K. It follows that the temperature limit for long-term stability of the microstructure is at the bottom of this range. The extent of crystallization may exceed 90% by volume and small crystal sizes of < 0.5 μm are produced (see Figure 4.32). Patterns indicating the positions and sizes of the heaters are fired into the top surface of the hob during the crystallization treatment. The major phases which

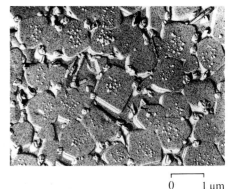

0   1 μm

Figure 4.32 Microstructure of a glass ceramic showing a small uniform grain size and an intergranular glassy phase

crystallize out are either β eucryptite or β spodumene, both of which have very low thermal-expansion coefficients and form solid solutions very readily with extra silica and with other oxides. Some degree of manipulation of the mechanical, optical, electrical and thermal properties is possible by changing the composition and hence the high number of oxides in the mixtures. One major advantage of the glass-ceramic fabrication route is the possibility of using continuous-glass-technology techniques to produce large quantities of highly reproducible items — with zero porosity.

4.3.4 Optical and radiant properties

Some of the commercial glass-ceramic hobs are white and translucent, while others are fairly transparent and dark brown in colour. Transparency is possible when the crystals are smaller than the wavelength of visible light. The transparency enables the power setting of the cooker to be estimated at a glance and the dark colour reduces the visibility of burnt-on stains.

The optical transmission properties are important in that they enable you to see whether or not the hob is operating and they also influence the efficiency of heat transfer to the cooking vessel. The intensity of radiation transmitted through a hob, I, varies with the wavelength of the radiation (see Figure 4.33). It depends on the thickness of the plate, t, according to the equation:

$$I = I_i \exp(-kt) \qquad (4.10)$$

where I_i is the intensity of the incident radiation and k is the spectral-

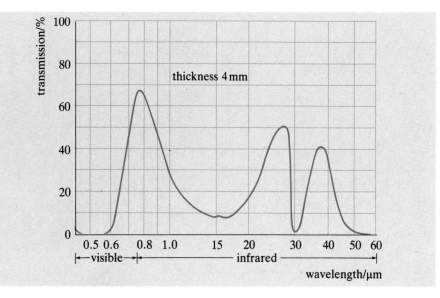

Figure 4.33 Light transmission through a 4 mm-thick glass ceramic as a function of wavelength

absorption coefficient. Physically, k is the reciprocal of the thickness that transmits the fraction 1/e of the incident radiation at a given wavelength. Notice that the material referred to in Figure 4.33 transmits best in the infrared part of the spectrum, which is where most of the radiation from a heater at 900 K is concentrated.

4.3.5 An assessment of the glass-ceramic hob

A good appearance appears to be the hob's main selling point and its smooth seamless surface fits in well with a smart modern kitchen. Also, it is easy to clean. The smooth flat one-piece construction ensures that there are no crevices to harbour spillages. Furthermore, the surface is not only a cooker and hot plate, but also a useful worktop when cold.

It is difficult to see a cost-effective alternative to glass ceramic for a seamless hob. The material can be formed at modest temperatures with zero porosity and it can be both hardened and toughened by the crystallization treatment. With no voids present and a very small grain size, it is transparent and therefore permits heat to be transmitted by radiation. Notice the wide range of properties that had to be considered in choosing the material: mechanical, thermal, optical and processing properties.

Compared to conventional hobs, there are some disadvantages. One is cost — they are more expensive than, say, enamelled steel. Another is fragility: it *is* possible to damage ceramic hobs with blows from a sharp pointed object, but this is very rare. A glass-ceramic hob is somewhat slower to boil water than some other electric or gas hobs. Finally, care must be taken not to scratch the hob. Special cleaning agents are necessary because cleaners containing coarse hard scouring powder would abrade and damage the surface quality.

SAQ 4.8   (Objectives 4.4 and 4.5)
Give two reasons why glass ceramics may be preferred to ordinary sintered ceramics for cooking hobs.

SAQ 4.9   (Objective 4.3)
If a glass-ceramic hob measuring 500 mm × 500 mm were heated to a uniform temperature of 900 K and then quenched along one edge by contact with cold water at 300 K, calculate the thermal strain, the possible dimensional change and the largest thermal stress that could occur in the material. Take $\alpha = 1 \times 10^{-6} \, \text{K}^{-1}$ between 300 K and 900 K, Young's modulus, E, to be 92 GN m^{-2} and Poisson's ratio, v, to be 0.24.

Why do manufacturers not use borosilicate glass? Take $\alpha = 3.2 \times 10^{-6} \, \text{K}^{-1}$, $E = 70 \, \text{GN m}^{-2}$, $v = 0.25$ and bending strength $\simeq 70 \, \text{MN m}^{-2}$.

SAQ 4.10 (Revision)
A glass-ceramic hob of length 500 mm, breadth 500 mm and thickness 8 mm is supported along two opposite edges. It is struck in the middle by a large full saucepan weighing 4 kg falling from a height h and fracture occurs from cracks on the underside of the hob. If the bending strength is 100 MN m^{-2} and the Young's modulus, E, is 92 GN m^{-2}, what minimum height h would cause fracture?

(Hint: balance the initial potential energy of the pan against the elastic energy, U, stored in a beam at the instant of fracture. For a simply supported, centrally loaded beam, $U = tbL\sigma^2/18E$, where σ is the maximum stress in the beam and L, b, t are its length, breadth and thickness, respectively.

SAQ 4.11 (Objective 4.4)
Many kitchen cleaning fluids contain alumina powder. Should these be used on a ceramic hob? Justify your answer.

SAQ 4.12 (Objective 4.11)
From Figure 4.33, what is the minimum value of the absorption coefficient, k_{min}, for this glass ceramic and to which value of wavelength does it apply? What is the value of $1/k_{min}$ and what is its significance?

4.4 Hip prostheses

4.4.1 Introduction to orthopaedic implants

Many of us have an elderly relative or friend who has undergone hip-replacement surgery. Generally, the reason for replacing a hip is because diseases such as arthritis have caused so much pain and stiffness that the person can no longer lead a normal life. Over 40 000 hip replacements are performed in Britain each year, so it is becoming a fairly routine surgical procedure.

The natural hip is a ball-and-socket joint, the ball being the head of the femur, the thigh bone, and the socket is the cup in the pelvis called the acetabulum. Arthritis causes the lubricating surfaces, made of cartilage, to degrade and eventually bone bears upon bone. This is the cause of stiffness and pain. In the replacement operation, all the damaged bone and cartilage is removed. The entire natural head of the femur is cut off and replaced by a metal head with a stem that fits down the central cavity of the femur. A stable fixation is a prime consideration and the prosthesis is often cemented into place using an acrylic cement packed into the femoral cavity before the replacement is inserted and then allowed to cure *in situ*. The socket is relined with an ultrahigh molecular-mass polyethylene (UHMPE) which is cemented into place.

This produces an artificial joint of a metal bearing on a polymer, which gives a coefficient of friction of about 0.001, some ten times higher than that of the original cartilage-on-cartilage bearing.

Of the conventional prosthesis, it might be asked, 'Where is the problem?' The main limitation to the use of conventional implants is the age of the patient. There are many people in the population between 30–60 whose lives would be greatly improved by this type of operation. Unfortunately, this is seldom successful. One of the main functional causes of failure is loosening of that part of the artificial hip inserted in the femur. This is aggravated by younger people being physically more active and the extra time the replacement hip needs to be in service. It is also thought that loosening of the new joint occurs if the bone around the implant does not undergo as near normal loading as possible so that it still feels 'needed'. The current thinking is to try to reduce the age of potential patients by improved material selection and component design. What kind of material is required for a better femoral head?

4.4.2 The properties required

The key functional requirements of a replacement hip are disparate and demanding. The ideal hip replacement would give pain-free normal mobility for the lifetime of the patient. The coefficient of friction affects the ease of movement of the joint and its wear rate and therefore it should be as small as possible. Friction also affects the rotational torque that is generated between the prosthesis and the bone, and thus the stability of the implant in the bone.

The joint may undergo fairly severe cyclic loading: for example, it may have to withstand occasional transient loads up to fifty times body weight when landing from a jump and cope with normal walking loads of $2\frac{1}{2}$ times body weight applied 10^6 times a year. In addition, there must be no possibility of catastrophic failure, such as that caused by rapid crack propagation from a crack-like defect.

The internal biological environment is chemically aggressive and any material chosen for long-term implantation must be able to resist these conditions without any reduction in mechanical properties or the release of undesirable degradation products into the body.

Consider the femur to be a tube constructed of a material with a Young's modulus of $20\,\text{GN}\,\text{m}^{-2}$. Since it is filled with marrow, a material that has a very low Young's modulus, the walls of the tube would have to carry virtually all the applied load. However, a rod is inserted down the tube and stress can be transferred from the bone to the rod.

Older rods were made from alloys that had a Young's modulus about ten times bigger than that of bone. The strain in the rod must equal the strain in the surrounding bone, so that the deformations of the rod and bone are compatible, and therefore the rod had to take a stress ten times larger than the stress in the bone. When the implanted bone is relieved of most of its usual loading it becomes thinner due to natural processes and this wasting of the bone is thought to aggravate the loosening of the implant. To reduce this effect, the implant material should therefore have a value of Young's modulus as close to that of bone as possible.

Wear of the prosthesis is another important consideration. Although the conventional metal on plastic prosthesis may wear very little over twenty years in an older patient, the wear rate would be much greater in a more active young person and such a patient requires their hip to last maybe fifty years. As you saw in Chapter 2, the potential mechanisms of wear on sliding surfaces are adhesion and abrasion, with possible contributions from chemical effects involving liquid lubricants. To minimize adhesion, it is generally advisable to use two dissimilar materials that differ significantly in hardness. Since the polyethylene used for the socket is soft, the ball should be hard. To reduce wear by abrasion, the harder surface, the ball, should have a good surface finish. To eliminate chemical wear, the implant should be unaffected by body fluid.

Finally, it is evident that the femoral part of the prosthesis, the stem of which in use is subjected to bending, must be tough in order to resist the occasional large bending loads without cracking.

You are looking for a material with the tensile strength, toughness and fatigue strength of stainless steel, but with a smaller Young's modulus, a higher hardness and good chemical stability in body fluid. Unfortunately, some of these requirements, for example, to reduce Young's modulus while raising the hardness, are mutually exclusive, so no single material can be found to meet all these needs. Combinations of two different materials — the 'composite approach' — may, however, be the answer. Toughness and fatigue strength are required by the stem, while hardness is required by the sliding surface of the ball: perhaps the stem and ball should be made of different materials.

4.4.3 A modular head

The conventional hip replacement comprises a cup made from a polyethylene with an ultrahigh molecular mass and a titanium alloy (Ti–6%Al–4%V) head. This is a highly passive alloy which has a very high degree of resistance to corrosion. At one time the heads were made from either stainless steel (18%Cr) or cobalt–chromium alloys, both of which have high strength and excellent fatigue properties combined with fair corrosion resistance. We shall examine the trend to titanium since it influenced the development of a whole new generation of hips with ceramic heads.

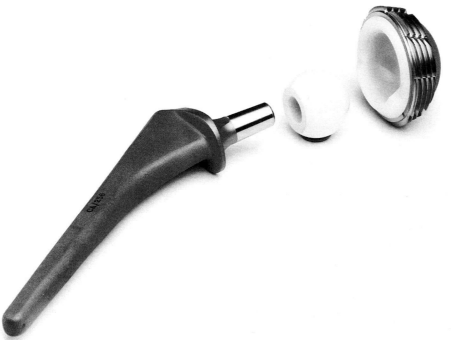

Figure 4.34 A modern hip replacement with a ceramic head

Titanium alloy has a Young's modulus of about five times that of the cortical bone of the femur. If this is compared with the much higher Young's moduli of the alloys used earlier, it would appear to be beneficial to use titanium alloys for more effective stress transfer into the bone, while maintaining adequate strength and fatigue properties. The release of metal ions from titanium alloys is lower than that of stainless steel or cobalt–chromium alloys and this may also be an advantage in the long term. Like all things, however, there is still a trade-off. Titanium alloys are not as hard as the other alloys and so the sliding surface would wear quicker. Ceramics would provide a better surface. Their high hardness and wear resistance, low friction and density, and good chemical stability are favourable properties. Their low toughness and tensile strength, high Young's modulus and large scatter in properties would, however, rule out their use for the stem.

The new generation of hips for long-term use in younger patients, have ceramic heads made of high-density alumina, which are attached to a titanium alloy stem (Figure 4.34). When this hip development work started in West Germany, alumina was already being used in dental porcelains. It was therefore known to be safe for implantation and so development centered on high-density aluminas (see ▼Alumina▲). The design standard ISO 6474:1981 requires the material to be capable of enduring a lifetime of thirty years, loaded one to two million times per year. These stringent requirements are met by the careful design and control of the processing conditions, which affect the microstructure and hence the properties of the final component.

▼ Alumina ▲

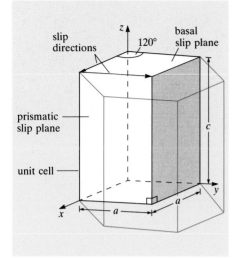

Figure 4.35 An hexagonal structure cell showing two slip systems

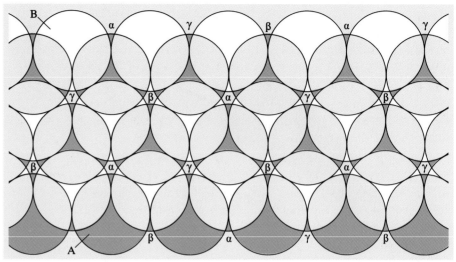

Figure 4.36 A plan view of the close-packed layer of oxygen ions in alumina. There are three sets of octahedral spaces (α, β, γ) just above the layer

Alumina contains two elements with different valencies: aluminium has a valency of three while that of oxygen is two. These elements also differ significantly in their electronegativities (Al 1.5, O 3.5), so they are expected to show some appreciable ionic bonding. (According to Table 4.2, you would expect the bonding to be about 50% ionic). When ions are formed, electrons are transferred from aluminium to oxygen. Each aluminium atom loses three electrons and each oxygen atom gains two, producing Al^{3+} and O^{2-} ions. It follows that in alumina, Al_2O_3, there must be three atoms of oxygen for every two of aluminium.

The structure of alumina is based on a hexagonal cell (Figure 4.35). If you imagine oxygen ions to be on the cell corners and at the point in the cell with coordinates ($2a/3$, $a/3$, $c/2$), where could you put the aluminium ions?

The ratio of the ionic radii, r_{Al}/r_O, is 0.31 and is less than the ratio of radii for the octahedral hole, 0.414, which suggests that

the aluminium ions might prefer the octahedral interstitial spaces to the smaller tetrahedral ones. This is the case and each aluminium ion occupies an octahedral site, where it is surrounded by six equidistant oxygen ions. The oxygen ions therefore have a one-sixth share in each aluminium ion. It follows that an oxygen ion needs four nearest aluminium ions to balance its charge (since $4 \times 3 \times 1/6 = 2$). The octahedral sites have this property.

The hexagonal cell 'owns' two oxygen ions — an eighth 'share' of each of the eight corner ions and a whole share of the ion within the cell — and two octahedral holes — at the points ($a/3$, $2a/3$, $c/4$) and ($a/3$, $2a/3$, $3c/4$), so only two-thirds of these spaces can contain aluminium ions if the formula Al_2O_3 is to be obeyed.

The structure is simply described by locating the one-third of the octahedral sites which are empty. Figure 4.36 shows a plan view of the close-packed layers of oxygen ions. Between adjacent layers, there are three 'close-packed' sets of octahedral sites, labelled α, β and γ by

analogy with the oxygen layers A, B, etc. One of these sets is unoccupied. Suppose that between A and B it is the set α, then in the next parallel layer of octahedral spaces, between B and A, the β set will be empty and in the next layer, between A and B, the set γ will be empty. The layers of octahedral spaces are unoccupied in the sequence α, β, γ, α, ... The overall stacking sequence of oxygen and metal ions is $A\alpha B\beta A\gamma B\alpha A$ . . . . Remember that the capital letters represent the positions of the oxygen ions while the Greek letters denote octahedral sites *unoccupied* by aluminium ions.

At temperatures below 1200 K, slip is restricted to the basal and prismatic slip planes shown in Figure 4.35. These do not provide five independent slip systems so polycrystalline alumina is not ductile at these temperatures. However, at higher temperatures, there is sufficient thermal stimulation available to make dislocations mobile on an additional slip system and polycrystalline alumina can be forged into complex shapes under these conditions.

Fabrication and microstructure

The manufacture of bulk ceramics usually starts from powdered raw materials and is described in ▼Fabrication▲. Ceramics, like any other polycrystalline material, have a characteristic microstructure, which can be seen by polishing, etching and magnifying (Figure 4.39). The most important features are the grain size and the level of porosity.

Grain size and strength are inextricably linked because the grain size is related to the pore size and the pores constitute defects which control the tensile strength. Reducing the grain size increases both the fracture strength and the toughness (Figures 4.40 and 4.41).

In the manufacture of the replacement hips the grain size of the ceramic head is therefore made as small as possible. This is done by starting with a very fine powder and controlling the sintering conditions so that grain growth is not significant. There is a conflict between the conditions required for sintering (high temperatures, long time) and those required to avoid grain growth during sintering (low temperatures, short time). A compromise is reached where the manufacturers use ultrapure (99.9%) alumina and start with a small particle size of less than 1 µm. After processing this gives a ceramic with a mean grain size that is less than 8 µm, typically 2 µm, and a porosity of less than 2%.

The sintering process is improved by the use of 0.05–0.2% magnesia, which prevents grain growth by segregating to the alumina grain boundaries — 'pinning' them. Hot pressing at 1350–1800 K may also be used where the energy supplied by pressing will enhance sintering, so the ceramic can be held for a shorter time at the sintering temperature of 1850–2000 K.

Manufacture of an alumina head

The very fine alumina powders are dispersed in a solution of a detergent-like deflocculant (see 'Gels', Chapter 8). This contains cations which are transferred to the surface of the particles causing them to repel each other. When a very small amount of organic binder, such as polyvinyl alcohol, is introduced to the suspension, every individual particle is then coated. The minimum binder is used to reduce both contamination of the product and the creation of void space when the binder is burned away. The suspension is spray-dried prior to pressing to shape. The binder is used to help the pressing process and to improve the green strength of the compact. These types of heads are usually isostatically pressed by encasing the powder mix in a flexible rubber bag and compressing through a pressurized fluid at 500 MN m^{-2}. The ball is then machined in its green state and a cylindrical hole introduced to accommodate the metal stem (Figure 4.34). Sintering is usually carried out in a vacuum or under hydrogen so that entrapped air and porosity are minimized. The sphericity of the ball is quite critical and should be correct to within a few micrometres. It is totally unacceptable for shrinkage, which may be as much as 20%, to be anisotropic and extremely careful control is needed during drying and sintering to avoid

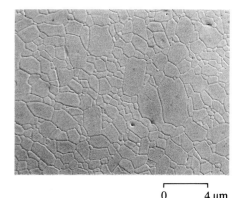

Figure 4.39 The microstructure of fine-grained 99.9% alumina

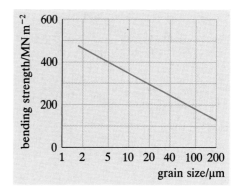

Figure 4.40 The dependence of the strength of alumina on grain size

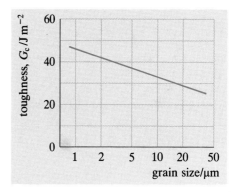

Figure 4.41 The dependence of the toughness of alumina on grain size

▼Fabrication▲

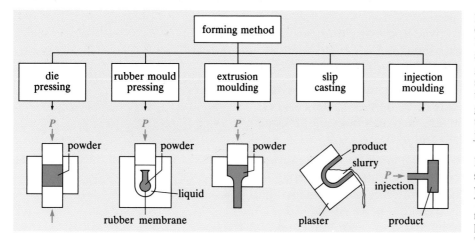

Figure 4.37 Processes for the fabrication of ceramics

The vast majority of ceramics are manufactured from fine powders. Some final grinding, blending or mixing is often carried out prior to the shaping process itself. Usually the shaping is achieved at low temperatures using one of the variety of forming methods described below (see Figure 4.37) and then the porous preform, sometimes known as a 'green' body, is brought to a higher density by the action of heat in the sintering process. Full compaction to the theoretical density is often not achieved and the product then contains porosity, which is usually expressed as a percentage of the total volume.

Die pressing is by far the most common fabrication process for small ceramic components and high production rates of up to 5000 pieces per minute can be achieved. Sizes can range from about 0.1 mm to 100 mm, with reasonably good dimensional tolerance on the final product ($\pm 1\%$). The main limitations are the complexity of the shape that can be achieved and the uniformity of the product's density.

Isostatic pressing overcomes the problem of achieving a uniform density by using a flexible membrane to transmit hydrostatic pressure from a compressed fluid to the component, thereby ensuring uniform compaction throughout the body. This process is best used for parts with cylindrical symmetry, such as spark plug insulators or tubes. Rates of production of around 1500 pieces per hour are possible. Quite often some final machining of the 'green' components will be required because of the imprecise control of dimensions, which arises from using a compressible membrane.

Extrusion through a suitable die is an excellent method of producing components with a constant cross-section, provided a batch material with adequate plasticity is available. For clay-based ceramics, the plasticity is achieved simply by controlling the amount of water present. For powders, suitable organic plasticizing agents are used to provide the correct consistency.

Injection moulding is a technique borrowed from the plastics industry. The batch used contains up to 60–70% by volume of fine ceramic powder in a polymeric matrix. The mix is heated to a plastic state and is then injected into a mould under pressure where it cools and sets prior to extraction. The next, delicate step is to eliminate the polymeric material without damaging or disrupting the preform. Long controlled burn-out treatments of, say, 36–48 hours are often required. The method is very good for mass production of complicated shapes, provided the maximum wall thickness of the component is not too great and the initial high capital cost of the injection machine and dies can be justified.

In **slip casting** the 'slip' is a colloidal suspension of ceramic powder in a liquid, which is usually, but not always, water. The slip is poured into a mould which is microporous and slowly draws out the liquid from the slip by capillary action. The result is that a layer of fairly solid material is built up against the wall over a period of a few hours. The excess slip can then be poured out to leave a hollow component of uniform thickness. The moulds are usually made of plaster of Paris (hydrated calcium sulphate) and are relatively cheap and easy to make. The process is slow, labour intensive and lacks precision, but is nevertheless used for some engineering components.

All the above processes are followed by drying and then a sintering process at an appropriate temperature, often around $0.7T_m$, to bring about densification. The temperature required is high because sintering occurs by solid-state diffusion, either along the external surfaces or through the crystalline interior. In the latter case, crystal defects such as grain boundaries are preferred paths. Sometimes impurities in the batch or deliberate additions (fluxes) can speed up the sintering processes or allow lower-than-normal temperatures to be used.

Liquid-phase sintering can occur at lower temperatures than solid phase sintering when a liquid phase is present to the extent of 2–20% by volume. The formation of this phase depends on having appropriate additives in the powder mix. In this mode of sintering, dissolution of

196

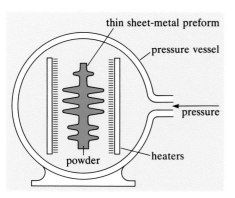

Figure 4.38 A hot isostatic press

atoms from the powder into the liquid occurs, followed by reprecipitation.

In **reaction sintering** the densification is helped by a chemical reaction that takes place, usually between a solid powder and a liquid or gas. The commonest examples are the production of silicon nitride by firing a silicon powder compact in nitrogen gas and the production of silicon carbide by reacting graphite with liquid silicon in the presence of some silicon carbide particles.

Sometimes both shaping and sintering are done simultaneously by die pressing at high temperatures, known as **hot pressing**. It is a slow and expensive process capable of producing only simple shapes of limited dimensions. A related process is known as **hot isostatic pressing**, in which a batch of the order of 10–100 ceramic components can be densified simultaneously in one high-temperature, high-pressure gas chamber (Figure 4.38). The gas, usually nitrogen or argon, is the medium by which pressure is transmitted to the preshaped, partially densified components. The economics of this process are much better than those of hot die pressing, but nevertheless it is used only for products of high value, such as ceramic cutting tools or turbocharger rotors.

this. After firing, the balls are put through a stringent polishing procedure using various grades of diamond powder. Typically a surface finish of less than 0.02 μm is achieved, thanks to the low porosity of the material. This is the finish that is also given to comparable metal heads, but with its higher hardness (see Table 4.3) alumina is found to give a seven times improvement in wear life.

In summary, the processing of alumina implants is carried out in such a way as to minimize the grain size and porosity, thereby enhancing the strength, toughness and surface finish of the product.

4.4.4 Prospects

Are these implants proving to be adequate for the younger patient? Not yet — engineers still have not overcome the problems of quality control of the component, and implants show a wide variation in performance. At the present time, however, the hip prosthesis with a ceramic head is the best solution available and its use is increasing steadily. With further development work based on extensive service experience, the prospects appear good for the service life to increase to that required for the thirty-year-old patient.

SAQ 4.13   (Objective 4.5)
How would you expect the bending strength, σ_b, of alumina to vary with the grain size, d? State clearly any assumptions made.

SAQ 4.14   (Objective 4.6)
Describe the fabrication method you would use to make a femoral head from alumina powder. Identify *two* microstructural features that must be controlled to within strict limits and briefly explain why.

SAQ 4.15   (Objectives 4.7 and 4.8)
By considering the active slip systems, decide whether a crystal of alumina at 1100 K can undergo the following changes of shape by plastic deformation:
(a) a change in length parallel to the *c*-axis;
(b) a change of shape within the basal plane.

4.5 Turbocharger rotors

4.5.1 Introduction

Currently, a major goal of ceramicists is to develop structural ceramics for use in heat engines. The following case is one of many steps in this direction.

The diesel engine was invented around 1890. By 1910, some diesel engines normally rated at 500 bhp were made to develop 1000 bhp for short periods by supplying air to their cylinders at an excess pressure of one atmosphere. Towards the end of World War I, military aircraft were being fitted with supercharged petrol engines containing blower turbines driven by hot exhaust gases. This effectively counteracted the loss in engine power that occurred in naturally aspirated engines due to the drop in atmospheric pressure with increasing altitude. These developments were the forerunners of turbocharged engines.

With the increases in fuel prices that occurred in the 1970s, there arose a greater awareness of the performance and economic advantages to be had from turbocharging small diesel and petrol engines. The advantages include increased engine power and torque, reduced exhaust pollutants, less exhaust noise and greater fuel economy. One major drawback with the existing turbochargers is 'turbo lag', the sluggish response of a small engine when the accelerator is depressed. The inertia of the rotating component, the rotor, plays a significant role in this effect and the replacement of a nickel-base superalloy rotor, of density $7.9\,Mg\,m^{-3}$, by a ceramic rotor with a density of about $3\,Mg\,m^{-3}$ would provide a significant reduction in the inertia of the system, as shown in Figure 4.42.

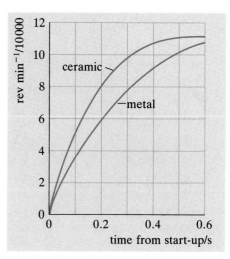

Figure 4.42 Step response characteristics of metal and ceramic rotors

4.5.2 What *is* a turbocharger?

Figure 4.43 shows a section through the moving parts of a turbocharger. A shaft connects an air-compressor wheel at one end to a high-temperature rotor at the other. The compressor has the function of feeding air to the engine at a pressure up to 3.5 times higher than atmospheric. The compressor wheel is usually made of a high-strength aluminium alloy which has adequate strength up to 520 K, the maximum temperature it should ever experience. The rotor is driven at high speed by the exhaust gases from the engine. The temperature of these gases from petrol engines can reach 1250 K on occasion, but the somewhat lower temperature of 1000 K is more typical of the exhaust gases issuing from a diesel engine. The device rotates at speeds up to $260\,000\,rev\,min^{-1}$ and therefore high-strength, creep-resistant materials are required to withstand the centripetal forces set up in the rotor (see ▼Centripetal stress▲). Nickel-based superalloys are conventionally used for this particular component (see Table 4.5).

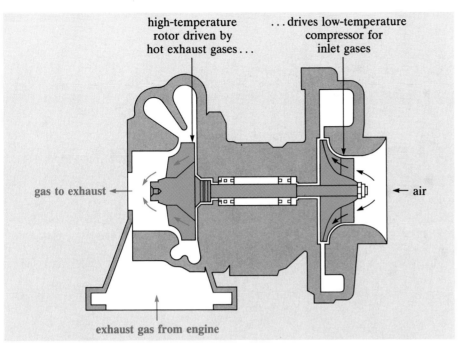

Figure 4.43 Internal construction of a turbocharger

Table 4.5   Composition by wt % of a typical nickel-based rotor alloy

alloy	C	Cr	Mo	Nb	Ti	Al	Zr	B
IN 713C	0.12	12.5	4.2	2.0	0.8	6.1	0.10	0.012

There are usually nine identical vanes on each of these wheels. The vanes on the compressor and rotor have complex three-dimensional shapes, which have been developed empirically over the years. The rotating shaft is located in lubricated bearings and complex scroll-shaped housings lead the air and exhaust gases to and from the wheels.

The designed life of a turbocharger for a petrol-engined vehicle is normally around 150 000 km or 5000 hours with little or no maintenance required during that period.

4.5.3 Suitable ceramics

We have already seen that to reduce turbo lag, a rotor ceramic should have a density less than that of the superalloy — the smaller, the better. There is a further benefit of a low density in that it reduces the centripetal stresses, σ_c, developed in blades. If the length of the blade is l and it is rotating at a given angular speed, ω, the expression for maximum centripetal stress is given in Equation (4.11)

$$\sigma_{max} = \rho \omega^2 l \left(R + \frac{l}{2} \right)$$

Failure will occur when this becomes equal to the strength, σ_f, of the

▼Centripetal stress▲

In order for you to corner in your car, a force towards the centre of the bend is required to enable you to change direction. It is applied *by* the road *to* the tyres and is due to the friction between them. This is a **centripetal force** and such forces are involved whenever a moving mass changes direction: for example, when a rotor spins.

We shall consider a simplified model of a rotor blade in order to deduce the factors that affect the centripetal forces acting and hence the stresses to which the blade is subjected. Consider the blade to be a straight-edged bar, ignoring its twist and the change of aerofoil section along its length. You want to calculate the stress at any point in the blade, which will be due to the motion of all the mass which lies further out along the blade.

The centripetal force due to the rotation of a mass m at angular velocity ω and at radius r is $m\omega^2 r$. If this force is supported normally by material of cross-sectional area a, then the tensile stress across the section will be

$$\sigma_c = \frac{m\omega^2 r}{a}$$

The problem is to use this formula for a distributed mass.

Figure 4.44 defines the symbols we are going to use. We shall let ρ be the density of the material used and a be the constant cross-sectional area of the blade. The mass of an element that is dx thick is $\rho a\, dx$ (density × volume) and it is rotating at radius $(R + x)$. The centripetal force acting on this element is $\rho a\, dx \omega^2 (R + x)$. The maximum force due to all such elements acts at the root and is

$$F_{max} = \int_0^l \rho \omega^2 (R + x) a\, dx$$

$$= \rho \omega^2 a \int_0^l (R + x)\, dx$$

$$= \rho \omega^2 a \left[Rx + \frac{x^2}{2} \right]_0^l$$

$$= \rho \omega^2 a l \left(R + \frac{l}{2} \right)$$

This force is carried by the cross-section of area a, so the maximum centripetal stress, at the root, is

$$\sigma_{max} = \frac{F_{max}}{a} = \rho \omega^2 l \left(R + \frac{l}{2} \right) \quad (4.11)$$

Note that its value depends on the density ρ of the material, the square of the angular velocity ω and the dimensions R and l. The variation in stress along the blade is shown in Figure 4.45.

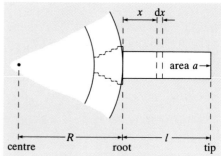

Figure 4.44 A model rotor blade

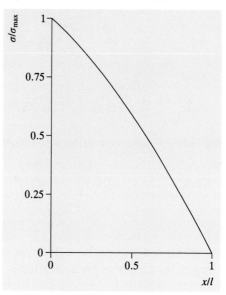

Figure 4.45 The variation in centripetal stress along the span of a blade with $R/l = 1.75$

material: that is, when

$$\sigma_f = \rho \omega^2 l \left(R + \frac{l}{2} \right)$$

For the case of fixed dimensions R and l, the maximum rotational speed, ω_{max}, that the rotor can have without breaking depends on the specific strength and is given by the expression

$$\omega_{max} \propto \sqrt{(\sigma_f/\rho)} \quad (4.12)$$

This can be used as a merit index in selecting the rotor material.

At start-up and when the throttle is opened or closed quickly, the rotor will be subjected to sudden changes of heat transfer with the attendant risk of thermal shock. You saw earlier how the resistance to thermal shock can be represented by three different parameters (R, R' and R''), depending on the condition that is kept constant.

In addition, the rotor material must be chemically stable and must not creep at the maximum operating temperature (1200 K). Table 4.3 contains properties for a range of four structural ceramics, all of which meet the temperature requirement. The oxides are obviously stable in oxygen, but so too are Si_3N_4 and SiC, due to the formation of a thin protective layer of silica on the surface, like the layer on a passive metal. Although these ceramics gain weight at 1270 K due to oxidation, it is small, of the order of $10\,g\,m^{-2}$ of the surface in 100 hours.

With these criteria in mind, consider which of these four ceramics would be the most appropriate for making a ceramic rotor.

> **SAQ 4.16** (Objective 4.4)
> Use the data in Table 4.3 to select the two most suitable ceramic materials for a turbocharger rotor. Assume that the values of given strength apply at the operating temperature of the rotor.

4.5.4 First steps with ceramics

In the 1970s, the Garrett Company in the USA carried out extensive testing of engineering ceramics for turbocharger rotors. The only ceramic materials with promising thermal and mechanical characteristics that were then available for production in these complex forms were reaction-bonded silicon nitride (RBSN) and reaction-sintered silicon carbide (RSSC). Rotors of 74 mm diameter were tested and found to burst at rotational speeds of less than half that required.

By 1982, progress had been made in improving the strength and reproducibility of these materials. The breakthrough was the identification of sintering additives and the development of pressureless sintering. The problem with the earlier materials had been the large variations in strength from one sample to another. The weakest of them failed at well below the average value. Ways have been devised of describing the variability of data (see ▼Scatter▲) and of extrapolating data in order to find stress levels with high probabilities of survival. In crude terms, the average tensile strength of the reaction bonded products had been doubled and the scatter in strength measurements had been halved.

These developments enabled the launch by Nissan in November 1985 of the *Fairlady Z* sports car which incorporated a ceramic turbocharger rotor (Figure 4.47). The ceramic chosen by Nissan was sintered silicon nitride (see ▼Silicon nitride▲). The properties of this material are compared to those of a nickel rotor alloy in Table 4.7. Based on *average* values of strength at the operating temperature, the specific strength of the ceramic is four times higher than that of the superalloy, but the scatter in strength values is larger in the ceramic. Widespread use of structural ceramics depends largely on reducing this scatter by careful control of the processing route and, hence, of the microstructure.

Figure 4.47 The Nissan *Fairlady Z* coupé

▼Scatter▲

The strength of a glass or ceramic varies significantly, even for supposedly identical test pieces, and this poses a serious problem for designers when using these materials to carry loads. Is there a minimum value of strength that can be assumed to apply and, if so, what is it?

In order to design with materials which show large variations in strength, you have to use a statistical approach. You can only say that a component of given size and shape made from a given material has a 99% chance, or whatever probability you choose, of having an actual strength greater than a given value. You may then use this value of strength in design and accept a small probability of failure.

This approach requires you to be able to extrapolate test results to large probabilities of survival and to do this you need a mathematical model or an equation that describes the observed scatter of strengths.

Table 4.6 shows a set of fracture strengths for test pieces of 95% alumina obtained by four-point bending carried out at room temperature. The values vary from 230 to 363 MN m^{-2} and can be described roughly in terms of the mean value, $\bar{\sigma}$ (304 MN m^{-2}) and the standard deviation, s (36 MN m^{-2}). The standard deviation (or root mean square deviation) describes the spread of the values and is defined as

$$s = \sqrt{\frac{\sum_i (\sigma_i - \bar{\sigma})^2}{n}} \qquad (4.13)$$

where n is the total number of measurements and σ_i is the ith value of the measured strengths ($1 < i < n$). For ceramics, the standard deviation is usually about 5–15% of the mean value, as in this case.

A model used widely to describe the scatter in the properties of ceramics is the Weibull distribution. First, the data are put into descending order with the largest value having rank $j = 1$ and the next largest $j = 2$, etc. A probability of survival, S_j, is then assigned to each value

of strength, where

$$S_j = \frac{j}{n+1} \qquad (4.14)$$

Again, n is the total number of values, which is twenty for the data in Table 4.6. For the highest value there is a small, but finite, chance of survival (1/21) and for the lowest value the chance is large (20/21), but this is not a certainty, because if more tests had been done, even smaller values may have been found. In this way, each measured strength, σ_j, is assigned a value of S_j.

To obtain a Weibull plot, ln ln (1/S_j) is plotted against ln σ_j, as shown in Figure 4.46. The plot produces a straight line, the slope of which, m, provides a measure of the scatter of the strengths and is called the **Weibull's modulus**. When m is large, the scatter is small and when m is small,

the scatter is large. For the data in Table 4.6, $m = 8.8$ and for the yield strength of a ductile metal, values of m of the order of 50 would be expected.

This plot indicates that the relation between S_j and σ_j can be written

$$S = \exp - (\sigma/\sigma_0)^m \qquad (4.15)$$

where σ_0 is the value of σ for which $S = 1/e$ ($\sim 37\%$). For the data in Figure 4.47, $\sigma_0 \simeq 320$ MN m^{-2}, a little higher than the mean value.

WORKED EXAMPLE According to Weibull's model, what is the probability that a test piece of the type referred to in Table 4.6 will survive a stress of 270 MN m^{-2}?

The probability of survival is found by putting the relevant values into Equation

Table 4.6   Fracture strengths, in MN m^{-2}, of a batch of alumina test pieces

230.5	243.0	257.5	273.5	274.5	283.0	287.0	296.3	303.7	306.0
313.0	315.2	315.6	315.7	320.0	333.5	341.6	351.6	353.0	362.6

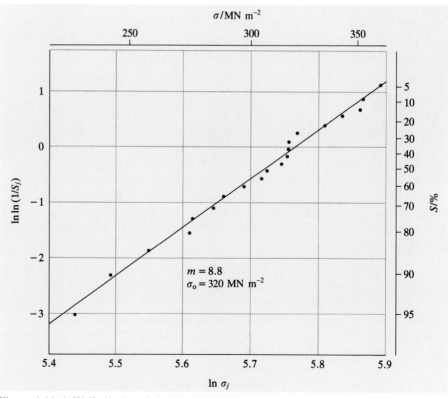

Figure 4.46 A Weibull plot of the data in Table 4.6

(4.15). In this case, $\sigma = 270\,\mathrm{MN\,m^{-2}}$, $\sigma_0 = 320\,\mathrm{MN\,m^{-2}}$ and $m = 8.8$. The probability is therefore

$$S = \exp - \left(\frac{270}{320}\right)^{8.8}$$

$$= 0.80 \text{ or } 80\%$$

SAQ 4.17   (Objective 4.9)
For what value of stress will there be a 96% probability of survival, according to Weibull's model? Assume the test piece is of the type referred to in Table 4.6.

▼Silicon nitride▲

Silicon nitride, Si_3N_4, is a crystalline ceramic and it has much in common with silica in that it is based on the packing of tetrahedra. Each tetrahedron has a silicon atom at the centre and a nitrogen atom at each corner, just like the SiO_4 tetrahedra in silica. Where the two ceramics differ is in the way that neighbouring tetrahedra join up. In silica, an oxygen atom is shared by *two* tetrahedra, whereas in silicon nitride a nitrogen atom is shared by *three* tetrahedra. The structure is hexagonal: that is, the unit cell is shaped like that shown in Figure 4.35. Within each cell, the tetrahedra are linked together in a ring as shown in Figure 4.48. This structure is strong and stable and is the only structure in which this composition is found.

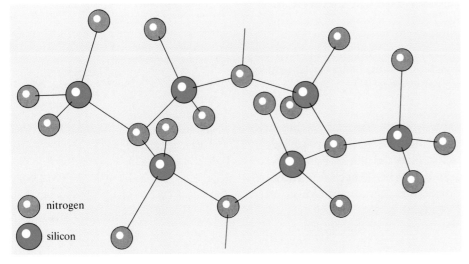

nitrogen

silicon

Figure 4.48 The structure of silicon nitride, Si_3N_4

Table 4.7   Turbocharger rotor materials

Property	Units	Silicon nitride	Nickel alloy
density, ρ	$\mathrm{Mg\,m^{-3}}$	3.0	8.0
fracture stress, σ_p (average at 1100 K)	$\mathrm{MN\,m^{-2}}$	300	200
Weibull's modulus, m	–	47	> 50
Young's modulus, E	$\mathrm{GN\,m^{-2}}$	230	250
toughness, G_c	$\mathrm{J\,m^{-2}}$	175	> 30 000
coefficient of thermal expansion between 0–1300 K	$\mathrm{\mu\,K^{-1}}$	3.3	16.4

4.5.5 Processing routes

The production of a 'green' body or preform of the required shape calls for slip casting or injection moulding.

A modified version of slip casting is used to form the rotor shape. A negative wax mould of the rotor is made and mounted on a plaster base. A slip of ceramic powder is poured in and the assembly is rotated at about 100 rev min^{-1} to speed up the capillary action of the plaster. Once cast, the wax mould is removed by a solvent and the casting is dried prior to sintering.

The process of injection moulding requires less careful skills, but its major disadvantage is the capital cost of the injection-moulding machine and dies, which can be of the order of £$\frac{1}{4}$ million and £30,000, respectively. Figure 4.49 shows the sequence of steps for the production of rotors by injection moulding. There are a great many material variables involved in optimizing this process: volume fraction of ceramic powder, particle-size distribution, choice of resin, the relationship between resin viscosity and temperature, resin burn-out characteristics. After moulding, the removal of the resin requires a very slow heating cycle in a low-pressure oxidizing enivronment. It is sometimes an advantage to have a mixture of resins present to avoid sudden burn-off within a narrow temperature range. Traces of residual carbon should be avoided in the case of silicon nitride preforms since these will be detrimental to the sintering process.

Once the dried 'green' shapes have passed the inspection stage they are ready for sintering. The temperature, time and environment required will depend upon the details of the ceramic material itself. Table 4.8 indicates the approximate sintering conditions for a range of materials. High-pressure gas sintering of silicon nitride has the advantage of suppressing the volatization of silicon nitride at high temperatures, therefore allowing higher sintering temperatures and shorter times to be used.

Table 4.8   Sintering conditions (24 hours)

Material	Preform	Temperature/K	Atmosphere/environment
RBSN	Si	> 1680	nitrogen
SRBSN	Si + Y_2O_3	> 1680	nitrogen
SSN	Si_3N_4 $+ Y_2O_3 (+ Al_2O_3)$ $+ MgO (+ Al_2O_3)$	> 2070	nitrogen
RSSC	C + SiC	> 1720	liquid silicon
SSC	SiC $+ B + C$	> 2270	argon

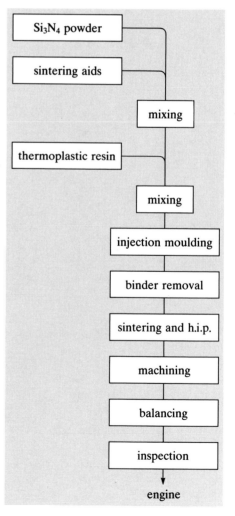

Figure 4.49 Steps for injection moulding

The partially sintered body may be subjected to a final densification treatment by hot isostatic pressing (h.i.p.) in which any isolated internal pores are eliminated by the applied pressure. Any sizeable surface cracks, however, are opened up by the process because the pressurizing gas can get between the crack faces, forcing them apart.

In the case of reaction-bonded silicon nitride and reaction-sintered silicon carbide, very little dimensional change takes place during sintering, but in the other sintering processes, linear shrinkage of anything from 8–25% can occur, depending upon the packing density of the starting materials. Some diamond grinding will therefore be necessary to achieve critical dimensions, after which the rotors are given a nondestructive inspection, which aims to detect strength-reducing cracks and pores.

The option of having an integral rotor and shaft was rejected for reasons of safety and reliability. If a ceramic shaft were ever to break in the region of the oil-lubricated bearings, then leakage of oil and fire might result. Thus it was necessary to join a ceramic rotor to a metal shaft. The temperatures encountered at the joint are not more than 570 K. The Nissan solution consists of a simple butt joint, as shown in Figure 4.50 made with a silver-copper brazing alloy. Several interlayer materials with different properties are used, one of which has a low Young's modulus and therefore deforms readily, while another has a thermal expansion coefficient close to that of silicon nitride.

After joining to the shaft, the rotors are subjected to spin tests which check the balance of the rotor and which also serve as 'proof tests'. If the assembly is out of balance then grinding of the material at the appropriate point is carried out. Proof testing involves spinning the rotor at a speed in excess of that which will be encountered in service. The inadequate specimens are broken during the test, thereby ensuring that the survivors all have adequate strength.

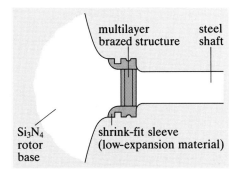

Figure 4.50 Joining the ceramic rotor to a steel shaft

4.5.6 Properties obtained

You have seen that two families of engineering ceramics have been evaluated for turbocharger rotors, one based on silicon nitride and the other based on silicon carbide. In this section we shall compare the properties obtained from the alternative processing routes. Reaction-bonded silicon nitride (RBSN) normally contains open porosity of at least 15% and a small proportion of unreacted silicon. With this high level of porosity, the strength is relatively low and the scatter large. For this reason, development of rotors based on this material was abandoned in about 1978.

An interesting variant of reaction-bonded silicon nitride was developed by the Ford Motor Company in the period 1978–82. The usual silicon

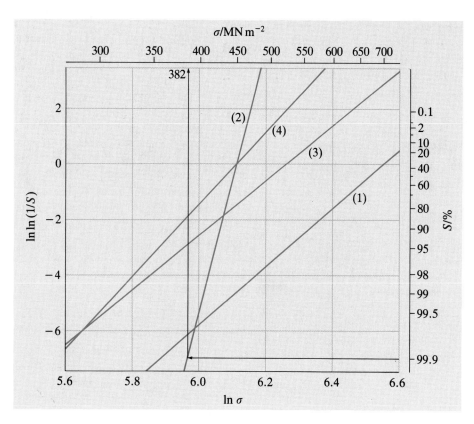

Figure 4.51 The spread of strength values for various ceramics

starting powder was doped with up to 14% by weight of yttria (Y_2O_3). After the normal nitridation treatment, the temperature was raised, whereupon the yttria formed a flux suitable for liquid-phase sintering.

The residual porosity was thereby eliminated. The mean strength of this material, sintered reaction-bonded silicon nitride (SRBSN), was good, but the scatter in strength remained rather high, as is shown by line 1 in Figure 4.51. This approach will no doubt receive further attention in the future.

Sintered silicon nitride (SSN) requires the presence of additives such as Y_2O_3 or MgO to the level of 3–8%, sometimes combined with Al_2O_3, to form silicate fluxes at the grain interfaces. The end product has a good mean strength and a narrow strength distribution (line 2). The silicates form either a glassy or crystalline phase at the grain boundaries and can lead to problems of creep deformation, but only at temperatures above 1570 K, higher than those in a turbocharger.

Reaction-sintered silicon carbide (RSSC) contains a proportion of residual silicon (8–40%, see Figure 4.52), but a little porosity ($< 1\%$) in volumes that were incompletely infiltrated by silicon. The grain size is between 2 μm and 50 μm. The strengths, although better than for reaction-bonded silicon nitride, are only moderately good and the scatter in strength is rather high (see line 3, Figure 4.51).

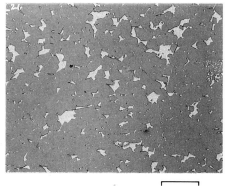

Figure 4.52 Microstructure of reaction-sintered silicon carbide (RSSC)

Finally, sintered silicon carbide (SSC) requires very fine starting powders of silicon carbide and additions of sintering aids such as 1% each of boron and carbon. Strengths are good, but not outstandingly so (see line 4), due to the difficulty of preventing large crystals from developing during sintering (see Figure 4.53).

In the case of engine components, highly reliable systems with fracture probabilities of less than 0.001 are required. Figure 4.51 shows the corresponding strengths for the various materials. The sintered silicon nitride (line 2) has the narrowest distribution of fracture stress and therefore should give the highest strength at a very low failure probability. This is the material used in the Nissan *Fairlady*.

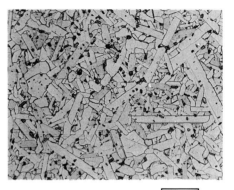

Figure 4.53 Microstructure of sintered silicon carbide (SSC)

> **SAQ 4.18** (Objective 4.4)
> Explain in terms of strength why sintered silicon nitride (SSN) was preferred to the other materials in Figure 4.51 for the rotor of the Nissan *Fairlady*.

> **SAQ 4.19** (Objective 4.10)
> What size of defect is responsible for the fracture of a test piece of sintered silicon nitride (SSN) of the type referred to in Table 4.7, at the average strength level?

4.5.7 Ceramic versus superalloys for turbocharger rotors: a summary

Ceramic rotors offer a number of advantages over the conventional metallic ones: they have a lower inertia, leading to less 'turbocharger lag'; they have smaller clearances and tolerances at blade tips, due to a lower coefficient of thermal expansion; and thus there is improved efficiency and fuel consumption. The lower thermal conductivity of the ceramic rotor keeps exhaust gases at higher temperatures, which improves the effectiveness of the catalytic converters used to clean-up the exhaust. Finally, a ceramic rotor can tolerate higher temperatures without damage due to creep occurring. This will be advantageous when adiabatic (insulated) diesel-engine cylinders come to be used.

Ceramic rotors have some disadvantages, however. Their performance, such as burst speed, is more variable than that of a metal rotor and this poses problems for the designer in achieving high levels of reliability. Another problem for the designer lies in the difficulty of joining a ceramic rotor to a metallic shaft. Finally, the ceramic rotor currently carries a cost penalty. High material costs, low volumes and multistage fabrication techniques all act to reduce their competitiveness. Judging by the large volume of research and development that is going on worldwide into ceramics for engines, there is a widespread belief that these disadvantages can be overcome.

4.6 Looking back

This concludes the three case studies in which you have seen how the attractive property profiles of some ceramics are being exploited. In the cooker hob, glass ceramics are able to transmit heat by radiation and to withstand the temperatures of the 'heater', while providing a high level of hardness and low thermal expansion which brings with it good scratch resistance combined with good thermal shock resistance. In the case of replacement hips, no high temperatures are involved — the environment is thermostatically controlled at 310 K! Here it is the ceramic's good chemical stability, wear resistance and low density that is sought. In contrast, the turbocharger rotor requires high strength and chemical stability at high temperature, good resistance to thermal shock and again a low density. Progress with ceramics in this application will, no doubt, lead to further applications in related products such as gas turbine stators and rotor blades.

You have seen that the typical ceramic can be described as hard, brittle, stable and solid up to high temperatures, with a high thermal conductivity at low temperatures and very low electrical conductivity, although a few semi-conducting ceramics conduct at high temperatures.

The hardness and brittleness are associated with the difficulty in moving narrow dislocations and, in some ceramics in the polycrystalline state, with having less than five independent slip systems. The brittleness is manifest as a low value of toughness (some 1000 times lower than that of ductile metals) and this leads to the strength of ceramics being controlled by very small crack-like defects in the microstructure. The distribution in location, size, shape and orientation of these defects leads to wide variations in the strength of ceramics. This scatter is a major problem in designing reliable load-bearing components made from ceramics.

Ceramics provide good examples of the 'PPP principle' that processing, properties and product design are inextricably linked. With their high melting points, ceramics are processed by powder and powder slurry methods rather than as melts and the main objective of processing, in addition to achieving a desired shape, is to refine the grain size and reduce the porosity, preferably to zero, because this reduces and refines the defect population and gives higher strength with less scatter. Indeed, the future success of structural ceramics will depend very largely on the progress that is made in developing further the methods of processing.

Objectives for Chapter 4

After studying this chapter, you should be able to:

4.1 Sketch the atom positions in the unit cells of magnesia and silicon/ diamond (SAQs 4.1, 4.3).

4.2 Describe how a glass is formed and two ways by which glass can be strengthened (SAQs 4.4, 4.6).

4.3 State the factors that affect the level of thermal stress in a ceramic containing a temperature gradient (SAQs 4.5, 4.6, 4.9).

4.4 Use property profiles in making choices between materials (SAQs 4.5, 4.7, 4.8, 4.11, 4.16, 4.18).

4.5 Explain what porosity is, how it is affected by processing and how it can be controlled. Describe how porosity affects the strength of a ceramic (SAQs 4.8, 4.13).

4.6 Describe five processes for shaping preforms from powders and two processes for densifying the preforms (SAQ 4.14).

4.7 Recognize sketches of the unit cells of the following ceramics: silica (high crystobalite); alumina; silicon carbide; silicon nitride (SAQ 4.15).

4.8 Explain qualitatively the brittleness of a typical polycrystalline ceramic in terms of the von Mises criterion (SAQ 4.15).

4.9 Explain why the strength values of a ceramic usually show a large scatter and show how this can be described by the Weibull distribution (SAQ 4.17).

4.10 Explain why ceramics in bulk, while containing high strength interatomic bonds, usually have low tensile strength (SAQ 4.19).

4.11 Define and use the following terms:
absorption coefficient
centripetal stress
chemical toughening
die pressing
dislocation width
electronegativity
fictive temperature
glass ceramic
glass-transition temperature
Hertzian stress
hot isostatic pressing
hot pressing
internal, residual or self-stresses
liquid-phase sintering
octahedral space
reaction sintering
slip casting
stoichiometry
thermal toughening
working, softening, annealing and strain points

Answers to self-assessment questions

SAQ 4.1 The coordination number is 4 because each atom can be considered to be at the centre of a tetrahedron, surrounded by four atoms, one at each corner.

Since the slip plane is the plane of maximum interplanar spacing it must pass through three nonadjacent corners of the cube, as it does for a face-centred cubic structure. Figure 4.54, which is based on Figure 4.2(a), thus shows the diamond cubic structure cell with one of the slip planes indicated by the shaded triangle.

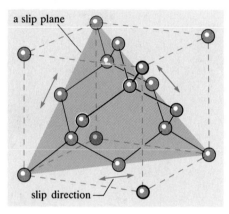

Figure 4.54

Since slip can occur only along a direction in which the atomic spacing is a minimum, within the slip plane there are only three possible slip directions, each one parallel to a side of the triangle defining a slip plane.

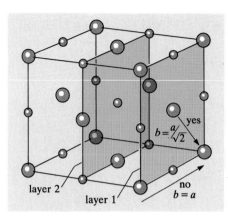

Figure 4.55 Potential slip directions on a cube face in magnesia

SAQ 4.2 The extent of ionic bonding increases with the *difference* in electronegativity between the elements. The electronegativity of oxygen is 3.5 and that of magnesium is 1.2, giving a difference of 2.3. This is greater than that of the H–F bond, which is 44% ionic. The bonding of magnesia is therefore over 44% ionic. (It's actually about 60%.)

SAQ 4.3 Figure 4.55 shows the positions of the ions on the cube faces of magnesia. Slip parallel to a cube edge makes ions of like charge become nearest neighbours, halfway through the slip displacement, b. This difficulty does not apply to slip parallel to the face diagonal, so this is the preferred slip direction.

SAQ 4.4 Generally, the molecular structure of a glass is not in true thermodynamic equilibrium. Rather, it is in a metastable equilibrium, with a structure frozen as it cooled through the glass transition. At room temperature its structure (e.g. its density) is the same as a true equilibrium structure at a higher temperature — the *fictive temperature*.

The fictive temperature increases as the rate of cooling increases, since, as can be seen in Figure 4.17, the greater the rate of cooling, the smaller is the density of the resulting glass.

SAQ 4.5 There are three different measures of the resistance to thermal shock — R, R' and R'', which vary with the heating or cooling conditions. These can be used as merit indices — 'the bigger, the better'! Their values for the three given glasses, derived by taking the appropriate properties from Table 4.3, are tabulated in Table 4.9.

On all three criteria, silica is best followed by borosilicate (e.g. Pyrex).

SAQ 4.6 A material whose coefficient of thermal expansion is zero over the temperature range from the softening point to ambient temperature could not be thermally toughened, since it would contain no residual stress after cooling. Other things being equal, the smaller the coefficient, the better the thermal shock resistance, but the less is the potential for thermal toughening.

SAQ 4.7 The metallic candidate (18/8 stainless steel) can be quickly eliminated; although it is tough enough to resist impact and thermal shock, it is opaque

Table 4.9 Merit indices derived from the data in Table 4.3

	Silica glass	Borosilicate glass	Annealed soda-lime silica glass	Toughened soda-lime silica glass	Glass ceramic	Silicon nitride	Magnesia	Alumina	Silicon carbide	Stainless steel
R/K	1210	234	93	334	909	583	40	93	188	122
$R'/W\,m^{-1}$	2180	351	168	601	2180	7290	996	2310	18750	18300
$R''/m^2\,\mu K\,s^{-1}$	1090	156	69	245	1010	3390	267	599	6660	4540
specific strength/ $N\,m\,g^{-1}$	32	29	28	100	>42	130	48	97	141	76

and therefore depends on the conduction of heat. Its high conductivity would result in the entire hob becoming hot, which would waste heat and pose a safety hazard. Its hardness is also not high and it would become scratched in use.

The glasses are all transparent and therefore allow heat transfer to occur by radiation. Therefore their low thermal conductivities are not a disadvantage. Indeed, they ensure that the parts of the hob away from the heaters remain cool in use. Annealed soda-lime silica glass lacks the required resistance to thermal shock — as you know, pouring boiling water into, say, a cold jam jar will crack it. Thermal toughening improves the merit indices R, R' and R'', but they are still no match for the values of silica or glass-ceramic. Furthermore, with an annealing point of about 750 K, this toughened glass would quickly lose its beneficial self-stresses on contact with the heater, so this material must be rejected. Also, neither the borosilicate nor the soda-lime silica glasses can meet the maximum temperature requirement.

This reduces the choice to silica glass and glass ceramic. Both have good thermal shock resistance and can tolerate the required temperature, 900 K. The glass ceramic, however, has a toughness some ten times higher than silica glass and should therefore prove more resistant to impacts. It is also cheaper and easier to process to shape than silica glass — even at 2300 K, silica glass has a viscosity of about 10^5 N m^{-2} s (see Figure 4.30). Glass ceramic is therefore the preferred choice.

SAQ 4.8

1 There is no porosity in a glass ceramic as there usually is in a sintered ceramic and porosity reduces the strength.

2 The glass ceramic is transparent to the radiation from the heaters, whereas a sintered (polycrystalline) ceramic is usually opaque, due to the presence of porosity and grains larger than the wavelength of the radiation.

SAQ 4.9

Thermal strain, ε, is given by the equation

$$\varepsilon = \alpha\Delta T$$

$$= (1 \times 10^{-6}) \times 600$$

$$= 6 \times 10^{-4}$$

Thus the thermal strain is 6×10^{-4}.

Since

$$\varepsilon = \frac{\Delta L}{L}$$

$$\Delta L = (6 \times 10^{-4}) \times 500\,\text{mm}$$

$$= 0.3\,\text{mm}$$

The dimensional change is therefore 0.3 mm.

The largest thermal stress occurs if the thermal strain at the surface is *completely* constrained by the underlying hot material. From Equation (4.4), this thermal stress is given by

$$\sigma = \frac{\alpha E \Delta T}{1 - v}$$

$$= \frac{(1 \times 10^{-6})(92 \times 10^9) \times 600}{1 - 0.24}\,\text{N m}^{-2}$$

$$= 72.6\,\text{MN m}^{-2}$$

This is less than the bending strength (see Table 4.3).

For borosilicate glass, the corresponding thermal stress would be

$$\sigma = \frac{(3.2 \times 10^{-6})(70 \times 10^9) \times 600}{1 - 0.25}\,\text{N m}^{-2}$$

$$= 179\,\text{MN m}^{-2}$$

This greatly exceeds the bending strength given in Table 4.3 (70 MN m^{-2}) and therefore a borosilicate hob may crack from thermal shock. Furthermore, the hob's operating temperature, 900 K, exceeds the material's temperature limit of 720 K, also given in Table 4.3.

SAQ 4.10

For fracture to occur, the initial potential energy of the weight must be equal to or greater than the energy stored by the hob when the maximum surface stress is the fracture stress, σ_f.

$$mgh \geqslant tbL\sigma_f^2/18E$$

$$4 \times 9.8 \times h \geqslant \frac{(0.008)(0.5)^2(100 \times 10^6)^2}{18(92 \times 10^9)}$$

$$h \geqslant 0.30\,\text{m}$$

The pan would have to fall through a distance of at least 0.30 m to fracture the hob.

SAQ 4.11

No! From Table 4.3, the hardness of alumina (1900 H_V) greatly exceeds that of a glass ceramic hob (600 H_V) and therefore would scratch the hob. A suitable cleaning fluid would contain a softer abrasive such as magnesia, MgO (600 H_V).

SAQ 4.12

The minimum absorption coefficient corresponds to the *maximum* transmitted intensity which is 67% (i.e. $I = 0.67I_i$) and occurs at a wavelength of 0.75 µm, which is on the borderline between the visible and infrared spectra. Equation (4.10) gives

$$I = I_i \exp\{-kt\}$$

If you rearrange this equation and take natural logarithms, you get

$$k_{min} = \frac{1}{t} \ln\left(\frac{I_i}{I}\right)_{max}$$

$$= \frac{1}{4} \ln\left(\frac{1}{0.67}\right)$$

$$= 0.100\,\text{mm}^{-1}$$

The reciprocal ($1/k_{min} = 10\,\text{mm}$) is the thickness that would transmit the fraction $1/e$ (about 37%) of the radiation with wavelength 0.75 µm.

SAQ 4.13

First you must assume that the maximum length of crack-like defects, $2a$, is approximately the same as the grain size, d. These cracks will become critical when the condition given by Equation (2.22) is fulfilled: that is

$$a_c = \frac{G_c E}{\pi\sigma^2} = d/2$$

From this, you can write

$$\sigma = \sqrt{\left(\frac{2G_c E}{\pi d}\right)}$$

Thus the bending strength, σ_b, varies inversely with $\sqrt{d}$. This is the type of dependence actually found and is shown in Figure 4.40.

SAQ 4.14 Submicrometre alumina powder → isostatically pressed to a sphere → machined → sintered → polished.

Two microstructural features have to be controlled: (a) the volume fraction of the pores; (b) grain size. Both should be as small as possible for minimum defect size and, hence, maximum strength.

SAQ 4.15 (a) No. Slip causes no change of length in directions either parallel to or perpendicular to the slip plane. Think of shearing a pack of cards; the thickness of the pack does not change and the size of each card does not change when the pack is sheared. The direction specified (the c-axis) is perpendicular to the basal plane and parallel to the prismatic planes, so both basal and prismatic slip systems are unable to cause changes of length along the c-axis.
(b) Yes. Although slip on the basal planes causes no change in shape of the basal planes (just as no card suffers a change of shape in the sheared pack), slip on a prismatic plane causes a change of shape (a shear) of the basal plane.

SAQ 4.16 If you consider resistance to thermal shock, each of the three merit indices R, R' and R'' (see Table 4.9) puts silicon nitride and silicon carbide either first or second, so these two materials are clearly the best from this point of view, largely on account of their small coefficients of thermal expansion. You also need the materials with the larger values of specific strength (σ_f/ρ). On this ground, too, the values for silicon nitride and silicon carbide (130 and 140 N m g^{-1}, respectively) are superior to those of magnesia and alumina (48 and 97 N m g^{-1}, respectively).

Silicon nitride and silicon carbide are the most suitable materials.

SAQ 4.17 You know from Equation (4.15) that the probability of survival is

$$S = \exp -(\sigma/\sigma_0)^m$$

The values given are $S = 0.96$, $\sigma_0 = 320$ MN m^{-2} and $m = 8.8$, so you can write

$$0.96 = \exp -\left(\frac{\sigma}{320}\right)^{8.8}$$

Taking natural logarithms

$$\ln 0.96 = -\left(\frac{\sigma}{320}\right)^{8.8}$$

Rearranging this gives

$$\frac{\sigma}{320} = (-\ln 0.96)^{1/8.8}$$

and therefore

$$\sigma = 222.5 \text{ MN m}^{-2}$$

This is the stress level for which there is a 96% chance of survival or a 4% probability of failure.

SAQ 4.18 Although sintered silicon nitride does not have the largest value of σ_0, the stress level corresponding to a probability of survival of 37%, it has the highest strength, σ, at large values of survival probability, S. You can see this from Figure 4.51 by reading off the values of stresses corresponding to a probability of survival of, say, 99.9%, or 1 part in 1000. The largest stress (382 MN m^{-2}) is that for sintered silicon nitride (line 2).

SAQ 4.19 The strength of ceramics is normally controlled by the presence of crack-like defects. On loading to the fracture stress, one of the defects becomes unstable and starts to grow rapidly. The relationship between the fracture stress and the critical crack length, a_c, of the defect was given by Equation (2.22),

$$\sigma = \sqrt{\frac{EG_c}{\pi a_c}}$$

This can be rearranged to give

$$a_c = \frac{EG_c}{\pi\sigma^2}$$

If you are taking the values for sintered silicon nitride from Table 4.7, you get

$$a_c = \frac{(230 \times 10^9)(175)}{\pi(300 \times 10^6)^2}$$

$$= 0.14 \text{ mm}$$

This is small, but larger than the size of a single pore. It is probably an array of closely spaced pores that join up on loading to form a critical crack.

Chapter 5
Polymeric Materials

by Peter Lewis
(consultants: Nigel Mills and Andrew Stevenson)

Chapter 5 Polymeric materials

5.1 Introduction to polymers

5.1.1 Bonding, structure and physical properties

In previous chapters we have seen that metals, ceramics and glasses possess particular property profiles that are closely related to their atomic or molecular bonding. Here we shall be examining two closely related materials — plastics and elastomers — which share a very different property profile. They are based on long-chain polymer molecules and it is these long chains that are responsible for their characteristic mechanical and thermal properties, which distinguish them from other materials. However, similar concepts to those used for metallic and inorganic materials can be applied to obtain general relationships between the structure and properties of polymers. The properties of some representatives of different classes of polymer are shown in Table 5.1.

We shall be exploring their peculiarities by looking at some common polymers and examples of design and failure. Their failures, from a plastic carrier rupturing at the handle, to plastic knobs coming away in

Table 5.1 Short-term mechanical (293 K) and thermal properties of polymers

Class	Polymer	$\rho/\mathrm{kg\,m^{-3}}$	$E/\mathrm{GN\,m^{-2}}$	$\sigma_{TS}/\mathrm{MN\,m^{-2}}$	Breaking strain/%	Izod impact strength/$\mathrm{J\,m^{-2}}$	$G_c/\mathrm{kJ\,m^{-2}}$	T_g/K	T_m/K
Thermoplastic polyolefins	high-density polyethylene (HDPE)	960	1.0	30	600	10	5	183	404
	low-density polyethylene (LDPE)	920	0.2	10	800	> 50	6.5	173	388
	polypropylene (PP)	910	1.5	33	400	6	8	263	449
Thermoplastic vinyl polymers	unplasticized poly(vinyl chloride) (UPVC)	1450	3.0	50	30	3.2	5	353	–
	polystyrene (PS)	1060	3.4	50	2.5	1.7	2	370	–
	acrylonitrile-butadiene styrene (ABS)	1050	2.5	42	80	15.0	5	373	–
	poly(methyl methacrylate) (PMMA)	1180	3.2	65	2.0	2.0	0.5	378	–
Thermoplastic polyesters	polycarbonate (PC)	1210	2.5	60	125	60.0	0.7	415	(540)
	poly(ethylene terephthalate) (PET)	1390	3.0	54	275	3.5	–	338	549
Thermoplastic polyamide	nylon (PA6,6)	1140	2.0	80	200	10.0	3	~273	540
Elastomer	natural rubber (NR)	930	~0.003	50	~700	–	–	200	298
Rigid thermoset	fabric-filled phenol formaldehyde (PF)	1400	8.0	50	0.7	7.5	–	–	–

your hand, are all too familiar! It is an important aspect of polymer technology to know how to design against such failures and how to work to the limit of the properties of polymers. Some of the ways in which a polymer's properties can be improved for particular functions will lead us to look at orientation, a theme that continues in the two subsequent chapters. Finally, we shall be looking at the unique properties possessed by elastomers, properties which engineers exploit in a host of applications, from car tyres to mountings for whole buildings.

Given the importance of bonding for property profiles, what can be said about bonds in polymers?

Consider a simple polymer such as polyethylene used, for example, in plastic shopping bags or washing-up bowls.

Its molecular structure or **configuration** (see ▼Structural terms for polymers▲) is shown in Figure 5.1 and comprises a sequence of carbon atoms, linked together in a mainly linear chain, with side bonds to hydrogen atoms. The covalent primary bonds that occur along the length of the chain, have energies of about $370 \, \text{kJ mole}^{-1}$. They are weaker than the carbon–carbon bonds in diamond ($750 \, \text{kJ mole}^{-1}$), due to the influence of the side bonds with hydrogen, but very much stronger than the van der Waals' bonds between the chains, which have a *maximum* value of about $5 \, \text{kJ mole}^{-1}$.

There is also an unexpected van der Waals' bond that occurs within the chain due to rotation about the axis of the carbon–carbon bond (Figure 5.1). The energy difference between different rotational states is only about $3 \, \text{kJ mole}^{-1}$, a value comparable with the thermal energy present in the material at 300 K ($RT = 2.5 \, \text{kJ mole}^{-1}$). The energy barrier to chain rotation, however, is higher (about $13 \, \text{kJ mole}^{-1}$) and affects the rate of chain rotation. As the temperature rises, the rate of rotation increases because the higher vibrational (i.e. thermal) energy in the molecules increases the probability of overcoming the barrier.

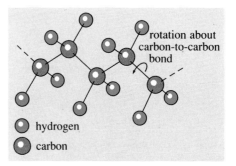

Figure 5.1 Planar zig-zag conformation of linear polyethylene: different conformations arise by rotation about the carbon–carbon covalent bond, as indicated by the arrow

EXERCISE 5.1   Compare the densities, ρ, of high-density polyethylene in Table 5.1 and diamond (Section 4.1.2). Explain the difference in terms of bonding.

Apart from their processability, the low density of polymers is one of the main reasons why they are so widely used for such items as containers, consumer durables and building products. Polymers have densities around $1000 \, \text{kg m}^{-3}$ because of the presence of light elements and their bonding, but it is the bonding which has the major influence on their mechanical properties and also their thermal behaviour (see ▼Thermal transitions in polymers▲).

Note that because the stress–strain behaviour of polymers is generally nonlinear and, as you will see, it also depends on time, the linear–elastic concept of Young's modulus does not strictly apply. Instead, the term 'tensile modulus' is therefore used throughout this chapter.

▼Structural terms for polymers▲

The structure of polymer chains is specified in two ways: by the shape of individual chains in space, **conformation**, and by the way in which each chain is constructed from its component covalently bonded atoms, **configuration**. The chain configuration is determined mainly during polymerization, when the monomer units are linked together to make chains. Several types of configuration arise as a result:

branched polymer   A linear chain to which one or more side chains are covalently bonded.

copolymer   A chain composed of two or more different repeat units, which may be randomly sequenced (random copolymer), or more regular (block and alternating copolymers).

crosslinked polymer   Covalent linking between individual chains to create a three-dimensional network.

homopolymer   A chain composed of identical repeat units.

Other useful terms include:

molecular mass   The number of repeat units in a chain multiplied by the molecular mass of each repeat unit. Statistical quantities such as the number-average molecular mass ($\bar{M}_n$) and weight-average molecular mass ($\bar{M}_w$) are important. At the entanglement molecular mass ($\bar{M}_c$), chains first start to entangle with one another, giving a solid with useful mechanical properties.

stereoisomers   Different spatial arrangements of the same atoms around a double bond in a repeat unit. There are two important types: **cis** and **trans** as shown in Figure 5.2(a).

tacticity   The arrangement of covalently bonded atoms around asymmetric carbon atoms (i.e. those with four different groups attached). Three different forms of polymer chain result (Figure 5.2b): **isotactic**, all the groups in identical positions; **syndiotactic**, alternately opposed groups; **atactic**, randomly placed groups.

aromatic   Structures based on the six-membered benzene ring (C_6H_6) and its derivatives (see Appendix 2).

aliphatic   Nonaromatic organic structures.

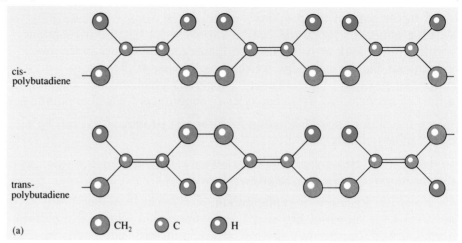

cis-polybutadiene

trans-polybutadiene

(a)   CH₂   C   H

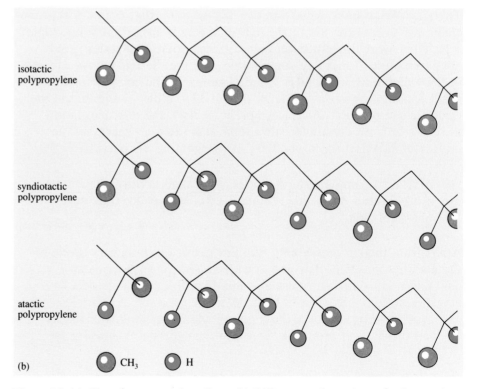

isotactic polypropylene

syndiotactic polypropylene

atactic polypropylene

(b)   CH₃   H

Figure 5.2 (a) *Cis* and *trans* polybutadiene. (b) Different configurations of polypropylene

▼Thermal transitions in polymers▲

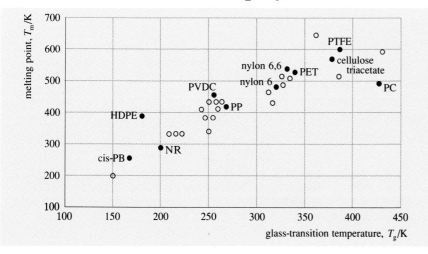

Figure 5.3 The melting points, T_m, of crystallizable polymers plotted against their glass-transition temperatures, T_g

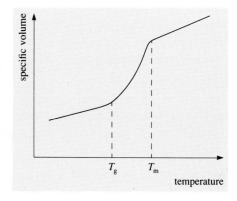

Figure 5.4 Transition curve for polymers

The melting of a thermoplastic kitchen utensil when it touches the hot surface of a cooker, or the softening of a polystyrene cup in boiling water, are readily observed examples of thermal transitions in thermoplastic polymers. Each represents a different type of thermal transition characterized by specific temperatures: the **melting point**, T_m, of crystalline polymers and the **glass-transition temperature**, T_g. They are important because their values are characteristic of the polymer's structure and they delimit the mechanical behaviour of polymeric materials. Partially crystalline thermoplastics show both transitions (Table 5.1) and there is a rough correlation between the two (Figure 5.3), with

$$T_g(\mathrm{K}) \simeq \tfrac{2}{3} T_m(\mathrm{K})$$

At the melting point, the ordered closely packed crystalline structure changes into the looser random arrangement of the liquid melt. Heat (the latent heat of fusion) must be supplied for melting to happen. Glass transition only occurs in amorphous structures and is not a

structural rearrangement requiring a defined amount of energy. Instead, it marks a change in the *rate* of energy absorption, due to a transition from localized oscillation and flexing of bonds to larger scale chain movements, once sufficient energy is available. These include rotation of chain segments (like a skipping rope being turned) and cooperative movements of neighbouring chains, and require more space. Thus as we saw in Chapter 4, Section 4.2, one way of determining the glass-transition temperature is from the change in slope of the graph of specific volume (or reciprocal density) against temperature (Figure 5.4). The melting point is shown by a sharp step in the same curve.

EXERCISE 5.2   Account for the difference between the glass-transition temperatures, T_g, of soda-lime silica glass ($\sim 970\,\mathrm{K}$) and that of a typical amorphous thermoplastic such as UPVC (Table 5.1).

Clearly, the melting point is related to structure since the chain conformation affects the ease of packing and, hence, the energy required to unpack chains, but what about the glass-transition temperature? If you return to the simple polyethylene chain shown in Figure 5.1, it is flexible at room temperature and has a very low glass-transition temperature of about 170 K. If a methyl group ($-CH_3$) is substituted for a hydrogen atom at alternate carbon atoms to form polypropylene, effectively the energy barrier to chain rotation is increased, so the glass-transition temperature should be higher. This is what is found. Polypropylene has a glass-transition temperature of about 260 K. UPVC and polystyrene have even bulkier sidegroups (Appendix 2), pushing their glass transition temperatures still higher to 353 and 370 K respectively.

EXERCISE 5.3   Why has natural rubber a glass-transition temperature of 200 K, while that of PET is 338 K?

5.1.2 Mechanical properties and structure

Compared with other materials, the moduli and strengths of bulk polymers are not very impressive. This is not always the case, however; some polymer fibres possess tensile moduli approaching that of steel as you will see in Chapter 6.

Using bond energies, it is possible to estimate the tensile modulus of an assembly of high-density polyethylene chains all of whose primary bonds are free to flex and rotate. It turns out to be very low — about $0.003\,\text{GN}\,\text{m}^{-2}$ compared to measured values of about $0.2\text{--}1.0\,\text{GN}\,\text{m}^{-2}$ (Table 5.1).

Why should there be such a difference?

The reason is that polyethylene crystallizes very easily in the **planar zig-zag** conformation shown in Figure 5.1. Neighbouring chains in the same conformation can pack together, so lowering the energy of the system (Figure 5.5). When strained along their lengths the chains can only respond by distorting covalent bonds and so have a high stiffness. The crystal structure of polyethylene is, however, more complex than any inorganic single crystal, with **lamellae** of **folded chains** (Figure 5.6) which combine together into a larger unit, the **spherulite** (Figure 5.7). Moreover, not all the material is crystalline. A significant portion is amorphous (i.e. noncrystalline) as indicated in Figure 5.6. Normally, spherulites are relatively large structural features and scatter light easily, so otherwise transparent polymers are translucent when crystalline.

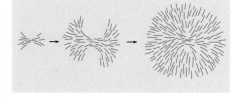

(a)

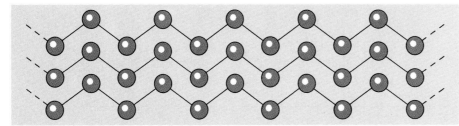

Figure 5.5 Crystallization of polyethylene by regular packing of chains in the zig-zag conformation

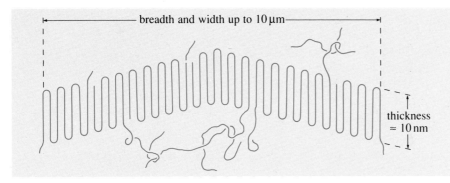

breadth and width up to 10 μm

thickness ≈ 10 nm

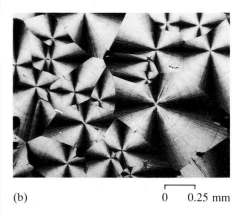

(b)　　　　　　　　0　　0.25 mm

Figure 5.7 (a) Stages in the formation of a spherulite from a stack of lamellae and (b) a polarized-light micrograph of spherulites in polyethylene oxide

Figure 5.6 Chain packing to produce lamellar crystal in polyethylene, with folds on crystal surfaces

Figure 5.8 shows schematic stress–strain curves for the main classes of polymer. From curves (c) and (d), you can see that some polymers, particularly the polyolefins and elastomers, can withstand enormous strains before breaking (see also Table 5.1). What both groups of material have in common is that, at room temperature, they are well above their glass-transition temperatures. The more open structure of the amorphous material makes it easier for chains to move past one another in response to a load and it allows larger movements (i.e. greater extensibility). Elastomers are wholly amorphous, but the chains are lightly crosslinked, so these large extensions are recoverable and they show **long-range elasticity**. In the polyolefins the behaviour is modified by the presence of crystallites and the absence of crosslinking. The large strains are not recoverable, the materials yield and show high ductility. Their moduli increase with increasing crystallinity from low-density polyethylene (50–60% crystalline—chain branching inhibits further crystallization), to high-density polyethylene (70–80% crystalline) and polypropylene (90–95% crystalline), but this is associated with decreasing breaking strains and impact strengths.

What about polymers below their glass-transition temperatures?

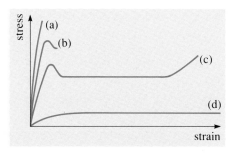

Figure 5.8 Stress–strain curves for polymers at 298 K: (a) a low-ductility polymer (e.g. PMMA); (b) a ductile polymer (e.g. PVC); (c) a ductile polymer showing cold drawing (e.g. polypropylene); (d) a polymer with long-range elasticity (e.g. natural rubber)

EXERCISE 5.4   Compare the tensile modulus, breaking strain, impact strength and glass-transition temperature of PVC and polystyrene. How do these values compare with the polyolefins? Explain the differences in terms of chain bonding.

Poly(methyl methacrylate) is a low-ductility polymer at 298 K, as shown by curve (a) in Figure 5.8, yet is widely used for products that are stressed in service, such as domestic baths. The reason is basically that not only can it be rapidly formed into shape by methods like thermoforming (heating sheet polymer), but the kind of stresses to which it is exposed are intermittent and relatively low. Being non-crystalline as a result of its atactic structure, it is transparent like PVC and polystyrene. In baths, it is pigmented and thus opaque, but is widely used for applications where transparency is at a premium. In lenses, for example, its low density gives it an advantage over inorganic glasses.

Where polymers with a higher tensile modulus are needed, an extra structural ploy, crosslinking the chains together, can be used. In fact, crosslinked rigid thermosets such as phenol formaldehyde (Table 5.1) were among the first synthetic polymers to be produced. Unlike thermo-plastics, which can be recycled after moulding, the structure of thermosets is locked in by chemical crosslinks. The resistance to rotation about the bonds is very high because of aromatic groups which are present in many thermoset chains, as well as the very high density of crosslinks, which further restrict chain mobility. Phenolic thermosets, as used, for example, in light fittings are normally filled with other materials (see Chapter 7) and have neither a melting point, because they are not crystalline, nor a glass-transition temperature, owing to the restricted chain rotation. Before any possible glass transition, degradation intervenes, evidence for which is given by the smell of overheated light fittings! In general, rigid thermosets exhibit the highest tensile moduli and lowest breaking strains of common polymers. Their stress–strain curves are typically similar to curve (a) in Figure 5.8, but they do *not* show the characteristics of curves (b)–(d), which typify the successive stages undergone by a thermoplastic polymer as the temperature is raised.

Since the properties of rigid thermosets are so different from those of thermoplastics, we shall not examine thermosets in any detail in this chapter, reserving them for Chapter 7. Because of their brittle nature, they are most commonly used in composites, the subject of that chapter.

SAQ 5.1 (Objective 5.1)
Using the data shown in Table 5.1, select the most appropriate noncrystalline thermoplastic polymer to make a beam of the same length, width and flexural stiffness as an existing beam made from fabric-filled phenolic, if the thickness of the new beam is to be 40% greater than that of the phenolic beam and if it is to withstand tensile strains in excess of 10%. Assume that the second moment of area, I, for a beam of rectangular cross-section is $bd^3/12$ (see Table 1.1).

5.1.3 Toughness and structure

For many applications, where both toughness and a high tensile modulus of $1-5\,\text{GN m}^{-2}$ are required, the chain structure of brittle polymers can be modified to produce substantial improvements. There are several different strategies employed, of which two important ones are:
1 Modification of the chain structure by copolymerization.
2 Modification of the repeat unit by synthesis of new polymers.

Copolymers include such materials as acrylonitrile-butadiene styrene (ABS) which has a complex structure involving random copolymer chains of acrylonitrile and styrene to which side branches of polybutadiene are attached (Appendix 2). The elastomeric polybutadiene phase separates to give globules of elastomer (0.1–1.0 μm in diameter) in the matrix of styrene acrylonitrile (SAN) copolymer (Figure 5.9). The tensile properties of styrene acrylonitrile itself (Appendix 2) are similar to those of polystyrene on its own, so why should the resultant copolymer be relatively tough despite being totally noncrystalline?

The answer lies in the effect of low-modulus inclusions in relatively rigid material: each inclusion raises the internal stress, with the stress concentration factor $K_t \simeq 2.0$ (see 'Stress concentrations' in Chapter 2). When stressed, defects grow from the inclusions, absorbing significant extra energy as they do so. The principle of **elastomer toughening** is now widely employed in other polymers — poly(methyl methacrylate), UPVC, thermoset resins and even relatively tough material such as

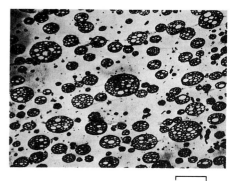

0   1 μm

Figure 5.9 Electron micrograph of acrylonitrile-butadiene styrene showing rubber particles stained black

nylon. The elastomeric inclusions, however, lower the overall tensile modulus of the material (Table 5.1).

Other tough polymers are based on different kinds of repeat unit: one example is transparent amorphous polycarbonate, where the backbone chain is composed of rigid aromatic groups (Appendix 2). The surprising features of the mechanical properties of polycarbonate are its high ductility and toughness well below its glass-transition temperature of 415 K, probably due to defect formation of the same kind that increases the toughness of acrylonitrile-butadiene styrene compared with polystyrene.

Two other polymers listed in Table 5.1 — poly(ethylene terephthalate) and nylon 6,6 — are well-known both in bulk form and as fibres for a wide variety of textile applications, from clothing to ropes and webbing (see Chapter 6).

Both have high melting points (about 550 K, see Table 5.1) and relatively low glass-transition temperatures: that of nylon 6,6 is sensitive to water content, but is generally accepted to be below room temperature at its equilibrium moisture content of about 4%. Drying the material increases its glass-transition temperature and reduces its toughness, unless special precautions are taken. From previous arguments, the polyethylene-like structure of the short chains connecting the amide bonds (Appendix 2) should make for a much lower glass-transition temperature, but the effect of hydrogen bonding *between* adjacent chains hinders rotation and so raises it. Poly(ethylene terephthalate) (PET) falls into the sequence already discussed with a rigid aromatic group in its polyester main chain (Appendix 2).

Both materials are crystalline, but the degree of crystallinity varies considerably in the range 0–70%, depending on molecular mass, conditions of processing and degree of chain orientation. Like the polyolefins, they can be cold drawn at room temperature following yielding, so they exhibit high breaking strains and are reasonably tough materials, as can be judged by dropping a full PET beer bottle!

5.1.4 Time-dependent mechanical properties

Up to this point, we have treated the mechanical properties of polymers as though they were independent of the rate of testing. The data in Table 5.1 are short-term data and were determined on samples strained at the same rapid rate to ensure comparability. However, the chain-rotation model suggests testing rate should be an important variable because chain molecules with small barriers between different rotational states will respond more quickly to heat or imposed strain than those with higher resistance to internal rotation. As we shall see in the next section, this needs to be taken into account in detailed design calculations, but for the moment we are mainly interested in relating broader aspects of mechanical properties to chain structure. Read through ▼ **Viscoelasticity and chain structure** ▲.

▼Viscoelasticity and chain structure▲

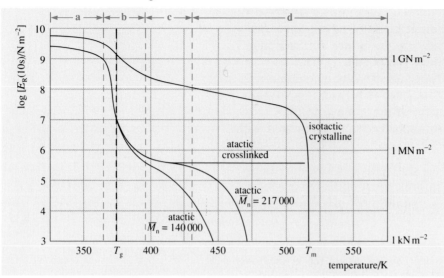

Figure 5.10 The relaxation modulus at 10 s plotted against temperature for different types of polystyrene. The viscoelastic states for thermoplastic PS: (a) glassy; (b) transition; (c) elastomeric; (d) viscous flow

The response of viscoelastic materials to an applied load is a combination of viscous and elastic components. The viscous component in polymers is caused by chains moving past one another in the noncrystalline zones and depends on time. The elastic component arises from the elastomeric behaviour of the amorphous chains plus, in partially crystalline polymers, the more nearly Hookean elastic response of the crystalline zones. Both elastic contributions are reversible so, on unloading, they drive the viscous component in reverse, producing time-dependent recovery. In practical situations, there are two particularly important observed effects, **creep** and **stress relaxation**. Creep is the continuing elongation of a sample under a constant load while stress relaxation is the decrease in stress for a sample held at constant deformation. Creep moduli, E_C, and stress relaxation moduli, E_R, are defined by the equations

$$E_C(t) = \frac{\sigma_0}{\varepsilon(t)}$$

and

$$E_R(t) = \frac{\sigma(t)}{\varepsilon_0}$$

where σ_0 and ε_0 are the constant stress and strain applied to the samples and $\varepsilon(t)$ and $\sigma(t)$ are the stress and strain at a time t.

Stress relaxation experiments performed on noncrystalline polymers will give a characteristic curve when $E_R(t)$ is plotted as a function of temperature (Figure 5.10). It shows a glassy region below the glass-transition temperature, T_g, a transition region around the temperature T_g, an elastomeric plateau for temperatures above T_g and a viscous flow region at temperatures much greater than T_g. The lower three curves are for different grades of atactic amorphous polystyrene. This material can also be prepared in isotactic form and when tested in a similar way produces a curve which shows a high modulus in the elastomeric region. This is due to the reinforcing effect of the crystallites on the elastomeric matrix. At T_m, melting occurs relatively rapidly and the modulus drops very quickly as the material flows.

Lightly crosslinking atactic polystyrene effectively prevents viscous flow: the elastomeric plateau above T_g continues to high temperatures, essentially because crosslinking inhibits large-scale creep of the material. Normal uncrosslinked polystyrene exhibits behaviour typified by the curves for the two different molecular masses of polystyrene. The grade with the higher molecular mass has a higher viscosity at comparable temperatures than the grade with the lower one and its elastometric plateau is significantly longer. The behaviour shown in Figure 5.10 is typical of most polymers.

One other feature of viscoelastic materials enables you to predict the way mechanical properties are affected by changing the rate of testing. In the chain rotation model, it is the energy differences between different rotational conformations that largely determine thermomechanical behaviour. Rates of change are affected by the energy barrier — the greater the energy barrier, the slower will changes occur at a given rate of testing. The time scale of an experiment is directly related to the rate of testing, so long times correspond to slow rates. Time and temperature are related: the more thermal energy is supplied to a given polymer, the easier will it respond to strain since its flexibility is increased and this is the same as the response at longer times. In other words, extending the time scale of an experiment is equivalent to raising the temperature: this is known as **time–temperature equivalence**. You will see later how the principle works for the dynamic properties of polymers.

SAQ 5.2 (Objective 5.2)
Polypropylene with a high molecular mass ($\bar{M}_n > 200\,000$) can be made in atactic form by changing the type of polymerization system. How will the stress-relaxation modulus, $E_R(10\,\text{s})$, vary with temperature (refer to Figure 5.10)? What will be the state of the polymer at 293 K? What will be the effect of changing the time scale of the stress-relaxation measurements from 10 s to 1000 s at 293 K?

5.1.5 Other factors

The polymers we have considered have low densities and relatively low tensile moduli, but varying degrees of toughness and ductility. There are two other key areas for polymers: ease of processing and raw material costs. Above T_g for an amorphous polymer, or above T_m if it is partially crystalline, thermoplastics are highly viscous liquids and flow when deformed. Increasing the temperature to about 100 K above T_g or to about 20 K above T_m, as appropriate, gives fluids of a viscosity suitable for processing. It is this property of polymers that is exploited in processing them to shape, most commonly by injection moulding and extrusion. The two processes are complementary in the sense that mouldings are compact products whose shapes can be complex, while extrudates are products of constant cross-section. Battery cases, for example, require injection moulding, but long lengths of pipe are always extruded.

EXERCISE 5.5 Compare the processing costs of polystyrene with those of soda-lime silica glass. Assume that soda-lime silica glass is shaped at 1073 K and that heating dominates processing. Take c_p (glass) $= 0.98\,\mathrm{kJ\,kg^{-1}\,K^{-1}}$ and c_p (polystyrene) $= 1.35\,\mathrm{kJ\,kg^{-1}\,K^{-1}}$.

Raw materials costs are less favourable for polymers because of the high cost of oil compared with inorganic raw materials like silica sand. There are also many more steps in converting oil to a monomer followed by polymerization. In common with commodities like steel and glass, such costs are offset by the scale of production. Typical costs of a range of polymers and metals are shown in Table 5.2. They can be used for comparison, although the prices of materials fluctuate and depend on the amount ordered for a given job.

For polymers, the price reflects the degree of chemical processing needed to make a monomer. The greater the number of steps (e.g. nylon, polycarbonate) the greater the price. This explains the lower prices of the polyolefins, which are made from very simple monomers. PVC ought to be more expensive, but isn't because of economies of scale and the low cost of chlorine, a cheap industrial chemical.

SAQ 5.3 (Objective 5.3)
A copolymer of polypropylene with 10 wt% ethylene is widely used in many products. The ethylene repeat units are randomly distributed in the polypropylene chain, Using molecular arguments about the flexibility of the backbone chain, describe qualitatively how the copolymer will compare with polypropylene in terms of: (a) density; (b) T_g, T_m; (c) tensile modulus; (d) impact strength.

Table 5.2 Raw material prices (Spring 1989)

	Price (tonne lots)/£ kg^{-1}
Thermoplastics	
polyethylene (low-density) (LDPE)	0.74
polyethylene (high-density) (HDPE)	0.73
PVC (plasticized)	1.10
PVC (unplasticized)	1.00
polypropylene homopolymer (PP)	0.80
poly(methyl methacrylate) (PMMA)	1.70
acrylonitrile-butadiene styrene (ABS)	1.60
nylon 6,6 (PA6,6)	2.90
polycarbonate (PC)	3.20
polystyrene (PS)	1.00
poly(ethylene terephthalate) (PET)	2.00
Metals (LME)	
nickel	6.90
copper	1.70
aluminium	1.34
zinc	1.05
lead	0.37
mild steel	0.30
Rubbers	
standard Malaysian rubber (NR)	0.70

SAQ 5.4   (Objective 5.1)
Select two polymers to satisfy the following criteria.

(a) One polymer should float on water, have a minimum tensile modulus of $1.0\,GN\,m^{-2}$ and be tough at temperatures below 223 K.
(b) The other should be transparent, rigid in boiling water and have a maximum density of $1200\,kg\,m^{-3}$.

5.2 Making the most of modulus

Selection of a material for a particular function calls for a mass of data, not just on thermal and mechanical properties like those in Table 5.1, but also derived properties like creep, stress relaxation, recovery, friction, wear and less easily quantifiable properties like ease of handling. Environmental resistance is also less easy to quantify, but often critical if a product is to last a long time. Above all, raw material and processing costs are vital. In this section, we shall be looking at not only the properties of different polymers, but also other factors that have to be considered in selecting the material for a specific application. Although short-term mechanical data are useful for preliminary sorting of materials, in products designed to last for a long time under loads, you will see that the viscoelastic properties of polymers must be taken into account.

5.2.1 Rainwater removal

With over forty million dwellings in this country, the market for products such as guttering and piping is very large indeed. New buildings and the renovation of older properties have created the potential for replacing traditional materials with cost effective alternatives. A basic function of a roof is efficient removal of rainwater and in our changeable climate, designers work to a maximum rainfall rate of $75\,mm\,hr^{-1}$, typical of a monsoon! The corresponding rate of flow into guttering is 2.1 litres per second per $100\,m^2$ of roof area, so it must be able to accommodate this high flow rate without spillage onto adjoining brick work. For the guttering and downpipe to fit the pitched roof shown in Figure 5.11, the optimum design solution is semi-circular (or 'half-round') guttering with a centrally located downpipe. Guttering systems may also need bends at corners of buildings, S-bends to bridge the eaves and seals at free ends. The variety of components is shown in Figure 5.12 including the support brackets. The spacing of these supports is critical, particularly to withstand sliding snow and ice, which can extend to the top of the roof and so put very high bending loads on the guttering.

5.2.2 Preliminary materials selection

What are the main requirements for guttering? They are:

(a) high stiffness in bending, for stability against rain and snow loading;
(b) strength to withstand installation stresses;

Figure 5.11 Guttering and down-pipe for a pitched roof

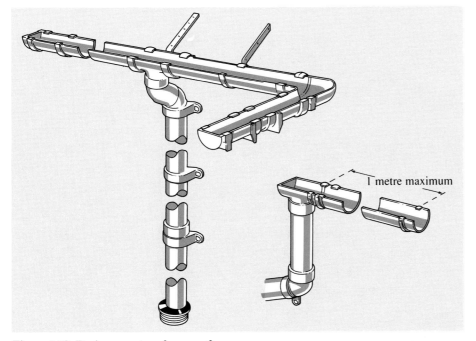

Figure 5.12 Drainage system for a roof

(c) low weight to aid installation and minimize the number of supports needed;

(d) low material cost and ease of manufacture;

(e) durability in the environment.

This list includes two installation criteria: strength and low weight to facilitate fitting the system together. Strength is particularly important due to the abuse that building materials receive on site. Weight is important since installation of guttering systems is often hazardous owing to the height of roofs and the difficulty of manipulating guttering lengths of up to 4 m into position. Weight has a knock-on effect because the greater the weight per unit length of guttering, the stiffer and stronger must be the supports. Since the product is for mass-markets, it must be of intrinsically low cost and be easy to process into its finished shape. A final criterion demands durability in the environment to which it is exposed: that is, resistance to low and high temperatures, rainwater and the ultraviolet radiation in sunlight.

EXERCISE 5.6   Examine the rainwater disposal systems on old, unrenovated dwellings. What materials have been used and what are their principal limitations?

Other materials used for guttering in modern houses include aluminium and thermoplastics. So what is the basis for materials selection? One way to rank the materials is in terms of merit indices and this is done in
▼Guttering materials▲

225

▼Guttering materials▲

The mechanical properties of the main materials suitable for guttering are shown in Table 5.3, clearly indicating the superiority of metals over the two thermoplastics, polypropylene and UPVC. These two commodity polymers are relatively cheap and easy to manufacture: high-density polyethylene is cheaper than polypropylene, but its tensile modulus is substantially lower (Table 5.1). Higher modulus polymers such as poly(methyl methacrylate) are much more expensive, while polystyrene is too brittle.

Straight comparisons of properties, however, do not do justice to the much lower densities of polymers. Low weight is highly desirable, so Table 5.3 also lists the merit indices E/ρ, and $E^{1/3}/\rho$ (see Section 2.4 'Merit indices'). The last represents the

flexural stiffness per unit mass for a beam of rectangular section and fixed thickness d. The ratio of one value to another gives the relative mass of two beams of given length and width, but with the same bending stiffness. It can also be shown that the thicknesses d_1 and d_2 of two beams made of materials with moduli E_1 and E_2 and of identical flexural stiffness are given by the simple equation

$$\frac{d_1}{d_2} = \left(\frac{E_2}{E_1}\right)^{1/3}$$

As Exercise 5.7 shows, a polypropylene beam of the same stiffness is only a little heavier than an aluminium beam, but considerably thicker owing to its much lower tensile modulus.

In terms of bending, aluminium, with the highest flexural merit index, is the best choice for rectangular beams. By dividing the specific flexural modulus by materials cost, however, gives a different ranking (column 9), with cast iron the best choice. The two thermoplastics are apparently poor contenders for guttering material.

EXERCISE 5.7 What is the relative mass of two rectangular beams of identical width, length and flexural stiffness made from (a) aluminium alloy and (b) polypropylene. How much thicker will the polypropylene beam need to be?

Table 5.3 Guttering materials

Material	Density, $\rho/\mathrm{kg\,m^{-3}}$	Tensile modulus, $E/\mathrm{GN\,m^{-2}}$	Tensile strength, $\sigma/\mathrm{MN\,m^{-2}}$	Specific modulus, $E\rho^{-1}/10^6$	Specific strength $\sigma\rho^{-1}/10^3$	Specific flexural modulus $E^{1/3}\rho^{-1}/10^3$	Cost/£ $\mathrm{kg^{-1}}$	Specific flexural modulus/ cost
mild steel	7860	210	460	26.7	58.5	0.76	0.30	2.5
cast iron	7150	100	100	14.0	14.0	0.65	0.18	3.6
aluminium alloy	2800	75	500	26.8	179.0	1.51	1.34	1.1
UPVC	1450	3.0	50	2.07	34.5	1.00	1.00	1.0
polypropylene	910	1.5	33	1.64	36.3	1.26	0.80	1.6

5.2.3 Guttering under load

There are several problems with using merit indices to select materials. Firstly, qualities such as corrosion resistance and ease of manufacture, which are inherently difficult to quantify, are neglected. Secondly, a single value of Young's modulus is assumed for a given material. This may be justified for metals, but not for polymers, which are not only significantly nonlinear, but also creep under continuous load (see ▼Creep in plastics▲). Finally, flat beams are not representative of more complex geometries. All of these objections apply to guttering: durability requires corrosion resistance, long-term loading can occur under adverse circumstances and guttering is not flat (Figure 5.16). So we shall now look briefly at traditional guttering materials to see why thermoplastics have become so widely used.

The rusting of steel was examined in Chapter 3 and even when it is galvanized for protection, it will rust over a period of years, so must eventually be replaced. The merit index approach yields the amount of material just to fulfil the mechanical needs. For steel, with its high Young's modulus, this is a relatively thin section (~ 1 mm) so little

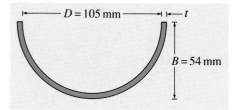

Figure 5.16 Dimensions of domestic guttering

corrosion is required before the guttering loses its integrity. Cast iron is much more resistant to rusting, but is difficult to manufacture in thin section and so is made with a thickness of about 3 mm. Owing to its high density, a 3 m length would weigh over 10 kg. Its apparently high strength (Table 5.3) is deceptive because it is a brittle material. Drop a length and it will shatter into pieces! Ductile thermoplastics like UPVC and polypropylene do not, of course, rust, but can suffer **ultraviolet degradation** in sunlight. Polypropylene is more sensitive than PVC, but in both cases, protection is given by 0.2–2% of carbon black in the raw polymer. This is why most plastic rainwater goods are either black or grey in colour.

To explore the effect of loading, we shall apply the relevant bending formula for half-round guttering. As we have already seen, the worst case is due to ice and snow on a pitched roof. The vertical component of the load is uniformly distributed along the length of the guttering

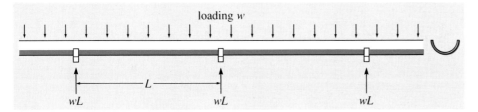

Figure 5.17 Loading conditions of domestic guttering

(Figure 5.17). If the guttering is fixed at the supports and cannot rotate about them the maximum deflection is given by

$$\Delta_{max} = \frac{wL^4}{384EI}$$

The second moment of area is given by

$$I = 0.0375D^3t$$

so

$$\Delta_{max} = \frac{wL^4}{14.4ED^3t}$$

where L is the length of gutter between supports, w is the imposed load per unit length, D the diameter of the gutter and t its thickness (Figure 5.16). The maximum deflection is five times larger if the guttering is free to rotate at the supports.

EXERCISE 5.10   Calculate the maximum deflection caused by a uniform loading of 334 N m^{-1} on cast iron guttering of thickness 3 mm and diameter 105 mm. Assume the supports are 2 m apart and allow rotation and that Young's modulus for cast iron is 100 GN m^{-2}.

▼Creep in plastics▲

Creep data for the many different grades of polymer marketed today are widely available from manufacturers' literature. One way of representing tensile creep data is in the form of linear strain–log time curves for different imposed stresses. Figure 5.13 shows the creep curves for a pipe-grade PVC at 293 K and it is clear that the creep rate increases rapidly with increasing imposed stress. The relationships between the different ways of presenting creep data are shown in Figure 5.14.

> EXERCISE 5.8   Using Figure 5.13, estimate the creep moduli for PVC at times of 10^2 s and 10^6 s at stress levels of $5 \, \text{MN m}^{-2}$ and $10 \, \text{MN m}^{-2}$. How will the creep moduli be affected by increasing temperature?

At stress levels low relative to the tensile strength, the creep moduli at different stress levels can be used in design calculations using classical formulae that assume Hookean elasticity, such as the engineers' bending formula. With all polymers, it is important to remember their temperature sensitivity, particularly for polymers with relatively low glass-transition temperatures, such as polyolefins. Since part of their structure is in the amorphous elastomeric state at room temperature, their creep rates will, in general, be considerably higher than materials like PVC, PMMA and PC, which have glass-transition temperatures well above 293 K. They will also be considerably higher than those of the highly crosslinked thermosets.

How important is creep for polymer products?

For constantly loaded products, such as pipes for water or gas, it is essential to know how the creep modulus decreases over the designed lifetime. Many polymer products, however, are only stressed intermittently, so how can creep moduli be adjusted for such situations? It is important to appreciate that when the stress is removed, creep recovery occurs, which, although frequently incomplete, will clearly help reduce the overall distortion. As might be expected, intermittent creep is dependent on the ratio of creep time to recovery time (lower curves of Figure 5.15).

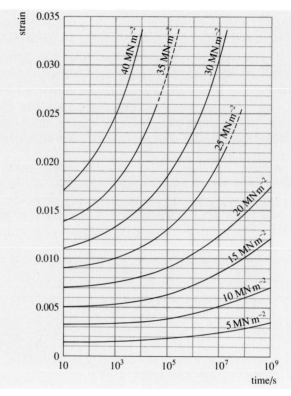

Figure 5.13 The tensile creep of pipe-grade PVC at 293 K

> EXERCISE 5.9   Estimate the creep moduli of polypropylene at 10^6 s and 293 K (a) under a continuous stress and (b) when recovery lasts three times longer than creep. Assume that the applied stress is $10 \, \text{MN m}^{-2}$.

Data such as those shown in Figures 5.13 and 5.15 are normally obtained from tensile creep tests conducted under carefully controlled conditions, but standards frequently specify tests for products such as plastic guttering to allow for the complexities of their geometry.

Even when the time dependence is removed, the stress–strain relationship is, in general, markedly nonlinear, as indicated in the isochronous plots in Figure 5.14, with the modulus decreasing with increasing strain. (Remember that all materials are intrinsically nonlinear; the deviations from linearity in, for instance, metals and ceramics, are negligibly small.) Thus, for polymers, the tensile modulus at a prescribed value of strain is usually

taken as the ratio of stress to strain at that strain (the **secant modulus**). The alternative, **tangent modulus**, the slope of the curve at a given strain, is less often used.

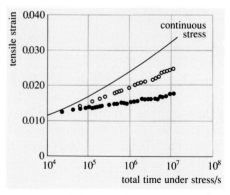

Figure 5.15 Intermittent tensile creep of polypropylene under a stress of $10 \, \text{MN m}^{-2}$ at 293 K. Open circles show results when the creep period is three times the recovery period; full circles show results when the creep period is one third of the recovery period

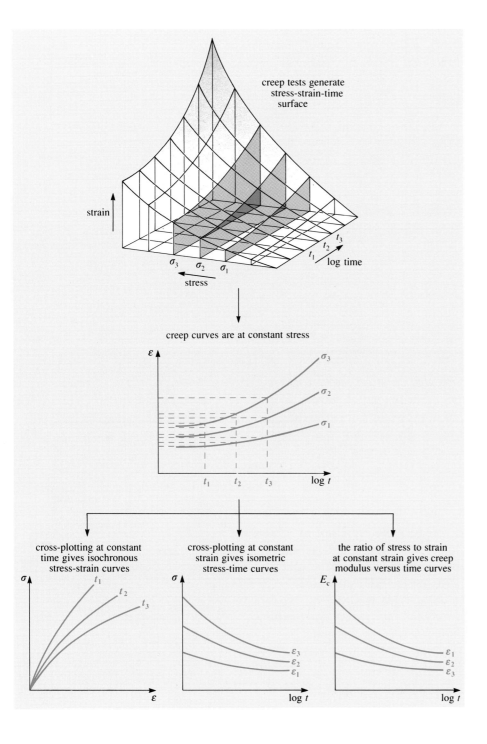

Figure 5.14 Presentation of creep data

Cast iron deflects only slightly, as expected given its high modulus. The loading conditions are severe: $334\,\mathrm{N\,m^{-1}}$ is equivalent to nearly $35\,\mathrm{kg}$ per metre of guttering! Cast iron guttering is thus over-engineered for a climate where such snow loading conditions are unlikely. What happens if thermoplastics are used instead? The maximum deflection is most sensitive to the distance between supports, L. If this is reduced, it should be possible to use a material with a lower modulus. The other problem with using thermoplastics is creep under the imposed load. This can be allowed for by using the appropriate creep modulus to calculate the deflection. Provided that the deflection is not excessive, it should be possible to use thermoplastic materials and so exploit their enivronmental resistance and ease of processing. The relevant standard (BS4576) stipulates that the maximum deflection after fourteen days should not exceed $19\,\mathrm{mm}$ under a uniform load of $334\,\mathrm{N\,m^{-1}}$ when tested at $293\,\mathrm{K}$.

SAQ 5.5 (Objective 5.1)
Calculate the maximum deflection that would be achieved in a British Standard creep test for guttering manufactured in (a) pipe-grade PVC and (b) polypropylene.

Assume the supports are designed to prevent rotation and are spaced $1\,\mathrm{m}$ apart. The thickness is $2.5\,\mathrm{mm}$ and the diameter is $105\,\mathrm{mm}$. Use Figures 5.13 and 5.15 to estimate relevant creep moduli, assuming a creep stress of $10\,\mathrm{MN\,m^{-2}}$. What would be the best material to use and what design changes could be recommended to give extra resistance to creep?

What emerges from the analysis is that the thickness, t, is not as critical a design parameter as D or L. In fact, in terms of merit indices, the appropriate index is the specific modulus (E/ρ) rather than the specific flexural modulus ($E^{1/3}/\rho$). The limits of the merit-index approach become apparent, however, because the thermoplastic guttering can simply be supported at more frequent intervals than would be required for an aluminium or steel beam. The manufacturers recommend a separation between the supports of $1\,\mathrm{m}$ where snow is not likely to be a problem and $0.8\,\mathrm{m}$ where snow loading could occur.

Most new dwellings use UPVC guttering systems for their ease of construction and relatively low cost, about one third of equivalent aluminium systems and a half that of cast iron guttering. They are also widely used for replacing cast-iron and steel guttering as both come to the end of their useful lives. Potable cold-water mains and waste-water pipes below ground are also made from PVC as is a wide range of other products (see ▼PVC — the versatile polymer▲).

▼PVC — the versatile polymer▲

PVC is probably the most versatile of the common thermoplastics in the way that its properties can be modified by suitable additives. Until about thirty years ago, PVC products were generally of low strength, due to thermal degradation during processing, and hence they had limited applicability. Improvements in methods of polymerization and the addition of stabilizing chemicals have much improved its heat stability and UPVC is now a commodity thermoplastic. Careful control of the processing temperature is still needed, however, in order to prevent the molecular mass being reduced. Extrusion grades of PVC typically have $\bar{M}_{\mathrm{w}} \approx 140\,000$ and, for moulding grades, $\bar{M}_{\mathrm{w}} \approx 70\,000$.

PVC with a low tensile modulus is familiar in many applications where flexibility is required, such as flooring (vinyl sheet), wire insulation, leathercloth upholstery (particularly in cars), bottles and sheet (e.g. PVC macs). A lower modulus implies higher molecular flexibility, and this is achieved by adding **plasticizers** to the polymer. They generally consist of polymers of low molecular mass, such as dioctyl phthalate (DOP), an

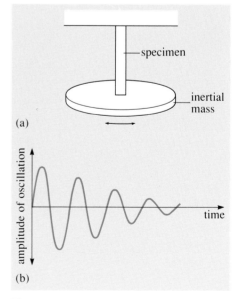

Figure 5.18 (a) A torsion pendulum; (b) the damping curve showing a reduction in the amplitude, A, of successive oscillations

oily clear liquid with a high boiling point (> 473 K). The plasticizer separates the chains and lowers the resistance to chain rotation, reducing the glass-transition temperature to 293 K or below. In effect, plasticizers are high-boiling solvents for polymers. The structure of the plasticizer needs matching to the end function, because if its molecular mass is too low, it can volatilize and the flexible product becomes brittle. You may have noticed this effect on your car windscreen, where the plasticizer condenses after evaporating from hot upholstery!

To study the effects of plasticizers, the torsion pendulum (Figure 5.18a) is commonly used. A strip of material, supporting a heavy disc, is twisted and then allowed to oscillate. As energy is absorbed by hysteresis, the amplitude of successive oscillations will be damped until the oscillations cease (Figure 5.18b).

Two useful quantities can be measured by such an experiment: the shear modulus, G, and the **damping** or **loss factor**, $\tan \delta$. The shear modulus, G, is given by

$$G = \frac{I}{Kt^2}$$

where I is the second moment of area of the disc, K relates to the geometries of the specimen and the disc and t is the period of oscillation. The damping factor depends on the ratio of the amplitudes of successive oscillations and is given by

$$\tan \delta = \ln \left(\frac{A_n}{A_{n+1}} \right) \times \frac{1}{\pi}$$

It is a measure of internal friction and is usually negligible for metals, but can rise to high values approaching unity for polymers, particularly in the region of the glass-transition temperature. Below this temperature, molecular vibrations are short ranged and localized, with little damping, but above the glass-transition temperature the material is elastomeric, so long-range molecular motion is relatively easy. In the transition zone, long-range molecular motion is hindered, so damping is great.

Figure 5.19 shows both the shear modulus and damping for PVC plasticized with different types of plasticizer. Each peak corresponds to the glass-transition temperature, but the value of T_g depends critically on the composition as well as the type of plasticizer.

> EXERCISE 5.12 The amplitude of each oscillation of a polymer rod suspended at a temperature of 298 K in a torsion pendulum was found to be 50% of the previous oscillation. Calculate the loss factor for the polymer.

Generally, plasticizers increase the elongation-to-break, lower the tensile strength and modulus, but increase impact strength. These can be controlled so that mechanical properties can be matched to a particular function.

Inert fillers, typically materials such as kaolin (china clay, a silicate mineral) and calcium carbonate, are also added in substantial quantities to PVC. Although fillers will be discussed in more detail in Chapter 7, they can modify the properties of plasticized PVC yet further. For example, plasticized PVC for wire insulation contains a filler to reduce its tear strength and make it easier to strip it from the wire when making electrical connections, an instance of deliberately *reducing* strength to suit the function of the product.

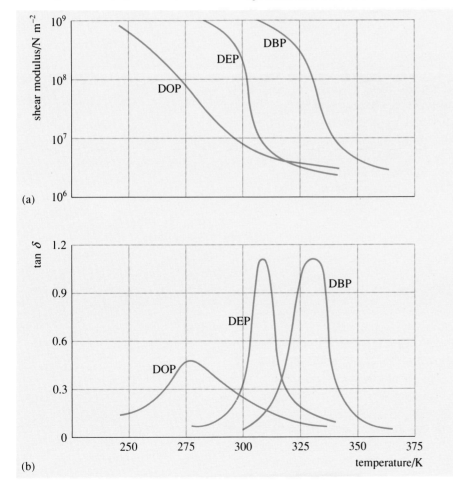

(a)

(b)

Figure 5.19 (a) The shear modulus and (b) the damping factor for PVC that has been plasticized with 40.2 vol.% dioctyl phthalate (DOP), 25.4 vol.% diethyl phthalate (DEP), 15.8 vol.% dibutyl phthalate (DBP)

Having satisfied the loading criterion, there are other factors which favour PVC for guttering: it is much lighter than metal guttering of equivalent thickness, so is much easier to handle in long lengths. Ease of handling is also aided by its low thermal conductivity (typical of all polymers) and low friction, a property which makes it easier to slide guttering between the supports. The low coefficient of friction at the supports helps in an unexpected way: polymers have high coefficients of thermal expansion ($\alpha_{PVC} = 70 \times 10^{-6} \, K^{-1}$) compared to steel ($\alpha_{steel} = 15 \times 10^{-6} \, K^{-1}$), so on a hot day when the temperature of the plastic rises to over 300 K, it will expand and so create unwanted stresses.

EXERCISE 5.11   The coefficient of thermal expansion is $\alpha = \Delta L / L_0 \Delta T$, where ΔL is the change in original length, L_0, which is produced by a temperature change ΔT. Calculate the change in length of a 10 m run of PVC guttering when heated by sunshine from 273 K to 303 K. How will friction at the supports affect the guttering and how could this be counteracted?

One potential problem remains. How long will guttering made of PVC last? It is a polymer highly resistant to water and most organic liquids, although it is soluble in cyclohexanone and THF, which are used to solvent-weld pipes, but is sensitive to ultraviolet light. The repeat unit (Appendix 2) suggests no sensitivity to ultraviolet light since it contains no aromatic or carbonyl groups, but the heat applied during processing can degrade the material and make it more sensitive. The problem is solved by the use of additives, of which carbon black is probably the cheapest and most effective.

5.3 Why plastics break

5.3.1 Types of failure

Understanding the mechanisms by which polymers fail, both at the molecular level and higher levels of organization, leads to positive ways of improving the toughness of polymers, subject to the ever-present constraint of processing to shape. Sudden failures of plastic articles are all too familiar, plastic knobs which come away in your hand, ice-cube trays which spring leaks when bent, bottles which leak when dropped and, of course, the inevitable failures of overloaded shopping bags! This last kind of ductile failure differs from brittle cracking and you will see how it can be predicted and prevented.

In the guttering example, one of the design criteria was that the material and the product should be strong enough to withstand installation

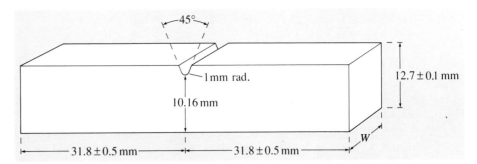

Figure 5.20 A specimen for an Izod impact test (BS 2782)

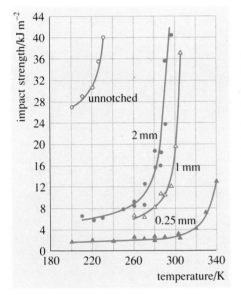

Figure 5.24 The effect that the radius of the tip of a notch has on the impact strength of PVC

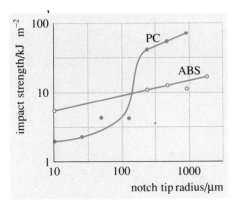

Figure 5.25 The effect of varying the radius of the tip of a notch on the impact strength of acrylonitrile-butadiene styrene and polycarbonate at 293 K

stresses. In practical terms, that implies some flexibility in order to enable the guttering to be manipulated into the supports, but it also covers other eventualities such as accidentally dropping the guttering or dropping objects onto the guttering. In both situations, impact strength is important and is affected by processing, the geometry of the product or other effects. Stress concentrations are particularly pernicious, lowering the strengths of apparently tough as well as brittle materials.

Pendulum impact tests on specially shaped and notched specimens (Figure 5.20) are popular because they are simple and quick to perform. The results of such tests, however, are only *qualitatively* useful in predicting the impact resistance of products. This is because the response of a product to an impact is complex. It depends on such factors as the local geometry of both the impactor and the product around the area of contact and the propagation of stress waves in the product and their interaction with each other and geometrical features of the product (e.g. edges, corners and holes). The very short-term (i.e. dynamic) mechanical properties of the materials involved also affect the response. Only for the simplest geometries is any analysis and prediction of performance possible.

Thus, increasingly, tests on complete products are being used to evaluate their response to impact, one of which is described in ▼Product impact testing▲. Pendulum impact tests are, however, useful in evaluating notch sensitivity in materials and the temperatures of the transition from ductile to brittle behaviour (Figures 5.24 and 5.25), as well as in quality·control: for example, comparing different batches of the same material.

Impact is not the only type of failure in plastics. Like other materials, plastics can fail at lower loading rates than those of impact, as well as under static loads and fluctuating loads (i.e. fatigue). There are three mechanisms of failure that reflect the singular nature of the structure and chemistry of plastics. These are **crazing, environmental stress cracking** and **creep rupture**, and we shall consider them in turn after we have looked as stress concentrations and their effect.

▼Product impact testing▲

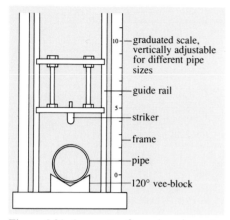

Figure 5.21 Apparatus for a drop impact test (BS 3505/4576)

Product impact tests evaluate a product in terms of its ductile or brittle behaviour and are preferred to specimen impact tests because they better reflect the effects of geometry and processing on a product's integrity. They are therefore often written into standards for plastic products. The drop test you will look at here is that used for PVC pipe, a product that can expect abuse on a building site. For example, when PVC pipes are unloaded from a truck on a cold day they are not handled gently! A drop of 2 m onto an uneven surface is quite probable. The tests are also used for quality control purposes because the processing of UPVC powder into pipes by extrusion must be carried out correctly to maximize the mechanical properties of the pipe. If the boundaries of the original powder survive into the pipe, the PVC is referred to as 'incompletely fused' and the impact strength will be lower.

Figure 5.21 shows the impact tester for PVC pipe, where 200 mm lengths cut from the pipe are supported on a 120° vee-block. A 25 mm diameter hemispherical-nosed striker falls a distance of 2 m onto the pipe, which typically deforms as

shown in Figure 5.22. The recommended striker mass, m, increases with the diameter of the pipe so, for instance, it is 2.75 kg when the mean pipe diameter is 114 mm. The test is quantified in terms of the impact energy, U, of the striker as it hits the pipe, which equals its potential energy before release

$$U = mgh$$

where h is the height the striker falls.

EXERCISE 5.13   What is the impact energy of the striker hitting a PVC pipe of 114 mm diameter under the recommended conditions? Estimate upper and lower bounds to the time scale of the impact from a knowledge of the velocity of the striker, v, at impact and the size of the striker and pipe. Assume $v = \sqrt{2gh}$.

Strain gauges are often attached to the rear face of the striker, the output being recorded on a storage oscilloscope. These show that impacts typically last 1–10 ms, depending on the elastic stiffness of the pipe.

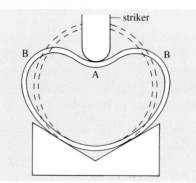

Figure 5.22 Deformation of a UPVC pipe during impact: at A there is biaxial tensile stress on the inside surface and at B there is radial tensile stress on the outside of the pipe

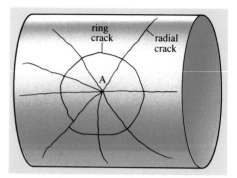

Figure 5.23 The pattern of brittle cracks in a PVC pipe tested at 273 K

What does this very short timescale imply?

From what we said earlier about the equivalence of time and temperature for viscoelastic materials, a faster rate is equivalent to a decrease in temperature. Since all polymers tend to become more brittle with decreasing temperature, the likelihood of brittle fracture will increase with higher impact velocities. If impacts on PVC pipe are conducted over a range of temperatures, a transition from ductile failure to brittle cracking occurs at about 273 K (Figure 5.23). Although the main glass transition of PVC occurs at 353 K, there is a subsidiary transition due to other molecular motions in PVC at about 225 K. It is thought that this peak may account for the brittle to ductile transition under impact at 273 K — chain molecular motion is inhibited and the chain molecules cannot easily respond to imposed stresses, so the pipe behaves in a brittle manner.

There may be other explanations for the brittle behaviour of UPVC, such as poor material, poor processing or the effects of notches. Nowadays, UPVC is toughened by copolymerization with elastomer to toughen it at ambient and low temperatures. This is used in products such as pipes and window frames.

5.3.2 Stress concentrations and strength

The notched impact test is an excellent example of putting a deliberate stress raiser in a sample so that failure will occur from a predictable point — the root of the notch where the stress is at a maximum. Such situations can be modelled theoretically, as shown by the stress-concentration diagram for a notch of hyperbolic shape in a flat plate of width w (perpendicular to the page), under a tensile force F (Figure 5.26). The distance of the notch tip of radius r from the base of the plate is d and the stress concentration factor, K_t, is the ratio of the stress at the notch tip (σ_{max}) to the nominal stress across the *remaining* section.

$$K_t = \frac{\sigma_{max}}{\sigma_{nom}} = \frac{\sigma_{max}}{F/wd}$$

so

$$\sigma_{max} = \frac{FK_t}{wd}$$

EXERCISE 5.14   Compared to the stress in an unnotched specimen, by how much is the stress raised by the notch in the Izod-test specimen in Figure 5.20, when loaded as shown in Figure 5.26?

Deliberate notches are widely used in brittle materials like inorganic glass to break them in a controlled way. Most polymer surfaces are soft compared with metals and glasses, so they are much more susceptible to surface damage by contact with harder materials. Scratch marks and wear can therefore weaken polymers substantially, but there are more pernicious stress concentrations which are frequently present as a result of design geometry, some inevitable, but many avoidable.

Stress concentration diagrams, such as Figure 5.26, show that notch effects can be minimized by several strategies:

(a) maximizing the radius of the notch tip;
(b) increasing the section over which the stress operates (i.e. increase w rather than d, since increasing the latter can increase K_t);
(c) repositioning inevitable notches away from areas likely to be highly stressed in use.

EXERCISE 5.15   Examine a number of everyday rigid polymer articles for inevitable and avoidable stress concentrations. What design steps have been taken to minimize the effects of stress concentration on the strength of the articles?

Another factor that makes stress-raising defects more serious is the interaction between different types of defect. If close enough together,

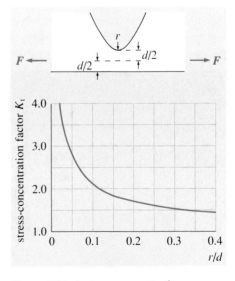

Figure 5.26 A stress-concentration diagram

the individual effects (K_1, K_2, etc.) are multiplied so that

$$K_t = K_1 \times K_2 \times \ldots$$

producing a much larger value of K_t. A sharp screw thread on a circular bottle top is a good example of this effect. A related problem occurs when the product design contains holes ($K_t \simeq 3$). During moulding, molten polymer must flow around them, dividing and rejoining on the opposite side. If the polymer is not hot enough, a visible **weld line** may occur where the material has not completely fused together. The weld line is then effectively a sharp crack at the edge of the hole and the resultant stress concentration can be very serious indeed (see ▼A failed bucket▲).

To resist both continuous stress or overload as well as occasional impacts, well-designed plastic products incorporate buttresses at highly stressed points with all the resultant corners well-rounded (Figure 5.30).

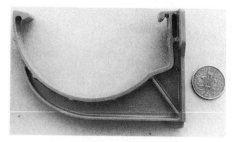

Figure 5.30 A support bracket for UPVC guttering buttressed on the underside

5.3.3 Crazing and molecular mass

Designing products to approach the strengths shown by plastics processed under optimum conditions (Table 5.1) can be achieved by careful analysis of the kind of stresses likely to be experienced in service, always backed up by rigorous tests on the prototype. Are there any correlations with molecular structure that might help in choosing the most suitable grade of a given polymer?

You need to look more closely at the effects of straining plastic products to failure. If a brittle transparent polymer such as polystyrene, widely used in rulers and the bodies of ballpoint pens, is broken in bending, shiny crack-like defects can be seen if the broken pieces are rotated in a bright light. They tend to occur close to the fracture surfaces of the broken parts. There will be few (less than fifty) and they will also be large (several millimetres in length). They are not cracks, but **crazes** and are still load bearing. They are the precursors to final failure in rigid polymers (Figure 5.31). Crazes are a form of highly

Figure 5.31 A polycarbonate tensile-test specimen containing several large crazes, produced by stressing while it is in an organic liquid

localized yielding and form on planes perpendicular to the tensile stress, usually starting from scratches on the surface. They are very thin initially, about 1 μm in width, and taper to zero at the advancing edge, with fibrils of drawn polymer within the bulk of the craze itself (Figure 5.32). The density of material in the craze is about half that of the bulk polymer, owing to the voids which have formed between the fibrils.

▼A failed bucket▲

Figure 5.27 The failed low-density polyethylene bucket

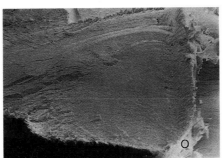

Figure 5.28 Fracture surface of the failed lug. Origin O, on the inside of the bucket

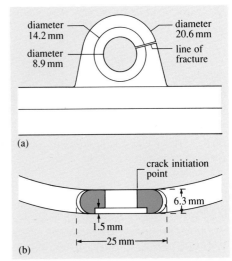

Figure 5.29 Details of the failed lug

Most fractures of plastic products are usually annoying, sometimes expensive, but rarely cause serious injury. Plastic buckets are such familiar domestic items that few of us would spare their design, material or manufacture a second thought. When they first appeared on the market, they were copies, in low-density polyethylene, of existing steel buckets, with little acknowledgement of any problems arising from the differences between the two materials. When full of water, a nine-litre bucket (the most common size) imposes a load of 45 N on each of the two handle lugs. The round lugs on a steel bucket each have a cross-sectional area of about 15 mm² and can easily withstand the imposed stress of 2.9 MN m⁻². With a Young's modulus of 210 GN m⁻², this is equivalent to a strain of only 1.4×10^{-5}. Low-density polyethylene has a short-term tensile modulus of 0.2 GN m⁻², so the plastic lugs were thickened substantially to reduce the stress on the section and reduce creep and the chances of failure.

Many such buckets failed, however, and in one case led to serious injury. At the time the bucket was nearly full with very hot water and, when one of the handle lugs

suddenly fractured (Figure 5.27), the woman carrying it was severely scalded. The failure was brittle with no yielding. The crack started at one corner of the section (Figure 5.28) and propagated straight across the lug at a slight angle to the horizontal (Figure 5.29). With a net cross sectional area on one lug of 57 mm², the nominal applied stress on the failed side was about 0.8 MN m⁻², corresponding to a strain of about 0.4%.

This applied stress was far short of the short-term ductile breaking strength of 10 MN m⁻², so how did *brittle* fracture occur?

The circular hole in itself represents a serious stress concentration and the lug walls constrain the imposed stress further so the net effect is to increase K_t. The investigators estimated that $K_t \simeq 5.9$, so that the maximum stress on the lug was about 4.1 MN m⁻², a value still less than the strength of the material. The temperature of the lugs was probably high due to the contents, but this could not explain the brittle failure, since most polymers increase in toughness as the temperature is raised, although their tensile modulus falls. The sharp inside corner on each lug probably increased the

stress further, but the crack did not initiate here (Figures 5.28 and 5.29), but rather from an external corner. There was, however, no evidence for fatigue, which can usually be detected by fine concentric striations centred on the initiation point (Figure 5.28) as described in Chapter 2, Section 2.4.

What other factors should be taken account?

An important variable is the way the material was processed to shape: the bucket was injection moulded from the centre of the base, so fluid polymer flowed up the sides and finally into the lugs. The molten front had to split and rejoin to create the lugs and if too cold would not fuse together, so forming a 'weld' line. These are very sharp cracks in the worst cases and concentrate the stress at the weld line tip more seriously than the factor of 5.9 estimated above (even a factor of 2 will push the total value of K_t to nearly 12). Several weld lines were found in the body of the bucket near the lug and it was concluded that it was probably a weld line in the lug that magnified the stress further and led to sudden catastrophic failure.

0   0.5 μm

Figure 5.32 A transmission-electron micrograph of a thin section through a craze in polystyrene showing bridging fibres

Crazing occurs at a relatively early stage in loading. It starts in polystyrene at less than $20 \, MN \, m^{-2}$, compared to its breaking strength of $50 \, MN \, m^{-2}$. A craze grows in size with further loading until a crack suddenly initiates and propagates within it and failure occurs.

Owing to the energy required to form crazes, a material can be toughened if the number nucleated can be increased and their growth inhibited. That is why acrylonitrile-butadiene styrene is so much tougher than polystyrene itself (Table 5.1), crazes forming at an early stage when it is strained. The crazes nucleate at the rubber–styrene acrylonitrile boundary in the two-phase structure due to the stress concentrating effect of the spherical elastomer particles (Figure 5.9) whose high extensibility inhibits craze growth. The smaller the particles, the more uniform their size, and the more evenly they are distributed in the rigid matrix, the tougher the ABS. A very high density of crazes is detected early in the stress–strain curve of polycarbonate, but the origin of the nucleation sites is still unknown. Because of the viscoelastic nature of polymers, craze formation depends on time. At very high testing rates they may be unable to form fast enough to relieve stresses at notch tips, so a tough material may crack in a brittle fashion. This is one reason for the brittle–ductile transition in impact behaviour.

EXERCISE 5.16   Describe what is seen when a thin strip of acrylonitrile-butadiene styrene is bent.

How is craze growth affected by molecular variables?

One critical variable is molecular mass: below the entanglement molecular mass, $\bar{M}_c$, the molecules are too short to bridge across crazes, so crazes become true cracks very quickly and the material possesses little strength. With increasing molecular mass, however, molecules are able to bridge across the craze and so strengthen it. The higher the molecular mass, the greater the improvement in strength. With most polymers a plateau usually occurs at a molecular mass of about 10^6 and Figure 5.33 shows this for two brittle amorphous plastics (PS and PMMA), as well as high-density polyethylene. The tensile strengths were measured at a high test rate and low temperatures so that the polymers

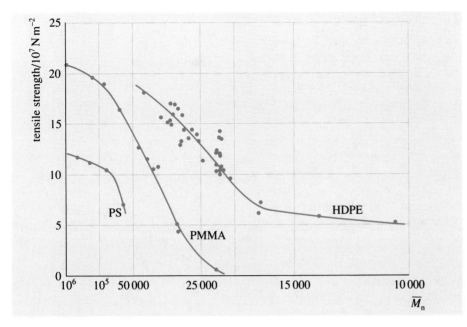

Figure 5.33 The tensile strength of three common polymers plotted against $\bar{M}_n$

were all brittle, but the effects of increasing molecular mass are quite general.

Craze formation in partially crystalline thermoplastics is more complex due to their spherulitic structure and the elastomeric nature of the amorphous regions. Stress whitening due to crazing, however, is familiar in polyolefins at stresses below the yield stress. The crazes probably form at the boundary between spherulites and the amorphous matrix.

The principle of controlling the molecular mass of polymers is applied extensively to products where strength is at a premium. Thermoformed baths made of poly(methyl methacrylate) have $\bar{M}_w$ approaching 10^6 to resist small impacts and repeated low-stress flexing. The effects of increased molecular mass are particularly dramatic where fatigue or wear may be a problem. That is why polyethylene with an ultrahigh molecular mass (UHMPE — see Chapter 4) of about 3.5–4 $\times$ 10^6 is used for the sockets of replacement hip joints. Figure 5.33 shows that to improve the strength of polymers, as high a molecular mass as possible is required, but this conflicts with another requirement, that of processability.

As the molecular mass of polymer increases, so does the viscosity of the melt and, hence, the difficulty of moulding. Special methods have been developed for grades of high molecular mass: PMMA is polymerized directly into cast sheets and then thermoformed to the shape of the bath; UHMPE is sintered to give the sockets of replacement hip joints. With injection moulding, however, the molecular mass is severely limited by process requirements, which in turn limits the maximum strength of the polymer. The strength may be further lowered by fillers, stress concentrations and damage in service.

5.3.4 Environmental stress cracking

The examples of thermoplastic products you have looked at so far are not those which involve exposure to fluids such as alcohol, dry cleaning fluids or other volatile organic fluids. Amorphous thermoplastics such as polystyrene, poly(methyl methacrylate) and polycarbonate can craze readily on exposure to such fluids, even without the application of stress (see Figure 5.31). The effect is essentially driven by the chain orientation produced by moulding which we shall discuss in the next section, and the fluid aids crazing by allowing localized relaxation of the polymer chains. The mechanism is analogous to plasticization, but is much more highly localized at craze tips. Eventually, a true crack will form inside a growing craze and, if the article is then stressed, it breaks easily. Just like a polystyrene ruler when bent, such crazes tend to grow from surface defects.

Environmental stress cracking can also occur in partially crystalline materials such as low-density polyethylene, although they are generally more resistant than amorphous thermoplastics. The first occurrence of environmental stress cracking in polyolefins was observed in the 1940s when low-density polyethylene cable insulation was greased and then flexed, for example, by being coiled. The imposed stresses were well within the breaking strength of the material, but the oil present in the grease was sufficient to start crazes, which grew slowly and, ultimately, led to brittle fracture. A similar effect occurred in washing-up bowls exposed to strong detergents, with long brittle cracks growing along the axis of orientation introduced by injection moulding.

How was the problem overcome?

Essentially the problem was solved by using material of higher molecular mass so that crazes were not so much suppressed as strengthened. Longer chains bridge the craze and effectively inhibit the growth of a crack within it. In more serious cases, the fluid causing the cracking must be removed or a different polymer used.

In general, the number of chemicals that can interact with polymer products is very large and manufacturers normally conduct both short-term and long-term tests to assess the sensitivity of their materials to particular chemicals. From what has already been said about plasticization, the higher the temperature of exposure, the greater the probability of such interaction, whether it be environmental stress cracking, crazing, swelling (where the polymer expands in the fluid) or dissolution (where the polymer goes into liquid solution).

Degradation in ultraviolet radiation is another hazard for many polymers, but is now usually prevented by the use of carbon black or special additives, which preferentially absorb ultraviolet radiation and convert it to less harmful thermal energy.

5.3.5 Creep rupture

With some problems of environmental stress cracking the rate of craze and crack growth is relatively slow and it can take years before sudden failure occurs. Failure can also occur, however, by **creep rupture**. No environmental agents are involved, although they would accelerate the process if they were involved. The phenomenon is important for pressurized water and gas pipes, but first look at a more familiar article, where rupture can occur as a result of overloading — a supermarket carrier bag.

A large amount of low-density polyethylene is used for packaging in the form of film. It is usually made by blowing air through and into an extruded hollow tube on a continuous basis — the blown-film process. It is an efficient way of converting polyethylene into tubular film for carrier bags with a thickness range of 0.01–0.5 mm. The bag needs to be flexible, to resist penetration by sharp corners of boxes, to take a load of 2.5–10 kg and yet use a minimum of material so that its cost is minimal.

A parallel-sided strip cut from such a bag can be pulled by hand to produce cold drawing. The specimen length increases by a factor of six to nine and the width and thickness decrease markedly. If the same test is carried out in a tensile-testing machine, with the crosshead moving at a constant rate, the load reaches a maximum just before the neck appears, then restabilizes at a lower level (Figure 5.34). During cold drawing, the neck, in which all the deformation is taking place, passes along the strip converting unoriented spherulitic low-density polyethylene into a highly oriented tape. It is this change of microstructure, in which the spherulites are destroyed and the polymer chains tend to line up in the direction of stretching, that makes drawn polymers stable. Fibre formation within crazes acts in a similar way in noncrystalline materials like acrylonitrile-butadiene styrene and polycarbonate. The neck propagates until it reaches the ends of the tensile specimen and only then does the load begin to rise further till failure occurs.

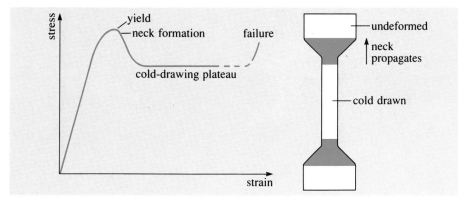

Figure 5.34 Stress–strain behaviour of low-density polyethylene at 293 K

When an overloaded plastic bag fails, the narrowest part near the handles usually necks down and the most serious defects in the edge propagate catastrophically across the handle. Some of the energy released by the crack propagating rapidly appears as heat at the crack tip and the rise in temperature will exacerbate the process since the tensile modulus decreases rapidly with increasing temperature. A loaded carrier bag left hanging on a peg may also fail after some time. This depends on the stress level and the creep properties of the material. The end point of creep curves is rupture— hence creep rupture — and this occurs at increasingly short times as the stress level is increased.

SAQ 5.6   (Objectives 5.2 and 5.4)
A carrier bag, 400 mm × 550 mm, is constructed of 0.05 mm thick blown film of low-density polyethylene. Elliptical holes are cut 40 mm from the top edge to form the handles. Given a one-hour creep-rupture stress of $10\,MN\,m^{-2}$ at 293 K, what is the *maximum* load that can be carried safely in the bag for a trip of one hour?

What load could be carried safely if the thickness of the bag was increased to 0.1 mm?

How would the safe load vary with (a) temperature of the environment and (b) nicks in the film near the handle?

Creep rupture is potentially more serious in UPVC water pipelines or polyethylene gas lines. Since the contents are under pressure, the wall of the pipe is stressed and designers must ensure that creep rupture is prevented over the designed lifetime of the pipe, normally fifty years. Rigorous creep-rupture tests on various types of polyethylene led to the development of a polyethylene of density $950\,kg\,m^{-3}$, roughly midway between that of low- and high-density polyethylene. This was achieved

SAQ 5.7   (Objectives 5.1 and 5.4)
Explain the following effects and suggest alternative polymers or grades for the failed product:
(a) PVC leathercloth plasticized with DOP is in contact with an acrylonitrile-butadiene styrene car dashboard. The acrylonitrile-butadiene styrene shows cracks in the contact zone.
(b) Garden furniture made of white pigmented polypropylene shows signs of cracking after prolonged exposure to sunlight.
(c) Domestic baths made from an injection-moulding grade of poly(methyl methacrylate) show fatigue cracks on the inner surface after normal use.
(d) Polycarbonate motor-cycle helmets show signs of crazing when treated with aerosol paint.
(e) Polycarbonate car bumpers crack from fixing holes.

by a particular kind of polymerization catalyst which introduced low controllable levels of chain branching. It lowered the degree of crystallinity and improved the toughness of the material. This medium-density polyethylene (MDPE) had better creep-rupture properties than high-density polyethylene and was thus selected for natural-gas pipes.

5.4 Making the most of chains

You saw in the previous section how increasing molecular mass generally improves the strength of polymers, and hence the strength of products, but brings problems of processability. This arises because the viscosity of the melt increases rapidly with rising molecular mass, so that a molecular mass of several millions usually means that special and, therefore, expensive methods have to be developed to shape the polymer. In this section you will see how methods of orienting chain molecules can be used to improve the mechanical properties of polymers, including both the tensile modulus and the tensile strength. However, elongation-to-break is usually sacrificed in the process.

We shall look first at the often uncontrolled effects of molecular orientation that are usually produced by moulding. This will lead on to the relatively simple methods that have been used to control orientation along a single axis, uniaxial orientation, which is exploited in net and mesh structures manufactured in thermoplastic polymer. The concept can be extended further by producing biaxially oriented products, of which the poly(ethylene terephthalate) bottle is the best example. Here, control of crystallinity is an important aspect of the full exploitation of the polymer property profile and can also shed light on the nature of crystallization in long chain materials.

5.4.1 Unwanted chain orientation

As you have already seen in Section 5.1.5, processing thermoplastics involves heating them to about 100 K above their glass-transition temperatures for amorphous polymers and for partially crystalline polymers to about 20 K above their melting points. The chains at these temperatures are sufficiently flexible to flow relatively easily to shape. Injection moulding involves pumping a molten polymer into closed steel moulds held at relatively low temperatures, 273–320 K, so that the solidification can occur before the shaped product is removed from the mould.

The conformation of a simple polymer chain is changed during flow to shape. The equilibrium conformation of a single polymer chain in the fluid state is a so-called **random coil** (Figure 5.35) where free chain rotation produces a shape that turns out to be, on average, roughly spherical. In a molten polymer, such randomly coiled chains will be highly entangled with one another. When strained during flow, the coils are distorted from their spherical shape to a greater or lesser degree depending on the strain.

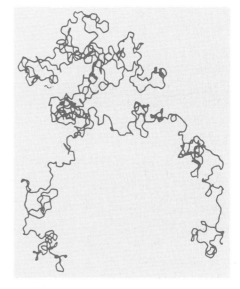

Figure 5.35 A random coil of a freely jointed polyethylene chain

The rates at which distortion or recovery of the polymer coils occur depend on the degree of flexibility of the chains and hence on temperature and time.

> EXERCISE 5.17   What will be the effect on the polymer chains of using very cool ($\sim$ 273 K) or very hot ($>$ 320 K) mould temperatures in the manufacture of a product composed of a noncrystalline polymer with a glass-transition temperature, T_g, greater than 410 K?

Since the bonding in polymer chains is highly anisotropic (Section 5.1), uncontrolled orientation produced by poor control of the mould (or melt) temperatures can lead to a product with a highly anisotropic resistance to crack propagation. The results of impact tests on injection-moulded bars made from ABS polymer are shown in Figure 5.36. Tests were carried out across and along the direction of flow and the effects of orientation produced by moulding at above normal temperature (500 K rather than 470 K) are clearly visible. The figure also shows the much more dramatic difference produced by using too cool a melt (440 K). The effects of orientation on strength are most severe with tough polymers, but they affect maximum achievable strength in all moulded polymer products (see ▼Orientation in polymer mouldings▲).

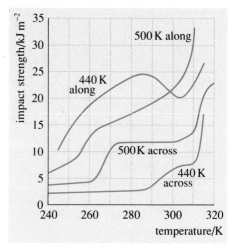

Figure 5.36 The effect of orientation on the impact strength of ABS polymer moulded at different melt temperatures

5.4.2 Uniaxial orientation

Molecular orientation can create problems in moulded thermoplastic products, but can it possibly be exploited? After all, very high impact strengths in acrylonitrile-butadiene styrene polymer can be obtained if the stress is applied parallel to, rather than perpendicular to, the chain axis (Figure 5.36). Provided some way can be found of orienting chains along a single axis, then the polymer should show a greatly improved tensile strength along this axis. It is really a question of matching product *shape* to the chain orientation within the product. When we discussed cold drawing in the previous section, the chains were oriented along the drawn bar. If the drawn bar is stressed along its length then a substantial improvement in tensile behaviour will be achieved. We therefore want to look more closely at cold drawing of partially crystalline polymers and how it can be exploited.

As we mentioned earlier (Section 5.3.3 and Figure 5.34), crazing occurs in partially crystalline thermoplastics before yield, but at larger strains spherulite deformation occurs. The lamellar crystals are distorted in such a way that they tend to line up at right angles to the tensile axis in the cold drawing region. Since the individual polymer chains are oriented at right-angles to the planar surfaces of the lamellae (Figure 5.6), the chains are therefore oriented along the drawing axis (Figure 5.39). The left-hand side of the diagram is an oversimplification of spherulite structure, since undeformed lamellae crystals will be oriented in all directions in the spherulite (Figure 5.7), but the main features of the model have been confirmed by detailed structural studies using X-rays and electron microscopy.

▼Orientation in polymer mouldings▲

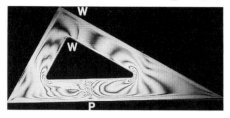

Figure 5.37 A polycarbonate set-square in polarized light showing the injection point at P and a weld line W–W

(a)   (b)

Figure 5.38 Polycarbonate battery cases in polarized light

The most useful method for examining chain orientation is by means of circularly polarized light. When oriented transparent polymers are placed between crossed polars (Polaroid sheet), the plane of polarization is changed because of the difference between the refractive indices (birefringence) of the polymer across and along the chain axis. Different wavelengths are affected to different extents, so that if a white light is used, coloured fringes are observed in the plastic object being examined. The sequence of colours observed can easily be demonstrated by flexing a thin piece of unoriented plastic between crossed polars, where the orientation is produced by low-strain elastic deformation. The birefringence can be measured quantitatively by using light of only one wavelength, usually yellow sodium light, and is directly related to the number of black fringes observed. The greater the number of fringes seen in a flat plane specimen, the greater the strain. In fact, engineers use the method widely to examine the effect of stress on polymeric models of structures such as buildings and dams. The zones that are likely to be most highly stressed in service are thus revealed and the design can be modified accordingly. Stress-concentration diagrams similar to that used in the previous section are determined in this way.

Relatively few polymers are sensitive

enough to give useful birefringence. They must also be optically transparent so excluding most highly crystalline polymers. Polycarbonate, polyurethane and epoxy resin are used widely for modelling. The sheet that is used must have very low initial birefringence, which means the chains must be close to their equilibrium conformations. Cast or extruded sheets are normally used because they possess little orientation compared with injection mouldings.

The second way in which polarized light is used in practical problems is the analysis of polymer orientation in mouldings. The object is to minimize the degree of orientation in areas likely to be impacted or highly stressed in service. Look first at a simple object such as a set-square (Figure 5.37). Although you would not expect it to be seriously stressed in service, it does show a high degree of orientation near the injection point, P, and at the two lower internal corners, judging by the density of fringes in these areas. A weld line, WW, is also visible at the top of the

set-square where two streams of the melt have only partially fused together.

The second example shows how the method can prove useful in monitoring orientation levels. The polycarbonate battery case in Figure 5.38(a) exhibits a very high fringe density, which was caused by a low mould temperature of less than 313 K. For the battery case in Figure 3.38(b) the mould temperature was in excess of 363 K and the rate of cooling was slower, allowing the chain molecules to come much closer to their equilibrium states. Although not so clearly visible in these plates, the highest fringe densities were found near internal corners and the orientation was such as to lower the impact strength drastically in places that already possessed severe stress concentrations. For products like this, which may suffer considerable abuse in service, polarized light analysis is a very valuable tool, not just for trouble shooting, but for routine quality control, since it is very easy to use and is entirely nondestructive.

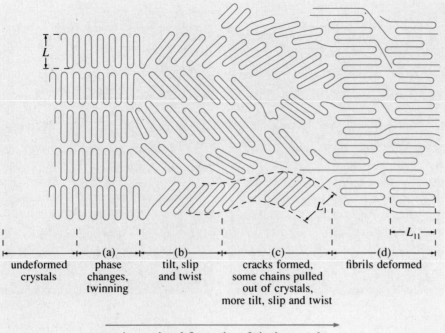

Figure 5.39 Deformation processes during cold drawing

As the strain increases, going from left to right, phase changes and twinning of crystals occur first, followed by gross tilting and slip of whole lamellae. Some unravelling of the crystals probably occurs, increasing the number of chains between crystals (so-called 'tie' molecules), so that in the cold drawn state the lamellar spacing has increased from its original value, L_I, to a larger value, L_II. The net effect is that single crystals are preserved, but in a substantially deformed state, and chain folding is maintained. The lamellae are stacked together in fibrils, which are oriented parallel to the drawing direction and are held together by a smaller number of lateral tie molecules than those along the drawing axis.

EXERCISE 5.18   How will cold drawing affect the tensile modulus of a partially crystalline thermoplastic?

The **draw ratio** of polymers is defined as the ratio of the cross-sectional areas before and after drawing: for most partially crystalline samples at 298 K, the ratio lies between about six and nine. Since the volume of the specimen is roughly constant, the length of drawn material is six to nine times the undrawn specimen length. For extrusion grades of high-density polyethylene, the tensile modulus increases from about $1.0\,\mathrm{GN\,m^{-2}}$ to values up to about $13\,\mathrm{GN\,m^{-2}}$ and the tensile strength from about $30\,\mathrm{MN\,m^{-2}}$ to several hundred. The elongation-to-break, however, is lowered substantially from several hundred percent to values of 1–10%.

246

Below a critical value of molecular mass, the draw ratio increases with decreasing molecular mass. For high-density polyethylene the critical value lies in the region of $\bar{M}_w = 100\,000$. Above this value, the draw ratio is constant, but below it, draw ratios up to about 15 are possible at normal strain rates (Figure 5.40).

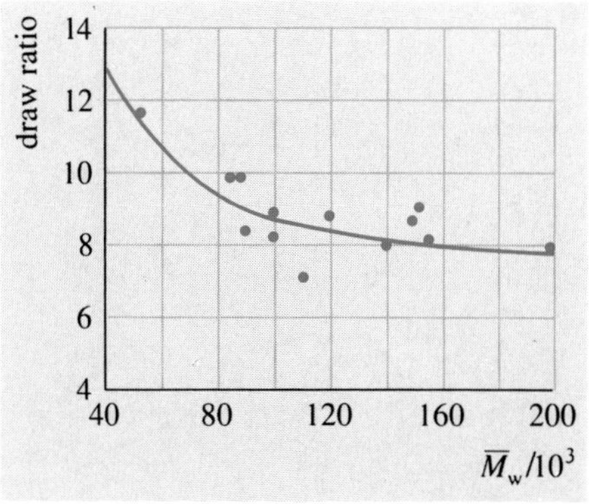

Since higher draw ratios give even greater chain orientation along the tensile axis, the tensile modulus increases yet further (Figure 5.41). As expected from our discussion of strength in the last section, however, the strength of the drawn material drops rapidly with lower molecular mass.

For partially crystalline thermoplastics with a molecular mass above about $\bar{M}_w = 100\,000$, drawing improves both stiffness *and* strength and can be usefully exploited in many products (see ▼**Exploitation of drawn thermoplastics**▲).

5.4.3 Biaxial orientation

If uniaxial orientation gives such big improvements in product stiffness and strength, are there any other ways in which orientation can be utilized? The second principal orientation method is termed biaxial orientation. If the random coil model for a freely jointed chain (Figure 5.35) can be represented as a sphere, then uniaxial orientation will distort the sphere into a cigar shape — a prolate ellipsoid. Orientation along *two* axes, however, will distort the sphere into a squashed sphere — an oblate ellipsoid — as shown in Figure 5.44.

How is this achieved and how can it be used in products?

The simplest way of producing biaxial orientation is simply by blowing up a rubber balloon. The material responds to the uniform air pressure by extending equally in all directions in the plane of the material. This is roughly what occurs during the process of blow moulding, where a tube of thermoplastic is heated past its melting point or glass-transition temperature, inserted into a steel mould of the desired shape and

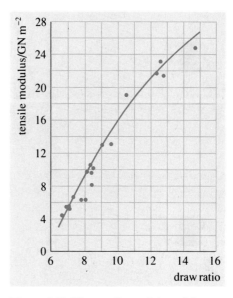

Figure 5.40 The variation in the draw ratio of high-density polyethylene with $\bar{M}_w$ at 293 K

Figure 5.41 The tensile modulus of drawn monofilaments of high-density polyethylene plotted against draw ratio

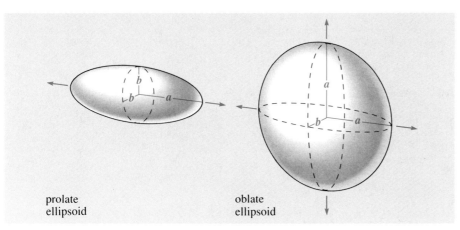

prolate
ellipsoid

oblate
ellipsoid

Figure 5.44 Uniaxial and biaxial distortions of a sphere

▼Exploitation of drawn thermoplastics▲

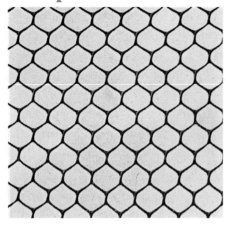

(a)

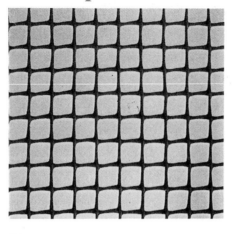

(b)

(c)

Figure 5.42 Extruded net structures

Thermoplastic nets (Figure 5.42) are widely used for packaging fruit, vegetables and many other products. Unlike conventional textile nets, which are produced from twisted fibres, thermoplastic net is extruded from a contrarotating set of split die holes to produce a tubular product which provides flat netting when cut open. It is used in both the normal unoriented form and in an oriented form where the individual filaments have been drawn uniaxially. Polymer of high molecular mass is used, so the draw ratio is relatively low, about six. Because the yield stress decreases with increasing temperature, the filaments are usually drawn by passing the extruded net through hot water at a temperature above 333 K and reeling-up under stress. Because of the enhanced molecular flexibility of the noncrystalline zones at this temperature, the drawing stress is also lower.

What are the main advantages of drawing to the user?

In stressed applications, like some packaging net, the filament modulus and strength are considerably higher and creep is much lower than unoriented polymer. It means that the net will be stable under a tensile load, but flexible enough in bending to conform to the shape of the contents. The stiffness of the filaments can be varied over a large range by using different polyolefins, low-density polyethylene giving a less stiff oriented net than high-density polyethylene or polypropylene owing to its lower degree of

crystallinity. The stiffness can also be varied by changing the filament diameter or cross-sectional shape. Hot drawing is not always performed to improve mechanical properties, however. Some forms of garden netting, such as pea and bean support netting, are drawn simply to change the size of the mesh.

There is an important limitation on oriented extruded net polymer. The nodes where filaments cross are relatively large and contain much more oriented polymer than is actually needed to hold the net together. To increase the specific stiffness and strength of heavy duty net used in civil engineering applications, an entirely different approach is taken. Extruded polyolefin sheet of highly uniform thickness is stamped with circular dies so as to produce a square array of round holes, as shown in the lower part of Figure 5.43.

When uniaxially oriented at high temperature, the material between the holes draws down so that the circular holes become highly distorted ellipses which can be seen in the top part of Figure 5.43. The side of a hole at right-angles to the direction in which the sheet is being drawn concentrates the stress by a factor of about three, so drawing starts here and progresses in both directions until the neck can progress no further. The mesh can also be uniaxially drawn at right-angles to the first drawing direction to make 'square' mesh. This 'Tensar' mesh made from 4 mm thick high-density

polyethylene sheet is typically used for fencing as well as supporting road and runway foundations. In fencing, it competes with steel wire mesh and can be tensioned in the same way to increase the bending stiffness. The mesh gives an improvement of 10–15 times the specific stiffness and strength of conventional filament extruded net and is virtually corrosion free. Protection against ultraviolet light is provided by addition of low levels of carbon black.

Figure 5.43 Uniaxially drawn high-density polyethylene mesh (top) produced from stamped sheet (bottom)

expanded internally by means of an applied air pressure. The process is used to make a wide variety of containers such as high-density polyethylene milk containers, PVC water bottles and low-density polyethylene detergent bottles. The materials are reasonably cheap, tough, and easily processable.

Suppose you wished to make a pressurized container, such as those needed for holding beer, cider and carbonated soft drinks. Would these materials be sufficient for the purpose?

To answer that question you really need to look at first, the stress system operating in a pressure vessel, secondly, the mechanical and other properties required and, finally, to see which polymers meet those requirements. You will see that biaxial orientation will be an important part of the answer and that crystallizable polymers have many advantages over noncrystalline polymers in the control they offer over mechanical properties (see ▼Polymer crystallization▲).

5.4.4 Pressurized containers

The product in which we are interested is the bottle for carbonated soft drinks. It comes in various sizes — 1, 1.5, 2, 3 litres — and is found increasingly on supermarket shelves. Marketing experts discovered that greater amounts of carbonated drinks could be sold if the containers were larger, so there was an incentive to develop safe packaging for large bottles. Traditional glass bottles are quite dangerous; if accidentally dropped onto the tiled floor of a supermarket, glass fractures into extremely sharp fragments. The stored energy of the compressed gas inside propels the fragments at high speed and several deaths have occurred in this way in American supermarkets.

The pressure inside is about four atmospheres, or $0.4\,\mathrm{MN\,m^{-2}}$. How is this translated into a stress in the wall of the bottle?

There are two stresses developed in a thin-walled cylindrical pressure vessel of cylindrical diameter D and wall thickness t. They are the hoop stress σ_θ and the longitudinal stress σ_z (see 'Stresses in the walls of a pressurized vessel', Section 2.4) and are related to the net internal pressure p and the bottle dimensions by Equations (2.2) and (2.3)

$$\sigma_\theta = \frac{pD}{2t} \qquad \text{and} \qquad \sigma_z = \frac{pD}{4t}$$

The hoop stress acting around the wall tending to split it lengthwise is *twice* the longitudinal stress and is therefore the most serious stress for a wall material to withstand.

EXERCISE 5.20   What is the hoop stress in a two-litre bottle with an internal pressure of four atmospheres, a wall thickness $t = 0.30\,\mathrm{mm}$ and diameter $D = 100\,\mathrm{mm}$? Would any of the polymers shown in Table 5.1 be suitable to withstand the calculated stresses?

▼Polymer crystallization▲

The crystallization behaviour of polymers is controlled by the nature of long chains and their time- and temperature-dependent properties. To explain their behaviour, you need to recall the nature of crystallization in polymers. The basic backbone chain must itself be ordered for crystals to form at all, but the *rate* of spherulitic crystallization varies enormously from polymer to polymer. Table 5.4 shows the maximum growth rate in moulding-grade polymers.

Table 5.4 Crystallization rates of polymers

Polymer	T_g/K	T_m/K	Maximum spherulitic growth rate/nm s^{-1}
high-density polyethylene	183	403	83 000
nylon 6,6	273 (dry)	540	20 000
acetal (POM)	198	448	6 700
polypropylene	263	449	340
poly(ethylene terephthalate)	338	549	170
isotactic polystyrene	373	513	5
polycarbonate	415	549	0.2

> EXERCISE 5.19 Is there any correlation between the rotational behaviour of chains and the rates of crystallization shown in Table 5.4? If so, explain it in terms of the repeat unit structures (Appendix 2).

In fact, materials like polypropylene have such low crystallization rates that nucleation agents are added to speed them up. Poly(ethylene terephthalate) crystallizes at about half the rate of polypropylene and, in practical terms, this means that at the optimum crystallization temperature, its spherulites will take about 1.6 hours to grow from 1 μm to 1 mm in size. This behaviour can be exploited, however, for it means that amorphous (i.e. noncrystalline) poly(ethylene terephthalate) can be produced very easily and it is widely used in thin-film form for audio- and videotapes. It can also be highly crystalline, however, and this form, which is white and opaque, is produced in sheets or mouldings. It is used as a drafting material which is stable when subjected to heat and in the presence of moisture. Crystallization rates are determined mainly by the rate of cooling during processing, fast rates giving amorphous material and slow rates giving crystalline material.

Is there any other way in which crystallization rates can be controlled?

One form of control comes through molecular mass. As Figure 5.45 shows, *reducing* molecular mass *increases* the crystallization rate. This behaviour can be understood by examining the single-crystal model (Figure 5.6). For a chain to crystallize with others requires long-range flexibility and it is easier with shorter

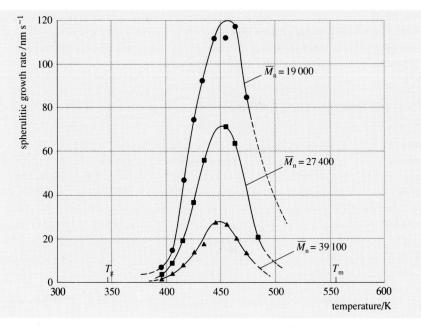

Figure 5.45 Effect of molecular mass on isothermal spherulitic growth rate of poly(ethylene terephthalate)

chains than with longer ones, where entanglements restrain the chain from moving into crystallites.

Another, somewhat different type of crystallization is that produced by orientation. If chains can be lined up in an orderly conformation, then crystallization will be aided. This is precisely what happens during orientation by strain, but the temperature must be such that the chains have some molecular flexibility so that they can pack together more easily. Orientation crystallization thus occurs above the glass-transition temperature: in the case of poly(ethylene terephthalate), it must be above 340 K. It is a form of

crystallization of polymers that does not generally involve spherulites, but chain folding does occur along the strain axis. If the orientation is performed on a polymer above its glass-transition temperature, but below its melting point, the crystallites are stable. If the orientation is performed above the melting point, the crystals will melt rapidly on removal of the strain. This phenomenon occurs in crystallizable rubber (such as natural rubber, which has a melting point of about 298 K) at high strains and the crystallites increase the strength and high-strain modulus substantially. As you will see in the next chapter, it is an important form of crystallization in fibrous polymers.

Table 5.5   Permeability of polymers to carbon dioxide at 293 K

Polymer	Permeability/$10^{-9}\,m^2\,day^{-1}\,bar^{-1}$
poly(vinylidene chloride) (PVDC)	0.12
polyacrylonitrile (PAN)	1.2
poly(ethylene terephthalate)	5–8
UPVC	8–16
polypropylene	59
high-density polyethylene	118

Nylon would be unsuitable due to its sensitivity to water, so what changes in design could be implemented to increase the strength of the bottle?

One design solution would be to thicken the walls considerably, but that would increase the mass and hence the cost of polymer used. Biaxial orientation offers a better solution involving no increase in the mass.

There is another consideration that is probably not very obvious and that is the permeability of the material to the diffusion of carbon dioxide gas from the bottle. Because of their low densities, polymers are poor barriers to gas diffusion, although there is considerable variation from polymer to polymer (see Table 5.5). Biaxial orientation lowers the permeability, which is why a range of values is shown for poly(ethylene terephthalate) and UPVC. The loss of carbon dioxide is proportional to permeability and to the ratio of surface area to volume of the bottle, so bigger bottles are better since they have a smaller ratio of surface area to volume. It is inversely proportional to the thickness of the wall. As a rough guide, for a two-litre bottle with a wall of normal thickness to have a ninety-day shelf-life, permeabilities down to that given for UPVC in Table 5.5 are acceptable. Polyacrylonitrile is ruled out because there is some concern that any residual acrylonitrile monomer in the bottle could be toxic.

That leaves PVDC, PET and UPVC. The first is an expensive speciality polymer so can be excluded, leaving the final choice as PET or UPVC. Since UPVC is practically noncrystalline, orientation crystallization is prevented, but PET is crystallizable and so it is the strongest candidate for pressurized containers.

5.4.5 The PET pressurized bottle

Unless the material is strong enough, the hoop stress created by pressurization will tend to create longitudinal cracks in the wall of a bottle. To resist hoop stress, the PET should therefore be more highly oriented around the bottle than along its length. This is what is effectively achieved during blow moulding because the noncrystalline preform (Figure 5.46) is stretched $2\frac{1}{2}$ times in length, but $3\frac{1}{2}$ times in diameter when blown to its final shape. This occurs at about 373 K, when the material is elastomeric: spherulite growth is suppressed (Figure 5.45), but orientation crystallization (15–20% crystallinity) occurs to

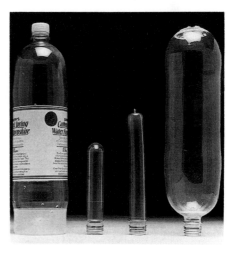

Figure 5.46 A complete 1.5-litre poly(ethylene terephthalate) bottle, with polyolefin base and cap; injection moulded preform; stretched preform; and blown bottle

251

improve the stiffness of the wall, its yield strength and toughness. Spherulite growth must be prevented because transparent walls are good for marketing the product and large spherulites embrittle the material. The molecular mass must be carefully controlled and $\bar{M}_n = 24\,000$ represents the optimum between the conflicting demands of strength, control of crystallization and processability.

The higher degree of orientation crystallization in the hoop direction is shown by the stress–strain curves of bottle material tested in the hoop direction and longitudinally (Figure 5.47). These curves may be compared with that for noncrystalline unoriented PET, the lowest curve. The yield stress increases from $40\,\mathrm{MN\,m^{-2}}$ to $80\,\mathrm{MN\,m^{-2}}$ in the longitudinal direction and to $165\,\mathrm{MN\,m^{-2}}$ in the hoop direction, more than sufficient to resist the stresses of $33.3\,\mathrm{MN\,m^{-2}}$ and $67\,\mathrm{MN\,m^{-2}}$ produced by pressurization. The relatively low level of creep (1% on diameter) mainly occurs within two to three days of pressurization. Although PET is expensive (£2000 tonne^{-1}), the improvement in properties allows a very thin wall of 0.25–0.30 mm, so the total mass of material in a two-litre bottle, 0.05 kg, is only about 2.5% of the total mass of the full bottle. PET now has a major share of the carbonated soft-drink market and a growing proportion of the beer and cider market. In the latter, a very thin PVDC coating is applied to the outer surface to inhibit oxygen diffusing inwards. This is needed to prevent oxidation of the alcoholic contents.

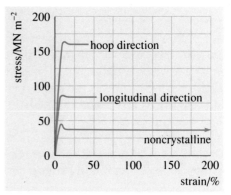

Figure 5.47 Stress–strain curves at 293 K for samples cut from the two principal directions in the poly(ethylene terephthalate) wall of the bottle compared with the stress–strain curve for the glassy preform

SAQ 5.8 (Objectives 5.1 and 5.5)
A three-litre bottle, made of poly(ethylene terephthalate) and with a diameter of 120 mm, is required to withstand a pressure of four atmospheres. Assuming that is must withstand the same hoop stress as a two-litre bottle, calculate the thickness of the wall needed.

The ratio of the surface area to the volume, S/V, of a cylinder with hemispherical ends of length L and of diameter D is given by the formula

$$\frac{S}{V} = \frac{12L}{D(3L - D)}$$

If the three-litre bottle has a total length of 340 mm and the two-litre bottle has a total length of 270 mm, estimate the shelf-life of the three-litre bottle compared with that of the two-litre bottle, assuming that permeation through the wall is the main cause of gas lost.

Could poly(ethylene terephthalate) be replaced by UPVC of the same thickness in a three-litre bottle?

5.5 Bouncing back: elastomers

We have referred at several points in this chapter to elastomers and the elastomeric 'state', but what is it exactly and how can it be exploited in engineering products? These two questions are addressed in this section and we shall look first at the basic nature of elastomers and how chain structure and crosslinking control their mechanical properties. If you look at Table 5.1, one of the most important elastomers, natural rubber, at first seems a very unlikely candidate for engineering products because of its very low value of tensile modulus. In fact, its properties can be improved by crosslinking, the judicious use of fillers, particularly carbon black, and by fibre reinforcement. Engineering products are often composed of comparatively large amounts of elastomer to achieve a respectable overall stiffness. This is particularly true of the massive elastomeric bearings used to support bridges and the similar mounts used to support buildings. There are also other strategies used to improve stiffness. Air under pressure, for example, is an excellent way of stiffening a flexible hollow product and the car tyre is probably the best example where pressurized air is used as an engineering material. Crosslinked elastomers are used because they show long-range reversible elasticity, but there is also a bonus. This comes in the form of damping, where the deleterious effects of vibration can be controlled by elastomers. As you will see, different elastomers possess different damping properties so that engineering with elastomers involves selecting the right elastomer for the intended function.

5.5.1 Mechanical behaviour of elastomers

The most surprising feature of elastomers is their long-range reversible elasticity. An ordinary rubber band can be stretched by over seven times its original length, yet it returns to its original length when released (Figure 5.48). By contrast, steel can be considered perfectly elastic only for strains of up to around 0.1%.

In metals and ceramics, elasticity is due to very small displacements of atoms from their equilibrium positions in the crystal: release the load and the atoms return to their original positions. Is there a simple molecular model to explain elasticity in elastomers?

Most certainly, yes. All elastomers are long-chain polymers with a glass-transition temperature that must lie below about 293 K and is usually very much lower (see Table 5.1).

EXERCISE 5.21   Examine the elastomeric repeat units shown in Appendix 2. What is the common bonding feature in the chains of polybutadiene, polychloroprene and natural rubber? What effect will it have on chain flexibility?

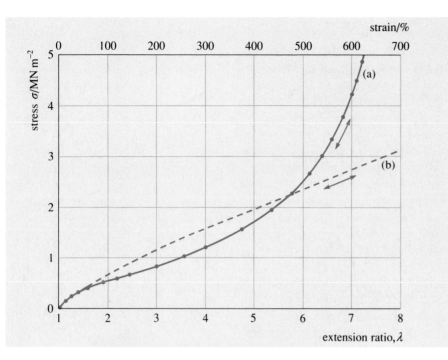

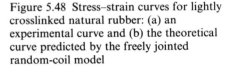

Figure 5.48 Stress–strain curves for lightly crosslinked natural rubber: (a) an experimental curve and (b) the theoretical curve predicted by the freely jointed random-coil model

Elastomer chains are greatly extended when strained from the random coils that exist in the unstrained state (Figure 5.35), but why should a crosslinked elastomer snap back to its original length? The basic reason is thermodynamic. In the strained state the chains become oriented along the tensile axis and, as you saw in the previous section, orientation is a form of chain ordering. The entropy, S, being a measure of disorder, will therefore *decrease* on straining. Provided the enthalpy, H, does not change

$$\text{free energy change on straining} = \Delta H - T\Delta S$$
$$= -T\Delta S$$

where T is the absolute temperature. Since T is always positive and ΔS is negative, the free energy change must be positive. The strained state is therefore energetically unfavourable. When the strained rubber band is released, it will return to a more favoured higher-entropy state where the chains have random conformations.

An analysis of elasticity in elastomers using a random-coil model of freely rotating repeat units with infinitely long chains shows that the stress–elongation behaviour is given by

$$\sigma = G\left(\lambda - \frac{1}{\lambda^2}\right)$$

where $\lambda = L/L_0$ and G is the shear modulus of the elastomer. The theory also predicts that the shear modulus is related to the number of network chains, N, by the equation

$$G = NkT$$

where k is Boltzmann's constant. A network chain is defined as that part of a chain lying between crosslinks, so *increasing* the number of crosslinks per unit volume of elastomer *increases* the number of network chains per unit volume. Since there is a constant number of polymer chains per unit volume, the molecular mass of chains between crosslinks *decreases* with increasing crosslinking. Therefore

$$G = \frac{\rho RT}{\bar{M}_n}$$

where ρ is the density of the elastomer and R the gas constant. The number average molecular mass of the network chains, $\bar{M}_n$, can be estimated from the degree of crosslinking, so the shear modulus, G, can be calculated at any temperature. Hence, the stress–strain curve can be derived using the equation above. The result is shown by the broken line in Figure 5.48.

The random-coil model shows reasonable agreement at low strains with the stress–strain curve of natural rubber, but diverges at high strains. This is attributed partly to the effects of entanglements and the finite length of real network chains and partly to energetic effects, all of which have been neglected in the simple random-coil model. Entanglements and finite chain length will limit the total elongation possible by interfering with orientation, so that real elastomers show a very steep rise in stress at high strain. Some elastomers can crystallize at high strains and these show an even steeper rise in stress at high strain owing to the reinforcing effect of crystallites.

EXERCISE 5.22 How will the modulus of a crosslinked elastomer change with increasing temperature? Is there anything unusual about this behaviour compared with metals or ceramics? What does this suggest about the bonding which controls the stress–strain behaviour?

For most elastomers, the change in volume when strained is very small indeed, so Poisson's ratio, v is about 0.5 and you can therefore say that the tensile modulus is about three times the shear modulus (see Chapter 1).

5.5.2 Engineering applications

For engineering applications, the modulus of elastomers can be controlled by the degree of crosslinking. Unfortunately, high degrees of crosslinking with sulphur, the most common crosslinking agent, produce a material where the long-range elasticity is lost. It is known as ebonite when the elastomer is natural rubber and is stiff and brittle.

Are there any other ways of changing the modulus of elastomers without compromising their long-range reversible elasticity?

We referred earlier to the use of fillers in the context of PVC guttering and other products, where the filler is added at relatively low levels (several percent). The main function of the filler was to inhibit degradation by ultraviolet radiation or to provide pigmentation. For elastomers the principal filler is carbon black, a relatively low-cost material which can be produced in a variety of particle sizes (see ▼Carbon black▲). With good mixing, most commodity elastomers will accept high levels of carbon black, up to about 40 weight %, increasing modulus without sacrificing reversible elasticity. The breaking strain, however, is drastically reduced.

EXERCISE 5.24   Make a list of elastomeric products and indicate whether or not they are filled with carbon black. To what approximate degree of strain are those products normally subjected? Is a high degree of reversible strain always required in those elastomeric products?

▼Carbon black▲

Carbon black is a widely used pigment, but a more important application is its use as a reinforcing agent for elastomers. As every schoolchild knows, it can easily be made by placing a metal spoon over a candle flame: the luminosity of the flame is caused by incandescent carbon particles, which are collected on the cold surface. The large-scale manufacture of carbon black produces particles of various sizes depending on the exact condition of production (Figure 5.49a). The basic unit of carbon black particles made using the oil furnace process, the most important industrially, is 20–30 nm in diameter. It consists of short lamellar graphite plates packed together into a roughly spherical shape (Figure 5.49b). Larger particles consist of accretions of this basic unit, but it is the smaller grades that are used for reinforcing elastomers. When added to thermoplastics for protection against ultraviolet light, much larger sizes (e.g. FT and MT blacks) are used, but they give little reinforcement.

The key to understanding the action of oil-furnace blacks lies not just in their small size, but also their chemical nature. The carbon black surface is very reactive to rubbers, primarily because the double bonds present in common elastomers can react easily with free radicals present on

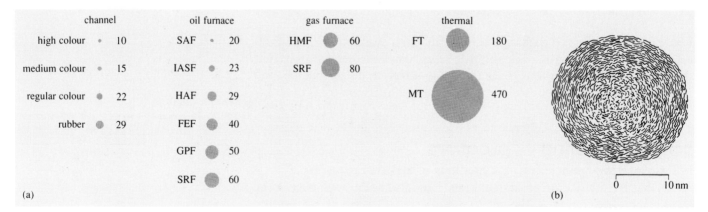

Figure 5.49 (a) The diameters, in nanometers, of particles of carbon black made by different methods and (b) the graphitic layer structure of oil-furnace black

the black surface. When a high-abrasion-furnace (HAF) black is mixed with natural rubber, for example, part of the rubber gels very quickly, indicating chemical crosslinking. Other rubber molecules are more weakly bonded to the carbon black surfaces.

Their reinforcing effect can be described by a relatively simple model, which helps to explain the mechanical properties of filled elastomers. Following formation (Figure 5.50a), the first elongation results in chain slippage so that the full strain is balanced between the chains connecting the particles (Figure 5.50b and c). On elastic retraction (Figure 5.50d), the chains between particles are roughly equal in length so that repeated straining is smoothly absorbed by the larger number of network chains in the reinforced material.

EXERCISE 5.23   What effects will increasing the HAF carbon black content have on the stress–strain behaviour of natural rubber in terms of: (a) its shear modulus, G; and (b) its ultimate strain?

The effects of increasing carbon black content on the modulus of natural rubber are shown in Figure 5.51: notice that the shear modulus has been plotted against the frequency of the test, since the modulus was determined by an oscillatory method. As the carbon black content is increased, the modulus at 10^{-4} Hz increases from about 0.25 MN m^{-2} to about 1.5 MN m^{-2} with a content of 50 phr carbon black. Carbon-black reinforcement affects the glass-transition temperature to a much smaller degree than crosslinking. Rather, it tends to 'smear' the transition out, as can be judged by the rise in the shear modulus at high frequencies of 10^4–10^6 Hz. High frequencies correspond to short times and hence low temperatures (from time–temperature equivalence). The smearing effect is due to decreased chain mobility, produced mainly by chain adsorption on the black particles. It is important that chain mobility is not too severely hindered, primarily because carbon-black reinforced elastomers should retain their elasticity at low temperatures (below 273 K) and high frequencies. Car tyres, for example, are often exposed to below freezing temperatures at driving speeds, where $\omega = 10$–30 Hz (68–204 km h^{-1}).

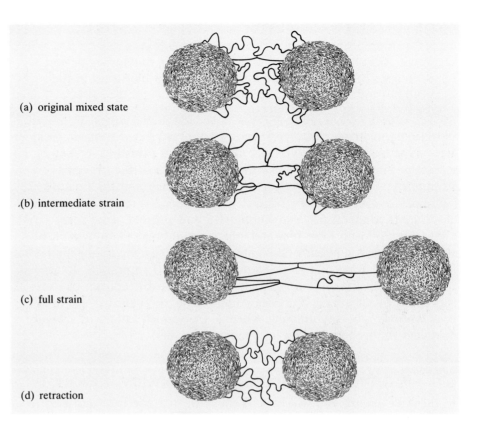

(a) original mixed state

(b) intermediate strain

(c) full strain

(d) retraction

Figure 5.50 A molecular model of carbon-black reinforcement

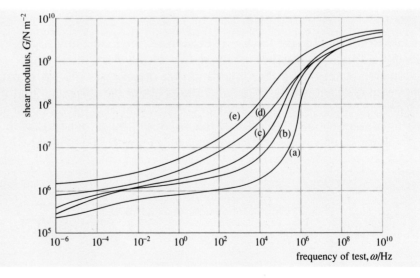

Figure 5.51 The effect on the dynamic shear modulus of natural rubber of reinforcing it with carbon black: (a) vulcanized with 1.5 phr sulphur, (b)–(e) 1.5 phr sulphur + 10, 20, 30 and 50 phr HAF black, respectively (phr = parts per hundred rubber by mass)

Many engineering products do not need the high strains that can be achieved in unfilled elastomers, although 100% reversible strain is still very high compared with other materials. With a maximum carbon black content of about 40% by weight, the reversible strain in the innertube or lining of a car's tyre, for example, is about 150%.

In the random-coil model, you were interested in elastomers in tension, but of course, many elastomeric products are subjected to more complex strains: balloons and car tyres are inflated by air pressure, so they are in a state of biaxial stress, just like the plastic bottle of the previous section; the walls of car tyres are flexed in bending and the tread is compressed and sheared by contact with the road surface. Many other elastomeric products are also compressed and sheared: shoe soles, elastomeric nonslip floor tiles and engine mountings. Compression of elastomers has a number of advantages in load-bearing applications, not least of which is that, for a given deflection, the compression modulus is three times the shear modulus. Very large imposed loads can be supported by elastomer blocks in compression and catastrophic failure in compression is very unlikely, whereas the sudden failure of a highly stretched rubber band is not an uncommon event!

5.5.3 Controlling stiffness in elastomeric products

In the next section you will be looking at elastomeric bearings used to support large structures, particularly bridges. The mass of a bridge can be very high, so if elastomeric bearings are used, they must be able to support the load without substantial deflection. In addition, they must be sufficiently flexible in shear so that thermal movements of the bridge can be accommodated. These apparently conflicting requirements of flexibility in one plane, but great stiffness in another, require a little explanation — how is it done?

The key to understanding the use of elastomers in engineering products boils down to an appreciation of the nature of the elastomeric state. Elastomers are very much like liquids in that they resist bulk hydrostatic compression, which is why Poisson's ratio is about 0.5. It means that if an elastomer block is put into *uniaxial* compression and the faces are fixed to prevent sliding, the sides of the block will bulge out so that the total volume of elastomer remains constant (Figure 5.52). In shear, the

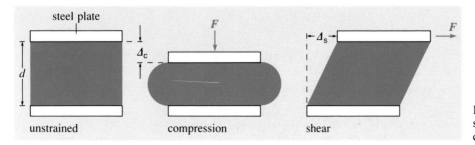

Figure 5.52 An elastomer block bonded to steel plates — unstrained, under uniaxial compression and under shear

sides of the block deform uniformly without any bulging. Notice the
block is bonded to steel plates, so that its top and bottom surfaces are
restrained and cannot deform laterally, whereas the centre of the block
is little restrained and so can bulge out. If the thickness of the block is
decreased, the restraining effect should *increase* since the degree of
bulging possible is reduced. A thin elastomer block will therefore be
stiffer than a thick block and this is the basis of controlling the
compressive stiffness of elastomer bearings. A simple measure of the
effect of thickness on stiffness can be calculated using a **shape factor**, S,
defined as the ratio of loaded area, A, to force-free area, A_1. For the
rectangular block shown in Figure 5.53, the top surface is the loaded
area and the *four* vertical surfaces are unloaded. Thus

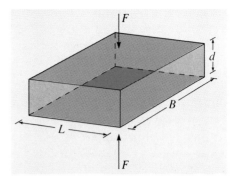

Figure 5.53 The dimensions of an
elastomer block used to determine its
shape factor

$$S = \frac{A}{A_1}$$

$$= \frac{LB}{2d(L + B)}$$

Thus if the thickness, d, is reduced, the shape factor increases. There is a
simple expression for the compression modulus of the block, E_c, in
terms of tensile modulus, E, and shape factor. For plane circular discs
or plane rectangular blocks, where B and L are not too different from
one another, then

$$E_c = E(1 + 2S^2)$$

The compression modulus of an elastomer bearing is thus very sensitive
to shape factor, S. The compressive stiffness, K_c, of a block is defined as
the compressive force, F, divided by the deflection, Δ_c (Figure 5.52), so

$$K_c = \frac{F}{\Delta_c} = \frac{FA}{A} \times \frac{d}{\Delta_c d} = \left(\frac{F}{A} \times \frac{d}{\Delta_c}\right) \times \frac{A}{d}$$

$$= \frac{E_c A}{d}$$

EXERCISE 5.25   What is the compression modulus, E_c, of a square
elastomeric bearing of side 0.2 m, thickness 50 mm and tensile
modulus 3.0 MN m^{-2}? Assume the bearing is bonded to steel plates
on its two largest surfaces. What deflection would occur if it were
loaded with a mass of five tonnes?

The deflection in the example given in Exercise 5.25 represents a
compressive strain of 13.6%, which is well within the capabilities of
carbon-black reinforced elastomers. The shape factor is low, but even
so, the compression modulus is substantially improved — by a factor of

three. Figure 5.54 shows that the compression stress–strain curves of elastomer blocks with a low shape factor are reasonably linear over a wide strain range, but become increasingly nonlinear as the shape factor increases.

The other factor that needs to be considered is the effect of changing dimensions on shear stiffness, K_s, which is defined, (Figure 5.52), in a similar way to compression stiffness, by the equation

$$K_s = \frac{F}{\Delta_s}$$

$$= \frac{GA}{d}$$

Clearly, the shear stiffness is inversely proportional to thickness, d, so that if the compressive stiffness is increased by decreasing d, then the shear stiffness will increase, too. The solution to the problem of controlling shear and compressive stiffness is simply to build the elastomer layers up into a composite block whose shear stiffness is given by the equation

$$K_s = \frac{GA}{nd}$$

where n is the number of layers of equal thickness d. Thus the shear modulus remains unaffected since it is dependent only on the total thickness of the block. The compression modulus of the block, however, is dependent on the shape factor of each individual layer.

EXERCISE 5.26 What is the horizontal deflection of the elastomer block described in Exercise 5.25 when subjected to a shear force of five tonnes? If the block is subdivided into two equal parts by insertion of a steel plate, which will be the effect on (a) deflection in shear and (b) deflection in compression? Assume that the shear and compression loads are identical.

5.5.4 Bridge support bearings

The ability to control two mutually perpendicular stiffnesses is widely exploited. A simple example where widely different horizontal and vertical stiffnesses are required is a bridge bearing. Bridges must accommodate the fluctuations in length caused by seasonal or daily temperature variations. Expansion or contraction of the long bridge deck causes horizontal movement which, if applied to the supporting piers, could generate dangerously high bending moments (Figure 5.55). The solution to the problem lies in finding a method whereby a large deflection can be accommodated without generating large reactionary forces.

A common method of reducing the bending moment on the supports is to mount the bridge deck on steel rollers. The major problem with this

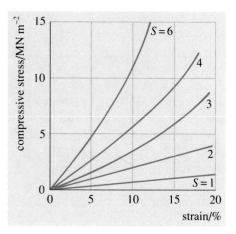

Figure 5.54 Compressive stress–strain curves for a natural rubber vulcanizate showing the effect of shape factor, S

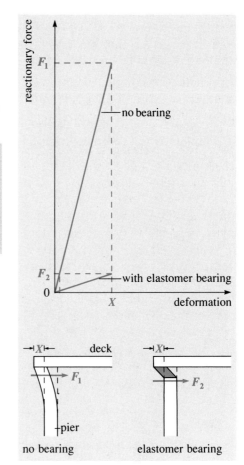

Figure 5.55 Relief of the stress in a bridge pier by using elastomer bearings

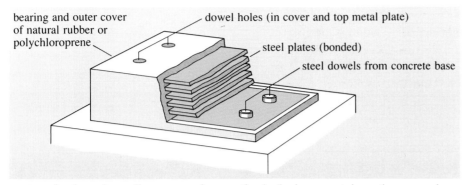

bearing and outer cover of natural rubber or polychloroprene

dowel holes (in cover and top metal plate)

steel plates (bonded)

steel dowels from concrete base

Figure 5.56 Construction of a typical elastomeric bridge bearing

system is that the rollers corrode, particularly in countries where roads are salted in winter. The deck movements also produce localized wear of the rollers, so they require regular maintenance to ensure good performance.

An alternative method, which became widespread in the 1960s and 1970s, is to use rubber bearings operating in shear to accommodate thermal motion of the bridge deck. Such units do not suffer from corrosion, require no maintenance and are easily installed. The loading requirements are three-fold: namely, to support the weight of the bridge and deck; to accommodate additional loads of traffic and wind; and to respond flexibly to thermal movements.

In practice, maximum permissible loads and deflections are specified by standards such as BS5400, Part 9. One reason for specifying maximum loads is to protect the concrete surfaces of the deck and pier: a typical stress limit of $5 \, \text{MN m}^{-2}$ may be recommended to prevent damage to the pier. There is clearly a considerable degree of freedom in designing with elastomeric bridge bearings: the total static load can be spread over many bearings, so that large motorway bridges for example may have twenty to forty bearings. The vertical deflection can be controlled by the number of steel plates in the block and typically there are three to ten plates inserted to create the individual elastomer layers (Figure 5.56). The layers are 2–5 mm thick so as to withstand all the imposed loads: compressive, shear and bending loads created by extra loads on the bridge. Corrosion of the steel plates is prevented by encasing the whole bearing with rubber; the bearing is prevented from moving across the pier surface by steel dowels.

SAQ 5.9 (Objective 5.6)
A small bridge of total mass 300 tonnes is to be supported by four natural-rubber bearings of square plan. The designed compression stress is $5 \, \text{MN m}^{-2}$. Assuming that, under a horizontal force of 60 kN, the maximum shear deflection is 25 mm, calculate the dimensions of the rubber bearing needed (the shear modulus of the rubber is $1.0 \, \text{MN m}^{-2}$). How many steel plates should be used in each bearing to restrict the vertical deflection to no more than 3 mm under the static deck load?

The natural-rubber bearings of many motorway and other bridges have now been in service for over twenty years in the United Kingdom and are monitored for signs of cracking, creep or other deterioration. Elastomers that have double bonds in their backbone chain, such as natural rubber, are susceptible to **ozone cracking**: the gas is highly oxidative and will split the chains at these points. Attack is highly localized, just like crazing, and ultimately produces deep cracks running into the rubber, which could lead to failure. The cracks grow at right-angles to the imposed tensile stress, which must exceed a low threshold for cracking to occur. With bridge bearings there is relatively little stress in the outer cover and, if cracks do develop, they are soon stopped by the internal compressive stress. Ozone cracking is also most likely to occur near electrical equipment, which generates low levels of the gas — an unlikely situation for bridge bearings in the open air! Polychloroprene elastomer is also widely used for bridge bearings, mainly outside the United Kingdom.

Another feature of elastomers that can be usefully exploited is their capacity for absorbing vibrations. In steel springs, damping is very low indeed and often controlled by external friction at the supports. That is why, in a car suspension system, hydraulic dampers are used in parallel with steel springs to damp driving vibrations.

Elastomeric springs, however, damp vibrations substantially, due to internal friction, and convert them into heat. Such bearings therefore combine the elasticity of steel springs with the damping characteristics of hydraulic pistons. The car tyre is itself an excellent example of a damping system, both for the elastomers used in its construction as well as the entrapped air. In fact, skill is needed in selecting the right combination of different elastomers with different damping properties to provide good driving characteristics (see ▼The car tyre▲). This is why bridge bearings can also be used to support whole buildings and reduce the effects of ground vibrations on the building and its contents.

At one extreme, such building mounts are used to provide protection from earthquakes since large shear strains of low frequency can be absorbed as well as low strains at much higher frequency. In the United Kingdom, rubber building mounts have found application in damping out the ground vibrations (20–50 Hz) from underground trains (Figure 5.59). The rubber mounts have been installed for over twenty years and although vertical creep has occurred, it has been minimal (less than 0.5 mm), just as in motorway bridge bearings. Natural rubber is the favoured elastomer, at present, although others could be used for their better damping properties.

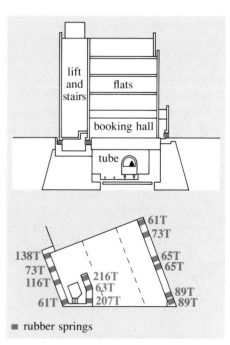

Figure 5.59 Section and plan of Albany Court flats in London showing distribution and capacity (in tonnes) of natural rubber mounts

▼The car tyre▲

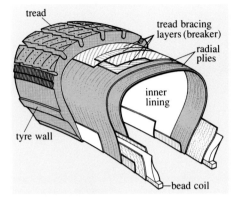

Figure 5.57 Structure of a radial car tyre

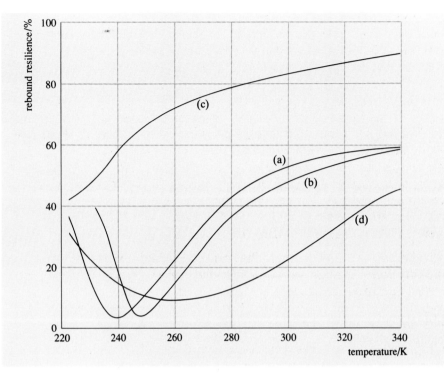

Figure 5.58 Rebound resilience as a function of temperature for: (a) natural rubber; (b) styrene-butadiene rubber; (c) cis-polybutadiene; and (d) polyisobutylene

To all outward appearances, a radial car tyre is composed of one uniformly black material. That couldn't be further from the truth! It is in fact a composite product composed of at least three different elastomers, together with steel wire and fibre plies (Figure 5.57). Although the pressurization of a tyre is relatively low (two atmospheres or $0.2\,MN\,m^{-2}$), the forces imposed on it by steering, braking and accelerating are relatively high, so reinforcement is essential. The key to the design of the car tyre, however, lies in the radial plies and the steel breaker under the tread. The plies are wound in the hoop direction to resist internal pressure, but are unidirectional in that they are flexible when the walls of the tyre bend. The tread, by contrast, is relatively inflexible because of the underlying steel breaker: it helps to keep the tread in contact with the road under most driving conditions. Under dynamic conditions, it is therefore the wall of the tyre that vibrates and deforms. Owing to internal damping, heat is evolved at driving frequencies (10–30 Hz) and it is essential to minimize this to prevent degradation of the material, which in extreme cases, such as in motor racing, can lead to the tyre catching fire. It is therefore essential to use elastomers in the walls of tyres that have low internal friction.

What elastomers, then, are suitable for tyre walls?

From what you have previously read about damping, an elastomer with a very low glass-transition temperature is essential, but this is not the only requirement, since damping can vary considerably at temperatures above the glass transition. The damping can be evaluated by comparing the bouncing behaviour of solid balls composed of different elastomers. The height of rebound divided by the height from which it is dropped gives the **rebound resilience** (Figure 5.58). There are large differences in rebound between a highly resilient elastomer such as cis-polybutadiene (c) — it's a favourite with children for super bouncing balls — and butyl rubber (d), which shows high damping over a wide temperature range. Rebound resilience is a direct measure of internal friction since it measures the energy absorbed by the material in flexure. Cis-polybutadiene is an excellent elastomer for tyre walls, but for cost and other reasons, is generally blended with natural rubber prior to crosslinking and manufacture of the tyre carcass.

The requirements for the tread are exactly the opposite. A high degree of damping is essential to give a high coefficient of friction at the road surface. Polyisobutylene would be an excellent choice and is, in fact, used in the treads of racing tyres, but processing problems limits its use in modified form to the inner lining, where its low permeability to the outward diffusion of air is utilized. In fact, a blend of natural rubber and styrene-butadiene rubber is used for treads, the random copolymer of styrene and butadiene being less resilient than natural rubber (Figure 5.58). It is also a relatively low-cost elastomer.

EXERCISE 5.27 For extra road grip, the styrene content of the tread is often increased to levels higher than the 25% used in normal grades of styrene-butadiene rubber. Explain, using the chain-flexibility theory, how the extra styrene repeat units improve the properties of the tread.

Objectives for Chapter 5

After studying this chapter you should be able to:

5.1 Select a polymer whose physical properties are sufficient to meet a specified mechanical function, given the appropriate data (SAQs 5.1, 5.4, 5.5, 5.7, 5.8).

5.2 Assess the effects of time and temperature on the mechanical properties of polymers (SAQs 5.2, 5.6).

5.3 Explain the thermomechanical properties of polymers in terms of chain flexibility and structure (SAQ 5.3).

5.4 Assess the strength of products and relate it to design geometry, polymer properties and environmental factors (SAQs 5.6, 5.7).

5.5 Calculate stresses in simple pressure vessels and relate stress to simple design parameters (SAQ 5.8).

5.6 Calculate the dimensions of elastomer bearings needed to meet compression and shear criteria using appropriate data (SAQ 5.9).

5.7 Define and use the following terms:

aliphatic structures	long-range elasticity
aromatic structures	melting point
atactic chain	molecular mass
branching	orientation of polymers
carbon black	ozone cracking
cis-stereoisomer	planar zig zag
configuration	plasticizer
conformation	random coil
copolymer	rebound resilience
crazing	secant modulus
creep	shape factor
creep rupture	spherulite
crosslinking	stress relaxation
damping (or loss) factor	syndiotactic chain
draw ratio	tangent modulus
elastomer toughening	time–temperature equivalence
environmental stress cracking	trans-stereoisomer
folded chains	ultraviolet degradation
glass-transition temperature	viscoelasticity
isotactic chain	weld line
lamella	

Answers to exercises

EXERCISE 5.1 The large difference in densities (diamond, $3500 \, \text{kg m}^{-3}$ and high-density polyethylene $960 \, \text{kg m}^{-3}$) is due to the compact highly ordered structure of diamond with strong covalent bonds between all the carbon atoms. Covalent bonds in polyethylene only occur along the chain; the bonds between chains are weak van der Waals' bonds. The chain spacing is therefore greater than in diamond so the density of polyethylene is much lower than that of diamond.

EXERCISE 5.2 From Table 5.1, the glass-transition temperature of UPVC is 353 K compared with that of about 970 K for soda-lime silica glass. The glass transition marks a change in molecular motion and its temperature depends on the strength of molecular bonds. For glasses with very strong ionic bonding between covalent silicate chains (chapter 4), the chains cannot rotate easily and the glass-transition temperature is high. In polymers, only weak van der Waals' bonds resist chain rotation, so the glass-transition temperature is low.

EXERCISE 5.3 Natural rubber has a double bond in its background chain at every fifth carbon atom (Appendix 2). Four hydrogen atoms have been replaced by one hydrogen atom and a methyl group. Thus the resistance to rotation should not be dissimilar to that of polyethylene, so that a low glass-transition temperature is expected. PET, however, contains a bulky aromatic group in the chain, locking the atoms into a rigid structure, so will possess a higher glass-transition temperature.

EXERCISE 5.4 Polystyrene has a higher tensile modulus ($3.4 \, \text{GN m}^{-2}$ compared with $3.0 \, \text{GN m}^{-2}$), but a much lower breaking strain than PVC (2.5% compared with 30%). PVC is more resistant to impact than polystyrene and has a lower glass-transition temperature (353 K compared with 370 K). Both have tensile moduli greater than the polyolefins, but their breaking strains are much lower. Their glass-transition temperatures are considerably higher than the polyolefins and this indicates a greater resistance to chain rotation. That resistance is reflected by the lower impact strength of PVC and polystyrene compared with the polyolefins.

Both are noncrystalline materials, so their relatively high moduli cannot be explained by the stiffening effect of crystallites. On the contrary, it is their higher resistance to chain rotation that accounts for their lower ductility compared with the polyolefins.

EXERCISE 5.5 If processed at 100 K above its glass-transition temperature, or about 473 K, the heat absorbed in heating polystyrene from 293 K to 473 K will be

$$\Delta H \simeq c_p \Delta T$$
$$= 1.35 \times 180$$
$$= 243 \, \text{kJ kg}^{-1}$$

The comparable heat absorbed by soda-lime silica glass heated from 293 K to 1073 K is

$$\Delta H \simeq 0.98 \times 780$$
$$= 764 \, \text{kJ kg}^{-1}$$

The heating costs for the glass are over three times those for polystyrene.

EXERCISE 5.6 The two main materials used for disposing of rainwater in the past were cast iron and thin sheet steel. Wood guttering was rarely used and copper guttering is a Tuscan idiosyncracy. Guttering was erected in short lengths of about 2 m, supported by steel or wrought iron brackets, often of crude design (e.g. a simple spike knocked into the wall). Putty was usually used to seal connections, but did it rather badly because it hardens and embrittles with age. The best cast-iron systems, found on country houses, for example, are superb examples of *over* design, being sturdy enough to be used by burglars! Apart from the great weight of cast-iron systems, they remain relatively free from corrosion, quite unlike steel guttering, which rusts through very quickly.

EXERCISE 5.7 Since the beams have the same flexural stiffness

$$\left(\frac{E^{1/3}}{\rho} \right)_{PP} \times m_{PP} = \left(\frac{E^{1/3}}{\rho} \right)_{Al} \times m_{Al}$$

Taking values for the specific flexural stiffnesses from Table 5.3

$$\frac{m_{PP}}{m_{Al}} = \frac{1.51}{1.26}$$
$$\approx 1.2$$

The polypropylene beam will therefore be about 20% heavier than the aluminium beam. The relative thickness will be given by

$$\frac{d_{PP}}{d_{Al}} = \left(\frac{E_{Al}}{E_{PP}} \right)^{1/3}$$
$$= \left(\frac{75}{1.5} \right)^{1/3}$$

Thus the polypropylene will be about 3.7 times thicker than the aluminium beam.

EXERCISE 5.8 From Figure 5.13, the strains produced by a stress of $5 \, \text{MN m}^{-2}$ acting for $10^2 \, \text{s}$ and $10^6 \, \text{s}$ are 0.0014 and 0.0021, respectively. Since the creep modulus is

$$E_C(t) = \sigma / \varepsilon(t)$$
$$E_C(10^2 \, \text{s}) = 5/0.0014$$
$$= 3.6 \, \text{GN m}^{-2}$$
$$E_C(10^6 \, \text{s}) = 5/0.0021$$
$$= 2.4 \, \text{GN m}^{-2}$$

Similarly, for $\sigma = 10 \, \text{MN m}^{-2}$

$$E_C(10^2 \, \text{s}) = 10/0.0033$$
$$= 3.0 \, \text{GN m}^{-2}$$
$$E_C(10^6 \, \text{s}) = 10/0.0045$$
$$= 2.2 \, \text{GN m}^{-2}$$

The creep moduli will *decrease* with increasing temperature, reflecting the greater molecular mobility of the polymer chains as the glass-transition temperature of 353 K is approached.

EXERCISE 5.9 By interpolation on Figure 5.15, at $t = 10^6 \, \text{s}$, the creep moduli are:

(a) continuous stress, $E_C(10^6 \, \text{s}) = 10/0.024 = 0.42 \, \text{GN m}^{-2}$
(b) intermittent stress, $E_C(10^6 \, \text{s}) = 10/0.015 = 0.67 \, \text{GN m}^{-2}$

EXERCISE 5.10 Since the guttering is free to rotate, the maximum deflection is five times greater than given by the equation, so

$$\Delta_{max} = \frac{5WL^4}{14.4ED^3t}$$

$$= \frac{5 \times 334 \times 2^4}{14.4 \times 100 \times 10^9 \times (0.105)^3 \times 3 \times 10^{-3}}m$$

$$= 5.3\,mm$$

EXERCISE 5.11 Since

$$\alpha_{PVC} = 70 \times 10^{-6}\,K^{-1}$$

$$\Delta L = \alpha L_0 \Delta T$$

$$= 70 \times 10^{-6} \times 10\,000 \times 30\,mm$$

$$= 21\,mm$$

If the frictional force is high, distortion of the constrained length of guttering will occur, particularly as the creep rate increases rapidly as the glass-transition temperature is approached. The coefficient of friction for PVC guttering is low, about 0.2, so the guttering will move through the supports. Provided one end is free, all is well, but if the system is constrained, then allowance must be made for the gross expansion. A joint with free surfaces would, however, solve the problem.

EXERCISE 5.12 Since

$$\tan \delta = \frac{1}{\pi}\ln \frac{A_n}{A_{n+1}}$$

and

$$\frac{A_n}{A_{n+1}} = \frac{1}{0.5} = 2$$

Then

$$\tan \delta = \frac{\ln 2}{\pi} = 0.22$$

Thus the loss factor is 0.22.

EXERCISE 5.13 The impact energy, $mgh = 2.75 \times 9.81 \times 2 = 54\,J$. The final velocity of the striker, v, is

$$v = \sqrt{2gh}$$

$$= \sqrt{2 \times 9.81 \times 2}\,m\,s^{-1}$$

$$= 6.26\,m\,s^{-1}$$

Upper and lower bounds can be estimated by simply calculating the time for the striker to move through its own length (12.5 mm) or the diameter of the pipe, depending on whether the product responds only at the point of impact or whether the whole product (114 mm) deforms.

lower bound to impact time

$$= \frac{12.5 \times 10^{-3}}{6.26} = 2\,ms$$

upper bound to impact time

$$= \frac{114 \times 10^{-3}}{6.26} = 18\,ms$$

EXERCISE 5.14 From Figure 5.20, $d = 10.16\,mm$ and $r = 1\,mm$, so $r/d \simeq 0.1$. Interpolation on Figure 5.26 gives $K_t = 2.1$, so the stress at the tip of the notch is raised by a factor of 2.1 over the *net* sectional stress. The unnotched area is 12.7/10.16 greater than the net section, so compared to the stress acting on it, the notch raises the stress by a factor of $2.1 \times 12.7/10.16 = 2.63$.

EXERCISE 5.15 Examples of stress raisers might include the following:

1 A support for a towel rail with a sharp corner at the attachment to the wall (put into bending by the load of a towel).
2 Sharp internal corners on the handles of winders on car windows (*not* visible when fitted!)
3 Sharp corners on thermoset electric plugs to conform with brass fittings.
4 Sharp edges to screw threads on plastic bottle tops.

The list is endless! All of these fairly obvious notches can of course be ameliorated by increasing the corner radii, but very often, like the window handle, they are not visible from the outside, so the consumer can make little judgement on the quality.

EXERCISE 5.16 The craze density becomes so high when acrylonitrile-butadiene styrene is strained that the surface appears white — an effect known as stress whitening. A microscope is necessary to see the individual crazes in ABS.

EXERCISE 5.17 Using a cold mould produces a high cooling rate in a product. The polymer chains are not able to respond quickly enough given that the material possesses a high glass-transition temperature, so that their fluid conformations will tend to be preserved. At a mould temperature greater than 320 K, however, the cooling rate from high temperature is much slower, and the chains will tend to return to their equilibrium conformation.

EXERCISE 5.18 Since the polymer chains are much more strongly bonded along the chain axis (Section 5.1) than laterally, the tensile modulus of drawn material is substantially higher than in the undrawn state. In spherulitic *undrawn* material, the net chain orientation in any direction is normally minimal, but cold drawing aligns the chains along the drawing (tensile) axis.

EXERCISE 5.19 The most flexible chain, that of high-density polyethylene, crystallizes most rapidly because chains can rotate very easily into a zig-zag conformation ready to pack together into crystals. Nylon and acetal also both have polyethylene-like chains (although modified by hydrogen bonding via oxygen and nitrogen atoms in the chain itself) so their crystallization rates are also high. Chains hindered by side groups, such as polypropylene and isotactic polystyrene, show severely reduced crystallization rates as the sizes of the side groups suggest. Both poly(ethylene terephthalate) and polycarbonate are hindered by aromatic groups in the chain itself, so have very slow crystallization rates.

EXERCISE 5.20 The hoop stress, σ_θ, is

$$\sigma_\theta = 0.4 \times \frac{100}{2 \times 0.3}\,MN\,m^{-2}$$

$$= 67\,MN\,m^{-2}$$

This value of hoop stress is considerably greater than the strengths of most of the common reasonably tough polymers shown in Table 5.1. The only material which has a higher tensile strength is nylon ($80\,MN\,m^{-2}$).

EXERCISE 5.21 All three polymers have a double bond in the repeat unit. The double bond means that there are fewer side atoms to interfere with chain rotation, so these chains will be flexible at 293 K. They should, as a consequence, have low glass-transition temperatures and natural rubber, for example, has a glass-transition temperature of 200 K.

EXERCISE 5.22 Since $G = \rho RT/\bar{M}_n$ for a cross-linked elastomer where R, ρ and $\bar{M}_n$ are constants, G will simply be proportional to T. In other words, the shear modulus of natural rubber will *increase* with increasing temperature. Increasing the temperature usually *decreases* the modulus of metals and ceramics. This occurs because higher temperatures make the atoms vibrate more rapidly about their equilibrium positions and easier to strain as the temperature rises. With elastomers, the effect is reversed. As the temperature rises, the chains can adopt a greater number of conformations in the unstrained state so the entropy change on straining becomes larger. The retractive stress therefore increases with rising temperature.

In metals and ceramics, the retractive stress arises from energetic rather than entropic changes, so will decrease with increasing temperature.

EXERCISE 5.23 Increasing the carbon black content increases the number of network chains, N, which contribute to the modulus of the natural rubber. Since $G = NkT$, the modulus will therefore increase. The ultimate strain, however, will be reduced, since the network chains are effectively shortened by reaction and adsorption on the filler particle surfaces (Figure 5.50). It will fall further as the carbon black content rises.

EXERCISE 5.24 Elastomeric products can be divided into those subjected to high strain (up to about 700%) and to low strain (up to about 100%).

High-strain products include 'rubber' bands, balloons and clothing elastic. High reversible tensile strains are an essential part of their function, but may only be used intermittently. None of these products are filled with carbon black.

Low-strain products cover car tyres, bottle bungs, hot-water bottles, tubing, antislip mats, shoe soles, wellington boots and a variety of seals and antivibration bearings (e.g. engine mounts). Most of these products are filled with carbon black.

EXERCISE 5.25 The shape factor, S, is given by

$$S = \frac{LB}{2d(L + B)}$$

Thus

$$S = \frac{0.2^2}{0.1 \times 0.4} = 1$$

The compression modulus is therefore given by

$$E_c = 3.0(1 + 2 \times 1^2)$$
$$= 9.0 \, \text{MN m}^{-2}$$

The deflection, Δ_c, is given by

$$\Delta_c = \frac{Fd}{E_c A}$$
$$= \frac{5 \times 10^3 \times 9.81 \times 0.05}{9.0 \times 10^6 \times 0.04} \, \text{m}$$
$$= 6.8 \, \text{mm}$$

EXERCISE 5.26 Since

$$K_s = \frac{F}{\Delta_s} = \frac{GA}{d}$$

$$\Delta_s = \frac{Fd}{GA}$$

Using the data in Exercise 5.25 and

$$G \simeq E/3$$

$$\Delta_s = \frac{5 \times 10^3 \times 9.81 \times 50 \times 10^{-3}}{1.0 \times 10^6 \times 0.04} \, \text{m}$$
$$= 61.3 \, \text{mm}$$

(a) If a steel plate is inserted in the middle plane of the original bearing, then the total thickness of rubber is unaffected, so the shear deflection will be identical to that of the original block (61.3 mm)

(b) The inserted plate will change the shape factor, so it must be recalculated for each layer:

$$S = \frac{0.04}{2 \times 25 \times 10^{-3} \times 0.4}$$
$$= 2$$

Thus

$$E_c = E(1 + 2S^2) = 3.0(1 + 8)$$
$$= 27 \, \text{MN m}^{-2}$$

Since

$$K_c = \frac{E_c A}{d} = \frac{F}{\Delta_c}$$

$$\Delta_c = \frac{Fd}{E_c A}$$

$$= \frac{5 \times 10^3 \times 9.81 \times 25 \times 10^{-3}}{27.0 \times 10^6 \times 0.04} \, \text{m}$$

So $\Delta_c = 1.13$ mm for one layer. For two equal layers, the total deflection will be 2.26 mm.

EXERCISE 5.27 Styrene chains are highly hindered by the large pendant benzene ring, so increasing the styrene level in SBR will *decrease* chain flexibility and increase the glass-transition temperature. The internal friction will increase, with the result that friction at the road surface is increased.

Answers to self-assessment questions

SAQ 5.1 From Equation (1.10), the engineers' bending equation, $MR = EI$. For the same flexibility in bending, both M and R are identical and therefore

$$E_1 I_1 = E_2 I_2$$

where 1 denotes the phenolic beam and 2 the unknown beam. Substituting for I gives

$$\frac{E_1 b d_1^3}{12} = \frac{E_2 b d_2^3}{12}$$

$$E_1 d_1^3 = E_2 d_2^3$$

Since $d_2 = 1.40 d_1$ and $E_1 = 8.0\,\text{GN m}^{-2}$,

$$E_2 = 8.0/1.40^3$$

$$= 2.92\,\text{GN m}^{-2}$$

From Table 5.1, four polymers have a tensile modulus of this order: PVC, PS, PMMA and PET. The last of these, PET, is partially crystalline and can therefore be excluded. PVC is the only noncrystalline polymer in Table 5.1 that has a breaking strain greater than 10%, so is the most appropriate one to replace the phenolic thermoset.

SAQ 5.2 The stress-relaxation curve of atactic polypropylene ($\bar{M}_n > 200\,000$) will be similar to that for atactic polystyrene of higher molecular mass (Figure 5.10), except that, with a lower glass-transition temperature, the temperature axis will be shifted so that the steep drop in the relaxation modulus will roughly coincide with the glass-transition temperature of 263 K. At 293 K, therefore, the atactic polypropylene will be elastomeric. Not being crosslinked it will show substantial relaxation at higher temperatures as it becomes a highly viscous liquid. Increasing the time scale of the relaxation experiments will be equivalent to increasing the temperature, so the material will behave like a highly viscous fluid at 293 K and have a much lower modulus.

SAQ 5.3 Polyethylene repeat units are less hindered than polypropylene, so they will improve the overall flexibility of the copolymer chain, but will affect crystallization behaviour. Since the polyethylene repeat units are scattered randomly along the polypropylene chains, order is reduced and the crystallinity is less. Thus, compared to polypropylene, the copolymer will have:

(a) a lower density;
(b) a lower glass-transition temperature, T_g, since chain flexibility is increased, and hence a lower melting point, T_m, since $T_g \simeq \frac{2}{3} T_m$;
(c) a lower tensile modulus, since it is related to both the degree of crystallinity and the chain flexibility;
(d) a higher impact strength because of increased chain flexibility and the copolymer will be tougher.

SAQ 5.4
(a) The first criterion immediately limits the choice to the polyolefins and natural rubber, all with bulk densities less than $1000\,\text{kg m}^{-3}$ (Table 5.1). Low-density polyethylene and natural rubber are excluded by the modulus criterion and polypropylene ($T_g = 263\,\text{K}$) is eliminated by the temperature limitation of 223 K. This leaves high-density polyethylene as the appropriate choice.

(b) Transparent polymers are limited to PVC, PS, PMMA, and possibly PET. To resist boiling water the glass-transition temperature must be greater than 373 K. On this criterion, PVC, PET and PS are excluded leaving PMMA and PC. The density limit of $1200\,\text{kg m}^{-3}$ eliminates polycarbonate, leaving PMMA.

SAQ 5.5 Interpolation onto the $10\,\text{MN m}^{-2}$ creep-stress curves shown in Figures 5.13 and 5.15 gives the following values of E_C for a creep time of 14 days or $1.2 \times 10^6\,\text{s}$.

$$E_C(\text{PVC}) = 10/0.005$$
$$= 2.0\,\text{GN m}^{-2}$$
$$E_C(\text{PP}) = 10/0.025$$
$$= 0.4\,\text{GN m}^{-2}$$

With the dimensions given, for PVC

$$\Delta_{max} =$$

$$\frac{334 \times 1^4}{14.4 \times (2 \times 10^9) \times (0.105)^3 \times (2.5 \times 10^{-3})}\,\text{m}$$

$$= 4\,\text{mm}$$

For PP

$$\Delta_{max} = 20\,\text{mm}$$

Thus PVC easily satisfies the British Standard creep condition and would be the best thermoplastic to use. Increasing the diameter, D, or the thickness, t, involves changing the profile of the product, so decreasing the length, L, is the easiest way to provide extra resistance to creep.

SAQ 5.6 The *minimum* cross-sectional area of each handle is

$$A = 40 \times 0.05 \times 10^{-6}\,\text{m}^2$$
$$= 2 \times 10^{-6}\,\text{m}^2$$

There are two handles and since each is looped around the hand to create two bands, the maximum safe load for a shopping trip, W, is $4A\sigma_R$, where σ_R is the creep-rupture stress, which is given at one hour. Hence,

$$W = 8 \times 10^{-6} \times 10 \times 10^6\,\text{N}$$
$$= 80\,\text{N or } 8.15\,\text{kg}$$

Using thicker film simply doubles the effective cross-sectional area, so the safe load using the new film will be 16.3 kg.

Since yielding tends to occur at lower stress levels when polymers are heated, at a given stress level the time to fail will decrease if the temperature rises. Nicks in the plastic film near the handle will become serious stress concentrators, multiplying the effective stress at the nick to several times the nominal stress calculated above. The safe load will be drastically reduced to a degree dependent on the severity of the nick.

SAQ 5.7
(a) The cracking is caused by DOP plasticizer migrating from the PVC into the acrylonitrile-butadiene styrene and initiating crazes. Since styrene-acrylonitrile, the matrix material, is highly sensitive to environmental stress cracking, the best solution would be to replace the acrylonitrile-butadiene styrene used in the dashboard by a polymer of similar modulus — high-density polyethylene or polypropylene. These are generally more resistant to environmental stress cracking than polystyrene-based polymers. An

alternative solution might involve selecting a less aggressive plasticizer for the PVC.

(b) Ultraviolet degradation is the problem here. The solution is to use a grade of polymer that has a high molecular mass together with a chemical or filler that absorbs ultraviolet light.

(c) Injection-moulding grades of PMMA have much lower molecular masses than the cast PMMA normally used for baths. Crazes are much weaker, so fatigue cracks will grow more easily. The solution is to use a grade of polymer that has, by direct polymerization, a much higher molecular mass.

(d) Environmental stress cracking of the amorphous polycarbonate from the aerosol paint is the basic problem. A solution would be to use a polycarbonate material with a higher molecular mass, but better solutions would involve using either a polypropylene copolymer of high molecular mass or a tough composite (Chapter 7).

(e) Cracking from the inherent stress concentration of a hole ($K_t = 3$) can be prevented by using a material that is less sensitive to notches, such as acrylonitrile-butadiene styrene or a polycarbonate that has a higher molecular mass.

SAQ 5.8 The main constraint is the hoop stress, σ_θ. Since

$$\sigma_\theta = \frac{pD}{2t}$$

$$t = \frac{pD}{2\sigma_\theta}$$

For the two-litre bottle you calculated the hoop stress to be $67\,\text{MN}\,\text{m}^{-2}$ and since it is the same for the three-litre bottle of diameter 120 mm

$$t = \frac{0.4 \times 120}{67 \times 2}\,\text{mm}$$

$$= 0.36\,\text{mm}$$

For a given polymer, permeation of carbon dioxide is controlled by the ratio of the surface area to volume, S/V, and the thickness of the wall of the bottle. The ratio of the shelf-lives of the bottles is given by

$$\frac{\text{shelf-life of three-litre bottle}}{\text{shelf-life of two-litre bottle}}$$

$$= \frac{(S/V)_{\text{2-litre}}}{(S/V)_{\text{3-litre}}} \times \frac{t_{\text{3-litre}}}{t_{\text{2-litre}}}$$

For the three-litre bottle

$$(S/V)_{\text{3 litre}} = \frac{12 \times 340}{120 \times 900}$$

$$= 0.038\,\text{mm}^{-1}$$

For the two-litre bottle

$$(S/V)_{\text{2 litre}} = \frac{12 \times 270}{100 \times 710}$$

$$= 0.046\,\text{mm}^{-1}$$

Therefore

$$\frac{\text{Shelf-life of three-litre bottle}}{\text{shelf-life of two-litre bottle}}$$

$$= \frac{0.046}{0.038} \times \frac{0.36}{0.3}$$

$$= 1.45$$

The three-litre bottle will thus have a shelf-life about 45% greater than that of a two-litre bottle because of its thicker wall and lower ratio of surface area to volume. If it is assumed that the shelf-life of the two-litre bottle is ninety days, the shelf-life of the three-litre bottle will be 131 days.

The replacement of PET by UPVC can be considered in terms of: (a) the effect on mechanical properties; (b) the effect on the permeation of carbon dioxide; (c) the effect of raw material costs.

In terms of the mechanical properties of unoriented material, UPVC, with a tensile strength of $50\,\text{MN}\,\text{m}^{-2}$ (Table 5.1), is clearly inadequate to withstand the hoop stress. During blow moulding, however, biaxial orientation will occur and the stiffness and strength will increase. Being a noncrystalline polymer, it will not show the same degree of improvement as PET, so thicker walls will be required.

The permeability of PVC to carbon dioxide compared with that of PET is higher by about a factor of two (Table 5.5), so a thicker wall will be required to keep losses of carbon dioxide to a low level. Since PVC also has a slightly higher density than PET (Table 5.1), the net effect will be to increase the weight of a bottle by at least a factor of two. This is offset by the lower raw-material costs of PVC, which are about half those of PET

(Table 5.2). UPVC could therefore be competitive with PET in terms of material properties, but to date has not succeeded in competing effectively with the polyester for biaxially oriented pressurized containers

SAQ 5.9 Since there are four bearings, each must support a load of 300/4, 75 tonnes, so if the designed stress under load is $5\,\text{MN}\,\text{m}^{-2}$, then the area, A, of each bearing can be calculated from the equation

$$\frac{F}{A} = \frac{75 \times 10^3 \times 9.81}{A}$$

$$= 5 \times 10^6\,\text{N}\,\text{m}^{-2}$$

Thus $A = 0.147\,\text{m}^2$

Since the bearings are square, if each side is of length a, then $a^2 = 0.147\,\text{m}^2$ and $a = 0.38\,\text{m}$. The thickness of the block can be calculated from the allowable shear strain, since

$$d = \frac{GA\Delta_s}{F}$$

$$d = \frac{1.0 \times 10^6 \times 0.147 \times 25 \times 10^{-3}}{60 \times 10^3}\,\text{m}$$

$$= 61.25\,\text{mm}$$

Now the compressive strain is

$$\varepsilon_c = \frac{\Delta_c}{d} = \frac{3}{61.25}$$

and the stress is

$$\sigma_c = 5 \times 10^6\,\text{N}\,\text{m}^{-2}$$

The compression modulus is therefore

$$E_c = 5 \times 10^6 \times \frac{61.25}{3}$$

$$= 102\,\text{MN}\,\text{m}^{-2}$$

The tensile modulus is

$$3G = 3.0\,\text{MN}\,\text{m}^{-2}, \text{ so}$$

$$E_c = E(1 + 2S^2)$$

$$= 3.0(1 + 2S^2)$$

and therefore

$$2S^2 = \frac{102}{3} - 1$$

$$= 33$$

Hence,

$$S = 4.06$$

For this shape factor, you can calculate the thickness of each rubber layer, since

$$S = \frac{a^2}{4ad} = \frac{0.147}{4 \times 0.38 \times d}$$

$$= 4.06$$

$$d = \frac{0.147}{0.38 \times 4 \times 4.06} \text{ m}$$

or

$$d = 24 \text{ mm}$$

Since the total thickness of 61.25 mm cannot be exceeded, it will be necessary to split it into three layers, each of thickness 20.42 mm. This will limit the vertical deflection to less than 3 mm, as required by the specification. Together with the two outer plates, a total of four steel plates will be needed for the bearing.

Chapter 6
Fibres and Fibre Assemblies

by Peter Lewis and George Weidmann
(consultant: Tom Frank)

Chapter 6 Fibres and fibre assemblies

6.1 Introduction

6.1.1 The nature of fibres

In this chapter and the next the emphasis turns to materials that are classified by their form and arrangement, rather than their chemical composition. We start with fibres and fibre assemblies.

Fibres are unique materials for two reasons. Firstly, their form endows them with an extraordinary degree of flexibility, which carries through to their assemblies. Secondly, whether natural or synthetic, they frequently have structures which confer on them significantly better mechanical properties than can be achieved in bulk form. They do, however, suffer from two drawbacks. Naturally occurring fibres, with the exception of silk, are only available in short, discrete lengths, known as **staple**. But even with continuous fibres, some means of holding fibres together is necessary to maintain the integrity of fibre assemblies when they are deformed in any direction other than along the fibre axis. Both of these problems are overcome by twisting fibres together to make a thread or yarn — the process of **spinning** (but note that 'spinning' is also used to describe the extrusion of synthetic fibres from the melt or from solution).

Spinning was one of the earliest technological skills acquired by human beings (the word 'spinster' is an indication of its importance to society). Products such as string and rope were used for joining animal skins or fastening wooden structures (Figure 6.1). The earliest textile materials, however, were probably made from felted animal hair. Archaeological evidence shows the subsequent development of **weaving** and the control of textile properties by varying the woven structure. Today, in addition to their uses in textiles and ropes, fibres are an essential component in composite materials, which are increasingly challenging the more traditional engineering materials in many areas.

What then is a fibre?

A fibre may be defined as a linear filament of material with a more or less uniform, small cross-section of thickness or diameter less than $100\,\mu m$ and **aspect ratio** (ratio of length to thickness or diameter) greater than 100.

Notice that this is a purely geometrical definition, and thus applies to *any* material. Although initially the only fibres available were those from plants and animals, fibres now encompass all classes of material. Useful fibres generally have aspect ratios considerably greater than 100, with synthetic ones such as polymeric fibres and inorganic glass fibres being

Figure 6.1 Bast fibre string from neolithic causeway site at Etton, Peterborough

produced in continuous form. When diameters exceed 100 μm they are known as **monofilaments** (polymers) or wires (metals).

In this chapter, we shall start by considering fibre flexibility. We then examine the structure and properties of fibres, with the emphasis on natural and synthetic organic fibres — the most important for textile and rope applications — but including the newer high performance ones. Finally, we deal with fibre assemblies, starting with spinning to produce yarns, and finishing with woven textile structures and twisted and braided rope structures.

6.1.2 Fibre flexibility

We saw in the context of elastomers in the previous chapter that flexibility can be a highly desirable attribute in a product — particularly where large movements or deformations have to be accommodated while maintaining product integrity. What was also important was not so much flexibility itself, but the ability to control it in different directions (in the case of elastomers by reinforcing with fibre plies or plates).

With fibres and fibre assemblies, the flexibility of the fibres is essential to the making of an assembly (by, for example, spinning or weaving), while the arrangement of the fibres within the assembly determines the directionality of stiffness.

EXERCISE 6.1   Identify fibre products requiring high tensile stiffness in one direction with flexibility in:

(a) all other directions,
(b) just one other direction.

Let's examine why a single fibre is so flexible and then investigate how and to what extent this flexibility is transferred to a simple array of fibres.

SAQ 6.1   (Objective 6.1)
Use the simple engineers' bending theory (Chapter 1) to compare the bending stiffness of a fibre with that of a monofilament of the same material whose stiffness in *tension* is 100 times greater than that of the fibre. Both are of circular cross-section with radii r_f and r_m, respectively.

So, it's because flexural stiffness varies with the fourth power of the cross-sectional dimension, while tensile stiffness only varies with its square, that flexibility increases so rapidly with decreasing cross-section.

What happens in an array of fibres?

Consider a simple, hexagonal array of parallel fibres around a central one, as shown in cross-section in Figure 6.2(a). What happens if we treat it as a beam in bending? To do this we need to use a theorem which states that the second moment of area of a section about any axis BB a distance y from the parallel axis AA through the centroid of the section (Figure 6.2b) is $I_{BB} = I_{AA} + Ay^2$ where A is the area of the section. Thus in Figure 6.2(b),

$$I_{BB} = \frac{\pi r^4}{4} + \pi r^2 y^2$$

If we apply this to the off-axis fibres in the array shown in Figure 6.2(a) and add their contributions to that of the central fibre, we get

$$I_{AA} = \frac{\pi r_f^4}{4} + 4\left(\frac{\pi r_f^4}{4} + \pi r_f^4\right) + 2\left(\frac{\pi r_f^4}{4} + 4\pi r_f^4\right)$$

$$= \frac{55\pi r_f^4}{4}$$

What then is the second moment of area of a monofilament of the same total cross-section as the seven fibres (Figure 6.2c)?

$$I_{AA} = \frac{\pi r_m^4}{4}$$

and

$$\pi r_m^2 = 7\pi r_f^2$$

or

$$I_{AA} = \frac{49\pi r_f^4}{4}$$

which is *less* than that for the seven-fibre array. In other words, these results predict that the bending stiffness of the array will be *greater* than that of a monofilament of the same total cross-sectional area.

This clearly runs counter to common experience and to what's expected of fibre assemblies. The problem lies in the difference between the assumptions of bending theory and the way fibre assemblies behave. Bending theory assumes that there is no relative shearing movement between adjacent layers of a beam to accommodate the difference in bending stresses through the section. On the other hand, although there are frictional forces between the fibres, they *can* move relative to one another when an array of them is put into bending. Thus, a stress gradient like that which resists bending in a monolithic beam cannot develop, and the resistance to bending is very much lower — though not quite as low as an equivalent array of nontouching fibres. We'll return to the flexibility of fibre assemblies in Section 6.3.

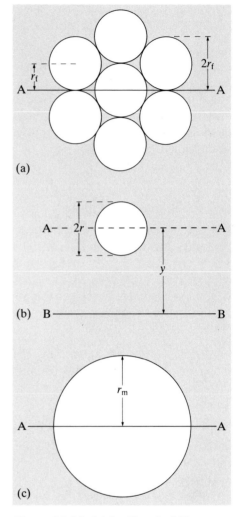

Figure 6.2 Model for fibre flexibility

6.2 Structure and properties of fibres

6.2.1 Common features

Fibre materials can be classified as natural or synthetic, and further sub-divided into organic, inorganic and metallic.

The vast majority of natural fibres are organic, the only significant inorganic one being asbestos (there is no natural source of metallic fibres). Both animal fibres, such as wool and silk (Figure 6.3), and plant fibres, like cotton and flax (Figure 6.4), were and still are widely used in a variety of textile products. Now they are often blended with synthetic organic fibres like nylon 6,6 and PET. Such synthetics are important in their own right, and offer a wide range of properties. Table 6.1 shows the mechanical properties of some of the more common natural and synthetic organic fibres, compared to those for steel piano wire (Chapter 3).

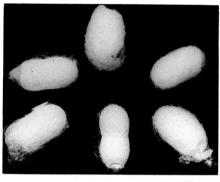

Figure 6.3 Cocoons of the cultivated silk moth *bombyx mori*

Table 6.1 Tensile properties of single textile fibres

Material	Density/ kg m^{-3}	Tensile modulus (bulk)/GN m^{-2}	Tensile modulus (fibre)/GN m^{-2}	Strength (bulk)/ MN m^{-2}	Strength (fibre)/ MN m^{-2}
natural polymer fibres					
cotton	1520	–	6–10	–	300–800
silk	1340	–	8–13	–	300–650
wool	1300	–	3–4	–	100–200
synthetic polymer fibres					
nylon 6,6	1140	2	1–5	80	400–750
PET	1380	3	12–19	54	600–800
PP	910	1.5	6.4	33	600
metallic wire					
steel (piano wire)	7860	210	210	460	3000

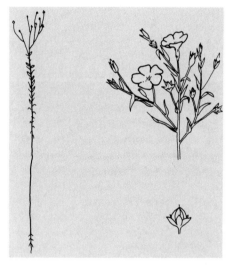

Figure 6.4 Flax plant carrying ripened seeds; head of flax in bloom; a seed capsule

It's apparent that the tensile moduli of the synthetic polymeric fibres are significantly higher than the moduli of the bulk materials. From Chapter 5, this immediately suggests that, relative to the bulk material, the polymer in the fibres is either more highly oriented, or more highly crystalline, or both. In fact, crystallinity is less important than *orientation*, and virtually all organic polymeric fibres, whether natural or synthetic, share a high degree of molecular orientation along the fibre axis. Progress towards higher strength, higher modulus fibres has relied as much on methods of achieving even higher degrees of orientation as on molecular engineering to produce stiffer and stronger polymer chains.

Table 6.2 Approximate dimensions of organic fibre features

Dimension	Feature	
100–10 mm	staple lengths	macrostructure
1 mm	spacing of crimps in wool	
100 μm	spacing of nodes in flax	
10 μm	fibre diameters	
1 μm	microfibrils molecular length of extended chains	microstructure
10 nm	folded chain lengths crystalline and noncrystalline regions	
1 nm	aliphatic and aromatic ring sizes	molecular structure
0.1 nm	atomic diameters (C, N, O, H etc.)	

Table 6.2 shows the dimensions of structural features of organic fibres. These range from the atomic scale of around 0.1 nm to staple lengths of up to 100 mm. In natural fibres, the hierarchy of structural features is more complex than in synthetic fibres.

The surface structure of fibres is also important since it affects the frictional characteristics between adjacent fibres. These, in turn, influence the ease with which a yarn can be spun from the fibre and the integrity of knitted structures and knots. Frictional characteristics are frequently related to the abrasion resistance of the fibre which, in turn, affects its durability in assemblies subject to repeated loading.

Partly because of history and partly because of the particular characteristics of fibres, textile technology has developed its own ways of describing the response of fibres and yarns to stress, as outlined in ▼Specific stress and tenacity▲. We'll be using descriptors such as LT, MT and HT in this chapter, although we'll use conventional engineering units as far as possible.

Let's start by looking at the synthetic polymeric fibres, since they tend to have simpler overall structures and many of their repeat units were discussed in Chapter 5. We'll then look at the more complex, natural polymeric fibres, before considering the newer, high-performance fibres.

6.2.2 Synthetic organic fibres

Although all synthetic polymers can form fibres, only about five have established a large market share in a variety of applications. The bulk commodity fibres include both nylon 6,6 and nylon 6, PET, PAN and polypropylene, but there are other fibres available for special uses, such as PVC and PTFE. Different kinds of fibre are derived from cellulose. These so-called man-made fibres include viscose rayon and acetate rayon. Rayon is produced by degrading natural cellulose (mainly from

▼Specific stress and tenacity▲

Determining the cross-sectional area of fibres is complicated by their small size, by their shapes especially in natural fibres, (Figure 6.5) and by their variability. This problem is even greater for yarns, so textile technology uses a measure based on the mass of a specified length of fibre or yarn: its **tex** is defined as the mass in grams of a 1 km length. A kilotex (ktex) is the mass in kilograms of a 1 km length. (Previously the **denier**, the mass in grams of a 9 km length, was used. Thus a 1 tex fibre would have a denier of 9.)

How is the tex of a fibre related to its cross-sectional area, A, and its density, ρ?

For a fibre of mass M kg, length L m, density ρ kg m^{-3} and tex T (g km^{-1}),

$$\text{tex, } T = \frac{M \times 10^3}{L \times 10^{-3}} \text{ g km}^{-1}$$

$$= 10^6 \rho A \text{ g km}^{-1}$$

The units of ρ and A must be SI units (respectively, kg m^{-3} and m^2) in this equation for T.

EXERCISE 6.2   Estimate the diameter of a two-tex, circular-section nylon fibre.

Tex can be measured relatively easily from the length and mass of a spool of fibre or yarn and, for fibres of similar density (see Table 6.1), it provides an approximate comparison of their cross-sectional areas.

Thus, textile technologists use N tex^{-1} (newtons per tex) for applied stress, a measure called **specific stress**.

Also widely used are the merit indices **specific modulus** or **stiffness** (for fibre assemblies) and **specific tensile strength** (or **tenacity** in textile parlance).

EXERCISE 6.3   Estimate the tensile strength (in N m^{-2}) of a single nylon fibre of tenacity 0.49 N tex^{-1}.

Fibres are frequently designated in terms of their tenacity, with LT, MT and HT referring to low, medium and high tenacity, respectively.

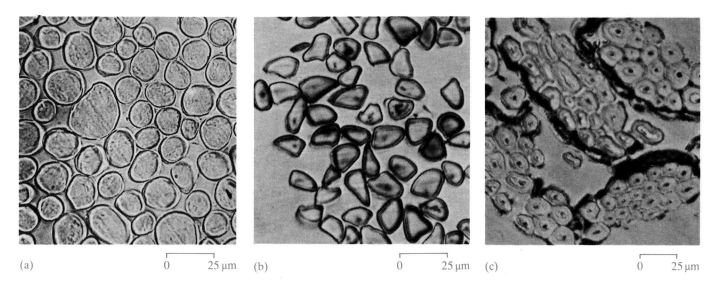

Figure 6.5 Cross-sections of natural fibres: (a) wool; (b) silk; (c) flax

wood) from a high molecular mass in excess of a million to much lower values ($\bar{M}_w \approx 60\,000$) so as to produce a solution which can ultimately be spun into fibre. The repeat unit of viscose rayon is identical to that of natural cellulose (Section 6.2.3), but acetate rayon has additionally been treated with acetic acid so as to convert the hydroxyl units into acetate groups. The crystallinity and orientation of rayon are lower than in cellulose.

The repeat units of synthetics are relatively simple (Appendix 2), but not all the commodity fibres are as simple as they might appear. Modifications are frequently made to the basic homopolymer backbone chain to manipulate fibre properties. Polyacrylonitrile (PAN) is often copolymerized with vinyl chloride to produce fire-resistant cloth, for example. With the exception of PAN, which, being mainly atactic, is largely noncrystalline, all the commodity synthetic fibres are partially crystalline, but to varying and controllable extents. Crystallinity is increased by orientation during processing, while spherulitic growth is suppressed, just as in the PET bottle (Chapter 5).

How is orientation achieved?

Simply by pulling the fibre from the **spinneret** hole (in effect, an extrusion die) at different speeds (Figure 6.6). The faster the speed, the greater the orientation along the fibre axis. The rates of crystallization associated with orientation follow roughly the same sequence as for spherulite growth rate (Chapter 5), so the final degree of crystallinity is controlled by careful control of temperature during drawing. Nylon and PET have about 60% crystallinity, polypropylene up to 90% crystallinity. The chains are folded along the fibre axis (Figure 5.39) and, in the amorphous zones between crystallites, are highly entangled. Molecular masses ($\bar{M}_w$) are about 20 000 for PET and nylon, but up to about 100 000 for PP and PAN.

In general, synthetic fibres don't have the range of macrostructures of most natural fibres, although spinning methods can mimic them. Simple circular sections are common, but a variety of other profiles are made (Figure 6.7). Other methods of spinning use bi- or multi-component fibres to produce crimped or kinked fibres when cooled, since each component polymer shrinks differentially. Crimping of staple fibre or texturizing of continuous filament is common where the fibre is to be used in clothing fabrics. A relatively new advance in polypropylene technology starts with thin film which is highly stretched and laterally split to give **fibrillated** film which can be spun further (Figure 6.8).

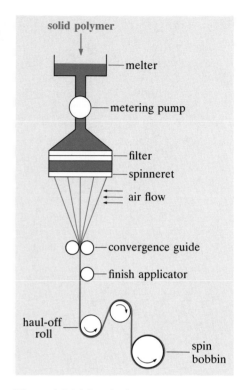

Figure 6.6 Melt spinning process

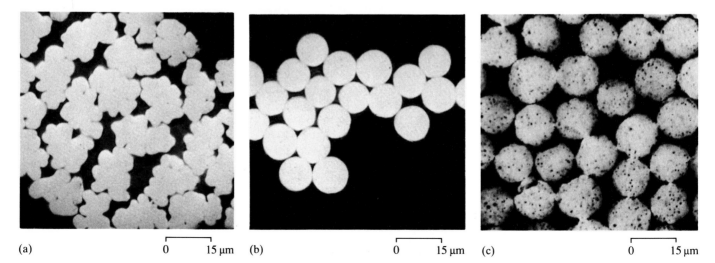

(a)   0   15 μm   (b)   0   15 μm   (c)   0   15 μm

Figure 6.7 Man-made and synthetic fibre shapes: (a) cellulose triacetate (Tricel); (b) PAN (Courtelle); (c) PET (Terylene)

Most synthetic fibre is simply chopped into shorter, staple lengths after spinning. Why should this be done when silk is regarded as the queen of natural fibres because of its continuity and continuity is seen as a virtue in a fibre? Synthetics are often blended with staple natural fibres and the spinning of such yarn assemblies is much easier when the synthetic fibre is also in staple form. In general, synthetic fibres used for staple have a low degree of chain orientation and high breaking strains. This is shown in Figure 6.9, which compares the tensile stress–strain behaviour of

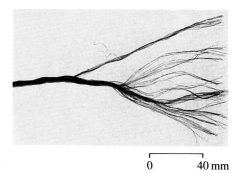

Figure 6.8 Fibrillated polypropylene baling twine

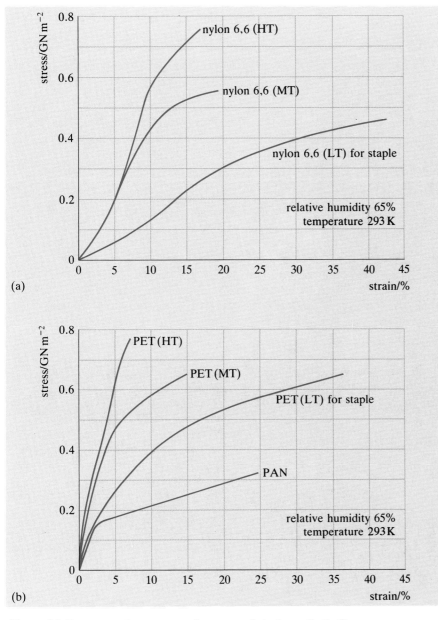

(a)

(b)

Figure 6.9 Representative stress–strain curves of single synthetic fibres

Table 6.3 Properties of single synthetic polymer fibres

Polymer	Some fibre trade names	Density/ $kg\,m^{-3}$	T_g/K	T_m/K	Initial modulus/ $GN\,m^{-2}$	Tensile strength/ $MN\,m^{-2}$	Specific strength/ $MN\,m\,kg^{-1}$
nylon 6,6	Bri-nylon	1140	323 (dry)	540	1.0	425 (staple)	0.37
					3.4	550 (MT)*	0.48
			273		5.0	750 (HT)	0.66
nylon 6	Perlon	1130	(4% wet)	498	0.7	330 (staple)	0.29
PET	Dacron, Terylene, Kodel	1400	338	549	12.3	660 (staple)	0.47
					14.8	660 (MT)	0.47
					18.5	780 (HT)	0.56
PAN	Orlon, Acrilan, Courtelle	1140	398	–	7.0	380 (staple)	0.33
PP	Meraklon	910	263	449	6.4	600	0.66

*MT = medium tenacity, HT = high tenacity

PAN fibre and different grades of nylon and PET fibre. Table 6.3 shows their mechanical and thermal properties.

The stress–strain characteristics shown in Figure 6.9 are nonlinear and, again, as with bulk polymers, a unique value of tensile modulus cannot be defined. A value that is frequently used for fibres, however, is the tangent modulus at low strain (see 'Creep in plastics', Chapter 5) and is known as the **initial modulus**.

Why should low-modulus staple fibre be advantageous?

This relates to the function of the fibre. In clothing, flexibility of the woven material is a prized virtue and, generally, the lower the fibre modulus, the greater the flexibility of material constructed from it. As we'll see in the next section, flexibility is also affected by assembly structure. There's a further reason: the breaking strains of staple PET, nylon and PAN are considerably greater than those for the more highly oriented fibres (Figure 6.9). Many textile applications involve high strains (think of the strain applied to clothing during use) so fibres of *low* degree of orientation are preferred here. Other applications, however, require greater tensile stiffness or strength (seat belts, safety webbing and some ropes), so that MT, HT or fibres of even greater orientation are used. PET (HT) has the highest tensile strength and modulus (Table 6.3) due to its inherently stiffer backbone chain.

6.2.3 Natural organic fibres

All plant fibres are based on **cellulose**, whose repeat unit is shown in Figure 6.10. It contains a cyclic unit, comprising a six-membered ring of one oxygen and five carbon atoms. The cyclic units are linked by oxygen atoms which have a specific configuration in space. The bridging oxygen atom determines many properties of cellulose, in particular, making it difficult for animals to digest. Starch possesses an identical repeat unit except in the configuration of this oxygen atom and can be

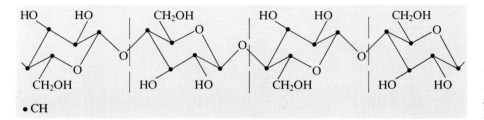

Figure 6.10 Repeat units of cellulose. Each successive ring is rotated by 180° about the linking oxygen atom to give a linear chain

easily broken down by animals for food. The cyclic unit is aliphatic, in contrast to the aromatic rings that occur in many synthetic polymers (Appendix 2). Aromatic rings are *flat*, with only one hydrogen atom attached to each carbon atom. The aliphatic ring in cellulose, however, is puckered and each carbon atom possesses *two* covalent bonds to which are attached hydrogen (–H) and hydroxyl (–OH) units or the –CH_2OH group (Figure 6.10). Natural cellulose has 10 000 or more repeat units per molecule.

EXERCISE 6.4   Estimate the molecular mass of natural cellulose (C = 12, O = 16, H = 1).

Hydrogen bonds formed by the hydroxyl units play a major role in the behaviour of cellulose. Of greater energy ($\approx$ 20 kJ mole^{-1}) than van der Waals' bonds, they stabilize the linear conformation of the chain to produce a ladder-like structure (Figure 6.11). In crystalline cellulose, the spare hydroxyl units are linked to adjacent chains via hydrogen bonds. Cellulose in plants is 60% to 90% crystalline, in contrast to starch, which is largely noncrystalline since the chains are difficult to align and the degree of hydrogen bonding is lower. Groups of 100 or more cellulose molecules are aligned in **microfibrils**, with individual molecules passing through crystalline and amorphous regions. These microfibrils are the principal constituent of cell walls in plants.

There are differences in the cell-wall structure depending on the plant species. In fibres from the stems of plants such as a flax (Figure 6.4), the microfibrils tend to form a spiral or helical conformation with a low degree of twist. In cotton fibres, which occur in the seed pod of the

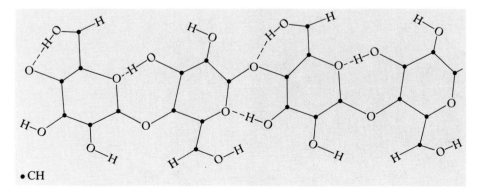

Figure 6.11 The intrachain bonding (broken lines) of a cellulose molecule

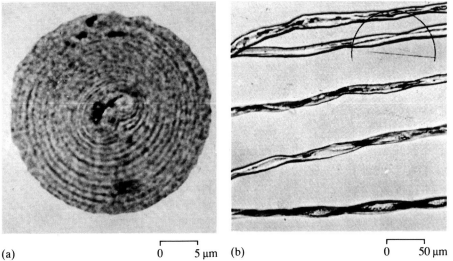

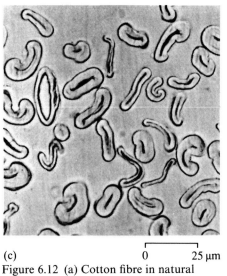

(a)　　　　　　　　　0　　5 μm　(b)　　　　　　　0　　50 μm　(c)　　　　　　0　　25 μm

Figure 6.12 (a) Cotton fibre in natural state showing daily growth rings. (b) Longitudinal micrograph of raw cotton. (c) After picking they show tubular structure when dry

plant, the helix is more twisted, making a more compact structure. Cotton, flax, manila and hemp fibres are relatively pure cellulose. Cellulose is also an important component of paper (Section 6.3) and wood (Chapter 7). As we've seen, it's also the precursor to rayon — the fibre often used in radial-tyre plies (Chapter 5).

Above the molecular level, the structure becomes more complex (Table 6.2), not least because the fibres are natural products whose growth is reflected in their structure. Cotton fibres of about 10 μm diameter in the natural state (Figure 6.12a) show daily growth rings similar to the annual ones in wood. After harvesting, however, they lose water and collapse to form fibres of irregular cross section and shape (Figures 6.12b and c). Intermediate between this gross fibre structure and the molecular level are the microfibrils which are revealed after fracture and wear (Figure 6.13).

While the main structural polymer in plants is cellulose, **proteins** are the

0　　　　25 μm

Figure 6.13 Scanning electron micrograph of flax fibres in a heavily worn modern linen sheet

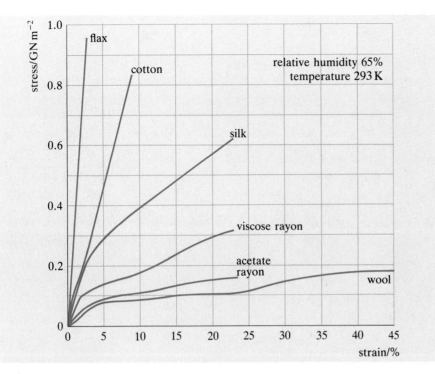

Figure 6.18 Representative stress–strain curves of single natural fibres

polymers used by animals. Their range of structures is much greater than that of cellulose and they are considered in ▼ **Protein-fibre structure**▲

Figure 6.18 shows the tensile stress–strain behaviour of single cellulosic and protein fibres. The differences can be accounted for in terms of their microstructure. Thus flax has a higher degree of crystallinity ($\approx 90\%$) than cotton and its microfibrils are more nearly aligned with the fibre axis. Hence it has a higher modulus and strength than cotton. Viscose rayon has a lower degree of crystallinity than cotton, and in acetate rayon it is even lower. Thus their moduli and strengths are lower than those of cotton, but they have enhanced extensibilities.

The behaviour of silk differs markedly from that of wool: silk has a much higher modulus and strength, but lower extensibility. In silk, the fibroin molecules are extended and highly oriented, but in wool the keratin is coiled into helices which are relatively easy to unwind. The high amorphous content in silk and wool (about 50% and 60%, respectively) modifies their tensile behaviour just as in the cellulosics. The high molecular mass of both fibroin and keratin ensures relatively high breaking strains. The effect of water on fibres and fibre assemblies (see ▼ **Water absorption**▲) is important both in terms of their properties and in terms of comfort for wearers of textile garments. A final point worth emphasizing is the considerable variation of tensile behaviour between different varieties of single natural fibres. It is the skill of the textile technologist to exploit not just the gross differences between fibres produced by different species, but also the different varieties within a particular species.

▼Protein-fibre structure▲

Protein fibres include collagen (skin, bone, tendon), keratin (hair, wool) and fibroin (silk, spiders' web). All possess very high molecular masses (in excess of 10^6), which maximizes strength, but they differ from one another in their repeat-unit structure, which in turn affects their microstructure. Unlike synthetic homopolymers, where each repeat unit is chemically identical, each protein chain possesses a *unique* sequence of *different* units, called 'chain units'.

There are, however, some simplifying features. Each chain unit is known as an **amino acid**, which, as the name suggests, possesses an amino group ($-NH_2$) and an acidic group ($-CO_2H$), both of which are linked to a central carbon atom. This carbon atom has two spare bonds, one of which is linked to a hydrogen atom (see Figure 6.14a). The comparison ends there because the other carbon bond is linked to a side group, R, and it is the presence of different side groups that makes each amino acid different.

During polymerization in plant and animal cells, the amino and acidic groups react together to form an amide link ($-CO-NH-$), which is the same link as in nylon (see Appendix 2). Figure 6.14(b) shows three amino acids joined in this way. R_1, R_2 and R_3, could be any one of twenty or more different side groups, the simplest of which are hydrogen in glycine and the methyl group in alanine. These amino-acid structures constitute the 'alphabet of life' as used by DNA.

What *conformations* can protein molecules adopt?

Let's consider just two proteins, fibroin in silk and keratin in wool fibre. In fibroin, large parts of the molecule are regular, just as in synthetic polymers. In the cultivated silk fibre, those parts consist of an almost regular, alternating copolymer sequence of glycine and alanine. The result is a regular crystal structure of extended chain molecules, effectively sheets held together by hydrogen bonds (Figure 6.15a) just like the sheets in nylon 6,6 (Figure 6.15b). The covalent backbone chains are oriented along the fibre axis. Although keratin is much more complex than fibroin, parts of each chain tend to curl into a helical conformation, where the hydrogen bonds are now *intra*- rather than *inter*molecular (Figure 6.16).

> EXERCISE 6.5 Which synthetic polymer crystallizes into a helix and why? How do extra bonds further stabilize natural helices?

Silk fibre is spun directly by the silk worm and consists of two filaments of triangular cross-section glued together by natural gum. Wool fibre is produced *in situ* by cell growth and as a result possesses a more complex macrostructure (Figure 6.17). The surface is scaly and, in addition, wool fibres show crimps (Table 6.2) along their length, produced by abrupt changes in fibre direction.

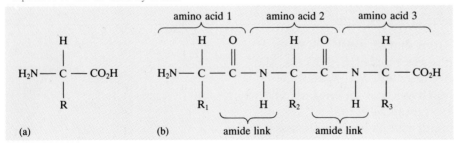

Figure 6.14 (a) An amino acid, with side group R. (b) Three amino acids linked together to show how protein molecules are constructed

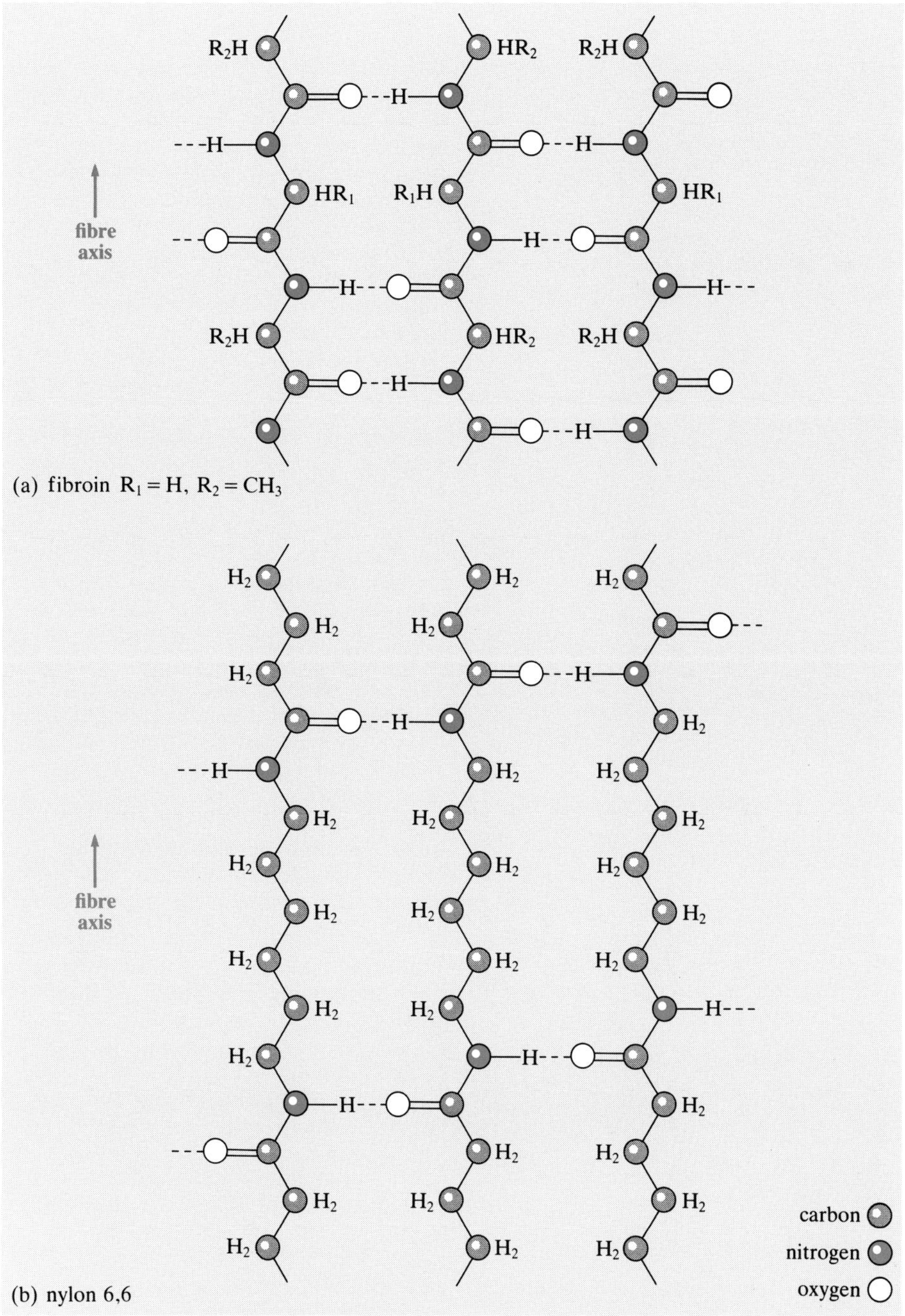
fibre axis
(a) fibroin $R_1 = H$, $R_2 = CH_3$
fibre axis
(b) nylon 6,6
carbon
nitrogen
oxygen

Figure 6.15 Sheet structures formed by hydrogen bonding

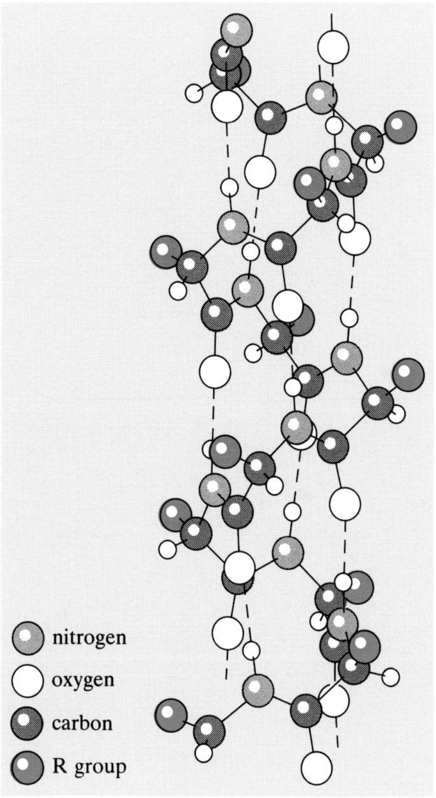
nitrogen
oxygen
carbon
R group

Figure 6.16 The α-helical structure

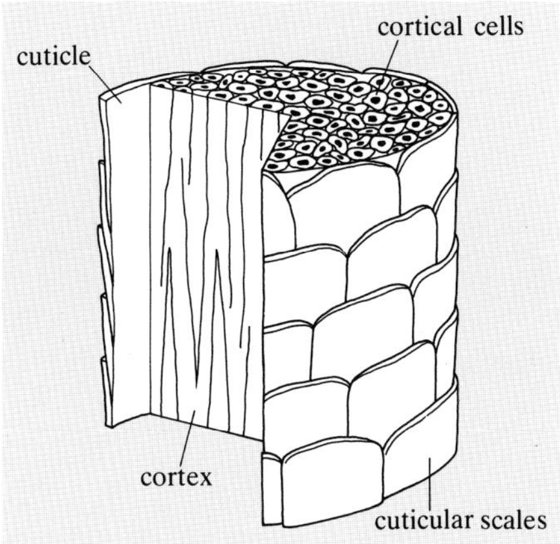
cortical cells
cuticle
cortex
cuticular scales

Figure 6.17 Cellular structure of wool fibre

▼Water absorption▲

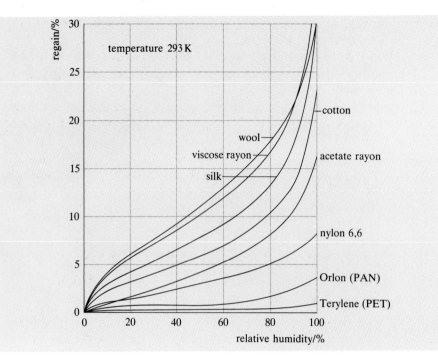

Figure 6.19 Equilibrium moisture content of a range of fibres as a function of relative humidity

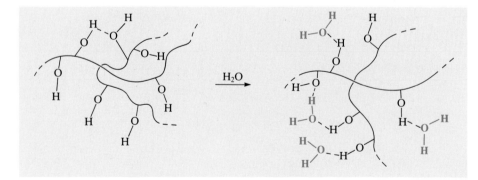

Figure 6.20 Water-absorption model for cellulosic fibres

In Britain's changeable climate, humidity plays a key part in environmental comfort. The absorption of water vapour from the body by fibres is an important factor in their use in clothing, mainly because such absorption helps to keep our skin dry and hence improves comfort. This is one of the main reasons why natural fibres have never been totally displaced by synthetics, which generally absorb much smaller amounts of water. The general relationship between fibre water content and relative humidity of the air for a range of fibres is shown in Figure 6.19 (**Regain** is the ratio of the mass of the water absorbed by a specimen to the mass of the specimen when dry.)

Increase in humidity and exposure to liquid water strongly affects the mechanical properties of natural fibres: wool, for example, becomes much softer when moist. In general, the tensile modulus of most single fibres of natural origin, both proteins and cellulosics, decreases with increasing moisture content, whereas their elongation increases. Tensile strengths usually decrease, but natural cellulosics like cotton and flax are exceptional in showing a slight *increase* in strength as the water content increases, perhaps because their structure tends to revert to that of their natural state (Figure 6.12a shows this for cotton).

How can these differing characteristics be explained in structural terms? The most important absorption mechanism is inwards diffusion of water from the surface, resulting in swelling. The thermodynamic driving force is the formation of hydrogen bonds with free hydroxyl groups as sketched in Figure 6.20. Hydroxyl units within cellulose crystallites are already hydrogen bonded (Figure 6.11) so it is principally the nonattached hydroxyl groups in the amorphous regions which interact with water molecules. The density of amorphous zones is lower than the crystallites, so the diffusion rate of the water molecules is higher. The degree of swelling in amorphous material controls the overall modulus of the fibre, so the modulus *decreases* and the breaking strain *increases* with increasing water content.

In polymers containing amide links (–CO–NH–), interaction with water occurs in a similar way: preferential absorption in the amorphous regions occurs via bonding with the amine units (–NH–) on unattached amide groups. In both wool and silk, the breaking strain roughly doubles and the fibre modulus drops substantially. The tensile strength and modulus of nylon drop slightly while fibres which do not form hydrogen bonds (PET, PAN, PP) are much less affected by water.

SAQ 6.2 (Objective 6.2)
Calculate the water regain at 100% humidity of (a) silk fibre and (b) nylon 6,6 fibre by estimating the mass of water which can hydrogen bond to all amine groups in the amorphous regions of the fibres. Assume that silk and nylon 6,6 are 50% and 60% crystalline, respectively. (Use these relative atomic masses: C = 12, N = 14, O = 16, H = 1.) Compare the answer with the measured values shown in Figure 6.19 and account for any discrepancy.

Other water transport mechanisms can operate with synthetic fibres — for example, by capillary action between fibres without water absorption by the fibre itself. Clearly, the ways in which the fibres are assembled into yarn and woven cloth are critical for such mechanisms.

6.2.4 High-performance organic fibres

Synthetic fibres have revolutionized the textile industry since nylon 6,6 first became available in the 1930s. Compared with natural fibres they offer greater breaking strains and absorb more energy up to breaking. They are more resistant to degradation and have the potential for property modification through the control of fibre diameter, length and molecular orientation. The most highly oriented, synthetic, conventional fibres still have tensile strengths lower than some cellulose fibres (see Figures 6.9 and 6.18). The high intrinsic strength and stiffness of the carbon–carbon bond, however, has encouraged research into synthetic organic fibres with chains much more highly oriented along the fibre axis. The first high-modulus, high-strength fibres were those made by carbonizing cotton, and this led directly to ▼Carbon fibres▲ Since then, other developments have culminated in aramid fibres ('Kevlar', du Pont, 1968) and high performance polyethylene fibres ('Spectra', Dutch State Mines, 1984). Their properties are shown in Table 6.4, which should be compared with those in Table 6.1.

Let's see how their structures are controlled to obtain such high performance.

Aramid fibre comprises linear chains of aromatic rings linked by amide groups (Figure 6.23). The amide groups are hydrogen bonded laterally just as in nylon 6,6 (Figure 6.15b), but the critical difference lies in the short hydrocarbon chain sequence between amide groups. In nylon 6,6 they are aliphatic polyethylene chains, but in aramid they are aromatic rings. The short aliphatic chains in nylon 6,6 form a planar zig-zag in crystals, but in amorphous regions their molecular flexibility causes a high degree of chain entanglement. Aramid chains, by contrast, behave like rigid rods owing to the inflexibility of their repeat unit, so entanglements are virtually absent.

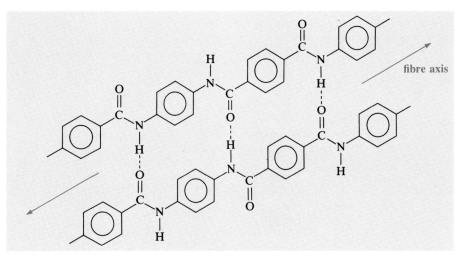

Figure 6.23 Chain structures of aramid fibre

▼Carbon fibres▲

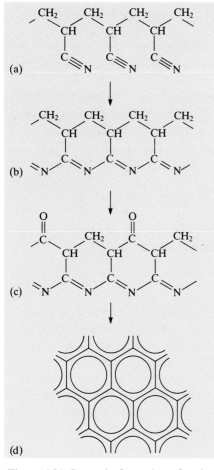

(a)

(b)

(c)

(d)

Figure 6.21 Stages in formation of carbon fibre: (a) hot stretch orientation of PAN yarn ($\approx 400\,K$); (b) formation of ladder polymer ($\approx 600\,K$); (c) chain oxidation ($\approx 600\,K$) followed by pyrolysis ($\approx 1300\,K$); (d) pyrolysis and carbonation (up to 2300 K) to give graphitic sheets

Table 6.4 Mechanical properties of single high-performance organic fibres

Fibre	Density/ $kg\,m^{-3}$	E/ $GN\,m^{-2}$	σ_{TS}/ $GN\,m^{-2}$	$E\rho^{-1}$/ $MN\,m\,kg^{-1}$	$\sigma_{TS}\rho^{-1}$/ $MN\,m\,kg^{-1}$
aramid (Kevlar 29)	1400	60	2.8	42.9	2.00
(Kevlar 49)	1440	124	3.1	86.1	2.15
polyethylene (Spectra 900)	970	120	2.6	124	2.68
polypropylene (gel-spun)	910	36	1.0	39.6	1.10
carbon fibre UHM*	1960	520	1.9	265	0.97
HM	1850	480	2.0	259	1.08
UHS	1750	270	5.2	154	2.97
HS	1760	265	2.8	151	1.59
steel piano wire	7860	210	3.0	26.7	0.38

*(U)HM = (ultra) high modulus, (U)HS = (ultra) high strength

It's somewhat surprising to learn that carbon fibres were first developed and used commercially over a century ago, by Swan and Edison in the first electric light bulbs. They carbonized viscose rayon or cotton to make their incandescent filaments, but the material was replaced by tungsten. In the intervening period, up to the late 1950s, carbon fibres were used for high-temperature insulation, but they did not possess outstanding mechanical properties. Subsequently fibres with strengths of $1.25\,GN\,m^{-2}$ and tensile moduli of $170\,GN\,m^{-2}$ were obtained by stretching viscose rayon at about 2300 K. An alternative route to carbon fibre used PAN continuous yarn. This is the fibre precursor used in the first commercial carbon fibres.

Why should PAN be a good precursor?

The reason is that the nitrile side groups ($-C\equiv N$) in PAN (Appendix 2) can react together to form a ladder-like polymeric structure (Figure 6.21). The first stage utilizes hot-stretched, oriented PAN fibre to give a ladder polymer by exposure to temperatures of up to 600 K. Main-chain carbon atoms oxidize in the same stage. The next stage involves elimination of nitrogen and oxygen by heating to 1300 K in an inert atmosphere, adjacent chains combining together to yield graphite sheets. They are, however, poorly oriented along the fibre axis, so the final stage involves heat treatment at temperatures up to 2300 K, when normally the fibres would contract substantially along their lengths. However, they are held at constant length, so the stress

generated in them orientes the graphite layers (Figure 6.22). Ultrahigh-modulus (UHM) fibres (Table 6.4) are formed by drawing in a further stage at temperatures up to 3000 K.

The fibres are only used in composite form, embedded in either a thermoplastic or thermosetting matrix (Chapter 7), and tendons have been made from 4 mm composite rod for potential structural application to moor floating oil rigs in the North Sea. The fibres are, however, more usually woven into cloth for incorporation into composite products, often of complex shape. This exploits the intrinsic flexibility of woven material yet maximizes the mechanical properties of the final crosslinked product.

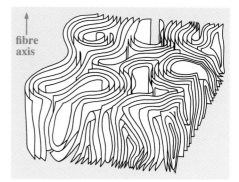

Figure 6.22 Structural model of a PAN-based carbon fibre

How is the aramid chain structure aligned with the fibre axis?

The rigid, rod-like monomer units are aligned in the flow direction in the spinneret. As fibre is formed by polymerization, pulled off and cooled, the chains are highly aligned. Aramid fibre possesses a fibrillar microstructure, not unlike that found in highly oriented cellulose fibres (Figure 6.13). The fibrils are aligned mainly along the fibre axis so the amorphous regions between fibrils do not seriously affect tensile stiffness. The higher-modulus form (Kevlar 49, Table 6.4) is hot stretched to align the fibrils further.

In high-performance polyethylene fibres (Spectra 900, Table 6.4), the planar zig-zag is also aligned along the fibre axis, but with relatively weak van der Waals' bonds between adjacent chains. Given the much greater intrinsic flexibility of polyethylene chains, how can they be aligned so effectively as to give tensile properties like those of aramid fibre?

The answer is a process called **gel spinning**. This starts with an HDPE solution in a suitable solvent at such a low concentration that the individual molecular coils do not touch, so they don't become entangled with one another. Rapid evaporation of the solvent produces a dry gel which can be hot drawn to no less than 72 times its original length to give high-modulus fibre! It turns out that ultrahigh-molecular-mass polymer ($\bar{M}_w = 1.5 \times 10^6$) can be drawn more easily by gel spinning than low molecular-mass polymer, so high strengths are obtained. The process results in chains extended along the fibre axis (Figure 6.24) without the folding present in conventional HDPE fibre. In principle, virtually any polymer can be gel drawn and Table 6.4 includes an experimental grade of gel-spun polypropylene, for example.

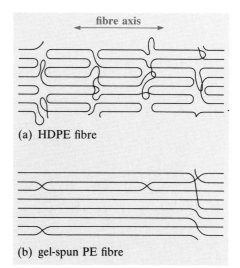

(a) HDPE fibre

(b) gel-spun PE fibre

Figure 6.24 Schematic model of oriented polyethylene fibre molecules

EXERCISE 6.6   What feature of the molecular structure makes the modulus of gel-spun isotactic polypropylene lower than that of high-modulus polyethylene?

6.2.5 Nonorganic fibres

Apart from the use of glass fibre and, previously, asbestos in thermal insulation and the growing importance of glass fibre in optical communications, the principal use of nonorganic fibres is as reinforcement in composite materials (see Chapter 7). The most commonly used reinforcing fibre is ▼ Glass fibre ▲. Table 6.5 lists the mechanical properties of a range of nonorganic fibres, again compared with those of steel piano wire.

'Saffil' zirconia and alumina fibres are examples of recent developments in ceramic fibres. Boron fibre, which is made by vapour deposition onto tungsten filament, and silicon carbide fibre are expensive, special-purpose materials. Asbestos, in contrast, is very cheap, but its use is limited by its associated health hazards.

▼Glass fibre▲

The first users of glass fibres were probably birds and small rodents living near volcanoes. They used 'Pele's hair' — filaments of molten lava, wind-blown from the lava surface and deposited on trees and bushes — for nest-building. Since the early days of glass technology, individual fibres drawn from heated glass rods have been used to decorate glassware (Figure 6.25), but methods of producing continuous filaments of controlled diameter, such as those used in reinforcement, were only developed in the 1930s. Although their diameters can range from about 3 μm to 30 μm, most reinforcing fibres lie between about 10 μm and 20 μm.

There are several types of glass available in fibre form, designated by different letter codes. Some of the more important of these are as follows.

● E-glass. A borosilicate glass (similar to Pyrex) originally developed for electrical applications, but which has a better resistance to water and to acids than the standard soda-lime silica glass as used in windows or containers (the latter is also known as A-glass).
● C-glass. A chemically resistant grade which withstands acidic, alkaline and aqueous environments better than E-glass.
● S-glass. An alumino-magnesium silicate glass of higher strength and modulus than other types, but also more expensive.

The most frequently used glass composition for reinforcement is E-glass. It's therefore available in the greatest variety of forms and is the least expensive. The last two are only used when enhanced chemical resistance or high specific strength and stiffness, respectively, are at a premium.

Glass fibres are made by drawing several hundred of them in parallel from the melt through platinum dies, the fibre diameter being determined by careful balance between viscosity and surface tension of the molten glass, die diameter and haul-off speed. The fibres are coated with **size** (which lubricates the contact between adjacent fibres and protects their surfaces from abrasion) and are then gathered into a strand.

Figure 6.25 Egyptian glassware from the fourteenth to the first century BC decorated with glass fibres

It's important to bear in mind that data such as these need some care in use. For instance, data on the strength of brittle materials such as the glasses refer to the pristine state. After they've been spun and woven their realizable strengths will be much lower. Also, the effective moduli and strengths will depend on the form of the reinforcement: rovings, yarns, woven structures and mats all have significantly different properties, which differ from those of single filaments, as you will see in the following sections.

Table 6.5   Mechanical properties of single nonorganic fibres

	Density/ρ, kg m^{-3}	Young's modulus/E, GN m^{-2}	Tensile strength, σ_{TS}/ GN m^{-2}
E-glass	2550	72	3.4
C-glass	2490	69	2.8
S-glass	2490	87	4.6
A-glass	2500	68	2.4
zirconia (Saffil)	5600	100	0.7
alumina (Saffil)	2800	100	1.0
boron	2600	380	3.8
asbestos (chrysotile)	2500	160	2.1
silicon carbide	2500	410	4.0
steel piano wire	7860	210	3.0

6.3 Fibre assemblies

6.3.1 Nonwoven assemblies

Textile structures can be produced from staple or continuous fibre in several different ways as shown in Figure 6.26. **Felts** are familiar nonwoven textiles which can be deformed easily into complex shapes, such as hats, without loss of integrity. Felts for blankets, mats, carpet underlay and so on are formed by compressing coarse wool fibres together. The scales on the wool-fibre surface (Figure 6.17) interlock with one another as the fibres are forced into contact, so preventing them slipping apart. The scales face in one direction, so only fibres where the scales are opposed will interlock. Substantial repeated compressive forces are needed to create such felts. Felting is aided by the crimp of wool, with fibre entanglements helping to maintain the structure.

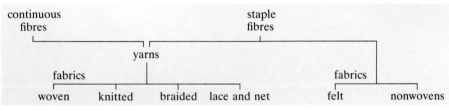

Figure 6.26 Main structural forms of textile fabrics and their relation to continuous and staple fibre

Fibre entanglements are also utilized to create felts from synthetic fibres, which have relatively smooth surfaces compared to wool. Artificially crimped (or texturized) staple fibres up to 50 mm in length are mechanically entangled with barbed needles to create the mat and the bulk density can be controlled by the amount of needle punching used. Relatively thick felts (up to 15 mm thick) can be produced and are increasingly used for road and rail underlays, and river bank and seashore revetments. Such 'geotextile' felts have bulk densities of 60 to 100 kg m^{-3} and tensile strengths from about 2 MN m^{-2} up to about 6 MN m^{-2}. Being mainly composed of interconnecting air spaces, they will allow water to percolate through very easily, but prevent soil particles from being swept away. They are often used in combination with drawn polymer grids (Chapter 5), which provide the inherent strength and stiffness to stabilize the structure. Polyester with a density of 1380 kg m^{-3} (Table 6.3) won't float in, and resists attack from, water (Figure 6.19), so is an ideal fibre for use in geotextile felts.

Another kind of nonwoven fabric is that represented by random assemblies of fibres laid in a flat plane and bonded at their points of contact. You are looking at a familiar example of such a structure as you read these words (see ▼Paper▲). Due to the random distribution of fibres in the plane, the mechanical properties of such fabrics are isotropic in their plane. Their stress–strain curves follow those of the single fibre more closely than woven structures, which are much more flexible than the fibres of which they are composed. This is illustrated in Figure 6.28.

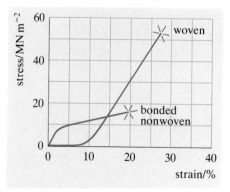

Figure 6.28 Comparison of stress–strain curves of typical woven and bonded fibre fabrics

▼Paper▲

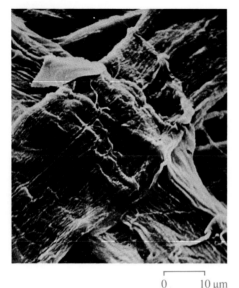

Figure 6.27 Cellulose fibre mat in paper showing fibre–fibre contact

It is very difficult to imagine a world without paper, in its multiplicity of applications from packaging to print and from absorbents to filters. Not many of these are structural, although the use of *pâpier-mâché* in eighteenth-century furniture and in aircraft fuel tanks in World War II shows what can be achieved with an appropriate binding agent. It might be thought, therefore, that the load-bearing ability of paper and paper-based materials is unimportant. However, most of us have cause to be grateful for the wet strength of tissue (probably daily!) and a reel of newsprint must be able to withstand the stress of being fed through printing presses at high speed without tearing. Bank notes ought to endure repeated folding without falling apart and again show the importance of the strength of paper in everyday life.

What, then, is paper?

Essentially, paper is a mat of cellulose fibres held together by hydrogen bonds.

Over 90% of paper is manufactured from wood pulp (See Chapter 7), but it can, in principle, be made from any material containing cellulose, such as cotton rags. Having obtained a pulp of fibres of the appropriate lengths (0.5 mm to 3 mm) by a combination of chemical and mechanical treatments, the next stage of papermaking is to break down partially the cell-wall structures. This is done by subjecting a suspension of about 2% of fibres in water to a high shear rate — a process known as **beating** or **fibrillation**. A combination of hydrolysis (swelling the fibres and breaking the cellulose–cellulose hydrogen bonds) and the shearing action in the fluid breaks the cell wall down into fibrils which splay out from the parent fibres. (This leads to a similar structure to that shown in Figure 6.13 for worn linen.) When the suspension is subsequently drained to make paper, the consolidating forces on the fibrous mat are provided by the surface tension of the draining water, both in bulk and in capillaries. The forces need to be quite high to ensure that intersecting fibres and fibrils are in close enough proximity to form hydrogen bonds. This is aided by the enhanced plasticity of the swollen fibres (Figure 6.27). It's been estimated that the fibre mat is bonded by 1.4% of the total hydrogen bonding in paper. The degree of beating is important because the amount and nature of the fibrillation produced critically affects the properties of the resulting paper.

Paper also incorporates a range of additives and coatings to control and/or modify not only mechanical properties, but environmental properties, such as water absorption and printability, and optical properties such as opacity, gloss and colour. Table 6.6 summarizes some typical paper properties. Paper also shows viscoelastic behaviour such as creep and stress relaxation.

Table 6.6   Typical paper properties

Property	Values
density*/$kg\,m^{-3}$	200–1400
tensile modulus/$GN\,m^{-2}$	1–5
tensile strength/$MN\,m^{-2}$	30–80
breaking strain/%	4–12

*A function of porosity and additives. The density of the fibre itself is about $1500\,kg\,m^{-3}$.

SAQ 6.3   (Objective 6.3)
Describe the structure of paper and explain the difference in properties between a cellulosic fibre, such as cotton (Table 6.1) and paper (Table 6.6). How will the bulk density of paper affect its properties?

6.3.2 Twisting of fibre into yarn

Like felting, spinning short natural staple fibres is one of the oldest textile processes. To spin, it's essential to produce a continuous thread (that is, a **yarn**) which can be built up in stages to the final structure. A minimum number of fibres is needed in the cross-section (about 30) for yarn integrity and an optimum degree of twist is needed to achieve maximum strength. The fibres are essentially held together by the friction generated by compressive forces induced by the twisting, which act across the yarn. The individual fibres are forced together and when the yarn is strained in tension, the stress is transferred evenly from fibre to fibre. The frictional forces increase with increasing twist, reaching a maximum value before the individual fibres are overstrained (Figure 6.29).

The maximum strength of a yarn is about half that of the individual fibres, irrespective of the type of staple fibre used. The full strength is never approached, simply because, when the yarn fails in tension, only a proportion of the fibres break, the rest being pulled out of the bundle. In continuous-fibre yarn, maximum strength is achieved without twist, and increasing the degree of twist lowers the tensile strength correspondingly. Yarn integrity is better maintained, however, especially after abrasive wear.

What is the effect on breaking strain of increasing the twist in staple yarn?

When a yarn is strained in tension, the first response is for the yarn to unwind, so the greater the twist, the greater the breaking strain (Figure 6.29). The type of staple fibre used in the yarn will affect the optimum degree of twist and the breaking strain, however. This is where synthetic

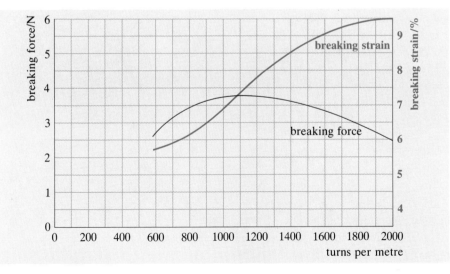

Figure 6.29 Effect of degree of twist on yarn strength and breaking strain (combed 25 tex cotton)

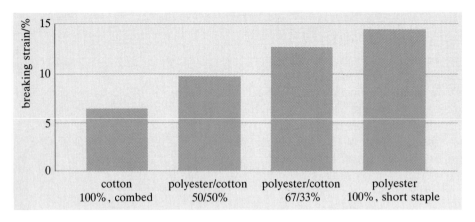
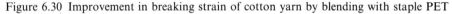

Figure 6.30 Improvement in breaking strain of cotton yarn by blending with staple PET

fibres generally have an advantage over natural cellulosics, and so are widely used in blends.

One of the most common blends is polyester with cotton. The textured-PET staple possesses a similar modulus to that of cotton so they will be mechanically compatible (Tables 6.1 and 6.3) and the PET fibre will reinforce the material by increasing the breaking strain. The higher the PET content, the greater the effect (Figure 6.30). The advantages of blended cotton include the possibility of higher process speeds during yarn manufacture and increased resistance to wear in fabrics composed of blended yarn. Clothing is often subjected to considerable strain and wear during use, particularly at knees and elbows, and blended cotton will give a more durable product.

Another factor favouring cotton–polyester blends is crease-resistance. Cellulosic materials crease badly during use and on hot washing due to viscoelastic creep in the low-modulus, swollen amorphous regions of the fibres. The creep will clearly depend on time, temperature and moisture content: sitting for a long time in one position is equivalent to a short time of exposure in a hot washing machine! Polyester is more resistant to moisture and creep, so blends with cotton will confer crease-resistance on garments: 100% nylon shirts (which are also crease resistant) were challenged in the late 1960s by the then recently developed blended garments. Polyester–cotton shirts have expanded their market share ever since, at the expense of both pure nylon (which tends to be uncomfortable in humid conditions) and pure cotton.

SAQ 6.4 (Objective 6.4)
Examine the clothing labels of a range of garments. What types of blended fibres are found? Why are polyester–cotton blends used instead of cotton? By using arguments based on their mechanical and environmental behaviour, explain why viscose rayon and staple nylon fibre are blended with wool.

▼Car seat belts▲

If you look at a seat belt, you will see a diagonal herringbone pattern running up and down the belt. The pattern is created by a twill-woven yarn (Figure 6.31b). Closer examination of the weave (for example, using a low-power magnifying glass) shows that the yarn has a relatively low degree of twist. Although not as apparent, the yarn is made from continuous fibre. Both the twill weave and the low degree of twist are designed to limit elongation under impact and maximize tensile strength. There are actually three grades of webbing for car seat belts, each with different breaking strains. Standard belts, for family saloon cars, have breaking strains of 6–8%; high-performance car belts have 16–18% breaking strains; and competition belting for racing cars (where cars can crash at speeds as high as 300 km h^{-1}) has 22–24% breaking strains. The design strains are specified to give maximum protection against deceleration forces well before failure, which of course should never occur. The reason for the higher breaking strains for high-performance and competition cars is to prevent the wearer's body being brought to a halt *too* quickly and so causing injuries. The greater the 'give' in the belting, the greater the energy the belt can absorb during deformation. The position of the driver in the cab must be matched to the webbing used since too much movement in a small cab could be disastrous.

SAQ 6.5 (Objective 6.4)
Using the data in Figure 6.9, select and justify the best fibre for the three grades of seat belt.

6.3.3 Textile fabrics

Textiles include both woven and knitted fabrics, the latter in effect being a two-dimensional array of knots. However, we'll concentrate on woven structures. The simplest possible structure in a flat textile is the plain weave (Figure 6.31a) where the yarns pass regularly over and under one another. **Warp** yarns are aligned along the length of the weave, with the **weft** yarns threaded across it to create the fabric. In the plain weave, the warp and weft yarns are usually woven at right angles to one another to create a wave-like conformation along both yarns (Figure 6.32). This increases the extensibility in both the warp and the weft directions because the yarn can straighten out before stretching. The **true bias** of the fabric (Figure 6.31a) represents the direction along which it will show greatest elongation for a given load.

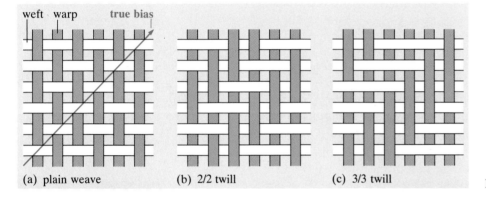

(a) plain weave   (b) 2/2 twill   (c) 3/3 twill

Figure 6.31 Structure of woven fabrics

EXERCISE 6.7   Test the flexibility of a plain-weave clothing fabric in different directions. What is the direction of greatest flexibility in the plane of the fabric?

A cursory examination of clothing will show the great variety of weaves that can exist. The twill structure (Figures 6.31b and c) exemplifies some of the complexities that can arise. If the sections along the warp for plain and twill weaves are compared (Figure 6.32), it's clear that more yarn will be aligned along the warp direction in the twill weaves than in the plain variety. However, more free yarn is exposed and there's a limit imposed by the increasing ease with which yarn can be hooked out (or 'pilled') as the spacing is increased. For structures designed to limit flexibility, such as linear webbing, the principal strain axis will be aligned along the warp direction. Woven-twill structures put a maximum amount of fibre into the main direction of strain without sacrificing fabric stability (see ▼Car seat belts▲).

For products which must perform reliably in an emergency, manufacturers prefer the predictable and consistent mechanical properties shown by synthetic fibres, particularly PET and nylon. They are also relatively unaffected by moisture and are resistant to the

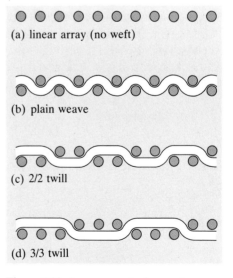

(a) linear array (no weft)

(b) plain weave

(c) 2/2 twill

(d) 3/3 twill

Figure 6.32 Arrangement of yarns in simple weaves

biological degradation, which attacks natural fibres. That's why synthetic fibres are so widely used in ropes — products, which, unlike car seat belts, are often subjected to high imposed strains in everyday use.

6.3.4 Rope

To distinguish rope from twine and string, it is defined as a linear flexible structure with a minimum diameter of 4 mm. The traditional rope structure is known as **hawser laid** and consists of three (or more rarely four) 'strands' twisted together in a helix (Figure 6.33). Each strand is itself a hierarchy of structures. Elementary base yarns are twisted together into 'primary' yarn. Primary yarns, in turn, are spun into 'roping' yarns, which again are twisted together to form the strand. The make-up of each structure in the hierarchy will determine the size of the rope: the greater the number of members in each class, the thicker the rope.

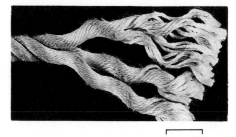

0   20 mm

Figure 6.33 Three-strand hawser-laid rope

EXERCISE 6.8   Make a list of the different ways in which rope is used. What functions involve: (a) low continuous loading; (b) high intermittent loading?

The strength of rope in tension is clearly a critical parameter for rope users and the relevant standards specify the minimum breaking loads that the rope must achieve (Table 6.7). The table includes the mass per unit length, called the **unit mass** (in grams per metre or ktex), for a range of rope diameters in three synthetic polymers (spun in continuous fibre form) and a natural cellulosic staple fibre, manila.

In many applications the unit mass of rope is important either because the rope is extensively manhandled or because very great lengths are needed (for example, buoy mooring lines). There's a simple measure of specific strength for ropes. This is the **breaking length**, which is defined as the length of rope which will just break under its own weight when freely suspended under gravity from a single point:

$$\text{Breaking length} = \frac{\text{total weight at break}}{\text{weight per unit length}} = \frac{\text{total mass at break}}{\text{unit mass}}$$

Table 6.7   Rope breaking loads (BS 4928/2052)

Nominal diameter/ mm	Nylon		Polyester		Polypropylene		Manila	
	breaking load/kN	unit mass/ktex	breaking load/kN	unit mass/ktex	breaking load/kN	unit mass/ktex	breaking load/kN	unit mass/ktex
4	3.14	10.5	2.9	14.6	–	–	–	–
8	13.2	41.9	10.0	65.0	9.41	30	5.35	54
12	29.4	93.7	22.3	116	19.9	66	10.4	105
24	118	373	89	460	73.5	260	44.8	400
48	412	1500	329	1850	267	1040	164	1580

EXERCISE 6.9 Which fibre material of those shown in Table 6.7 has the greatest breaking length when used in 12 mm diameter rope?

The ranking of rope breaking lengths is similar to the specific-strength ranking of the corresponding single fibres (Table 6.3), although their absolute strengths are lower due to their multiple twisted structures. In addition, as the diameter for hawser-laid rope increases, the breaking length tends to decrease owing to the higher twist levels needed to pack the yarns together. Conventional rope represents an interesting example of an engineering structure where strength is sacrificed for flexibility and fibre cohesion.

What are its tensile properties?

Ropemakers prefer a simple form of presentation of tensile behaviour: the load–strain curve. The initial loading curves for 24 mm, three-strand, hawser-laid rope in nylon and polyester are shown as curves A in Figures 6.34(a) and (b), respectively. The curve for nylon 6,6 rope shows

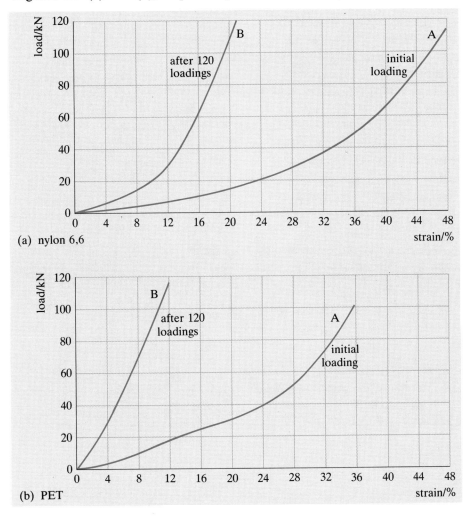

(a) nylon 6,6

(b) PET

Figure 6.34 Load–strain curves for 24 mm diameter three-strand hawser-laid ropes

that it extends much more easily than a single fibre of low tenacity nylon (Figure 6.9a).

Why should this be?

The answer lies in the rope's construction. Just like yarn or fabric when a load is applied (Figure 6.28), the coiled rope will act like a spring, so it extends easily under low loads. At high loads, the 'spring' is fully extended and individual fibres take the strain. If the rope is loaded repeatedly (120 times in this case) to half the rated strength, creep reduces the breaking extension from 48% to 20% (curve B). Part of the creep is structural as the fibres, yarns and strands pack closer together. Superimposed on this structural creep is viscoelastic creep of the fibre material itself. In line with the tensile properties of individual PET fibres (Figure 6.9b) the elongation-to-break on first loading of PET hawser-laid rope is considerably *less* than that of nylon rope, but it shows qualitatively similar behaviour to nylon on repeated loading.

Are there any disadvantages in hawser-laid ropes?

One problem with ropes twisted together from stranded yarn is that torque applied in the *opposite* direction to the twist angle will unwind the strands, which can snarl against projections, so weakening the structure. A way of overcoming this problem involves weaving the strands together to make **plaited** rope (Figure 6.35). An extension of this idea leads directly to woven tubes of strand, or **braided** rope, which for large mooring ropes, can enclose a similarly braided tube of smaller diameter.

Ropes are frequently required to be bent to small radii, for example when they're coiled up or when they're tied into a knot. It's in such cases that their flexibility becomes most important. Although the structure of the rope plays a role here, we can use a simple model to demonstrate how the flexibility depends on the diameter of the fibres.

Consider two ropes of the same material and of the same tensile strength. The cross-section of one contains n_1 fibres of radius r_1, while the other has n_2 fibres of radius r_2. Because their strengths are the same

$$n_1 \pi r_1^2 = n_2 \pi r_2^2$$

or

$$\frac{r_1}{r_2} = \sqrt{\frac{n_2}{n_1}}$$

If we say that, in bending, the maximum fibre stress in each must not exceed the *same* value of σ_{max}, we can use our engineers' bending theory to relate this to the minimum radius R_{min} to which the ropes can be bent, assuming friction is absent and the fibres can slip over one another. We have

$$\frac{\sigma_{max}}{y} = \frac{E}{R_{min}}$$

Since σ_{max} and E are the same for both ropes

Figure 6.35 Loop in plaited rope created by an eye-splice

$$\frac{y_1}{y_2} = \frac{R_{1min}}{R_{2min}}$$

Now, y is the distance from the neutral axis to where the stress σ is acting — in this case the surface of the fibre. Thus y is equal to the fibre radius, and

$$\frac{r_1}{r_2} = \frac{R_{1min}}{R_{2min}}$$

$$= \sqrt{\frac{n_2}{n_1}}$$

The minimum bending radius in this situation is therefore inversely proportional to the square root of the number of fibres in the cross-section and directly proportional to their diameter — the finer the fibre, the more flexible the rope and the easier to tie knots (see ▼Getting knotted and spliced▲).

When aramid fibres first became widely available in the 1970s, ropes seemed an obvious area of application. Their very high specific modulus and strength compared to that of steel wire (Table 6.4) made them a serious contender for demanding engineering applications. These include stays for guyed radio masts and mooring cables for oil rigs.

▼Getting knotted and spliced▲

One problem faced by the ropemaker is joining rope to itself or other structures, which usually creates severe bending strains at the joint (for example, Figure 6.35). How will the strength of the rope be affected? In effect these are stress concentrations and the strength of the rope will be decreased because of them. Is there any way of minimizing the stress concentration? One way is to use the relation between rope flexibility and fibre diameter, and use smaller diameter fibres in the rope.

Our modelling neglected friction between the fibres. The frictional effect will depend on the type of polymer: polyolefins with a lower coefficient of friction will be less affected than nylon or PET, for example. New rope will be more flexible in bending than rope which has been repeatedly strained in tension, since the latter will be more closely packed. Although friction will modify the relation between fibre size and bending stiffness, there still remains a qualitative effect.

Repeated loading increases the stiffness of ropes and reduces the energy required to

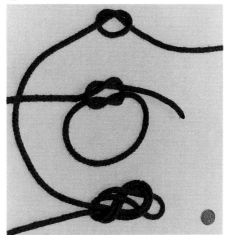

Figure 6.36 Simple knots: from the top, overhand, reef and double figure-of-eight

break them (Figure 6.34). Connections and knots are more important in limiting ropes' durability because of the high bending strains involved. Simple knots, like the overhand, are more serious than those where the strain is spread more evenly (Figure 6.36). The breaking load of

a rope with an overhand knot or reef knot is about 40% of that of unknotted rope, but that of the double figure-of-eight knot is about 74% of the unknotted rope. Splices are less serious in their stress concentrating effects, an eye-splice (Figure 6.35) typically reducing the breaking load by only about 15%.

SAQ 6.6 (Objectives 6.5 and 6.6) Select from polyester (PET), nylon and manilla (see Table 6.7) the best fibre for construction of a three-strand hawser-laid rope, 24 mm in diameter, to fulfil the following function (assume that the maximum imposed load should not exceed 20% of the breaking strength of the rope). The rope is a gymnasium climbing rope fixed by an eye splice to a roof hook and is 10 metres long, hanging freely to the floor. The rope is knotted with overhand knots at 0.5 metre intervals to aid climbing. The maximum load is ten pupils (each of mass 50 kg) with the new rope extending by not more than 8%.

EXERCISE 6.10 Compare the changes in specific modulus and strength of aramid (Kevlar 29) fibre with those of steel wire going from air to a subsea environment. Assume the density of sea water is 1030 kg m^{-3} and use Table 6.4.

The first ropes constructed from aramid fibre were twisted, hawser-like cables with very high breaking loads, but early experience with these ropes showed their strength to be seriously affected by unforseen factors (see ▼Failure in the Gulf▲).

The weakness of conventionally constructed cables using the new aramids led directly to the use of the much simpler parallel-fibre construction, where the properties of the rope are much closer to those of the individual fibres. Such aramid ropes are protected from the harmful effects of ultraviolet light (particularly noticeable in aramids

▼Failure in the Gulf▲

The use of aramid ropes for mooring drilling ships in deep water was tried for the first time in the Gulf of Mexico in mid-1983. The construction derrick ship, *Ocean Builder 1*, was employed to build an oil-drilling tower. Twelve lines were used for mooring it to the 300 m deep seabed, each line being a 60 mm diameter aramid cable (Figure 6.37). On first tensioning the system, 4–6 weeks after deployment, four of the ropes parted at 20% of their 200 tonne rated strength.

Examination of the ropes after the failure showed extensive kinking of single filaments in the aramid parts of the bundle. The problem lay in the low compressive strength of the fibres. Such kink bands form at 0.5 to 0.8% compressive strain, so can occur in mild

0   10 μm

Figure 6.38 Formation of kink bands on the compression side of a bent Kevlar 49 fibre

bending or direct compression (Figure 6.38). Tests on fresh cable after the accident showed that only about ten compressive cycles were needed to form kink bands and, once formed, they acted as a hinge about which the fibre could fatigue further when subjected to repeated strain. Stress is clearly localized and concentrated at these points, drastically weakening the overall structure. Where did these stresses originate? It was found that the loads imposed by tidal and wave action caused the rope to twist and unwind cyclically. When twisted tightly, the rope and hence the fibres were put into compression, and so formed kink bands. Low-load fatigue thus led directly to fibre failure. In the unwinding part of the cycle, the braided jacket surrounding the rope was eventually split, creating a

'birdcage' (Figure 6.39). Further tests showed that the higher the twist given to the rope, the greater the probability of failure, and it was also found that wetting the fibres lowered the fatigue strength. The water absorption of aramid is lower than nylon 6,6, but still relatively high (6% for Kevlar 29 and 3.5% for Kevlar 49). It is probably due to water diffusing into amorphous regions between crystal fibrils, leading to swelling and high compressive stresses between the fibres. The main recommendations of the failure analysis were to employ a stiffer, extruded thermoplastic jacket to inhibit flexing and prevent hinge points from developing. Torque-free construction such as braided rather than hawser-laid aramid rope was also recommended, but this can involve fibres crossing one another. When fatigued in tension, such ropes can fail by the high frictional forces developed at fibre–fibre contacts.

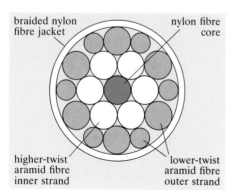

Figure 6.37 Aramid mooring line construction

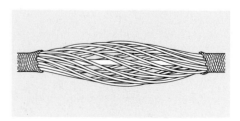

Figure 6.39 'Birdcage' formed in braided aramid cable

because of the ultraviolet absorbing properties of the aromatic group in the repeat unit) and from surface abrasion by an extruded polyethylene tube. Parallel-lay ropes of this type had been developed in the 1960s using high tenacity PET fibre, and so were an obvious way to use the new aramid fibres when they became available ten years later.

The simplicity of parallel-lay-rope construction avoids the stress concentrations arising in conventional ropes. One key to their successful use has been the design of high-strength terminations (Figure 6.40). The load supported by the rope is smoothly redistributed by spreading the fibre core in a conical steel sheath locked into place by a steel spike and sealed with silicone resin.

Owing to their more recent introduction, experience of high-performance polyethylene ropes has been more limited. One typical example of their application is in yachts where they replace aramid-fibre rigging rope. Here the clear advantage lies in the higher specific stiffness and strength of polyethylene (Table 6.4) owing to its substantially lower density, which also makes it buoyant in sea-water. Because of the lower coefficient of friction at fibre–fibre contacts, it can be braided or twisted into rope and the fatigue strength is substantially higher than that of equivalent aramid ropes.

There is now considerable engineering experience with the parallel-lay aramid ropes (for example Figure 6.41). They show linear stress–strain

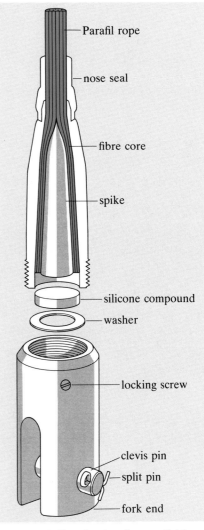

Figure 6.40 Termination for parallel-lay ropes

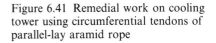

Figure 6.41 Remedial work on cooling tower using circumferential tendons of parallel-lay aramid rope

behaviour and break without yielding, like Hookean materials (Figure 6.42). Their advantage lies, of course, in their much lower unit mass (Figure 6.43a), as well as greater corrosion resistance. They have comparable breaking loads to steel ropes of the same nominal rope diameter (Figure 6.43b). Their ease of handling is thus much better and they should show greater longevity in severe environments than steel rope.

SAQ 6.7 (Objectives 6.5, 6.6)
What are the breaking lengths in air of 30 mm diameter rope constructed from: (a) parallel-lay Kevlar 49; (b) galvanized steel?

What is the best choice of material for tethering an oil-drilling rig in 2000 metres depth of sea water using 30 mm rope? (Assume the density of sea water ρ_s is 1030 kg m^{-3}, and that the maximum permitted load is 10% of the breaking load.) Comment on possible problems that might be encountered with the rope material selected.

Most applications of parallel-lay aramid ropes to date have exploited the tensile strength to full advantage by keeping the rope in tension and not allowing it to go into severe bending, where compressive stress can

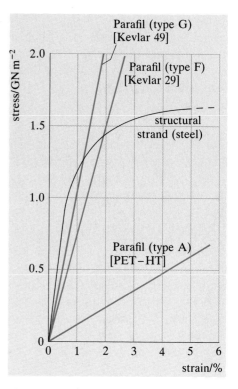

Figure 6.42 Stress–strain curves of parallel-lay and steel ropes

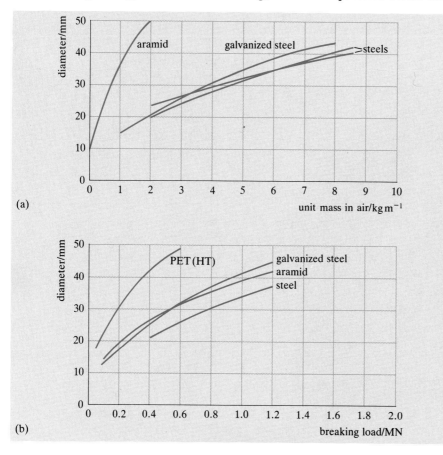

(a)

(b)

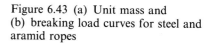

Figure 6.43 (a) Unit mass and (b) breaking load curves for steel and aramid ropes

cause fibre kinking. One further factor needs to be considered. The creep rate is higher than in equivalent steel ropes, but this can be circumvented by prestressing the rope before installation. There is undoubtedly an interesting future for high-performance fibre ropes complementing, but never replacing, steel (apart from their sensitivity to fatigue they are much more expensive than steel, by a factor of about 200). However, the versatility of their fibres is such that they also find application in rigid matrices, that is in composite materials — which is the subject of the next chapter.

Objectives for Chapter 6

After studying this chapter you should be able to:

6.1 Account for the effect of fibre diameter on bending flexibility (SAQ 6.1).

6.2 Assess the effect of changing molecular structure on such physical properties as water absorption by fibres (SAQ 6.2).

6.3 Describe and account for the structure and properties of paper (SAQ 6.3).

6.4 Justify the use of a particular fibre or blend in a textile application (SAQ 6.4, 6.5).

6.5 Calculate the breaking load of rope and assess the effects of stress concentrations on breaking load (SAQ 6.6, 6.7).

6.6 Select a material for a rope of given function (SAQ 6.6, 6.7).

6.7 Define and use the following terms and concepts:

amino acid	protein regain
aramid	size
braided rope	specific stress
cellulose	spinneret
denier	spinning
felt	staple
fibrillation	tenacity
gel spinning	tex
hawser laid	true bias
initial modulus	unit mass
microfibril	warp
monofilament	weaving
parallel laid	weft
plaited rope	yarn

Answers to exercises

EXERCISE 6.1 (a) Monofilaments, threads, ropes of all kinds, from simple string up to large boat cables, are stiff only along the longitudinal axis of the product and highly flexible in all other directions.
(b) Webbing is composed of woven fibres and is stiff along the long axis and along the axis at right angles to it, but flexible in bending along the third axis.

EXERCISE 6.2 From Table 6.1, the density of nylon in fibre form is $1140 \, \text{kg m}^{-3}$, so

$$A = \frac{T}{10^6 \rho} = \frac{2}{1.14 \times 10^9} \, \text{m}^2$$
$$= 1.75 \times 10^{-9} \, \text{m}^2$$

Since $A = \pi d^2/4$ (where d is diameter)

$$d^2 = \frac{4A}{\pi} = 2.23 \times 10^{-9} \, \text{m}^2$$

Thus $d = 4.73 \times 10^{-5}$ m or $47.3 \, \mu$m.

EXERCISE 6.3

$$1 \, \text{N m}^{-2} = 1 \, \text{N tex}^{-1} \times \text{tex m}^{-2}$$
$$= 1 \, \text{N tex}^{-1} \times 10^6 \rho$$

Thus

$$\sigma_{\text{TS}} = 0.49 \times 10^6 \times 1140 \, \text{N m}^{-2}$$
$$= 559 \, \text{MN m}^{-2}$$

EXERCISE 6.4 From Figure 6.10, the atomic composition of each repeat unit is 6 carbon atoms, 5 oxygen atoms and 10 hydrogen atoms. The repeat unit molecular mass is therefore $(72 + 80 + 10) = 162$. The molecular mass of natural cellulose is therefore at least $1\,620\,000$.

EXERCISE 6.5 Among common synthetic polymers, isotactic polypropylene crystallizes in a helical conformation. It does so because the side group ($-CH_3$) inhibits the formation of a regular zig-zag conformation (as in HDPE). The side groups twist out of one another's way in a regular way so as to form a helix. In natural polymers, there are extra *intra*molecular hydrogen bonds which provide extra stability for the helix.

EXERCISE 6.6 The reason is that the PP chain is coiled into a helix. Since the intramolecular bonds in the helix are weak van der Waals' bonds, straining will unwind the helix first. The fibre stiffness is thus much lower ($36 \, \text{GN m}^{-2}$) compared to that of high-performance PE fibre ($120 \, \text{GN m}^{-2}$).

EXERCISE 6.7 Plain weave fabrics are most resistant to strain along either warp or weft, and most flexible at $45°$.

EXERCISE 6.8
(a) Low continuous loading: clothes line, tent guy rope, children's swing, boat rigging, rope ladder, buoy mooring rope.

(b) High intermittent loading: car tow rope, climbing or caving rope, mooring ropes for ships.

EXERCISE 6.9

$$1 \, \text{ktex} = 1 \, \text{kg km}^{-1}$$

For nylon, from Table 6.7, the breaking load of 12 mm diameter rope is 29.4×10^3 N and *weight* per unit length is $93.7 \times 9.81 \, \text{N km}^{-1}$. Thus

$$\text{breaking length} = \frac{29.4 \times 10^3}{93.7 \times 9.81} \, \text{km}$$
$$= 32 \, \text{km}$$

For the other materials we obtain:

polypropylene	30.7 km
polyester	19.6 km
manila	10.1 km

So nylon is the strongest with a minimum specified breaking length of 32 km.

EXERCISE 6.10 For subsea applications, the specific modulus and strength will be $E/(\rho - \rho_s)$ and $\sigma_{\text{TS}}/(\rho - \rho_s)$ where ρ_s is the density of seawater. So the specific modulus of Kevlar 29 improves from 43 to 162 and steel from 26.7 to 30.75. The specific strength of Kevlar 29 changes from 2.00 to 7.57 and steel from 0.38 to 0.44. The specific modulus of Kevlar is thus improved about three times over that of steel when they are used in sea water.

Answers to self-assessment questions

SAQ 6.1 From Chapter 1 we have

$$\frac{M}{I} = \frac{E}{R}$$

so that, for the same M and same E, the relative stiffness of the fibre (subscript f) and monofilament (subscript m) is the ratio of the radii to which they are bent.

$$\frac{R_f}{R_m} = \frac{I_f}{I_m}$$
$$= \frac{r_f^4}{r_m^4}$$

We know that in tension the monofilament is 100 times stiffer, so that its cross-sectional area (πr^2) must be 100 times greater, or

$$r_m^2 = 100 r_f^2$$

Thus

$$\frac{R_f}{R_m} = \frac{r_f^4}{10^4 r_f^4}$$
$$= \frac{1}{10^4}$$

That is, the fibre is 10 000 times more flexible in bending than the monofilament.

SAQ 6.2 The first step in calculating water regain is to determine the repeat unit molecular mass M_R for each fibre.

(a) Silk fibroin (see Figure 6.15a)
$M_R = 5C + 2N + 2O + 8H = 128$ and there are *two* free amine groups, so they can bond with *two* water molecules (each of molecular mass 18). Hence water regain at 100% humidity is $(36/128) = 28\%$.

(b) Nylon 6,6 (see Appendix 2)
$M_R = 12C + 2N + 2O + 22H = 226$ with two free amine groups, so water regain is $(36/226) = 16\%$.

Assuming that only amorphous material absorbs water, then the water regain of the fibres will be

(a) fibroin, total water regain

$$= 28 \times 0.5 = 14\%$$

(b) nylon 6,6, total water regain

$$= 16 \times 0.4 = 6.4\%.$$

The water regain for silk should therefore be over twice that of nylon, but in fact it is up to about three times that of nylon over most of the regain curve (Figure 6.19). This may be because more than one molecule of water attaches to each free amine unit in silk; or because of penetration of water into the crystalline zones of silk; or a combination of both effects.

SAQ 6.3 Paper consists of a flat mat of cellulose fibrils, about 20 μm wide and 0.5 to 3 mm long, bonded together at their intersections by hydrogen bonds on the surface of the fibrils. The fibrils are randomly oriented with respect to one another in the plane of the paper, giving the maximum number of fibre–fibre intersections. If the paper is strained in one direction, only a small proportion of the fibres will be aligned along the strain axis, so the tensile modulus of paper will be substantially lower ($1-5\,\mathrm{GN\,m^{-2}}$) than that of the cellulose fibre ($6-10\,\mathrm{GN\,m^{-2}}$). Increasing the degree of orientation along the strain axis will increase the modulus. The tensile strength will be considerably lower than that of the single cotton fibre ($300-800\,\mathrm{MN\,m^{-2}}$) since the fibres are short and held together only by relatively weak hydrogen bonds. The strength of paper is thus typically $30-80\,\mathrm{MN\,m^{-2}}$, about 10% of the strength of individual fibres.

Bulk density will affect properties substantially: the lower the bulk density, the fewer the fibre–fibre intersections, so the weaker the paper will be. The modulus of paper both in tension and bending will also be lower, depending directly on the number of fibres in the cross section. This is essentially the difference between a lightweight, low-strength tissue paper and a dense, higher-strength writing paper.

SAQ 6.4 A typical selection might include

- 50% wool/35% viscose rayon/15% nylon (duffle coat),
- 55% PAN (acrylic)/45% cotton (coat lining),
- 67% polyester/33% cotton (raincoat),
- 50% polyester/50% cotton (coat lining),
- 65% polyester/35% cotton (shirt),
- 80% wool/20% nylon (sweater).

Polyester–cotton blends have higher breaking strains than pure cotton and are more resistant to wear, moisture, creep and creasing.

Although staple nylon and viscose rayon fibre possess greater tensile moduli than wool (Figure 6.9 and 6.18), they are closer to wool than most other synthetic fibres. They both possess greater strengths than wool, so they will help to reinforce blended yarns. Neither viscose rayon nor nylon will seriously affect comfort since both are hydrogen bonded and will absorb water, although not nearly to the same extent as wool itself (Figure 6.19).

SAQ 6.5 Since the weave itself, as well as yarn twist, will increase the elongation-to-break, those fibres with very low breaking strains are best suited for car seat belts: high tenacity polyester and high tenacity nylon 6,6 with breaking strains of 6% and 16%, respectively. So HT continuous PET fibre is the best choice for standard seat belts and HT continuous nylon 6,6 fibre for high performance belting. MT nylon 6,6 fibre is the best choice for competition belts with its lower initial modulus and greater breaking strain than MT PET.

SAQ 6.6 Consider manila, the weakest rope, first (Table 6.7). Its rated breaking strength is 44.8 kN which must be derated by 80% to give a maximum load of 8.96 kN. But the rope will be further weakened by the eye splice and overhand knots. The derating factor will be that of the worst joint, namely the overhand knot with a knot efficiency of 0.40. The maximum load it can withstand is therefore 3.6 kN. Since the required rope must withstand a load of 500 kg (4.905 kN), manila can be rejected as a choice.

The choice therefore narrows to nylon 6,6 or PET both of which possess breaking

strengths well in excess of the specification (Table 6.7). The choice will be determined by the strain characteristics shown by curves A in Figure 6.34. By inspection of the curves, the extensions under this load for first use are

PET 6%
nylon 10%

Polyester (PET) rope would therefore be the best choice given a maximum specified elongation of 8% on first loading.

SAQ 6.7 From Figure 6.43, we have

(a) For 30 mm diameter aramid, the breaking load is about 0.5 MN. Unit mass is $0.7\,\mathrm{kg\,m^{-1}}$. The breaking length L_B is therefore

$$L_B = \frac{0.5 \times 10^6}{0.7 \times 9.81}\,\mathrm{m} \approx 73\,\mathrm{km}$$

(b) For 30 mm diameter galvanized steel, the breaking load is again 0.5 MN. The unit mass is about $4\,\mathrm{kg\,m^{-1}}$. The breaking length L_B is then

$$L_B = \frac{0.5 \times 10^6}{4 \times 9.81}\,\mathrm{m} \approx 13\,\mathrm{km}$$

In sea water of density ρ_s, L_B will increase for both materials by $\rho/(\rho - \rho_s)$, assuming complete wetting. Thus the permitted lengths are

$$L = \frac{0.1 L_B \rho}{(\rho - \rho_s)}$$

Table 6.4 gives $\rho = 1440\,\mathrm{kg\,m^{-3}}$ for aramid (Kelvar 49), so from the equation above, $L \approx 26\,\mathrm{km}$. For steel, Table 6.4 gives $\rho = 7860\,\mathrm{kg\,m^{-3}}$ so $L \approx 1.5\,\mathrm{km}$, which is less than the specified depth of water.

Aramid is the best choice for such a tethering system, provided it's kept in tension. Steel, despite galvanizing, is susceptible to corrosion especially at its most highly stressed points. In water, aramid fibres can swell, so the aramid rope will be sensitive to fatigue. If water is prevented from entering the rope, then it will be buoyant in sea water (due to the entrapped air between the fibres), so making anchoring to the seabed critical. An alternative is to provide a matrix polymer to change the bulk density of the rope to match that of sea water, making the breaking length infinite!

Chapter 7
Composite Materials

by George Weidmann   (consultant: Martin Buggy)

Chapter 7 Composite materials

7.1 Introduction

In the previous chapter, and elsewhere, we've seen that materials in fibrous form have some of the highest moduli and greatest strengths available. Such materials frequently have significantly lower densities than those of many other structural materials. Given that fibres and fibre assemblies have little resistance to flexural and compressive deformations, the problem is how to take advantage of their outstanding properties in applications involving bending or compression. The answer, of course, is to embed the fibres in a suitable matrix material to form a composite.

However, that's not the only reason for making composites. Materials which are brittle, such as thermosetting resins and ceramics, can have their toughness significantly enhanced by incorporating fibres or fillers (in a similar way to the crystallites in glass-ceramics in Chapter 4). Fillers can also be used to enhance such properties as fire retardance or electrical conductivity, as well as to reduce the overall cost of the material. A further important class of composite material is that in which the second phase is a gas, namely, foams. Although most synthetic composites in use today are based on polymeric matrices, their relatively low maximum service temperatures has led to growing interest in metallic- and ceramic-based composites for high temperature applications.

Composite materials have been used for many thousands of years, starting with natural ones such as wood, bone and horn (based on fibres of cellulose, collagen and keratin, respectively). Also, the advantages of deliberately combining materials to obtain improved or modified properties were appreciated by some of the earliest civilizations. Examples of this are mixing the clay for air-dried bricks with straw to reduce the effects of shrinkage stresses (Figure 7.1), the use of wattle and daub in walls (Figure 7.2), the Egyptians' development of plywood and their use of linen–plaster composites for mummy cases (Figure 7.3).

Nowadays, especially with the growth of the plastics industry and developments in fibres, a vast range of combinations of materials is available for composites. The selection of a particular combination for a composite depends not only on the properties required in service, but also on the processing route. This in turn is conditioned by the form of the materials (for example, liquid or solid, particulate or fibrous, discrete or continuous) and by economic factors such as the quantity required. As an example, ▼*Hunt* class MCMV▲ and ▼Dunlop tennis racket▲ examine two contrasting composite products, both based on polymeric matrices filled with fibres. But that's where their similarity ends.

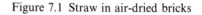

Figure 7.1 Straw in air-dried bricks

Figure 7.2 Wattle and daub

Figure 7.3 Linen–plaster composite (*cartonnage*) mummy case

▼ *Hunt* class MCMV ▲

Figure 7.4 A *Hunt* class MCMV

Like all navies, the Royal Navy needs a capability for detecting and clearing mines. These two functions of minehunting and minesweeping are combined in a mine countermeasures vessel (or MCMV). A prime requirement of an MCMV is that it should have the least possible chance of being detected magnetically. Materials that are inert magnetically must not only be non-magnetic, but also non-magnetizable and non-conducting (to avoid the magnetic fields associated with induced currents).

Previous vessels (the *Ton* class) had a double-skinned mahogany hull over an aluminium framework, but in the 1960s the Navy was looking for a replacement design with a higher explosive shock resistance and improved magnetic properties, coupled, if possible, with lower hull maintenance costs. Contending materials were wood, stainless steel and glass fibre reinforced plastic (GFRP). All-wood construction was ruled out because of inadequate resistance to explosions and its susceptibility to rot and attack by marine organisms. Stainless steel was eliminated because, being conducting, it did not satisfy the magnetic requirements,

and, despite its name, was still liable to corrosion in seawater. A 12-year development programme with GFRP culminated in the building and launch in June 1978 of HMS *Brecon*, the first of the *Hunt* class of MCMVs (Figure 7.4).

So far, 13 of this class have been built, and with a length of 60 m, beam of 10 m and displacement of 625 t, they are amongst the world's largest GFRP structures. Each one contains about 300 t of GFRP, with equal masses of E-glass fibres and polyester resin (see Section 7.2.2). (Epoxy resin was also a contender, but costs a lot more than polyester, and needs an elevated temperature treatment to gain a property advantage.) The fibres are in the form of continuous rovings (that is, lightly twisted yarns) woven in a plain weave (Figure 7.5). This form of reinforcement, together with the size of the hull and the structural benefits of having a one-piece moulding, meant that the only feasible production route was by hand lay-up into an open mould. The hull was built in successive layers by impregnating the glass cloth with resin, placing it in position in the mould and consolidating it by hand rolling. At its thickest section, the hull consists of 40 such layers, each about 1 mm thick.

It's estimated that the *Hunt* class has a hull life of some 60 years, and the savings from this and its ease of repair more than outweigh the higher initial costs in materials and labour compared with wooden and metal hulls. A further advantage of GFRP is that the woven glass cloth provides an effective fire barrier — a sensitive area for the Navy after its experience in the Falklands.

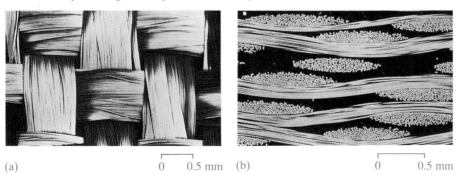

(a)   0   0.5 mm   (b)   0   0.5 mm

Figure 7.5 Woven glass roving (a) without resin (SEM) (b) section through composite (optical)

▼Dunlop tennis racket▲

The manufacturers of sports equipment are highly competitive, mirroring the sports they cater for. In the late 1970s, Dunlop Sports Company, a major manufacturer of wooden tennis rackets, found its business seriously threatened by rackets made from glass fibre and carbon fibre composites (aluminium alloy rackets, introduced in the 1960s, were suffering a similar threat).

To compete, a composite racket needs to be of similar size and mass to a wooden one; and since GFRP and CFRP have higher densities than wood, this means it needs a hollow or low-density core. The contemporary composite rackets had the uncured composite material laid by hand around an inflatable or expandable mandrel, and this assembly was inserted in a mould and hot cured for several minutes. The process was both labour intensive and time consuming. The rackets also had to be over-engineered to cope with the variability of hand fabrication, and were thus wasteful of material. Wooden rackets similarly require a substantial amount of hand work and the material is intrinsically variable.

The company was thus looking for a composite process route which was faster, more automated and more consistent. What the design team came up with was a variant of the traditional lost-wax process for casting — but applied successfully for the first time to injection moulding. A diecast, bismuth–tin eutectic alloy core is positioned in the moulding tool, and a melt of nylon 6,6 with 30% by volume of carbon fibres is injected around it (Figure 7.6). In a separate operation, the core is melted out by immersion in an oil bath and the alloy recycled into further cores.

For the process to work, the alloy ($T_m = 411.5\,\mathrm{K}$) must not melt significantly when the hot nylon ($T_m = 540\,\mathrm{K}$) is injected around it, yet should melt out easily at a temperature below the melting temperature of the nylon. This apparent paradox is resolved by a combination of the low thermal conductivity of nylon, the high thermal capacity and conductivity of the alloy core, and a short ($\sim 3\,\mathrm{s}$) injection time. Together these ensure that the heat input *rate* to the alloy is too low for any significant melting to occur.

A finished racket and details of it are shown in Figure 7.7. Volume manufacture started in November 1980, and it has been a highly successful product since its inception, growing to an annual production rate of several hundred thousand. The choice of injection moulding limited the reinforcement to short fibres, but nevertheless the racket is some three times stronger than a wooden one, and comparable with those of the hand-laid, continuous fibre composite rackets.

In this chapter we will first consider the nature of composites — how they can be classified and what they're made from. We'll then examine their more important mechanical properties, how these reflect those of the constituent materials, and to what extent they can be modelled.

7.2 Types of composite

7.2.1 Make-up and morphology

At its simplest, a composite material consists of a matrix material which encapsulates discrete elements of one or more different materials. These elements can be particles, fibres, assemblies of fibres, sheets or, in the case of foams, gas- or liquid-filled voids. Some examples of these are sketched in Figure 7.8 (overleaf), where they are ranked vertically according to whether they are equiaxed, extended in one or two dimensions, or assemblies, and divided horizontally according to whether the components of the composite are discrete or continuous.

For a given combination of second-phase and matrix materials, the form of the second phase strongly affects the properties of the composite. As we saw from the MCMV and tennis racket examples, the form of the second phase can also control the process route. However, another very important factor is how much of each component material is present in the composite. This is usually characterized by its volume fraction (see ▼Expressing concentration▲).

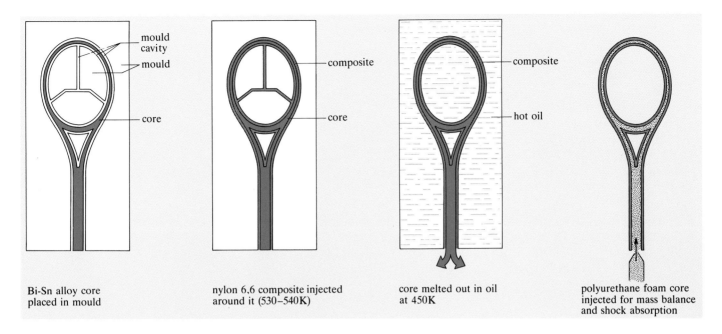

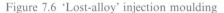

Figure 7.6 'Lost-alloy' injection moulding

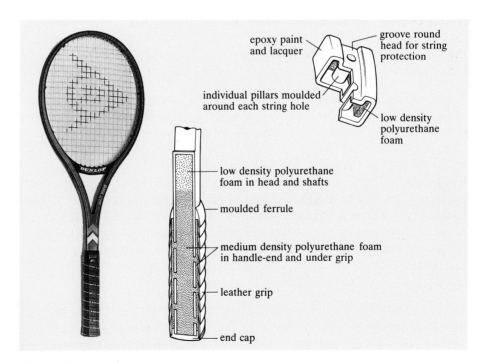

Figure 7.7 The finished racket

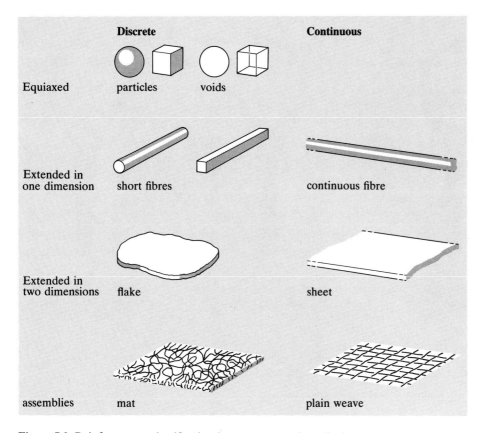

Figure 7.8 Reinforcement classification by geometry and continuity

As with the packing of atoms in a crystal, the *maximum* possible concentration, or volume fraction, of a given arrangement of a filler or reinforcement in a matrix is known as the **packing factor, P_f**. The value of P_f depends on the shape, size distribution and arrangement of the second phase. In the same way as a material in crystalline form has a higher density than in its amorphous form because of the more efficient packing of its atoms, a regular array of, for example, fibres has a higher packing factor than a random array (see ▼ **Packing and spacing** ▲). A broad distribution of sizes will have often a higher packing factor than narrow or monodisperse distributions, since smaller particles can fit in the interstices between larger ones. Hexagonal and square cross-sections can pack together more efficiently than circular ones.

So, what sorts of material are used in composites? We've already noted that most composites available today for load-bearing applications are, with the possible exception of concrete, based on polymeric matrix materials. The use of polymeric matrices means that advantage can be taken of both their easy processability and their relatively low densities in the resulting composites. Practically the whole range of polymeric materials can be used to encapsulate fibres or fillers or both. Table 7.2 sets out representative properties of some of them, together with those of other matrix materials. The first two are examples of the more

Table 7.2   Properties of some matrix materials

Temperature	Density $\rho/\mathrm{Mg\,m}^{-3}$	Tensile modulus/ $\mathrm{GN\,m}^{-2}$	Tensile strength $\sigma_{\mathrm{t}}/\mathrm{MN\,m}^{-2}$	Upper limit $T_{\mathrm{max}}/\mathrm{K}$
polypropylene	0.91	1.5	33	270
nylon 6,6	1.14	2.0	80	380
PES	1.37	2.4	80	450
PEEK	1.32	1.1	90	530
polyimide	1.36	1.4	80	590
polyester	1.3	3.3	65	490
epoxide	1.2	3.5	75	460
aluminium	2.7	71	80	370
titanium	4.5	116	180	770
silicon nitride	3.2	300	8000	1300

common thermoplastics (see Chapter 5) whilst the next three, polyethersulphone (PES), poly(ether ether ketone) (PEEK) and polyimide represent the newer, high temperature thermoplastics. Polyester and epoxide are thermosetting plastics, and these are considered separately in Section 7.2.2.

▼Expressing concentration▲

The concentration of a component in a composite is most frequently expressed in terms of its **volume fraction** V. In this chapter we'll be using lower case letters to denote actual weights (or masses) and volumes of components, and upper case letters for their respective fractions. The volume fraction V_i of the ith component is then

$$V_i = \frac{v_i}{v_\mathrm{c}}$$

where the volume of the composite v_c is

$$v_\mathrm{c} = v_1 + v_2 + \ldots = \sum_i v_i$$

and

$$V_1 + V_2 + \ldots = \sum_i V_i$$
$$= V_\mathrm{c} = 1$$

by definition.

When making a composite it's often convenient to measure out the components by weight. The **weight fraction** W_i of the ith component is related to V_i as follows:

$$\begin{aligned} W_i &= \frac{w_i}{w_\mathrm{c}} \\ &= \frac{m_i g}{m_\mathrm{c} g} \\ &= \frac{v_i \rho_i}{v_\mathrm{c} \rho_\mathrm{c}} \\ &= \frac{V_i \rho_i}{\rho_\mathrm{c}} \end{aligned} \tag{7.1}$$

We also have that the density of the composite ρ_c is

$$\begin{aligned} \rho_\mathrm{c} &= \rho_1 V_1 + \rho_2 V_2 + \ldots \\ &= \sum_i \rho_i V_i \end{aligned} \tag{7.2}$$

so that

$$W_i = \frac{V_i \rho_i}{\sum_i \rho_i V_i} \tag{7.3}$$

By a similar argument

$$V_i = \frac{W_i/\rho_i}{\sum_i W_i/\rho_i} \tag{7.4}$$

SAQ 7.1   (Objective 7.1)

(a) Verify Equation (7.4).
(b) A composite is made up of equal weights of nylon, chalk powder and glass spheres, whose densities are 1.1, 2.7 and 2.5 $\mathrm{Mg\,m}^{-3}$ respectively. What are the volume fractions of each component and what is the density of the composite?

▼Packing and spacing▲

Like atomic packing factors, those for composites can be derived for different model arrays. Two of the simplest are the hexagonal and square arrays of parallel fibres in contact (Figures 7.9a and b) of diameter d. The packing factors are

$$P_f = \frac{\pi}{2\sqrt{3}} \quad \text{for the hexagonal case}$$

and

$$P_f = \frac{\pi}{4} \quad \text{for the square one}$$

If, instead of touching, the fibre centres are a distance D from each other (Figures 7.9c and d), for both arrays

$$\frac{V_f}{P_f} = \frac{d^2}{D^2} = \frac{d^2}{(d+s)^2} \tag{7.5}$$

from which the fibre spacing s is

$$s = d\left[\left(\frac{P_f}{V_f}\right)^{1/2} - 1\right] \tag{7.6}$$

The packing factors for a selection of cases are listed in Table 7.1 (those for random packing were determined empirically — Figure 7.10).

This shows the marked effect that aspect ratio (Chapter 6) has on randomly packed fibres, but the value of $P_f = 0.11$ is very much *less* than the 0.20–0.30 volume fractions typical of injection moulding grades of thermoplastic containing fibres, whose aspect ratios are usually well over 100.

Why should this be?

EXERCISE 7.1 Suggest reasons for the discrepancy between V_f and P_f for fibre-filled thermoplastics.

Figure 7.11 shows a section transverse to the fibres in an aligned-fibre composite with $V_f = 0.75$, showing that the correspondence to a regular array is only approximate, but, if anything, the arrangement is hexagonal.

EXERCISE 7.2 If the composite in Figure 7.11 can be modelled as a regular array, estimate the mean fibre spacing.

Table 7.1 Packing factors P_f for different shapes and arrangements of filler

Filler	P_f
parallel fibres, same size, hexagonal array	$0.91 \ (= \pi/2\sqrt{3})$
parallel fibres, same size, square array	$0.79 \ (= \pi/4)$
spheres, same size, hexagonal close packing	0.74
spheres, same size, random loose packing	0.60
straight fibres, random packing	
aspect ratio 2	0.68
aspect ratio 10	0.42
aspect ratio 50	0.11

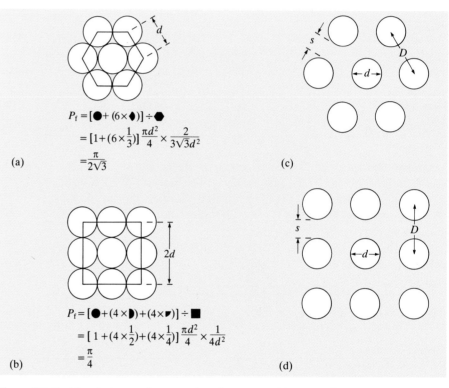

(a)

$$P_f = [\bullet + (6\times\blacklozenge)] \div \bullet$$
$$= [1 + (6\times\tfrac{1}{3})]\frac{\pi d^2}{4} \times \frac{2}{3\sqrt{3}d^2}$$
$$= \frac{\pi}{2\sqrt{3}}$$

(b)

$$P_f = [\bullet + (4\times\blacktriangleright) + (4\times\blacktriangledown)] \div \blacksquare$$
$$= [1 + (4\times\tfrac{1}{2}) + (4\times\tfrac{1}{4})]\frac{\pi d^2}{4} \times \frac{1}{4d^2}$$
$$= \frac{\pi}{4}$$

(c)

(d)

Figure 7.9 Packing factors and geometries of square and hexagonal arrays

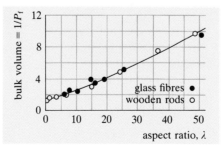

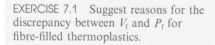

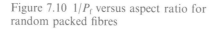

Figure 7.10 $1/P_f$ versus aspect ratio for random packed fibres

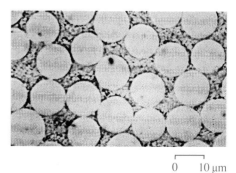

0   10 μm

Figure 7.11 Section of unidirectional glass fibre–polyester composite (optical)

Table 7.3   Typical properties of reinforcing fibres

	Density $\rho/\mathrm{Mg\,m^{-3}}$	Young's modulus $E/\mathrm{GN\,m^{-2}}$	Tensile strength $\sigma_{TS}/\mathrm{GN\,m^{-2}}$	Linear expansion coefficient $\alpha/10^{-6}\,\mathrm{K^{-1}}$	Typical diameter/μm	Melting or softening point/K	Relative price/kg^{-1}
E-glass	2.55	72	3.4	5.1		1110	1
C-glass	2.49	69	2.8	7.2	10–20	1030	5–10
S-glass	2.49	87	4.6	2.7		1220	5–12
A-glass	2.50	68	2.4	16.2		970	1.5
aramid (*Kevlar 49*)	1.44	124	3.1	$-4(\parallel)$ $59(\perp)$	12	770	10
UHMPE (*Spectra*)	0.97	120	2.6	—	38	420	30
carbon UHM	1.96	780	2.3				
HM	1.85	480	2.0	$-2(\parallel)$	8	3920	20–100
UHS	1.75	270	5.2	$17(\perp)$			
HS	1.76	265	2.8				
zirconia (*Saffil*)	5.6	100	0.7	—	3	2770	25
alumina (*Saffil*)	2.8	100	1.0	—	3	2270	
boron	2.6	380	3.8	—	100–200	2570	450
asbestos (chrysotile)	2.5	160	2.1	30	0.02	1790	0.1
silicon carbide	2.5	410	4.0	—	140		
steel piano wire	7.9	210	3.0	15	(up to 5 mm)	1670	2

One of the problems with polymeric matrix materials has been their relatively low maximum service temperatures (see Table 7.2), yet some of the available fibres can remain load-bearing at very high temperatures, so matrices are needed that more nearly match this performance. This has spurred the development of materials such as PES, PEEK and polyimide (but at some sacrifice of processability) on the one hand, and, on the other, to composites based on metal and ceramic matrices. Bear in mind, though, that the presence of reinforcing fibres frequently raises the upper service temperature of the composite *vis-à-vis* the matrix on its own. Thus nylon 6,6 with 30% by weight of glass fibres can perform adequately at about 470 K, whilst titanium–silicon carbide fibre composite can withstand some 1200 K.

Some typical properties of the most important, high performance reinforcing fibres are shown in Table 7.3. Aramid, high modulus polyethylene (HMPE), carbon and glass fibres were described in Chapter 6. Of the others in Table 7.3, the *Saffil* zirconia and alumina fibres are examples of recent developments in ceramic fibres; boron (which is made by vapour deposition onto a 13 μm diameter tungsten filament) and silicon carbide are expensive, special purpose materials; and asbestos, in contrast, is very cheap, but its use is severely restricted by the associated health hazards. Steel piano wire is included for comparison, although it is available only in much larger diameters than those listed for the fibres.

▼Fillers▲

Many of the more common and cheaper fillers are based on ground minerals, such as limestone, or on some of the many varieties of silicate (kaolin, talc, feldspar, mica or silica itself as quartz). Other low-cost fillers are derived from natural products, such as wood, nutshells and various fibrous plants. The last category, which includes cotton, jute, coconut, sisal and hemp is more expensive than wood flours and nutshell flours, but the fibres act more effectively as reinforcements. Apart from its use in fibrous forms (both chopped and continuous), glass is also used as a filler as solid spheres, as flakes and, more recently, as hollow microspheres.

Table 7.4 summarizes the characteristics of some of the more common fillers for plastics. The particle size ranges shown in the second column of the table represent the extremes of size, and generally there will be several grades of each filler type available, with different, narrower size distributions in between these extremes.

Flake-shaped reinforcements, such as the glass flakes mentioned above, talc and particularly mica can also reinforce if their aspect ratios are sufficiently high. The last two can, therefore, offer a low cost alternative to glass fibres. In terms of the

Table 7.4   Characteristics of some common fillers for plastics

Filler	Size range/μm	Density/kg m^{-3}	E/GN m^{-2}	Linear expansion coefficient α/10^{-6} K^{-1}	Relative cost/kg^{-1}
ground limestone	0.1–600	2710	26	10	1
calcium carbonate	0.1–2	2650	26	10	5
feldspar	0.3–90	2600	30	6.5	2
ground quartz	0.1–75	2650	30	10	4
kaolin	0.03–50	2580	20	8	
talc	0.05–60	2800	20	8	4
mica	4–400	2820	30	8	6
glass spheres (ballotini)	5–700	2480	60	8.6	15
hollow glass spheres	10–150	280	0.2	8.7	70
wood flour	100–1000	600	10	5.5	4
alumina trihydrate	0.05–150	2420	30	4.5	8

relative costs in the last column of Table 7.4, E-glass fibre would have a value of about 50, which is the same order as an inexpensive grade of a polyolefin such as polypropylene.

Another important function of fillers is fire retardance. Although mineral fillers are generally non-combustible, and some, like limestone and calcium carbonate, have a fire-retarding effect, there are special fillers which are used for this purpose. Alumina trihydrate is one of these. It decomposes endothermically at temperatures above 493 K to produce alumina and about 35% by weight of water. It is the heat absorbed by this decomposition which is the chief mechanism in suppressing combustion. It has the added benefit of increasing the resistance to electrical arcing and tracking in a polymer.

Apart from fibres, there still remains a very wide range of solid materials which are added to plastics in significant volume fractions (greater than 0.05, say) in order to modify one or more properties (especially if cost is regarded as a property). These types of additive are usually called **fillers**. Fillers may be distinguished from fibrous reinforcements by their much lower aspect ratios, so their effect on such mechanical properties as modulus, fatigue strength and toughness are generally smaller. They are reviewed in ▼Fillers▲.

The introduction of boron fibres and lower cost silicon carbide and alumina fibres and whiskers (see Chapter 6) has increased development work on metal-matrix composites using matrices such as aluminium (for example the Toyota piston ring groove in Figure 7.12 with alumina–silica fibres in aluminium alloy), magnesium alloys and titanium. To some extent this has been led by metal manufacturers who were losing aerospace markets to plastics-based composites. Boron reinforced aluminium is commercially available, its first major use being in the

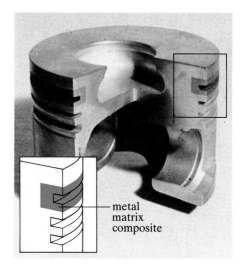

Figure 7.12 Toyota piston ring groove

space shuttle (Figure 7.13), and current programmes should lead to significantly increased usage of metal-matrix composites. The main technical challenge in producing suitable materials is to ensure that the metal matrix adequately wets, but does not react with, the reinforcement, thus achieving a satisfactory interface. Ceramic-matrix composites, either fibre or whisker reinforced, are also under active development to produce impact-resistant, damage-tolerant, high-temperature materials.

SAQ 7.2   (Objective 7.2)
What are the main reasons for making and using composite materials? What is the rôle of the matrix in composites based on high-performance fibres?

Figure 7.13 B–Al tubes in the mid-fuselage of a space shuttle

Figure 7.14 PF saucepan handles

7.2.2 Thermosets

Consideration of thermosets has been included here, rather than in Chapter 5, for two reasons. Firstly, as a class, they are much more brittle than thermoplastics, and so are very rarely used as load-bearing solids without some form of reinforcement (although this usually doesn't apply to their use as foams — see next section — adhesives and surface coatings). Examples of thermoset moulding materials, which, typically, are wood-flour filled, include phenol formaldehyde (PF) saucepan handles (Figure 7.14), urea formaldehyde (UF) electrical fittings (Figure 7.15), and melamine formaldehyde (MF) kitchen utensils (Figure 7.16).

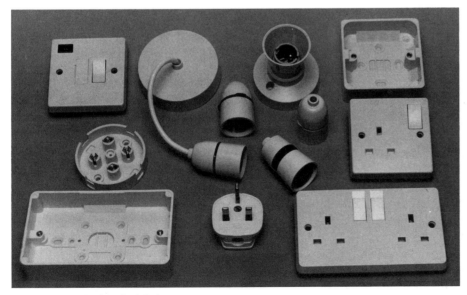

Figure 7.15 UF electrical fittings

Figure 7.16 MF kitchen utensils

The second reason is that, because they tend to perform better to higher temperatures and creep less than thermoplastics, and because it's easier to incorporate higher volume fractions of longer fibres and to achieve greater orientation of them, they are preferred as matrix materials for high-performance composites.

Thermosets differ from thermoplastics in that their molecular chains are crosslinked to one another by primary bonds, a characteristic they share with the majority of elastomers (Chapter 5). However, the crosslink density is far higher than in elastomers, leading to a rigid, rather than a rubbery, solid. The generic term **network polymer** includes both elastomers and thermosets, and gives a better indication of their structure than their more common names. In fact 'thermoset', an abbreviation of 'thermosetting polymer', is misleading. When it was coined, the only network polymers available required heat to set (or cure) them, but this is no longer the case. Nowadays it is used in the sense that, once cured, the materials do not undergo a melting transition on heating. Instead, they remain solid until they decompose. This places a constraint on their processing to shape compared with thermoplastics.

EXERCISE 7.3   Why should this limit the processing of thermosets?

There are other major differences between thermosets and thermoplastics arising out of their network structure. The arrangement of crosslinked molecules is most commonly random, leading to amorphous, glassy materials (one exception is radiation-crosslinked polyethylene, where the starting material is already partially crystalline — but the crosslink density is low compared with the usual thermosets). Since the molecular network is nominally continuous, the whole of any thermoset sample is one giant molecule, so chain molecular mass is meaningless as a characterizing parameter. Of more importance are the **crosslink density** and the associated, mean molecular mass between crosslinks. Varying these allows control of, in particular, mechanical properties. For example, natural rubber (NR), which is crosslinked with sulphur, might have a mean molecular mass between crosslinks of 8000 and a crosslink density of $4 \times 10^{25}\,\mathrm{m}^{-3}$. Increasing the latter by a factor of 20, by adding more sulphur to the mix, produces ebonite — a hard, rigid thermoset — with a mean molecular mass between crosslinks of about 400 (that is, some 5–6 NR repeat units).

The brittleness of thermosets is also a reflection of their degree of crosslinking. Mechanisms such as plastic flow and crazing, which operate in thermoplastics to increase the work required to propagate a crack, are severely restricted in thermosets, leading to their low toughness. Despite their crosslinking, thermosets still exhibit viscoelastic behaviour such as time-dependent creep and recovery, although on a smaller scale than thermoplastics. An example of this is shown in Figure 7.17 for four different epoxy resins at 293 K.

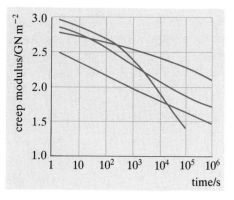

Figure 7.17 Tensile creep modulus at 1% strain for four different epoxy resins

The chemistry of thermosets tends to be more complex and capable of more variation than that of thermoplastics. The amorphous network structures in fully cured materials are also less amenable to physical and chemical analysis, so there is more uncertainty about their actual structures. However, common to them all is the necessity of having sufficient bonding sites available to form the network. This is usually expressed in terms of the **functionality** of a molecule, which, in simple terms, is the number of other molecules, of either the same or different species, to which it can bond. The sites at which such bonding occurs can be double bonds (unsaturated molecules), reactive side groups (such as hydroxyls or amines) or anywhere where a free radical can be formed. For network formation, all the species present must be at least difunctional and at least one species must be trifunctional or greater (Figure 7.18).

Thus, a trifunctional monomer can form a network by repeated linking to itself. Examples of this class of thermoset include the three formaldehyde-based materials, PF, UF and MF (see Appendix 2 for their molecular structures) introduced at the beginning of this section. Typically these come in the form of moulding powders already compounded with their reinforcement. Heating in a closed mould first melts the powder and then initiates the crosslinking reaction. Bakelite, a proprietory name for a PF moulding powder, was introduced in 1909 and was the first synthetic polymer composite material. One of the earliest products made from it is reputed to have been a gear lever knob for a Rolls–Royce. In liquid form, PF and MF are also used for impregnating paper laminates, widely used for kitchen units and as interior cladding materials.

Polyester, epoxies and polyurethanes differ from the formaldehyde-based thermosets in that they are each characterized by the presence of a particular chemical group — ester, epoxide and urethane respectively (see Appendix 2). However, since these groups can be part of a large variety of different molecules, each of these thermoset types represents a family of materials, whose properties can be tailored over a broad spectrum through appropriate formulation.

We've already met polyesters in Chapters 5 and 6. However, these were saturated polymer molecules with no double bonds available in the chain to act as potential crosslinking sites. *Unsaturated* polyesters (UP) for thermosets are made by reacting together a mixture of saturated and unsaturated species to make a **prepolymer** polyester with a molecular mass of about 2000 (Figure 7.19). The ratio of saturated to unsaturated segments in the prepolymer determines the spacing of the crosslinks. The prepolymer is then mixed with a **reactive diluent**, usually styrene, (which reduces the viscosity and forms bridging crosslinks) and a free-radical **initiator**. In the case of room-temperature curing systems for hand lay-up (as in the '*Hunt* class MCMV', Section 7.1) the initiator starts to react immediately on mixing, and the resin must be applied within its useful pot-life (typically 2–3 h). Moulding grades, such

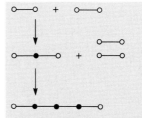

difunctional molecules can combine only to form chains

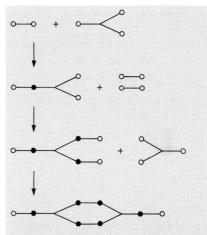

to form a network they require tri- (or higher) functional molecules

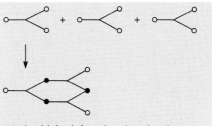

tri- (or higher) functional molecules can form a network with each other

Figure 7.18 Functionality and crosslinking

prepolymer (8–9 repeat units; S_1, S_2 – saturated; U – unsaturated)

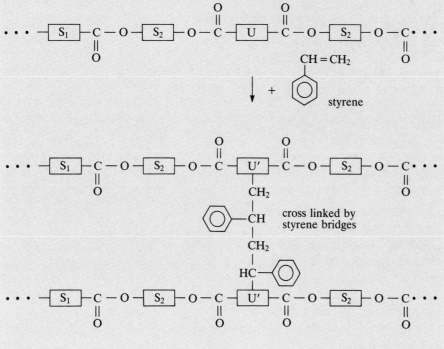

Figure 7.19 Polyester formation

as those used in **dough moulding compound (DMC)** — known in the USA as bulk moulding compound (BMC) — and **sheet moulding compound (SMC)** use initiators which produce free radicals only at elevated temperature, so these compounds normally have a shelf-life of about 3–6 months.

Both DMC and SMC incorporate glass fibres. DMC, as its name suggests, is a compound which has been mixed like a dough, and which is suitable for injection and compression moulding. It has been used successfully for products such as headlamp reflectors of complex shape (Figure 7.20). However, the mixing process breaks the fibres down more than in SMC. Polyester SMC was the GFRP roofing material in Chapter 1. It's made by spraying chopped glass fibres onto a moving layer of polyester mix and covering with a further layer of the mix. The compound is sandwiched between polyethylene or nylon film, consolidated by passing between rollers, and is left to congeal. Once it is handleable, an appropriate pattern of strips is cut from the sheet and, with the film removed, compression moulded.

In many compounds such as SMC, getting the basic thermoset chemistry right and adding the appropriate reinforcement, although important, are not enough. ▼SMC▲ typifies some of the extra complexities involved in formulating one grade for a particular

▼SMC▲

A typical formulation in wt % for an SMC for automotive purposes might be:

Unsaturated polyester resin in styrene	23.0
Chopped glass fibre (14 μm E-glass)	25.0
Tertiary butyl perbenzoate	0.4
Polystyrene in styrene	6.0
Magnesium oxide	0.8
Zinc stearate	1.1
Calcite (ground limestone)	7.5
Brominated aromatic compound	6.0
Antimony trioxide	3.7
Alumina trihydrate	26.5

Apart from the polyester and glass — the basic composite — what function do all the other ingredients have?

The *t*-butyl perbenzoate is the initiator. It decomposes at about 410 K to give free radicals which initiate the crosslinking of the polyester with the styrene. Polystyrene in styrene reduces the shrinkage during crosslinking from about 7% by volume to about 0.15% and improves the surface finish.

Magnesium oxide is a thickening agent which increases the viscosity of the SMC by a factor of about 1000 over a period of days following mixing by forming weak ionic bonds with acid groups in the polyester prepolymer. The glass fibres are added when the viscosity of the mix is low, so they are more easily wetted and impregnated; subsequent thickening makes the SMC tack-free and handleable.

Zinc stearate is a mould release agent. It melts at about 390 K. Calcite is an inert, low-cost filler. The brominated aromatic compound and the antimony trioxide are both added as fire retardants. Alumina trihydrate is a filler which also contributes to the fire retardance (see 'Fillers', Section 7.2.1).

In addition to these, the SMC might also contain a pigment (~2 wt %), the polyester resin could be chlorinated to aid the fire retardance, and the glass fibres would be not only coated with size but also with an antistatic agent to avoid clumping as they are sprayed on to the SMC mix.

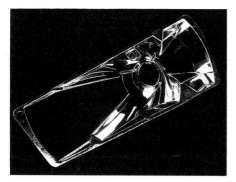

Figure 7.20 Lucas homofocal headlamp reflector

Figure 7.21 London Underground 'C' train

application. Phenol formaldehyde SMCs (PF SMC) are also now available and, because of their lower smoke emission on combustion, are preferred to polyester materials in some applications. One example is on the London Underground, where PF SMC is specified for seating as well as for the cab fronts (Figure 7.21) of the trains.

Unlike polyesters, in which the ester group plays no part in the crosslinking, the highly strained epoxide rings in epoxies are the reactive sites for network formation. Epoxy prepolymers usually have one epoxide group at each end, and each is difunctional. Di- and/or tri-functional crosslinking agents are used — typically these are amines or acid anhydrides. The nature of both the segment between the epoxide groups and the **crosslinking agent** determines the crosslink density. Although originally developed as adhesives, epoxies are now widely used in high-performance composites, especially in aerospace applications. Like the polyesters, they are available as liquid impregnating systems, but are more useful in **prepreg** (for pre-impregnated) tapes. In these, woven reinforcement of glass, aramid or carbon fibre is impregnated with an epoxy resin, and the resin is partially cured (or 'B-staged') to form flat, handleable tapes. These are laid up, according to the mix of fibre orientations and laminate thickness required, and then heat cured under pressure in a mould.

Like the polyesters, the polyurethane (PU) family includes both thermoplastics and thermosets, but if anything, it is even more versatile. Polyurethanes are made by reacting two difunctional species, a di-isocyanate ($O=C=N-R_1-N=C=O$) and a glycol ($HO-R_2-OH$) to form a trifunctional urethane (Appendix 2). Isocyanate is an extremely reactive group. If the two species are in equimolar proportions, a thermoplastic is produced, but if there is an excess of isocyanate, it can form crosslinks with the urethane group and other reaction products to make a thermoset. (Trifunctional amines and alcohols are sometimes also used as crosslinking agents). Once again, the nature of R_1 and R_2 controls the crosslink density and properties of the material, which can range from elastomeric to rigid solid.

Although polyurethanes are perhaps best known for foams and surface coatings, there is a growing market for products made by reaction injection moulding (RIM), in which the di-isocyanate and glycol are reacted together directly in the mould, with no prepolymer stage. Fibre or flake reinforcement can be added to one of the reactants (giving reinforced RIM or RRIM) and the process can include foaming. Car and truck bumpers and body panels are typical products (see the Pontiac *Fiero* in Figure 3.35).

One advantage of polyurethanes is the very low viscosities of the reactants. However, in general, thermoset processing involves much lower initial viscosities than the melt processing of thermoplastics. This, coupled with the usually simpler shapes, means that processing pressures can be lower, with savings in tooling and machinery. Cheaper tooling means smaller production runs are economically more feasible, and lower pressures also mean that larger area mouldings are easier to make. Figure 7.21 showed one example of this, and ▼ERF truck cab▲ describes the materials selection route which arrived at another.

The distinctions between the properties of thermosets and those of thermoplastics have, of necessity, been generalized, but are becoming increasingly blurred. Thermoplastic elastomers rely on secondary crosslinks which break down and reform reversibly as the temperature is raised and lowered. Crosslinked polyurethanes can be formulated to contain a mixture of hard, crystalline segments and soft, rubbery segments. A glassy thermoset known as CR-39, which is used for spectacle lenses, has a higher impact strength than PMMA (though less than PC). On the other side of the coin, *thermoplastic* prepregs, containing continuous fibre reinforcement, are becoming available based on matrices such as PEEK with carbon or aramid fibres.

> SAQ 7.3 (Objective 7.3)
> How, in general, do thermosets differ from thermoplastics in terms of structure, properties and processing? What are the requirements for thermoset formation and how are these achieved?

7.2.3 Foams

Foams are a special type of composite, in which the component bound by the matrix is not a solid, but a gas. In addition to their uses in thermal insulation and in providing buoyancy, foams are also widely used in load-bearing applications such as cushioning in furniture, energy-absorbent packaging and padding, and as the filling in sandwich panels to provide enhanced bending stiffness with minimum weight penalty (see ▼Sandwich beams▲).

Foams can be made in several ways:

(a) Mechanically. Gas (usually air) is entrapped in a liquid by beating the liquid (like beating an egg white), which then sets.

(b) Physically. Either an inert gas (such as N_2 or CO_2) is blown at high pressure into a melt just before moulding, or a low-boiling-temperature liquid, which volatilizes on heating to the process temperature, is mixed with the base material.

(c) Chemically. A chemical which decomposes on heating to produce gases (for example azodicarbonamide, which breaks down to $N_2 + CO + CO_2$) is added, or reactive chemicals are used to produce the gas (for example CO_2 in RIM and RRIM polyurethanes).

(d) Using hollow fillers. So-called **syntactic** foams are obtained when fillers such as hollow glass microspheres (see 'Fillers', Section 7.2.1) constitute the voids in the foam.

(e) By dissolution. A particulate composite is made from which the particles are dissolved or melted.

▼ERF truck cab▲

In common with the rest of the automotive industry, truck manufacturers are increasingly turning to plastics and plastics composites for such components as body panels, bumpers and interior trim. However, to date, there is only one manufacturer, ERF, that produces a complete cab bodyshell in a plastics composite material (Figure 7.22). The cab consists of 24 separate SMC mouldings (Figure 7.23) which are bolted to a welded steel frame. It was first introduced in 1975, and the company is currently producing some 5000 units annually.

Before selecting SMC as the truck cab material, ERF considered a number of alternative cladding materials including steel, aluminium, RIM polyurethane foam, vacuum-formed ABS, cold press moulded GFRP and hand lay-up GFRP. The tooling costs for steel would have been

Figure 7.23 SMC mouldings for truck cab

Figure 7.22 ERF truck cab

higher, and both steel and aluminium would have involved considerably higher assembly costs. In addition, there is the problem of corrosion with steel. The bolt-on method of construction adopted by ERF also means that replacement of damaged panels is easier than with a welded steel construction. Although the PU foam was rigid enough, its impact strength was inadequate, whilst the problem with any vacuum-formed thermoplastic is the high level of frozen-in strain. Apart from their deleterious effects on strength, any paint stoving process would lead to a relaxation of these strains with attendant changes in shape. Cold press moulding (CPM) of polyester–glass is about four or five times slower than SMC moulding and the tool lives are relatively short. It would not have been possible to mould in the integral attachment bosses either with CPM (moulding pressures too low) or with hand lay-up. The latter is also by far the slowest production process and an excessively large floor area would have been required to accommodate the multiple series of moulds that would have been necessary to achieve the desired production rate. An added bonus of SMC is its improved sound- and vibration-damping (Chapter 5) compared with steel.

An important parameter for characterizing foams is the ratio of their bulk density to that of the parent, unfoamed material (ρ_f/ρ_s), and this can vary over a wide range, from about 0.005 to 0.80. Another variable is the cellular structure of the foam. The foam can be either open-cell or closed-cell, or some mixture of the two. In a closed-cell foam (which includes syntactic foam), the gas bubbles are discrete and not interconnected — they form discrete elements in a continuous matrix (Figures 7.27a and b). In contrast, the bubbles (or voids) have coalesced and are interconnected in an open-cell foam, so both gas and matrix are continuous (Figure 7.27c). Clearly the cell walls in a closed-cell material must have continuous, unbroken surfaces, but in open cells the walls

▼Sandwich beams▲

We saw in Section 6.1.2, that the second moment of area I_{BB} of a body with respect to an axis parallel to, but offset by y from, the axis through its centroid is given by

$$I_{BB} = I_{AA} + Ay^2$$

Or, the minimum value of I is I_{AA}, that about the axis through its centroid, and I increases as the square of the distance y between the two axes.

Bending stiffness is given by the product EI, so one way of increasing this for a given quantity of material is to increase I by distributing material further away from the centroidal axis (in effect, the neutral axis in a beam of symmetrical cross-section), as sketched in Figure 7.24. However, to function as a beam, the two blocks of material in Figure 7.24 must be joined together to create a structure to resist bending. To avoid incurring an undue weight penalty, the joining material should be as light as possible.

The I-beam (Table 1.1) is one way of achieving this, in which the upper and lower plates are linked by a web of the same material. But another way is to make up a sandwich containing a low density 'filling' such as a honeycomb (Figure 7.25) or a foam (Figure 7.26(a)). An idealized bending-stress distribution through the thickness of such a sandwich is shown in Figure 7.26b, showing the discontinuity in stress at the interfaces between the core and the two outer layers (or 'skins'), with the latter bearing the highest stress.

The bending stiffness of a sandwich beam of width w (using subscript c for core and subscript s for the two skins) is

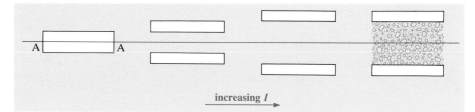

Figure 7.24 Increasing I in sandwich beam

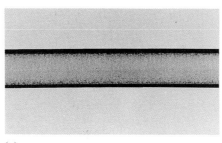

(a)

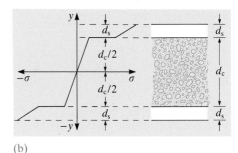

(b)

Figure 7.26 (a) 9 mm thick sandwich moulding (glass-filled nylon skins, nylon foam core) (b) stress distribution through sandwich thickness in bending

$$(EI)_{beam} = 2E_s I_s + E_c I_c$$

$$= 2E_s w\left[\frac{d_s^3}{12} + d_s\left(\frac{d_s + d_c}{2}\right)^2\right]$$

$$+ E_c w\left(\frac{d_c^3}{12}\right) \quad (7.7)$$

For a beam with high-modulus, thin skins and a low-modulus, thick core (say $E_s \geqslant 10E_c$ and $d_s \ll d_c$) this approximates to

$$(EI)_{beam} \approx \frac{E_s w d_s}{2}(d_s + d_c)^2 \quad (7.8)$$

(see SAQ 7.8).

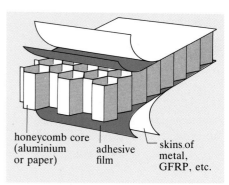

honeycomb core (aluminium or paper)   adhesive film   skins of metal, GFRP, etc.

Figure 7.25 Honeycomb sandwich

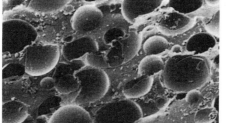

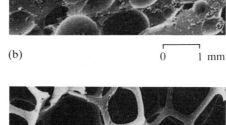
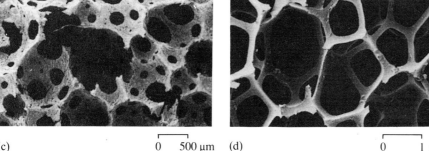
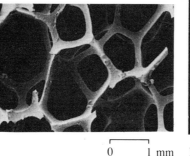

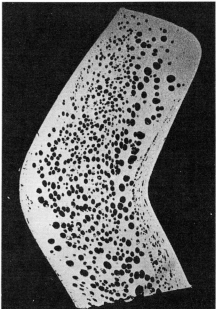
Figure 7.28 Section through a 5 mm thick PS structural foam moulding

Figure 7.27 (a) Closed-cell PS foam, $\rho = 115\,\mathrm{kg\,m^{-3}}$ (SEM) (b) syntactic foam, glass spheres in epoxy resin, $\rho = 475\,\mathrm{kg\,m^{-3}}$ (SEM) (c) open-cell SBR/NR foam, $\rho = 78\,\mathrm{kg\,m^{-3}}$ (SEM) (d) reticulated open-cell PU foam, $\rho = 36\,\mathrm{kg\,m^{-3}}$ (SEM)

can vary from a surface pierced by one or more holes, to the so-called **reticulated** cell (Figure 7.27d) where the cell structure is defined by more-or-less linear ties and struts.

Most synthetic foams are polymeric, and in volume terms, polyethylene, polyurethane, polystyrene and PVC are probably the most widely used. However, **structural foams** and **sandwich mouldings**, based on such materials as HDPE, polypropylene, poly(phenylene oxide), ABS and polycarbonate, are more important in load-bearing applications. These differ from other foams in that they have solid skins covering a foamed core. In both cases, the gas pressure in the foam allows thicker wall sections than are normally feasible in injection moulding, so increasing the bending stiffness per unit weight. Their densities are in the range 60–80% of that of the parent material. Figure 7.28 shows a section through a structural foam moulding.

It's not just polymers that can be foamed. The second illustration in this book, Figure 1.2(a), is of a foam. Many other natural materials are cellular (for example wood, bone, sponge), and earthenware and cement-based products are frequently porous. Breeze blocks used in building are, in effect, foams made from ash and clinker, and, at the dining table, bread, meringues, omelettes and mousses are all foams. Figure 7.29 is a section through a specimen of bone, which bears a striking resemblance to structural foam. Figure 7.30 overleaf shows examples of foams in other materials, some of which are very familiar.

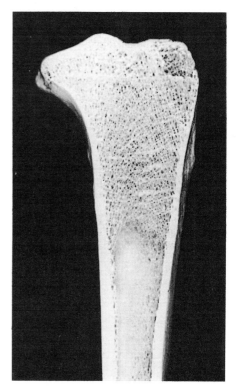
Figure 7.29 Section through tibia

325

7.2.4 A natural composite: wood

Wood, or timber, is one of our oldest structural materials. Its use in buildings, boats, bridges, furniture and similar artefacts stretches far back into antiquity. Its annual rate of consumption is similar to that of iron and steel, which, in terms of volume, means about ten times greater (and that's excluding its use as fuel).

Being a natural material, wood's properties cannot be manipulated to anything like the extent of those of many of the other materials we've considered in this and the previous chapters. Since there are some 30 000 species of tree, each with its own property profile, properties can be selected over quite a wide range through choice of species. But, within any given species, there is a large, and to some extent unpredictable, variability. Design with timber, and in our context, load-bearing design, has to bear this variability in mind.

There's little in the external appearance of a tree to suggest that wood is not only a cellular composite material, but also highly anisotropic. Essentially, it's a foam with anisotropic cells, whose walls are themselves an anisotropic composite. Wood is also used to make a range of composites (see ▼Wood composites▲) as well as paper (Chapter 6). Being based on cellulose, the cell wall structure of wood shares common features with cotton (Chapter 6), but there are significant differences.

Since it's made up of a very large number of cells, wood has a more complex structure than single-celled cotton, and at different levels of scale. To explore this, let's look at a typical softwood, since the

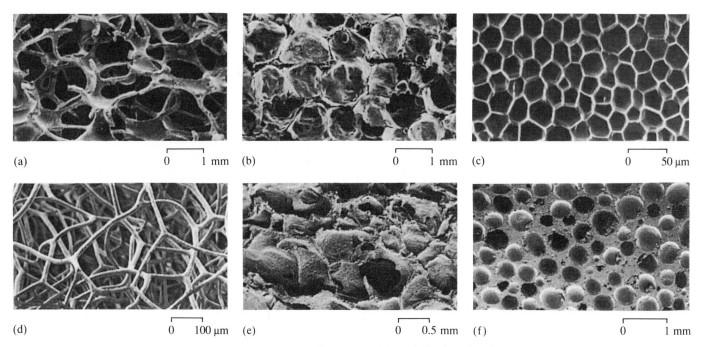

(a)   0   1 mm   (b)   0   1 mm   (c)   0   50 μm

(d)   0   100 μm   (e)   0   0.5 mm   (f)   0   1 mm

Figure 7.30 A range of foams: (a) nickel (b) glass (c) cork (d) sponge (e) bread (f) chocolate bar

structure is simpler than in a hardwood. Figure 7.34 shows some of the main structural features apparent at scales between a complete tree ($\approx 50\,\text{m}$) and the unit cell in crystalline cellulose ($\approx 1\,\text{nm}$).

The bulk of a tree is made up of dead cells, the live and growing region being confined to the **cambium** immediately beneath the bark. Wood, of course, displays the well-known, annual pattern of growth rings (in cotton they are daily ones). The cell wall structure contains about 40–50% by weight of cellulose and 20–25% of amorphous materials known as hemicelluloses. The remaining material is largely composed of **lignin**, a complex, tarry, aromatic compound. Whilst the hemicelluloses mainly act as the cement for the cellulose, the bulk of the lignin is found around the outside of the cell wall as the final layer to be deposited before death of the cell.

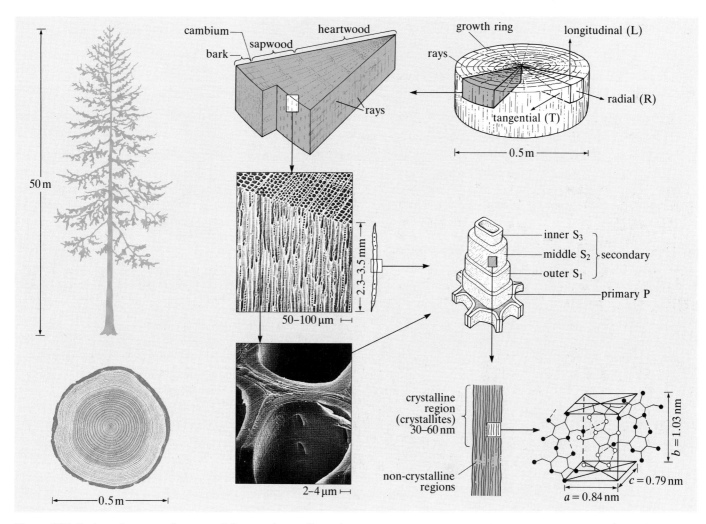

Figure 7.34 Scale and nature of structural features in a softwood

▼Wood composites▲

Versatile and useful a structural material that it is, wood nevertheless suffers from some deficiencies. Its variability is one, as is its propensity to structural flaws such as knots. Its very high anisotropy can also be a problem, not only for mechanical properties, but also because moisture-induced swelling and shrinkage are anisotropic, resulting in warping. Finally, its maximum width is limited by the diameter of the tree from which it's cut. Several wood composite materials have been developed which reduce wood's deficiencies. Most of these are in the form of sheet or board.

Plywood as we've seen, was known to the Egyptians, but wasn't exploited on a large scale until the 1930s. In this, thin layers (veneers or plies) of wood are laminated together with an adhesive so that the grain direction runs at right-angles in successive layers (Figure 7.31a). Plywoods always have an odd number of plies, so that they are balanced about the mid-plane of the central layer. Increasing the number of plies reduces the anisotropy, but increases the cost, so 3-ply and 5-ply are most commonly used. The adhesives are UF (Section 7.2.2) for indoor grades and PF for outdoor ones (such as marine ply). A variant of plywood has the veneers impregnated with a low viscosity UF solution before laminating and curing. The resulting material is used to support the windings in high voltage power transformers, whilst suitably profiled short lengths are used as electrically insulating fishplates in railway track (Figure 7.32), an indicator of its durability.

Figure 7.32 Insulating fishplate

Chipboard was developed to make use of waste timber. Resin-coated chips are compression moulded between flat platens to produce sheets of particulate composite (Figure 7.31b) with a matrix V_m of about 0.1. UF is the most commonly used matrix material. Chipboard's properties, although isotropic, are markedly inferior to those of plywood. But it is very much cheaper.

Fibreboard (or **hardboard**) (Figure 7.31c) is another inexpensive wood composite that starts with wood chips. The lignin, which is thermoplastic, is softened by pressurized steam so that the chips can be separated into fibres. These are formed into a mat, which, when hot-pressed, becomes a board bonded by lignin (a little UF can be added to assist this).

Finally, prefabricated **laminated timber** (Figure 7.31d) is used for such products as shaped beams in the construction industry and fan blades for wind tunnels (Figure

(a)　　　　　　　0　3 mm

(b)　　　　　　　0　3 mm

(a)

(c)　　　　　　　0　3 mm

(d)

(b)

Figure 7.33 Wind tunnel fan

Figure 7.31 Wood composites and their microstructures: (a) 15 mm plywood (b) 12 mm chipboard (c) 3 mm fibreboard (d) laminated timber (Stowe School swimming hall)

7.33), and is used to make tennis rackets and skis. Although UF, once again, is often used, epoxies are preferred in more demanding applications.

Table 7.5 lists some typical properties of the above materials compared with one of the softwoods. Notice how the number of plies reduces the anisotropy of strength and stiffness in plywood. Notice also the effect of resin type on the properties of chipboard, and the effect of density (reflecting moulding pressure) on the properties of fibreboard.

Table 7.5 Properties of wood composites (from J. M. Dinwoodie, *Timber, its Nature and Behaviour*, Van Nostrand Reinhold, 1981)

	Density $\rho/\mathrm{kg\,m^{-3}}$	Flexural strength		$E(\|)/\mathrm{GN\,m^{-2}}$	$E(\perp)/\mathrm{GN\,m^{-2}}$
		$\sigma(\|)/\mathrm{MN\,m^{-2}}$	$\sigma(\perp)/\mathrm{MN\,m^{-2}}$		
Douglas fir	590	120	3.2	16.4	1.1
plywood, 3-ply	520	73	16	12.1	0.9
9-ply	600	60	33	10.8	3.3
chipboard, UF	720	11.5		1.9	
PF	680	18.0		2.8	
MF/UF	660	27.1		3.5	
fibreboard, tempered	1030	67		4.6	
standard	1000	53		–	
medium density	680	19		–	

A horizontal slice through the trunk reveals the annual growth rings, and removing a 'pie-slice' from this and examining it at higher magnification shows that the main structural elements are an array of hollow, vertical cells, known as **tracheids**, and these are connected by a much smaller number of horizontal, radial cells located in the **rays**. Tracheids have aspect ratios of the order of 100, and a closer look at their cell walls shows that these have a layered structure.

The outermost primary (P) layer is the first one to form as the cell is growing. In this layer the microfibrils of cellulose are arranged in an irregular network. When a cell has stopped growing, wall thickening starts and the three secondary layers are formed. In the outer (S$_1$) layer the microfibrils lie almost perpendicularly to the long axis of the cell. The middle layer (S$_2$) is very much the thickest one and has the microfibrils running at about 20° to the long axis. Finally, the inner layer (S$_3$) has the microfibrils oriented almost at right-angles to the long axis again. In each of these layers the cellulose microfibrils are embedded in a mainly hemicellulose matrix.

Finally, the individual cells are joined together with an amorphous, lignin-rich material which contains practically no cellulose. The relative thicknesses of the different components of the cell wall are approximately:

primary	P	~1%
secondary	S_1	10–20%
	S_2	70–90%
	S_3	~5%

The microfibrils themselves, in a similar way to those in cotton, are composed of more-or-less parallel cellulose molecules, which lie alternately in crystalline and amorphous regions of the microfibrils. In the crystalline regions the cellulose molecules form a regular array held together by hydrogen bonds, with the unit cell dimensions some 2×10^{-11} times smaller than the tree of which they're part (Figure 7.34).

As mentioned earlier, since wood is a natural material, the properties obtained from a set of samples of any one species are likely to show a much greater variation than those from a set of a synthetic material. Standard deviations of 20% are typical. Density is one important variable, and both this and other properties are significantly affected by the moisture content. A further complication is that wood, like polymers, is a viscoelastic material: it creeps with time under load and recovers after unloading.

With the above factors in mind, let's look at the properties of some hardwoods and softwoods. Table 7.6 shows for a range of these the densities (ρ), Young's moduli (E) parallel ($\parallel$) and perpendicular ($\perp$) to the grain, and the strengths in tension (σ_t) and compression (σ_c) parallel to the grain, all at a moisture content of 9–12%. This shows the variation in properties across species, and the very high degree of elastic anisotropy.

Table 7.6 Mechanical properties of some woods

Species	$\rho/\mathrm{kg\,m^{-3}}$	$E(\parallel)/\mathrm{GN\,m^{-2}}$	$E(\perp)/\mathrm{GN\,m^{-2}}$	$\sigma_t(\parallel)/\mathrm{MN\,m^{-2}}$	$\sigma_c(\parallel)/\mathrm{MN\,m^{-2}}$
hardwoods					
balsa	200	6.3	0.2	23	12
mahogany	440	10.2	0.8	90	46
walnut	590	11.2	0.9		
birch	620	16.3	0.9		
ash	670	15.8	1.2	116	53
oak	690	13.6	1.0	97	52
beech	750	13.7	1.7	100	45
softwoods					
Norway spruce	390	10.7	0.6		
Sitka spruce	390	11.6	0.7	70	35
Scots pine	550	16.3	0.8	90	47
Douglas fir	590	16.4	1.1	120	50

7.3 Composite stiffness

7.3.1 Modelling composite behaviour

A considerable amount of work has been devoted to devising models of the behaviour of composite materials. The bulk of this work has concentrated on their mechanical properties. Other areas, such as thermal properties and dielectric properties, have received less attention. The aim of such modelling is to express the behaviour of composites in terms of the properties of the constituent materials, the proportion of each present in the composite and, in more sophisticated models, their geometry.

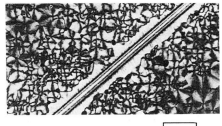

0   50 µm

Figure 7.35 Transcrystallinity around aramid fibre in nylon 6,6

However, the task of *accurately* predicting composite behaviour is straightforward in only a limited number of cases. Thus, the well-known 'rule of mixtures', which is a volume-weighted average of the moduli of the phases present (see Section 7.3.2), provides an exact solution for the tensile modulus in the fibre direction of an aligned, continuous fibre composite only when both fibres and matrix have the same Poisson's ratio. Even here, though, there may be some problem in assigning a correct value to the modulus of the matrix, since this can be affected by the presence of the fibres. For example, it has been suggested that hydroxyl ions on the surface of glass fibres could modify the crosslinking reaction of an epoxy or polyester matrix, and there is evidence that partially crystalline thermoplastics crystallize in a different manner when in contact with fibre surfaces (an example is shown in Figure 7.35). Both of these factors are liable to produce a matrix material whose mechanical properties differ from those of nominally the same material but which is unfilled.

A similar caveat is thought to apply to polymeric foams, in their case because the lower the density, the thinner and more highly drawn the cell-wall material is likely to be (to say nothing of possible differences in crystallization). Hence the properties of the bulk polymer are less likely to represent those of the foam matrix.

The problems of modelling composites intensify as the fibres become more discontinuous, as their lengths decrease, and as the distributions of length and of fibre orientation broaden. Similarly, with particulate composites (that is, those with more-or-less equiaxed second phase geometry) exact solutions are available only for some special packing arrangements. However, although it is important to bear these provisos in mind, it will be seen in the following sections that it's still possible to use simple models to establish limits on the behaviour of composites, and to help gain an insight into the processes of reinforcement.

7.3.2 Modelling stiffness

There are two related elementary models for the stiffness of composites (see ▼**Simple models for stiffness**▲) and these differ in their predictions of composite modulus and its variation with volume fraction of the components. If we use subscripts 'f' for fibre or filler and 'm' for matrix, Figure 7.37 shows the variation in moduli with volume fraction predicted by the models for $E_f = 70\,\mathrm{GN\,m^{-2}}$, with $E_m = 1\,\mathrm{GN\,m^{-2}}$ (lower pair) and $E_m = 35\,\mathrm{GN\,m^{-2}}$ (upper pair).

It can be seen that the parallel, homogeneous strain model predicts a higher value of composite modulus than the series, homogeneous stress model for all volume fractions other than zero and unity. Also the difference between the two models increases with increasing difference in modulus between the components. Since with most composites the modulus of the polymeric matrix is very much less than that of the filler or reinforcement, this would appear to restrict severely the usefulness of the models. However, by considering the assumptions made in setting up each model, it's possible to be a bit more specific about the type of composite to which they should apply.

Thus the assumption that the blocks are equally strained (the parallel model) is a good starting point in the description of a continuous, aligned fibre composite stressed in the fibre direction. On the other hand, the assumption of equal stress in the series model should correspond more closely to the case of transverse loading in an aligned fibre composite, and to that of particle-filled composites. To illustrate this, Figure 7.38 shows the moduli of a range of experimental copper matrix composites, containing either tungsten wires or tungsten

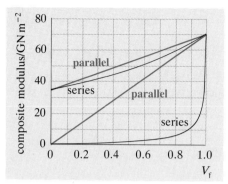

Figure 7.37 Predicted modulus versus V_f for the parallel and series models for two values of E_f/E_m

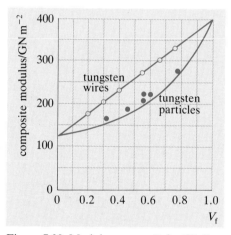

Figure 7.38 Modulus versus V_f for W–Cu composites. The tungsten wires are continuous and discontinuous

▼Simple models for stiffness▲

The two simplest models for predicting the elastic behaviour of composite materials are really two different variants of the same model. This considers the composite as being equivalent to a combination of elementary, parallel-sided blocks of the component materials, the number of such blocks equalling the number of components. The two models differ in their assumptions as to how these blocks are arranged in relation to the direction of the applied loading.

Figure 7.36 illustrates the two alternatives for a three component model. For loading in the x-direction, and assuming perfect bonding between the blocks, each component must be *equally strained*. The blocks are scaled so that the ratio of their areas a_i in the yz-plane perpendicular to x is the same as the ratio of the volume fractions of the components, that is

$$a_1:a_2:a_3 = V_1:V_2:V_3$$

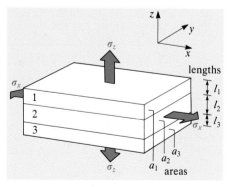

Figure 7.36 Geometry of simple models for composite stiffness

By expressing the load on each component and on the composite in terms of Young's modulus, strain and area, and imposing the above two conditions, it is straightforward to obtain for E_c, the modulus of the composite, that

$$E_c = E_1 V_1 + E_2 V_2 + E_3 V_3$$
$$= \sum_i E_i V_i \qquad (7.9)$$

This, of course, is the well known **rule of mixtures**, or **homogeneous strain** model for a composite.

If the composite model is loaded in the z-direction, each component is *equally stressed*, since each has the same area in the xy-plane. The component blocks are now scaled so the ratio of their lengths l_i in the loading direction equals the ratio of the volume fractions. By expressing, this time, the extensions of the components and composite in terms of Young's modulus, stress and length we obtain

$$\frac{1}{E_c} = \frac{V_1}{E_1} + \frac{V_2}{E_2} + \frac{V_3}{E_3}$$
$$= \sum_i \frac{V_i}{E_i} \qquad (7.10)$$

as the **homogeneous stress** model for E_c.

particles, together with the calculated values (lines) from the two models. It can be seen that for both the continuous and the discontinuous aligned wires the experimental measurements show excellent agreement with the rule of mixtures (parallel) solution. The data for dispersed particles of tungsten in copper lie closer to the series model solution, although they do not quite coincide with it.

It is important to appreciate that these two models are both very simple, are based on a bare minimum of information (moduli and volume fractions) and make no distinction between the matrix and the filler or reinforcement. They thus neglect any effects due to, amongst other things:
- the geometry of the individual phases
- the arrangement and distribution of the phases
- imperfect bonding
- differences in Poisson's ratio
- any viscoelasticity of polymeric components.

In effect, Equations (7.9) and (7.10) provide approximate upper and lower limits to the behaviour of composites. More sophisticated bounds have been derived, which take some of the above effects into account, and which should therefore correspond more closely to reality. It turns out that, for the tensile modulus in the fibre direction of a unidirectional fibre composite, these improved models differ very little from the rule of mixtures solution. However, they do provide better solutions for other moduli and along other axes. Neither foams nor wood are amenable to the same treatment as solid composites, as is discussed in ▼Stiffness of foams▲ and ▼Stiffness of wood▲.

The effects of fibre length and fibre misalignment require modifications to the modelling procedure, whilst viscoelasticity — which can clearly be important in polymer matrix composites — is usually taken into account by substituting the appropriate creep modulus at a given strain and time into the elastic equations. Before examining these modifications, the anisotropy of fibre composites will first be considered.

SAQ 7.6   (Objectives 7.6 and 7.7)

(a) A polypropylene foam has a short-term tensile modulus of $0.2\,\mathrm{GN\,m^{-2}}$. Estimate its density and compressive modulus.

(b) A manhole cover measuring $1\,\mathrm{m} \times 0.5\,\mathrm{m}$ is made in polypropylene structural foam with a blow ratio β (that is, ratio of foamed to unfoamed volume) of 1.25. It must support a central mass of $100\,\mathrm{kg}$ for short periods without deflecting more than $5\,\mathrm{mm}$. Estimate the minimum mass of polypropylene required. The deflection Δ at the centre of a simply supported rectangular plate whose length is twice its width w, whose thickness is t, under a central load F is given by the following equation

$$\Delta = \frac{2Fw^2}{11Et^3}$$

▼Stiffness of foams▲

At first sight it might appear that all we need to do to estimate the modulus of a foam is to use one of our simple composite models and take the modulus of the second phase (the gas) as effectively zero. Unfortunately, this rules out the homogeneous stress, series model (Equation 7.10) since we obtain zero for the composite (foam) modulus (which is what you might expect if one of the elementary blocks used in deriving the model had zero modulus). We do rather better with the rule of mixtures. Using subscripts f for foam and s for solid material we get $E_f = E_s V_s$.

> EXERCISE 7.4 Relate V_s to the density ratio ρ_f/ρ_s and to the porosity p of a foam.

Experimentally it's been found that for many uniform density foams

$$\frac{E_f}{E_s} = \left(\frac{\rho_f}{\rho_s}\right)^n \tag{7.11}$$

where $n \approx 1.5$ for tension and $n \approx 2$ for compression (Figure 7.39), whilst the rule of mixtures solution corresponds to $n = 1$.

The difference arises because the ways in which the cell walls of foams deform are very unlike the starting assumptions of the rule of mixtures model. The walls of the cells have virtually nothing — only a gas — to constrain them when they start to deform in response to a load. Thus, depending on their geometry and orientation with respect to neighbouring walls, their initial deformation will be some mixture of in-plane and out-of-plane bending, shear and tension or compression (which includes buckling), and the overall response of the foam is the aggregate of these.

In structural foams and sandwich mouldings, the problem is complicated by the presence of the solid skins, and in structural foams, by the non-uniformity of the core (see Figure 7.28). Here, for

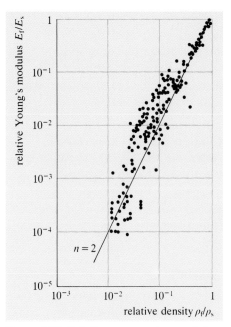

Figure 7.39 Relative compressive modulus versus relative density for a range of foams

flexure, the rule of mixtures solution (that is, $n = 1$) often provides a better approximation than the values for uniform density foams.

A further complication is added when the cells are liquid-filled, since changing the volume fraction of solid can change the connectivity of the cells, and thus the freedom of the liquid to move in response to load. This is exemplified by hardened cement, which contains water-filled pores. Here a value of $n = 3$ better fits the experimental variation of compressive modulus with density (or porosity).

Finally, returning to the lower density, more flexible foams, in many applications such as packaging, padding and upholstery, the deformation mode of interest is compression. The way in which foams deform can be tailored not just by varying (ρ_f/ρ_s) but also by varying the distribution of cell sizes and their wall structures. Figure 7.40 shows four examples of this.

Curve A is more typical of a foam used in cushioning, where the continual resistance to increasing load provides the person sitting or lying on it with the feeling of being supported. In contrast, foams like those represented by curves C and D would give the impression of 'bottoming-out' and thus would not rate highly in the comfort stakes! Further, as the plateaux are associated with irreversible deformation due to buckling collapse of the foam, they are even more unsuitable for furniture. On the other hand, the energy absorption associated with this deformation is very useful in packaging applications.

> SAQ 7.4 (Objective 7.4)
> Describe how foams can be made, and the different structures that can result. How is the stiffness of foams related to their structure?

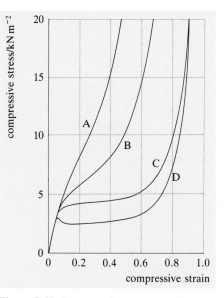

Figure 7.40 Compressive stress–strain behaviour of different foams: A, latex rubber, open-cell, broad size range; B, PU, non-reticulated, broad size range; C, D, PU, reticulated, narrow size range

▼Stiffness of wood▲

One feature of woods is that, despite the large variation between and within species, the composition, structure and properties of the cell *walls* remain very much the same. Typically they have a density of 1500 kg m^{-3}, and moduli of 35 GN m^{-2}

and 10 GN m^{-2} parallel to and transverse to the cell axis, respectively. This suggests that, in the same way as foams, the overall density of the material relative to that of the cell wall must be an important determinant of the mechanical properties.

Figure 7.41 shows the moduli of the woods in Table 7.6 plotted against their density. Here $E(\perp)$ has been split into its radial (R) and tangential (T) components (see Figure 7.34) as listed in Table 7.7. Also plotted on Figure 7.41 are two lines of slopes $n = 1$ and $n = 2$ which pass through the data points for the cell-wall material.

Despite the scatter, it's apparent that data for the longitudinal (or axial) direction are a reasonable fit to the line with $n = 1$, whilst for the two transverse moduli $n = 2$ is a better approximation. Thus in terms of our earlier Equation (7.11) for foams, the modulus of wood can be related to its density by the rule of mixtures ($n = 1$) along its grain, and by the same equation as for foams ($n = 2$) across its grain. Figure 7.41 also shows that the anisotropy of wood increases with decreasing density.

Table 7.7 Moduli of the woods of Table 7.6

Species	$E_L/\mathrm{GN\,m^{-2}}$	$E_R/\mathrm{GN\,m^{-2}}$	$E_T/\mathrm{GN\,m^{-2}}$
hardwoods			
balsa	6.3	0.30	0.11
mahogany	10.2	1.13	0.51
walnut	11.2	1.19	0.63
birch	16.3	1.11	0.62
ash	15.8	1.51	0.80
oak	13.6	1.28	0.66
beech	13.7	2.24	1.14
softwoods			
Norway spruce	10.7	0.71	0.43
Sitka spruce	11.6	0.90	0.50
Scots pine	16.3	1.10	0.57
Douglas fir	16.4	1.30	0.90

SAQ 7.5 (Objective 7.5)
Summarize the main structural features of softwoods, starting with the packing of cellulose molecules, and indicate the rôles these features play in the stiffness and strength of wood.

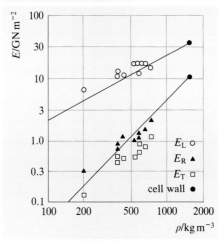

Figure 7.41 Moduli of woods versus density (log. axes)

▼Westland rotor blade▲

The blades of a rotating rotor of a helicopter in flight are subjected to a complex system of time-varying forces. These include centrifugal forces due to the rotation, bending forces supporting the weight of the machine and torsional forces due to alternating the aerofoil angle during each revolution.

Fatigue failure is clearly a potential problem in this situation, and, in the case of aluminium alloy blades, limits their service life to about 5000 flying hours. Since composite materials can offer significantly enhanced fatigue performance compared with homogeneous materials (see Section 7.4.2), Westland Helicopters Ltd developed a composite blade, initially for their Sea King helicopter.

The make-up of the blade is shown in Figure 7.42. It consists of a leading-edge spar bonded to a trailing-edge fin. The spar, which is the main load-carrying member, consists of composite layers around a thermoplastic foam core, the whole encased in a titanium erosion shield. (The fin also has a lightweight core, made from an aramid paper honeycomb.) All the composites are made from epoxy prepreg tapes — and a special machine was developd by Westland for laying these in the mould. Most of the reinforcement (about 80%) is directed along the axis of the spar to withstand the centrifugal and bending loads. The upper and lower sidewall slabs, containing equal proportions of glass and carbon fibre, are stiffer than the backwall slab and nose mould. The inner and outer wraps are both made from a 45° weave of carbon fibre prepreg to counter the torsional loads.

The initial cost of these blades was about double those of the metal ones, but their fatigue lives are at least a factor of four greater.

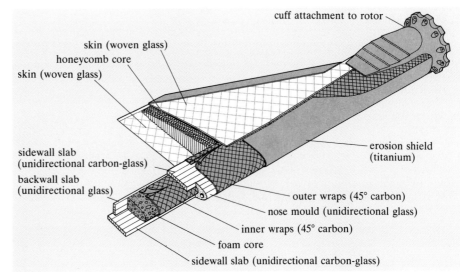

Figure 7.42 Construction of composite rotor blade

7.3.3 Anisotropy

Many materials can have some anisotropy conferred on them by processing (for example piano wire, Chapter 3, polymers, Chapter 5) and, in particular, polymer and polymer-based fibres are usually highly oriented (Chapter 6). Thus anisotropy in a composite arising from a preferential orientation of fibres can be enhanced by contributions from both these sources. One of the advantages of composites is the opportunity of tailoring the anisotropy to suit the application. In the case of complex loading, this can involve laminating layers of material with different fibre orientations, and sometimes different types of fibre. ▼Westland rotor blade▲ is one example of a critical application, involving complex and time-varying stresses.

Although anisotropy can be manifested in many physical properties (such as the easy cleavage directions in wood and slate, optical birefringence, thermal expansion), let's confine ourselves here to stiffness. An isotropic elastic material, by definition, has the same properties in every direction, and only requires any two out of the four elastic constants, E, G, v and K to characterize its elastic behaviour (Chapter 1). In contrast, a fully anisotropic material has properties that are different in *all* directions — and has 21 independent elastic constants! Fortunately, composite materials are either isotropic, or have axes or planes of symmetry which make them less than fully anisotropic Even so, analytical descriptions of them are beyond our present scope, and we'll confine ourselves to a very simple measure of anisotropy.

Consider the experimental values of modulus in Table 7.8 for a unidirectional E-glass fibre–epoxy composite with $V_f = 0.65$. The x-direction is taken as the fibre direction and the axes and moduli are as defined in Figure 7.43.

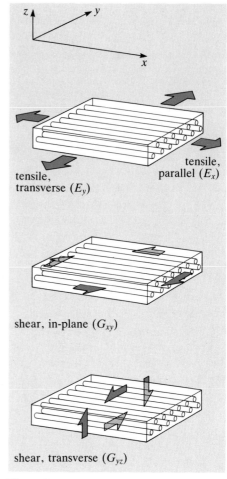

Figure 7.43 Axes and moduli in an aligned fibre composite

Table 7.8 Moduli (GN m^{-2}) of an epoxy-glass composite

Modulus	Measured	Equation (7.9)	Equation (7.10)
E_x (tensile, parallel)	46.5		
E_y (tensile, transverse)	16.4		
G_{xy} (shear, in-plane)	7.7		
G_{yz} (shear, transverse)	5.7		

EXERCISE 7.5 Using G (epoxy) $= 1.35$ GN m^{-2} and G (glass) $= 29.5$ GN m^{-2} together with data from Tables 7.2 and 7.3, complete the last two columns of Table 7.8. How well do Equations (7.9) and (7.10) predict the moduli? (Assume that the equations apply to G in the same way as to E.)

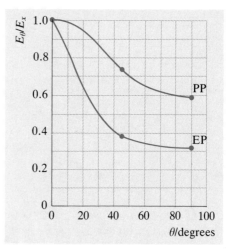

Figure 7.44 Orientation dependence of 100 s isochronous tensile moduli of PP and EP composites with glass

The **degree of anisotropy** κ of a material can be defined in terms of the moduli along two of the principal axes in the material. Thus, in the above example, and using E_x and E_y

$$\kappa = \frac{E_x}{E_y} - 1 \qquad (7.12)$$

which equals 1.84 for the values quoted in Table 7.8. When $E_x = E_y$, $\kappa = 0$, but this does not necessarily mean that the material is isotropic. For instance, the reinforcement might be in the form of a woven cloth with $E_x = E_y$, although the modulus in any other direction would be lower.

In a composite with discontinuous fibres or one in which the fibre alignment is distributed over a range of directions, the rule of mixtures expression, Equation (7.9), is no longer adequate in its simple form to predict the composite properties in the nominal fibre direction. Also the degree of anisotropy is reduced. Figure 7.44 illustrates this for an injection moulded, polypropylene–glass fibre material (PP), whose anisotropy of tensile modulus is compared with that of the unidirectional epoxy–glass fibre (EP) composite of Table 7.8. The vertical axis is normalized with respect to the modulus E_x of both materials in the fibre direction, and E_θ is the modulus at an angle θ to the x-direction in the xy plane. (Note that these are 100-second tensile creep moduli — see Figure 5.14.)

The degree of anisotropy κ for the PP is 0.62, while that for the EP, which we obtained above, is 1.84. This difference arises because, not only did the PP contain a lower proportion of glass than the EP (20% and 80% by weight respectively) but also because the fibres in the PP were short and imperfectly aligned. A typical example of such a distribution of fibre orientations in an injection moulding is shown in Figure 7.45; the non-uniform dispersion of the fibres is also apparent.

The effect of fibre aspect ratio for three different volume fractions of fibre is shown in Figure 7.46 in terms of the composite modulus as a fraction of the unmodified rule of mixtures solution. For λ greater than about 1000, there is very little effect of fibre aspect ratio, that is very little difference between continuous and discontinuous fibres. However, the aspect ratios of the fibres in most short-fibre composites tend to lie in the range between 10 and 1000, and it is just in this range that the modulus is changing rapidly with aspect ratio. It's important to bear in mind that the aspect ratio is defined relative to the direction of applied stress. Thus, if a fibre's aspect ratio was 100, say, when stressed parallel to its long axis, it would only be 0.01 when stressed at right-angles to this.

To measure an orientation distribution like the one shown in Figure 7.45, or to quantify it theoretically, is by no means straightforward, although the extremes of total alignment and total randomness are much easier to describe. The effect of fibre misalignment is frequently incorporated into the rule of mixtures solution by an **efficiency factor B**,

flow direction

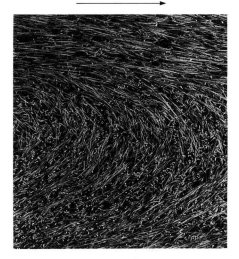

Figure 7.45 Contact microradiograph of a section through the thickness of an injection moulding containing 22% by volume of 13 µm diameter glass fibres

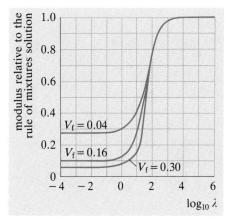

Figure 7.46 Calculated variation of composite modulus with aspect ratio of fibres

so that $\quad E_c = BE_f V_f + E_m V_m \qquad\qquad$ (7.13)

The efficiency factor has the following values:

- $B = 1$ complete alignment, stressed parallel to the fibre axis,
- $B = 1/2$ fibres aligned in two directions at right-angles, stressed in one of these directions,
- $B = 3/8$ random distribution in a plane, stressed in the plane,
- $B = 1/5$ random distribution in three dimensions.

> SAQ 7.7 (Objectives 7.1 and 7.6)
> Using data from the text and Tables 7.2, 7.3, 7.4 and 7.6 as appropriate, estimate the tensile moduli of the following:
>
> (a) the unidirectional glass fibre–carbon fibre hybrid composite in the Westland rotor blade in the fibre direction if the matrix volume fraction $V_m = 0.3$ and it contains equal volumes of the two fibres (E-glass and HM carbon);
>
> (b) the material in one layer of the MCMV hull along one of the fibre directions;
>
> (c) the material in the Dunlop tennis racket along the fibre direction in an aligned fibre specimen, if the average fibre aspect ratio is 200 (assume HM carbon);
>
> (d) a 5-ply plywood, with outer plies of walnut whose thicknesses are half that of each of the three inner plies of Douglas fir, stressed in the direction of the outer-ply grain;
>
> (e) an automotive grade of SMC if the fibres have $\lambda > 1000$ and are randomly directed in a plane, and the specimen is stressed in the plane (assume that the components not listed in Tables 7.2–7.4 have an aggregate density of $2200 \, \text{kg m}^{-3}$ and a modulus of $20 \, \text{GN m}^{-2}$).

7.3.4 Stress transfer and the interface

When a load is applied to a composite, it is almost invariably applied to one component and is transferred to the others by some combination of shear and tensile (or compressive) stresses acting across the interface. This must be so for any component which is not continuous. For composites containing continuous fibres the loads are typically applied to the outside of the material. The inner components are stressed by shear (Figure 7.47) and it is only at some distance away from the loading point that the strains across the cross-section become uniform.

Without a mechanism of stress transfer there would be no reinforcement. The composite properties would be more like those of a foam. This means that the nature of the bond at the interface between the matrix material and the reinforcement plays a critical rôle in determining the properties of the composite. Fibres are frequently given surface treatments to enhance the interfacial bond. One example of this is considered in ▼Washing machine tank▲.

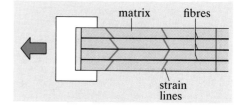

Figure 7.47 Displacement distribution in externally loaded continuous fibre composite

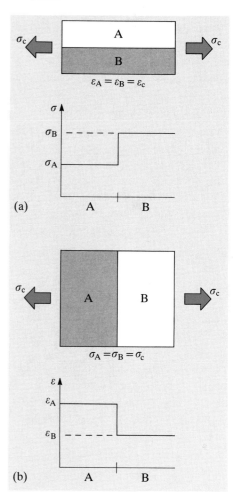

Figure 7.51 Idealized stress and strain distributions at an interface

Our simple composite models have implicitly assumed that the proximity of an interface does not affect the component's response to load, which in turn implies that there is a discontinuous (or step function) change in stress (or strain) at the interface. This is sketched in Figure 7.51 for the parallel and series models, for the case of component

▼Washing machine tank▲

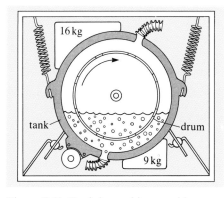

Figure 7.48 Tank in washing machine

The tank of a front-loading washing machine supports, and provides a water-tight container for, the rotating drum, in which the washing is done (Figure 7.48). Typically it has to support up to 30 kg of washing and water, carry 25 kg of balance weights, withstand the dynamic loadings of the drum spinning at up to 1000 r.p.m., and be unaffected by an aqueous detergent solution at up to 100 °C.

Usually the tanks are made from stainless steel, or the cheaper, vitreous-enamelled mild steel. In the Philips Variatronic 082 washing machine this involved welding together 16 separate components. Potentially a one-piece injection moulding offered significant savings in tooling and assembly costs, provided adequate stiffness and strength could be achieved. Philips selected a polypropylene (for low cost) structural foam (SF) reinforced with 30% by volume of glass fibres.

The fibres are treated with a silane **coupling agent** to improve the bond between them and the matrix. The effect of this can be seen in Figure 7.49, which compares fracture surfaces of untreated and treated specimens. The untreated fibres show greater evidence of pull-out from the matrix, and have no polymer adhering to them. The treated material is about 25% stiffer and twice as strong as

the untreated material. The reinforced SF tank (Figure 7.50) has the added advantage of better sound damping than its precursor.

SAQ 7.8   (Objectives 7.6 and 7.7) What thickness of steel would have the same bending stiffness as the polypropylene–glass SF, if the latter were modelled as a 6 mm thick sandwich with two 1.5 mm thick skins ($E = 6.6\,\mathrm{GN\,m^{-2}}$), and a core whose uniform density is 50% that of the skins?

Figure 7.50 Washing machine tanks in (a) vitreous enamelled mild steel, and (b) glass-reinforced structural foam PP

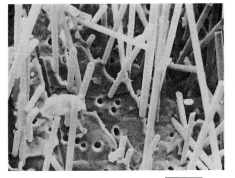
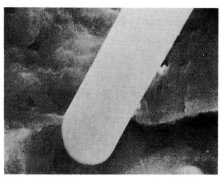
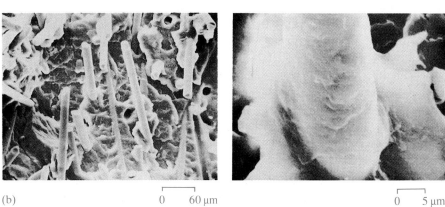

Figure 7.49 Fractured specimens in PP–glass composites (SEM) (a) untreated fibres (b) treated fibres

339

A having half the Young's modulus of component B. In real composites there are very many more than just one block of each component. Their effect is to constrain each other and thus to modify the system of stresses acting on each component.

The implication of the disparity in strains (or stresses) between the matrix and non-continuous components is that there is some minimum length below which they will not be fully stressed (or strained) by shear transfer. Consider, for example, a discontinuous fibre composite. This minimum length can be estimated by examining the balance of forces acting on a single fibre embedded in the matrix material. A model for this is developed in ▼Minimum fibre length▲, from which it emerges that the minimum fibre length depends on the ratio of maximum fibre stress to interfacial shear stress (Equation 7.17). The higher the shear stress, the shorter the length of fibre needed to attain the maximum fibre stress. Thus we need to consider the interfacial shear stress.

As can be seen from Figures 7.52(c) and (d), the shear stress τ_i varies from a maximum value at the ends of the fibre to zero when $\sigma_f = \sigma_{f(max)}$. The maximum attainable value of τ_i is determined by the characteristics of the interfacial bond. In many composites, the available evidence suggests that there is a substantial contribution to the shear stress from interfacial friction. The two principal sources of the stresses normal to the interface required to generate the frictional forces are the residual compressive stress due to differential thermal contraction following processing, and the stress due to lateral contraction of the matrix (Poisson's ratio effect) when the composite is stressed. The interfacial shear stress is just the coefficient of friction μ times the sum of these two contributions (see 'Wear' in Chapter 2). Typically, μ might be about 0.2 to 0.3 and the total normal stress about $30\,\mathrm{MN\,m^{-2}}$, leading to a maximum shear stress of less than $10\,\mathrm{MN\,m^{-2}}$.

The significance of this is in the effect on λ_{min}, and hence on the efficiency with which the greater stiffness or strength of the fibres can contribute to the composite properties. However, this depends not only on the value of τ_{max} but also on the condition for τ_i to reach zero. This is given approximately when the strains in the matrix and fibre are equal, namely

$$\varepsilon_m \approx \varepsilon_f$$

or

$$\frac{\sigma_m}{E_m} \approx \frac{\sigma_f}{E_f}$$

and this is when $\sigma_f = \sigma_{f(max)}$ (Figure 7.53). Hence

$$\sigma_{f(max)} \approx \frac{\sigma_m E_f}{E_m} \qquad (7.18)$$

As can be appreciated from Figure 7.52(d), even when the tensile stress in the fibre has reached its maximum value $\sigma_{f(max)}$, the *average* stress in

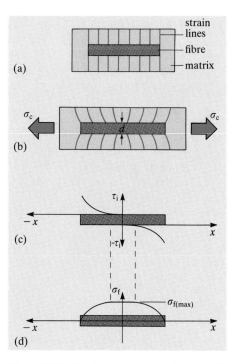

Figure 7.52 Displacements and stresses of a single fibre in a matrix: (a) unstressed (b) stressed (c) interfacial shear stress, τ_i (d) tensile stress in fibre, σ_f

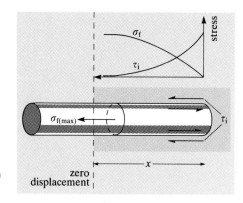

Figure 7.53 Model of fibre half-embedded in matrix

the fibre $\bar{\sigma}_f$ is still much less than this. It is only when aspect ratio $\lambda \gg \lambda_{min}$ that $\bar{\sigma}_f$ approaches the rule of mixtures value. It is clearly important, especially when expensive fibrous reinforcements are employed, to ensure that their lengths are such as to make certain that the maximum benefit is derived from them.

For a complete picture, modifications to the stress distribution around the fibre, due to tensile stresses at the fibre ends and due to stress concentrations in the matrix around the fibre ends, ought to be taken into account. Stress concentrations arise whenever there is discontinuity of elastic modulus in a stressed component, and their magnitude depends not only on the geometry but also on the relative values of modulus. In a composite, these stress concentrations have two consequences.

The first is that the stress in the matrix at, for example, the end of the fibre shown in Figure 7.52, can be increased locally to beyond its yield stress or fracture stress. The second is that the stress fields associated with such stress concentrations can interact with each other (Figure 7.54) to produce an average stress in the matrix which is higher than that implied by a simple rule of mixtures model. The net effect is to reduce the apparent stiffness of the matrix materials (assuming $E_f > E_m$).

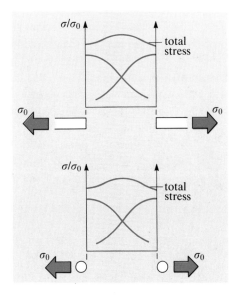

Figure 7.54 Stress fields around stress concentrations in a composite

▼Minimum fibre length▲

A model for evaluating the minimum fibre length is shown in Figures 7.52 and 7.53.

A single fibre of diameter d is embedded symmetrically in a matrix which is subjected to a stress σ_c. Because of the symmetry, the centre line of the composite in Figure 7.52 is not displaced on loading, and thus you need only consider the equilibrium of forces on one half of the fibre (Figure 7.53). The other half of the fibre can be treated as being free of the matrix, but exerting the same distribution of stresses on the right-hand half as shown in Figure 7.52 for the fully embedded case.

What, then, is the minimum value of embedded length x_{min} of the fibre such that the fibre stress can attain its maximum value $\sigma_{f(max)}$ through shear transfer? For the model depicted in Figure 7.53, the force on the fibre must balance the force on the matrix for equilibrium. So

(tensile stress) × (cross-sectional area of fibre)

= (average shear stress) × (surface area of fibre)

or

$$\frac{\sigma_f \pi d^2}{4} = \bar{\tau}_i \pi dx \quad (7.16)$$

where τ_i is the interfacial shear stress and the bar over τ_i indicates its mean value. This equation reduces to

$$\frac{x}{d} = \frac{\sigma_f}{4\bar{\tau}_i}$$

The minimum embedded length, x_{min}, for σ_f to attain $\sigma_{f(max)}$ is then

$$\frac{x_{min}}{d} = \frac{\sigma_{f(max)}}{4\bar{\tau}_i}$$

However, this is for the model of Figure 7.53. In a composite *both* ends of the fibre are embedded (Figure 7.52), so the minimum fibre length l_{min} is twice x_{min} and

$$\lambda_{min} = \frac{l_{min}}{d} = \frac{\sigma_{f(max)}}{2\bar{\tau}_i} \quad (7.17)$$

defines the minimum aspect ratio, λ_{min} to reach the maximum stress in the fibre through the shear transfer. In other words, the larger the value of the interfacial shear stress, τ_i, the shorter a given fibre needs to be to achieve a given value of $\sigma_{f(max)}$ in the fibre.

EXERCISE 7.6   If 15 μm diameter E-glass fibres in an aligned fibre composite were found to break when stressed in the fibre direction at lengths exceeding 5.1 mm, estimate the mean interfacial shear stress.

7.4 Composite failure

7.4.1 Tensile strength

The failure of composites frequently involves different mechanisms or combinations of mechanisms from that of homogeneous materials. Because of the need to consider the details of the different kinds of mechanisms involved, the plasticity and strength of composite materials are not as amenable to the simple generalized modelling which has been used to examine their elastic properties. Nevertheless for fibrous composites the same sort of models as were used for elastic behaviour still have some utility, particularly in describing their tensile strength. Let's start with a continuous, aligned-fibre composite stressed in the fibre direction. The rule of mixtures applies, as fibres and matrix are equally strained, so the moduli in Equation (7.9) can be replaced by stresses (but evaluated at the same strain).

EXERCISE 7.7   If the fibres in the epoxy–glass composite in Table 7.8 had a fracture strain of 1%, estimate the load at which they would start to break in a specimen of the composite loaded in the fibre direction and of cross-section 20 mm × 5 mm.

If the fibres are more brittle than the matrix, as is frequently the case, they will start to fail first. These failures will occur at different positions in different fibres, producing what is effectively a discontinuous fibre composite. The failure of discontinuous fibre composites can be analysed, using the ideas developed earlier for stress transfer to fibres. Equation (7.17) gave the minimum fibre aspect ratio for full load transfer as

$$\lambda_{\min} = \frac{\sigma_{f(\max)}}{2\bar{\tau}_i}$$

If the breaking strength of the fibres, σ_f^*, is put equal to $\sigma_{f(\max)}$ then the equation defines a critical fibre aspect ratio λ_{crit} for the stress on the fibre just to reach its breaking stress

$$\lambda_{\text{crit}} = \frac{\sigma_f^*}{2\bar{\tau}_i} \tag{7.19}$$

The strength of a short-fibre composite will clearly depend on whether the aspect ratios of its fibres are less than, equal to or greater than λ_{crit}. Expressions for the strengths of composites under these three different conditions are developed in ▼λ_{crit} and strength▲ These equations assume that the fibres in the composite are all aligned parallel to the tensile stress direction. For other orientation distributions the efficiency factor B in Equation (7.13) can be used to modify the terms containing V_f as before, so that σ_c^* can be estimated.

▼ λ_{crit} and strength ▲

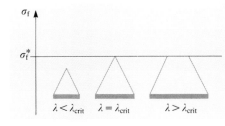

Figure 7.55 Assumed stress distributions in fibres for three fibre length conditions

In order to estimate the strength of a composite containing discontinuous fibres it is necessary to take into account that the stress varies along the length of the fibre. To do this the simplest approximation is to assume that the fibre stress distribution in Figure 7.52 can be represented by a linear increase in stress from both ends as shown in Figure 7.55.

So when $\lambda = \lambda_{\text{crit}}$, the average stress in the fibre $\bar{\sigma}_f$ is simply half the fracture stress, σ_f^*.

If the aspect ratio of the fibres is greater than λ_{crit} then the average stress in them will be given by a combination of σ_f^* and the contribution from the fibre ends

$$\bar{\sigma}_f = \left(1 - \frac{\lambda_{\text{crit}}}{2\lambda}\right)\sigma_f^*$$

If, on the other hand $\lambda < \lambda_{\text{crit}}$, then from Equation (7.17), the maximum stress in the fibres is

$$\sigma_{f(\text{max})} = 2\bar{\tau}_i\lambda$$

so the average stress is half of this,

$$\bar{\sigma}_f = \bar{\tau}_i\lambda$$

For all three length conditions the composite strength can be estimated from

the average fibre stress using the rule of mixtures

$$\sigma_c^* = \bar{\sigma}_f V_f + \sigma_m V_m$$

(remembering to evaluate σ_m at $\varepsilon = \bar{\varepsilon}_f$).

Substituting the three values of $\bar{\sigma}_f$ into this equation for $\lambda = \lambda_{\text{crit}}$ we have

$$\sigma_c^* = \frac{\sigma_f^*}{2}V_f + \sigma_m V_m \qquad (7.20)$$

For $\lambda > \lambda_{\text{crit}}$

$$\sigma_c^* = \sigma_f^* V_f\left(1 - \frac{\lambda_{\text{crit}}}{2\lambda}\right) + \sigma_m V_m \qquad (7.21)$$

For $\lambda < \lambda_{\text{crit}}$

$$\sigma_c^* = \bar{\tau}_i\lambda V_m + \sigma_m V_f \qquad (7.22)$$

This treatment implicitly assumes that the fibres all have the same length, which is extremely unlikely in practice. Most processing methods will leave a distribution of fibre lengths so that Equations (7.20) to (7.22) need to be modified to take this into account. In principle, this can be done by summing the separate conditions for those fibres with $\lambda < \lambda_{\text{crit}}$ and those for $\lambda > \lambda_{\text{crit}}$, but this is possible only when the form of the distribution is known. However, although the presence of a significant proportion of fibres with $\lambda < \lambda_{\text{crit}}$ can have a deleterious effect on the stiffness and strength of the composite, it can bring benefits in the form of increased toughness, as we'll now explore.

EXERCISE 7.8   Estimate λ_{crit} for the nylon–carbon fibre composite in SAQ 7.7(c), and the tensile strength of a sample with a random planar fibre distribution (take $\bar{\tau}_i = 10\,\text{MN}\,\text{m}^{-2}$).

7.4.2 Crack propagation and toughness

We've seen that the tensile failure of composite materials depends very much on the detailed mechanisms involved. Let's examine more closely one important aspect of their failure characteristics, namely, their resistance to crack propagation or their toughness. Table 7.9 shows some typical values of the toughness G_c for a range of materials.

Any process which leads to an increase in the amount of elastic strain energy absorbed when a crack propagates will increase the toughness of the material. In the majority of materials the amount of plastic work

Table 7.9   Values of toughness G_c for a range of materials

Material	$G_c/\text{kJ}\,\text{m}^{-2}$
mild steel	100
glass fibre–epoxy	40–100
woods, across grain	10–200
polypropylene	8
ABS	5
woods, with grain	0.5–2
epoxy resin	0.1–0.3
oxide glass	0.01

performed in a zone immediately ahead of the crack tip is the dominating mechanism of energy absorption. Even in very brittle materials such as the oxide glasses, their toughness is about an order of magnitude higher than their surface energy, and there is evidence that this is due to very small-scale yielding at the crack tip.

However, plastic work alone is insufficient to explain why a continuous fibre reinforced composite, such as epoxy resin with glass fibres, has a maximum G_c of around $10^5 \, \text{J m}^{-2}$ when that of glass on its own is about $8 \, \text{J m}^{-2}$ and that of the epoxy resin is about $500 \, \text{J m}^{-2}$ (it's clear that the toughness of this material doesn't follow the rule of mixtures!).

What happens in such a material depends very much on the strength of the bond between the fibre and the matrix. If the bond is strong and both fibre and matrix are brittle, a crack propagating in the matrix at right-angles to the fibre orientation will pass through the fibres with very little hindrance. The toughness of the composite will probably lie somewhere between those of the two components.

If, on the other hand, the bond is relatively weak, debonding can occur between the fibre and the matrix (Figure 7.56), under the action of increased stresses *in the crack direction*, σ_x, which reach a maximum a short distance ahead of the crack tip. One can either say that this effectively makes the crack tip very blunt, thus increasing the stress required to drive it forward, or one can say that extra strain energy is required to do the work of debonding. Both are ways of saying that the toughness of the material is increased. This also explains why the fatigue performance of composite materials is frequently very much better than homogeneous materials (you saw this in 'Westland rotor blade', Section 7.3.3).

There is a further contribution to the toughness from these debonded fibres. For the composite to fail completely, the fibres must fracture, and after fracturing, their ends have to be pulled out of their hole in the matrix, requiring work. A combination of these two mechanisms is responsible for the large increase in toughness in this class of material. It should be borne in mind, however, that these processes cannot operate for cracks running parallel to the fibres, and the material's toughness is very much less in this direction. The strategy of providing weak interfaces is a powerful way of ensuring a high toughness. In fact it is the reason for the high toughness of fibrous composites (including, as can be seen in Table 7.9, wood). One application of this is described in ▼Composite foil▲.

However, the presence of weak interfaces under normal stress can conflict with the requirements of good shear bonding for high stiffness and strength, as was noted in the previous section. This means that short-fibre composites have to be formulated so as to achieve the optimum compromise between stiffness and strength on the one hand, and toughness on the other. Such an optimization is in fact assisted by a

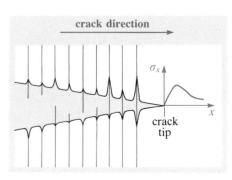

Figure 7.56 Debonding and fibre pull-out

distribution of fibre lengths brought about by processing, since the short fibres with $\lambda \leqslant \lambda_{\text{crit}}$ will tend to enhance the toughness whilst the longer ones will contribute to the stiffness and strength.

The situation is somewhat different with particulate composites and no comparable very large increases in toughness (relative to the matrix toughness) are attained. In fact, if high-modulus particles are blended with, for example, a tough, partially crystalline thermoplastic such as nylon to increase its stiffness, the toughness of the material decreases with increasing volume fraction of particles. On the other hand, if the matrix is brittle some increase in toughness can be obtained. Finally, we've already seen in Chapter 5 that one important method of improving the toughness of brittle plastics is to incorporate within them extremely low modulus particles, namely an elastomer.

SAQ 7.9   (Objective 7.8)

(a) What are λ_{min} and λ_{crit}?
(b) How do fibre length and orientation distributions affect the tensile strength of composites?
(c) Why does the toughness of many fibre composites *not* obey the rule of mixtures?

▼Composite foil▲

Although the sport of fencing grew out of men's attempts to kill one another in duels, the emphasis nowadays is very much the converse: every effort is made to prevent injuries and deaths, and those that do occur, although infrequently, still make the news (Figure 7.57).

The problem arises because the specification for the blades needs a steel with a yield stress of $2\,GN\,m^{-2}$ or more if the blade is to bend the required amount and still remain elastic. Steels of this level of yield strength have a concomitantly low toughness (Chapter 3), and are thus liable to break in a brittle manner if overstrained. When this happens, the whiplash as the stored elastic strain energy is released, together with any continuing forward thrust of the blade, can be sufficient for the broken tip of the blade to penetrate an opponent's protective clothing, especially if the fracture produces a sharp point.

A new type of blade has recently been introduced which relies on the crack-stopping mechanism in fibre composites to overcome this danger. It's made from a parallel array of nickel-coated steel wires which are drawn and then heat treated. On quenching, the nickel-rich areas remain austenitic, and hence soft, whilst the cores of the wires form hard martensite. Figures 7.58(a) and (b) show longitudinal and transverse sections of this metallic fibre composite material, known as *Paul-Steel*, whilst Figure 7.58(c) shows how a crack in the martensitic steel is halted at the interfaces with the austenitic regions. As Figure 7.59(a) shows, when this type of blade fails, it delaminates (absorbing a lot of extra strain energy) but does not break in two. In contrast, Figure 7.59(b) shows the brittle failure of a conventional blade.

Figure 7.57

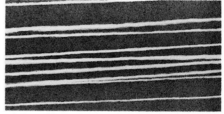

(a)   0   240 µm

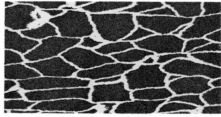

(b)   0   240 µm

(c)   0   30 µm

Figure 7.58 (a) Longitudinal and (b) transverse sections of *Paul-Steel* (c) crack stopping at interface

Figure 7.59 Failure of (a) *Paul-Steel* blade (b) conventional blade

Objectives for Chapter 7

After reading this chapter you should be able to:

7.1 Interrelate weight and volume fractions and densities of composites and their components. (SAQs 7.1, 7.7)

7.2 Account for the use of composite materials. (SAQ 7.2)

7.3 Explain the essential differences between thermosets and thermoplastics in terms of their structure, properties and processing. (SAQ 7.3)

7.4 Describe how foams and their structures are produced, and relate these to stiffness. (SAQ 7.4)

7.5 Outline the different levels of structure of softwoods and their effects on mechanical properties. (SAQ 7.5)

7.6 Apply the appropriate simple model(s) to estimate the elastic moduli of a variety of composite types, including foams. (SAQs 7.6, 7.7, 7.8)

7.7 Perform simple design calculations on composite products (SAQs 7.6, 7.8)

7.8 Relate the strength and toughness of composites to fibre length, orientation and interfacial bonding. (SAQ 7.9)

7.9 Define and use the following terms:

cambium	network polymer
chipboard	packing factor, P_f
coupling agent	plywood
crosslink density	prepolymer
crosslinking agent	prepreg
degree of anisotropy	rays (in wood)
dough moulding compound (DMC)	reactive diluent
efficiency factor, B	reticulated cell
fibreboard	rule of mixtures
fillers	sandwich mouldings
functionality	sheet moulding compound (SMC)
homogeneous strain model	structural foams
homogeneous stress model	syntactic foams
initiator	tracheids
laminated timber	volume fraction, V_i
lignin	weight fraction, W_i

Answers to exercises

EXERCISE 7.1 The assumptions underlying P_f are that the fibres are straight, of the same aspect ratio, and randomly packed. In fibre-filled thermoplastics the fibres are flexible so can easily be bent, processing can break up fibres, reducing their aspect ratios, as can clumping, and, as we'll see, the arrangement of the fibres is almost certainly not random.

EXERCISE 7.2 From the scale of the micrograph, the mean fibre diameter is about 12.5 µm, so that from Equation (7.6) with $P_f = 0.91$ for a hexagonal array

$$s = 12.5\left[\left(\frac{0.91}{0.75}\right)^{1/2} - 1\right]\mu m$$

$$= 1.27\,\mu m$$

EXERCISE 7.3 Many thermoplastics process routes involve melting and remelting, e.g. injection moulding of granules of chopped-up extrudate, recycling of scrap material, friction welding of moulded components. Thermosets can be processed only once from the liquid state, and then only from their precursors.

EXERCISE 7.4

$$V_s = \frac{v_s}{v_s + v_g}$$

where 'g' represents gas. If bulk density is ρ_f

$$V_s = \frac{m_s}{\rho_s} \times \frac{\rho_f}{m_s + m_g}$$

$$\approx \frac{\rho_f}{\rho_s}$$

since $m_g \approx 0$. Porosity p is given by

$$p = V_g$$

$$= 1 - V_s$$

$$= 1 - \frac{\rho_f}{\rho_s}$$

Table 7.10

	Equation (7.9)	Equation (7.10)
E_x	48.0	9.2
E_y	48.0	9.2
G_{xy}	19.6	3.6
G_{yz}	19.6	3.6

EXERCISE 7.5 The moduli (in $GN\,m^{-2}$) predicted by the two equations are given in Table 7.10.

Comparing these with Table 7.8, it's apparent that only E_x, the tensile modulus parallel to the fibres, is reasonably predicted (by the rule of mixtures). However, both equations provide upper and lower bounds to the other moduli.

EXERCISE 7.6 From Table 7.3 we have that the tensile strength of E-glass fibres is $3.4\,GN\,m^{-2}$. Hence, from Equation (7.17)

$$\bar{\tau}_i = \frac{\sigma_{f(max)}}{2} \times \frac{d}{l_{min}}$$

$$= \frac{3.4 \times 10^9}{2}$$

$$\times \frac{15 \times 10^{-6}}{5.1 \times 10^{-3}}\,N\,m^{-2}$$

$$= 5.0\,MN\,m^{-2}$$

EXERCISE 7.7 For loading in the fibre direction, the rule of mixtures applies, and since matrix and fibres are equally strained,

$$\sigma_c = \sigma_f V_f + \sigma_m V_m$$

From Table 7.3, $E_f = 72\,GN\,m^{-2}$ so at $\varepsilon = 0.01$, $\sigma_f = 720\,MN\,m^{-2}$ (note that this is less than the strength of pristine fibres shown in Table 7.3). From Table 7.2, for the epoxy at 1% strain, $\sigma_m = 35\,MN\,m^{-2}$. Thus

$$\sigma_c = (720 \times 10^6 \times 0.65)$$

$$+ (35 \times 10^6 \times 0.35)$$

$$= 480\,MN\,m^{-2}$$

Since the cross-section is $20 \times 5 \times 10^{-6}\,m^2 = 10^{-4}\,m^2$, this corresponds to a load of 48 kN.

EXERCISE 7.8 From Equation (7.19) and Table 7.3, for HM fibres we have

$$\lambda_{crit} = \frac{2.0 \times 10^9}{2 \times 10 \times 10^6}$$

$$= 100$$

We're given $\lambda = 200$ in SAQ 7.7 so $\lambda > \lambda_{crit}$, and Equation (7.21) applies modified by $B = 0.375$ from Equation (7.13). But first we need $\bar{\varepsilon}_f$ to evaluate σ_m.

$$\bar{\varepsilon}_f = \frac{\bar{\sigma}_f}{E_f} = \left(1 - \frac{\lambda_{crit}}{2\lambda}\right)\frac{\sigma_f^*}{E}$$

$$= \frac{0.75 \times 2 \times 10^9}{480 \times 10^9}$$

(from Table 7.3)

$$= 3.1 \times 10^{-3}$$

So $\sigma_m = \bar{\varepsilon}_f E_m = 6.2\,MN\,m^{-2}$ and, using Equation (7.21),

$$\sigma_c^* = (0.375 \times 2 \times 10^9 \times 0.3$$

$$\times 0.75) + (6.2 \times 10^6$$

$$\times 0.7)\,N\,m^{-2}$$

$$= 0.17\,GN\,m^{-2}$$

(The material suppliers give $\sigma_c^* = 0.25\,GN\,m^{-2}$ with fibre orientation unspecified, so our estimate is pretty good.)

Answers to self-assessment questions

SAQ 7.1

(a) Since

$$V_i = \frac{v_i}{v_1 + v_2 + \dots}$$

and

$$v = m/\rho$$

$$V_i = \frac{m_i/\rho_i}{m_1/\rho_1 + m_2/\rho_2 + \dots}$$

$$= \frac{m_i/m_c\rho_i}{m_1/m_c\rho_1 + m_2/m_c\rho_2 + \dots}$$

$$= \frac{W_i/\rho_i}{\sum_i W_i/\rho_i}$$

(b) Since $W_1 = W_2 = W_3 = \frac{1}{3}$ from Equation (7.1)

$$\frac{V_1\rho_1}{\rho_c} = \frac{V_2\rho_2}{\rho_c} = \frac{V_3\rho_3}{\rho_c}$$

Hence

$$V_2 \text{ (chalk)} = \frac{V_1\rho_1}{\rho_2} \text{ (nylon)}$$

$$= \frac{1.1V_1}{2.7}$$

and

$$V_3 \text{ (glass)} = \frac{1.1V_1}{2.5}$$

Since

$$\sum_i V_i = 1$$

$$V_1 + \frac{1.1V_1}{2.7} + \frac{1.1V_1}{2.5} = 1$$

whence

$$V_1 = 0.541$$
$$V_2 = 0.221$$
$$V_2 = 0.238$$

Since

$$W_1 = \frac{V_1\rho_1}{\rho_c} = \frac{1}{3}$$

$$\rho_c = 3V_1\rho_1$$

or

$$\rho_c = 1.786 \, \text{Mg m}^{-3}$$

SAQ 7.2 The main reasons are:

- to strengthen or toughen an otherwise weak or brittle matrix material using fibres,
- to improve or modify such properties as stiffness, cost, fire retardance and shrinkage using fillers,
- to benefit from the highest specific strength and specific modulus materials which are available only in fibrous form (Chapter 6).

The rôle of the matrix in the last case is to hold an assembly of fibres together, transfer stresses to them, protect their surfaces from damage or environmental attack, and produce a solid material that can withstand flexural and compressive loading.

SAQ 7.3 The outstanding structural difference is that in thermosets the polymer chains are crosslinked by primary bonds, whilst in thermoplastics the crosslinks are secondary, easily reversible, bonds. Because of the primary-bond network, thermosets are predominantly glassy materials, and, since plastic flow and crazing possibilities are thus more limited, they tend to be brittle materials. The crosslinking also restricts melt processing to that of their precursor materials. However, since the viscosities of their precursors tend to be lower than those of molten thermoplastics, larger area mouldings, cheaper tooling and smaller production runs are economically feasible.

Thermoset formation requires the reacting species to have a minimum functionality of two, with at least one component to be trifunctional or higher. These are achieved either by starting with trifunctional reactants (for example PF, UF, MF) or by using multifunctional crosslinking agents (for example UP, EP).

SAQ 7.4 Process routes include (a) mechanical beating, (b) physically adding a gas or volatile liquid, (c) chemical decomposition of a component, (d) hollow fillers and (e) dissolving or melting out a component. Foam structures are either open-celled (hence, interconnected) or closed-celled. The former includes reticulated foams (ties and struts) and the latter, syntactic foams in which the cell walls are formed by a second-phase material.

The main parameter linking stiffness to structure is ρ_f/ρ_s. Empirically, for uniform density foams,

$$\frac{E_f}{E_s} = \left(\frac{\rho_f}{\rho_s}\right)^n$$

with $n \approx 1.5$ for tension and $n \approx 2$ for compression.

SAQ 7.5 The cellulose molecules are aligned in microfibrils, each molecule passing through crystalline and amorphous regions of the fibrils. The fibrils in a matrix of mainly hemicelluloses form the layers which make up the cell walls as shown in Table 7.11 (starting from outside and in order of formation).

The cells (or tracheids) have $\lambda \approx 100$, and are bonded to each other by a lignin-rich matrix to form a parallel array whose axis is aligned with the trunk or branch of the tree. The cell *walls* of all trees have very similar properties, which are largely a reflection of those of the dominant S_2 layer. Thus the variation within and between species is determined by the overall density of the wood (in the same way as foams). The oriented cell array with anisotropic cell wall leads to highly anisotropic properties (very much more so than in fibre composites).

Table 7.11

Layer	Thickness/cell wall	Fibril arrangement
P	~1%	irregular network
S_1	10–20%	perpendicular to cell axis
S_2	70–90%	20° helically to cell axis
S_3	~5%	perpendicular to cell axis

STRUCTURAL MATERIALS

SAQ 7.6

(a) For tension we have from Equation (7.11)

$$\frac{E_f}{E_s} \approx \left(\frac{\rho_f}{\rho_s}\right)^{1.5}$$

$$\rho_f = \left(\frac{E_f}{E_s}\right)^{1/1.5} \rho_s$$

$$= \left(\frac{0.2}{1.5}\right)^{1/1.5} \times 910\,\text{kg m}^{-3}$$

(from Table 7.2)

$$\approx 240\,\text{kg m}^{-3}$$

Thus, for compression, with $n = 2$

$$E_f \approx \left(\frac{\rho_f}{\rho_s}\right)^2 \times 1.5\,\text{GN m}^{-2}$$

$$= 0.10\,\text{GN m}^{-2}$$

(b) Rearranging the equation for Δ,

$$t^3 = \frac{2Fw^2}{11E\Delta}$$

where, for flexure,

$$E = E_f \approx \left(\frac{\rho_f}{\rho_s}\right)^1 E_s$$

The blow ratio β is

$$\beta = \frac{v_s + v_g}{v_s} = \frac{1}{V_s}$$

$$= \frac{\rho_s}{\rho_f} \text{ (see Exercise 7.4)}$$

Therefore

$$E_f = \frac{E_s}{\beta}$$

$$t^3 = \frac{2 \times 100 \times 9.81 \times 0.25 \times 1.25}{11 \times 1.5 \times 10^9 \times 5 \times 10^{-3}}\text{m}^3$$

$$t = 0.0195\,\text{m}$$

Hence mass of polypropylene m is

$$m = \rho_f(v_s + v_g)$$

$$= \frac{910 \times 1 \times 0.5 \times 0.0195}{1.25}\text{kg}$$

$$= 7.1\,\text{kg}$$

SAQ 7.7

(a) This is a straightforward rule of mixtures calculation with $V_{f1} = V_{f2} = 0.35$. From Tables 7.2 and 7.3 we have

$$E_c = (0.3 \times 3.5) + (0.35 \times 72)$$

$$+ (0.35 \times 480)\,\text{GN m}^{-2}$$

$$= 194\,\text{GN m}^{-2}$$

(b) We're told the MCMV material contains equal masses of E-glass and polyester, or $W_m = W_f = 0.5$. From Equation (7.4) and Tables 7.2 and 7.3,

$$V_f = \frac{W_f/\rho_f}{W_f/\rho_f + W_m/\rho_m}$$

$$= \frac{0.5/2.55}{0.5/2.55 + 0.5/1.3}$$

$$= 0.338$$

So

$$V_m = 1 - V_f$$

$$= 0.662$$

The fibres are aligned in two directions at right-angles, so $B = 0.5$ in Equation (7.13) and

$$E_c = (0.5 \times 0.338 \times 72)$$

$$+ (0.662 \times 3.3)\,\text{GN m}^{-2}$$

$$= 14.4\,\text{GN m}^{-2}$$

(c) We know that $V_f = 0.3$ and the fibres are HM carbon. Applying the rule of mixtures

$$E_c = (0.3 \times 480)$$

$$+ (0.7 \times 2.0)\,\text{GN m}^{-2}$$

$$= 145.4\,\text{GN m}^{-2}$$

However, the fibre aspect ratio is 200, and from Figure 7.46, the expected modulus is only about 85% of the above value, or

$$E_c = 124\,\text{GN m}^{-2}$$

(d) Here again, we assume equal strains and apply the rule of mixtures using the data in Table 7.6. For the walnut, $V_f = 0.25$ for both layers, whilst for each layer of pine, $V_f = 0.25$. Using $E(\parallel)$ and $E(\perp)$ in alternate layers

Table 7.12

$$E_c = (0.25 \times 11.2)$$

$$+ (2 \times 0.25 \times 1.1)$$

$$+ (0.25 \times 16.4)\,\text{GN m}^{-2}$$

$$= 7.45\,\text{GN m}^{-2}$$

(e) Estimating the SMC modulus is a bit more complex. From 'SMC' and Tables 7.2–7.4 we get Table 7.12 below.

The modulus of the composite is given by Equation (7.13) with $B = 0.375$, but to evaluate this we need V_m and E_m. Thus we start by treating the matrix on its own as a particulate composite. By applying Equation (7.4) to the weight fractions *of the matrix* in the last column of the table, we get for components 1–4

$$V_1 = 0.447$$

$$V_2 = 0.070$$

$$V_3 = 0.276$$

$$V_4 = 0.207$$

From the homogeneous stress model (Equation 7.10)

$$\frac{1}{E_m} = \frac{0.447}{3.3} + \frac{0.070}{26} + \frac{0.276}{30}$$

$$+ \frac{0.207}{20}$$

or

$$E_m = 6.34\,\text{GN m}^{-2}$$

From Equation (7.2)

$$\rho_m = 1894\,\text{kg m}^{-3}$$

Then, applying Equation (7.4) again but this time to the SMC as a whole,

$$V_f = \frac{0.25/2550}{0.25/2550 + 0.75/1894}$$

$$= 0.198$$

Finally, from Equation (7.13)

$$E_c = (0.375 \times 0.198 \times 72)$$

$$+ (0.802 \times 6.34)\,\text{GN m}^{-2}$$

$$= 10.4\,\text{GN m}^{-2}$$

Component	$\rho/\text{kg m}^{-3}$	$E/\text{GN m}^{-2}$	W_f	$W_f/0.75$
1 polyester	1300	3.3	0.23	0.307
2 calcite	2710	26	0.075	0.100
3 alumina trihydrate	2420	30	0.265	0.353
4 others	2200	20	0.18	0.240
E-glass	2550	72	0.25	

SAQ 7.8 For equal bending stiffness

$$(EI)_{steel} = (EI)_{SF}$$

and, assuming we're comparing rectangular sections of the same width w, we need to use the full version of Equation (7.7) to determine $(EI)_{SF}$. We also need a value for the core modulus. Using Equation (7.11) with $n = 1.5$ we obtain $E_f = 2.33\,GN\,m^{-2}$, and with $n = 2$ we get $E_f = 1.65\,GN\,m^{-2}$. Since we're considering bending (involving both tension and compression), a better estimate is probably the mean of these two, or $E_f = 2.00\,GN\,m^{-2}$. Then, with $d_s = 1.5 \times 10^{-3}\,m$ and $d_c = 3.0 \times 10^{-3}\,m$, we obtain from Equation (7.7)

$$(EI)_{SF} = 1.98 \times 10^{-7}w[(1.88$$
$$\times 10^{-7}) + (5.06$$
$$\times 10^{-6})] + 4.50w$$
$$= 108w\,N\,m^2$$

$$(EI)_{steel} = \frac{210 \times 10^9 wd^3}{12}$$

$$(d^3)_{steel} = \frac{12 \times 108}{210 \times 10^9}\,m^3$$

$$= 6.17 \times 10^{-9}\,m^3$$

So $(d)_{steel} = 1.83 \times 10^{-3}\,m$ or $1.83\,mm$.

SAQ 7.9

(a) λ_{min} is the minimum fibre aspect ratio for the tensile stress in a fibre to reach the maximum value that can be achieved by shear transfer from the matrix. λ_{min} is inversely proportional to $\bar{\tau}_i$ (Equation 7.17). λ_{crit} is the value of λ_{min} at which the maximum stress in the fibre equals the tensile strength of the fibre (Equation 7.19).

(b) The tensile strength of a composite depends on the value of the aspect ratio of its fibres relative to λ_{crit} (Equations 7.20 to 7.22), and increases with increasing fibre length to a maximum value predicted by the rule of mixtures in an aligned, continuous fibre composite (Exercise 7.7). As the fibre-orientation distribution becomes increasingly random, the tensile strength of the composite falls. The effect can be simply modelled for some defined orientation distributions by modifying the rule of mixtures expression for strength with the factor B in Equation (7.13) as exemplified in Exercise 7.8.

(c) The toughness of many fibre composites does not obey the rule of mixtures because, for cracking perpendicular to the fibre direction, the array of fibres in a matrix provides mechanisms of energy absorption (namely, debonding and fibre pull-out) which are not available in either fibre or matrix materials on their own.

Chapter 8
Cement, Concrete and Reinforced Concrete

by George Weidmann

Chapter 8 Cement, concrete and reinforced concrete

8.1 Introduction

In terms of the old playground spoofing game 'Scissors, paper, stone', we have covered steel (scissors) in Chapter 3 and paper (briefly) in Chapter 6. Now comes the third, stone, but in the guise of an artificial stone — concrete, and its reinforcement with steel bars and wires. Although cement, concrete and reinforced concrete are all, in a sense, composite materials, only concrete is amenable to the composite modelling of the previous chapter. This and the intimate relationship between the three materials provide the justification for considering them together in this separate chapter.

Concrete is made from cement, aggregates (sand and crushed rock or gravel) and, crucially, water. The water is important for two reasons. Firstly, it is the water added to the mix that makes the concrete mouldable and, secondly, it is the hydration reaction between the water and the components of the cement that leads to the cement's hardening, so binding the aggregate and giving concrete its stone-like properties. Thus we have a material that can be readily moulded into complex shapes, yet subsequently acquires properties similar to those of natural stone. It is also generally cheaper than stone because its raw materials are more readily extracted and it can be formed to the required shape *in situ*, whereas stone has to be quarried and subsequently cut to shape.

This century has seen concrete and reinforced concrete become two of our most important civil-engineering materials, surpassing in tonnage the amount of wood and steel used. Their range of application is large and diverse, from garden paths to airport runways, and from sound-insulation partitions to massive engineering structures such as dams, bridges and high-rise buildings. Their extensive use in the construction industry has made the discovery of problems in concrete and reinforced-concrete structures a matter of considerable public concern, both from the point of view of safety and the costs of rectification.

Thus, in what follows, we shall first consider cement powder, what it is and how it interacts with water to form hydrated cement. We shall then examine concrete — hydrated cement plus aggregate — and reinforced concrete, with particular reference to how they can be modelled and how they can go wrong. Finally, we shall look at prestressed concrete and some possible future trends.

Figure 8.1 Reinforced-concrete structures:
(a) Clunie Dam, Tayside;

(b) Redheugh Bridge,
Newcastle-upon-Tyne;

Figure 8.2

(c) Water Tower Place, Chicago, which, at
260 m is the world's tallest reinforced-
concrete building

8.2 What is cement?

8.2.1 The ingredients

The term cement is a generic one and applies to all binders that are used in concrete. However, we shall confine discussion to **Portland cement**, which is by far the most important of these binders.

The starting material is a mixture of about 75% limestone, $CaCO_3$, and 25% clay, principally aluminosilicate, but with a significant iron and alkali oxide content. These are ground together and fed into a rotary kiln against a counterflow of air and burning powdered coal. As the temperature of the mixture increases, a series of reactions takes place. Between 500 K and 900 K the clay minerals break down and then, between 1180 K and 1300 K, the limestone calcines to **quicklime**, CaO, and carbon dioxide. Finally, at up to 1800 K, the different components partially melt and react together to give a mixture of the main constituents of cement. The high surface energy ($0.5–0.6\,J\,m^{-2}$), assisted by the low viscosity ($0.1–0.2\,N\,s\,m^{-2}$) of the molten material, drives the mixture to coalesce into nodules of clinker around 5–10 mm in size.

After cooling, the clinker is mixed with 3–5 wt %**gypsum** $CaSO_4.2H_2O$, which helps control the initial setting rate, and is then ground to give cement powder. The distribution of particle sizes is important since this determines the specific surface area of the cement grains. The hydration reactions occur on the surfaces of these grains, so the higher the specific surface area, the faster the rate of hydration. Ordinary Portland cement has a specific surface area in the range $300–350\,m^2\,kg^{-1}$, while that of a rapid hardening cement is $400–450\,m^2\,kg^{-1}$. A distribution of particle sizes for ordinary Portland cement is shown in Figure 8.3, from which it can be seen that approximately 90% of the particles measure more than 5 µm and 99% less than about 90 µm.

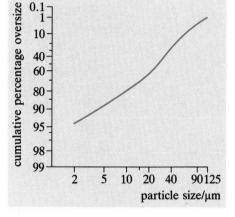

Figure 8.3 The distribution in the size of particles of ordinary Portland cement

> EXERCISE 8.1   A powder sample contains equally sized spherical grains of specific surface area $400\,m^2\,kg^{-1}$. If the density of the material is $3000\,kg\,m^{-3}$, what is the diameter of the grains?

What then are the constituents of cement? Apart from the gypsum, they are more or less impure minerals from the clinker. In decreasing percentage order these are **alite** (C_3S), **belite** (C_2S), **aluminate** (C_3A) and **ferrite** (C_4AF). The nomenclature of the constituents is considered in ▼**Naming the constituents of cement**▲. In addition there are small quantities, less than 1.5%, of alkali oxides and other impurities. An optical micrograph of a piece of cement clinker is shown in Figure 8.4 in which the darker areas are alite crystals fringed with belite, while the lighter areas are aluminate (dark) and ferrite (light). With so many ingredients, none of which is pure, it is evident that the reaction of

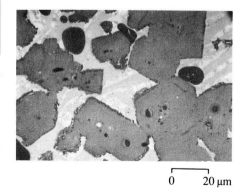

Figure 8.4 Cement clinker made up, in the darker areas, of alite crystals fringed with belite and, in the lighter areas, aluminate (dark) and ferrite (light)

▼Naming the constituents of cement▲

The constituents of cement are minerals, which exist as multicomponent solid solutions. The solutes or impurities are important because they can affect the crystal structures or the reactivities, or both, of the minerals. Despite this, it is usual to treat the main constituents as chemical compounds and to label them as if they were mixtures of oxides, but using a shorthand nomenclature. This involves replacing the usual chemical formula for an oxide with a single letter, as shown:

Oxide	CaO	SiO$_2$	Al$_2$O$_3$	Fe$_2$O$_3$	H$_2$O	SO$_3$
Symbol	C	S	A	F	H	$\bar{\text{S}}$

Notice we have used a different type face to emphasize the difference between the chemical formulae and the symbols for various oxides.

Table 8.1 summarizes the various ways of naming and characterizing the different phases in Portland cement, together with the typical percentage of each present. Bear in mind that the mixed oxide representation carries no structural information: for example, the presence of silicate ions, SiO$_4^{4-}$, in C$_3$S is not apparent, despite their vital role in the hydration of cement.

Table 8.1   Constituents of Portland cement

Compound name	Compound chemical formula		Equivalent oxide formula	Shorthand nomenclature	Mineral name	Density/ kg m^{-3}	Typical % by weight
tricalcium silicate	CaO.Ca$_2$SiO$_4$	≡	3CaO.SiO$_2$	C$_3$S	alite	3150	55
dicalcium silicate	Ca$_2$SiO$_4$	≡	2CaO.SiO$_2$	C$_2$S	belite	3280	20
tricalcium aluminate	2CaO.Ca(AlO$_2$)$_2$	≡	3CaO.Al$_2$O$_3$	C$_3$A	aluminate	3030	12
tetracalcium aluminoferrite	4CaO.Al$_2$O$_3$.Fe$_2$O$_3$			C$_4$AF	ferrite	3770	8
hydrated calcium sulphate	CaSO$_4$.2H$_2$O	≡	CaO.SO$_3$.2H$_2$O	C$\bar{\text{S}}$H$_2$	gypsum	2320	3.5
alkali oxides and other constituents	K$_2$O, Na$_2$O CaO						1.5

cement with water is going to be pretty complex. Each ingredient will not only react with water, but will also influence the way the others react. What happens to turn a powder into a paste and then into a stone-like material is considered in the next section.

8.2.2 Cement hydration

When water is added to a cement or a concrete mix, two things happen. The mix is turned into a paste: that is, it acquires cohesion with some rigidity, but is workable and pourable. At the same time, a considerable amount of heat is evolved. This constitutes the first stage of what is called **setting**. The setting period is the time for which the cement remains workable and is a vital factor in the usefulness of cement. (It corresponds to the pot life of thermosetting plastics discussed in Section 7.2). Setting ends when **hardening** starts and the cement begins to solidify.

The physical and chemical changes

The macroscopic changes involved can be followed by measuring the compressive strength and the rate at which heat is evolved as functions of time, as shown in Figures 8.5(a) and 8.5(b). What is remarkable is that, although the initial compressive strength is barely measurable, as might be expected from a workable paste, the rate at which heat is evolved is initially very high (it rises to about $200\,W\,kg^{-1}$ at 30 s), followed by a rapid fall to a low value during the setting period. After

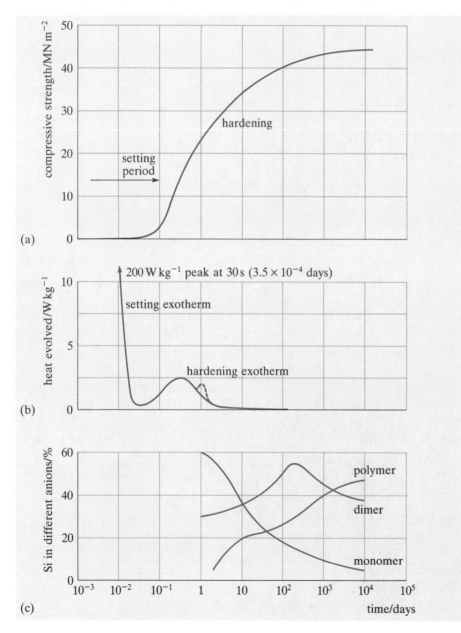

Figure 8.5 Cement hydration: changes with time of (a) compressive strength, (b) heat evolution rate, and (c) distribution of Si in anionic species

an hour or so, hardening starts and is accompanied by an increasing heat output, which rises to a maximum at around ten hours. This second maximum is, however, only about 1% of the initial exothermic peak. Thereafter, as the compressive strength steadily increases, the rate at which heat is evolved decreases to virtually nothing at ten days or so, although there is frequently a subsidiary peak. Notice that the time scale in Figure 8.5 goes from about fifteen minutes to almost thirty years.

What do these changes signify?

One clue to this is shown in Figure 8.5(c). It is well known that two silicate ions can combine together to form a **dimer**.

$$2SiO_4^{4-} \rightarrow Si_2O_7^{6-} + O^{2-}$$

This is achieved by two tetrahedra joining corner to corner.

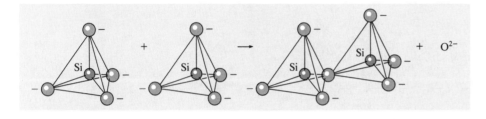

This combining together can continue to give higher-order anions, culminating in polymers. Figure 8.5(c) shows the results of measurements on cements whose ages ranged from one day to over twenty years and gives the proportion of silicon present in different anionic species. Here 'polymer' refers to anions with five silicon atoms (Si_5) or more, since anions with three or four silicon atoms (Si_3 and Si_4) were not detected. Note that each linking of silicate tetrahedra releases an O^{2-} ion.

EXERCISE 8.2   By extending the above diagram for the dimeric silicate anion, sketch the trimeric silicate anion, work out its formula and write down the equation for the reaction between the dimeric and trimeric silicate anions that produces the pentamer (Si_5). Deduce the general formula for a straight chain, silicate anion containing n atoms of silicon.

It is clear that the hardening process has associated with it an increasing proportion of dimeric and higher-order silicates. Although these are exothermic reactions this is not the whole story. Other processes are involved. Another clue is that the pH value of the mixture rises very rapidly within the first minute of water being added (see ▼pH▲), which must reflect the rate at which the concentration of hydroxyl ions is increasing.

▼pH▲

The acidity or alkalinity of an aqueous solution is characterized by its pH number, which is a measure of the concentration of hydrogen ions, $[H^+]$. Square brackets are the chemist's shorthand for 'concentration of'. The pH number is defined by

$$pH = log_{10} \frac{1}{[H^+]}$$

or

$$pH = -log_{10}[H^+]$$

In an aqueous solution, there is a fixed relationship between $[H^+]$ and $[OH^-]$: at 298 K, this is

$$[H^+] \times [OH^-] = 10^{-14}(mol\,l^{-1})^2$$

or

$$log_{10}[H^+] + log_{10}[OH^-] = -14$$

So, in terms of pH, you have

$$[H^+] = 10^{-pH}\,mol\,l^{-1}$$

and

$$[OH^-] = 10^{-(14-pH)}\,mol\,l^{-1}$$

Thus pH 1 is extremely acidic, while pH 14 is extremely alkaline. Pure water, which is neutral, has a pH value of 7.

All the starting ingredients of cement, given in Table 8.1, are ionic solids. Water is a highly polar liquid and will thus readily dissolve most ionic species. Information on the concentration changes with time for different ions should help establish what reactions are occurring and when, especially if you know something of the heats of hydration of the different ingredients.

Figure 8.6 shows the results of chemical analysis of the liquid phase of a sample of hydrating cement paste at different times. This shows the concentration of hydroxyl ions, $[OH^-]$, rising from about $0.020 \, mol \, l^{-1}$, at around one minute, to a peak of about $0.075 \, mol \, l^{-1}$, at around three hours, and then falling. The concentrations of calcium ions, $[Ca^{2+}]$, and sulphate ions, $[SO_4^{2-}]$, both show an early peak at about two minutes, with the former falling off and then rising again to a subsidiary peak at a similar time to the $[OH^-]$ peak. For potassium and sodium, $[K^+]$ and $[Na^+]$ both rise slowly, with $[K^+]$ more than four times $[Na^+]$. Finally, $[Si]$, $[Al]$ and $[Fe]$ all remain very small (about $2 \times 10^{-6} \, mol \, l^{-1}$) over the time scale of the measurements.

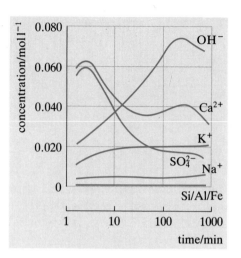

Figure 8.6 Typical graphs of concentration against time for ions in solution in a paste of Portland cement

EXERCISE 8.3   From Figure 8.6, estimate the pH of the solution at 1 min, 10 min and 3 h.

The pH number therefore rises in the first minute from about that of the starting water, 7, to a pH of 12.3 and continues rising to a maximum pH of 12.9 after about three hours. A *saturated* solution of $Ca(OH)_2$ has a pH of 12.6 ($[OH^-] = 0.04 \, mol \, l^{-1}$ and $[Ca^{2+}] = 0.02 \, mol \, l^{-1}$), so this must mean that $[OH^-]$ and $[Ca^{2+}]$ reach nearly double their saturation limit before $Ca(OH)_2$ starts to precipitate out of solution at a time corresponding to the end of the setting period. Thereafter, the cement settles down to a constant value of pH 12.6, due to the reversible reaction

$$Ca^{2+} + 2OH^- \rightleftharpoons Ca(OH)_2(solid)$$

If the pH drops slightly, some $Ca(OH)_2$ dissolves to restore the level, while if it increases, more $Ca(OH)_2$ precipitates from solution. As you will see later, this highly alkaline environment is very important for keeping the steel in reinforced concrete corrosion-free.

The high initial peak of $[Ca^{2+}]$ and $[SO_4^{2-}]$ must be mainly due to the gypsum, $C\bar{S}H_2$, dissolving, and you would expect $[Ca^{2+}] = [SO_4^{2-}]$. The fact that $[Ca^{2+}]$ is slightly larger than, but parallel to, $[SO_4^{2-}]$ in the first few minutes suggests a further source of Ca^{2+} ions — either unreacted CaO in the clinker, or the C components in C_3S, C_3A and C_4AF, or both. This source becomes more important at longer times when the two curves diverge. The reduction in $[SO_4^{2-}]$ means that there must be a mechanism for removing SO_4^{2-} ions from solution.

The potassium and sodium ions present clearly come from the minor alkali oxide constituents which also provide an additional source of

Table 8.2   Heats of hydration of cement constituents

Cement mineral	Heat of hydration/ $kJ\,kg^{-1}$	Hydration rate
alite C_3S	490	medium
belite C_2S	225	slow
aluminate C_3A	1200	fast
ferrite C_4AF	400	medium–slow
gypsum $C\bar{S}H_2$	110	very fast
lime C	1168	extremely fast

hydroxyl ions. The main sources of hydroxyl ions are the O^{2-} ions produced by the linking of silicate tetrahedra and by the dissolution of CaO. The O^{2-} ions react rapidly in the presence of water:

$$O^{2-} + H_2O \rightarrow 2OH^-$$

This emphasizes that oxygen ions from any source will be unstable in water and will instantly react to make hydroxyl ions.

The puzzling thing is, that while all this is going on during the setting period, there is virtually no trace of Si, Al or Fe in solution. The mechanisms involved are thought to be crucial to the setting period, and we shall therefore return to this after we have considered the heats of hydration.

The heats of hydration of the different cement minerals are given in Table 8.2 together with their relative rates of hydration. With this information you can start accounting for the form of the curve in Figure 8.5(b).

Although gypsum hydrates rapidly, its low heat of hydration and small content will contribute little to the initial exothermic peak. Any unreacted lime, although only little is likely to be present, would make a bigger contribution, though not nearly enough to account for the maximum of $200\,W\,kg^{-1}$. This means that the initial hydration of the other four constituents must be largely responsible. Of these C_2S and C_4AF will only play a minor role because they have lowish heats of hydration and are slow to hydrate, leaving C_3S and C_3A as the most likely candidates. The gypsum retards the hydration rate of C_3A so that it becomes comparable to that of C_3S. Thus it appears that most of the initial peak of $200\,W\,kg^{-1}$ is due to the initial hydration of C_3A and C_3S, with the lowish content of C_3A being balanced by its large heat of hydration. The second, hardening peak is also largely due to C_3A and C_3S, with the former being responsible for any subsidiary peak.

Mechanism of hydration

While the above accounts for the contributions to the peaks in the rate at which heat is evolved, it does not explain why the initial hydration rate slows down to a very low level during the setting period and then

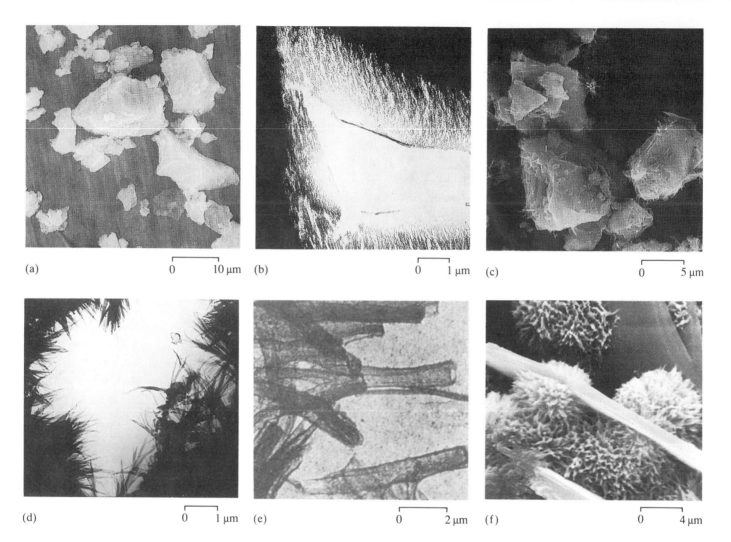

(a)　　　　　　0　　10 μm

(b)　　　　　　0　　1 μm

(c)　　　　　　0　　5 μm

(d)　　　　　　0　　1 μm

(e)　　　　　　0　　2 μm

(f)　　　　　　0　　4 μm

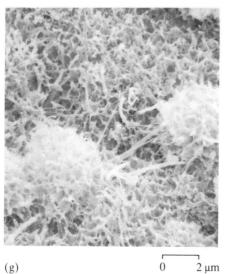

(g)　　　　　　0　　2 μm

(h)　　　　　　0　　5 μm

Figure 8.7 (a) Unreacted cement grains (SEM). (b) Colloidal gel coating on C_3S grain after immersion in water for five minutes (TEM). (c) Early formation of ettringite crystals on cement grains (SEM). (d) Gel foils and fibrils (TEM). (e) Tubules formed from C_3A gel (TEM). (f) Cement grains covered with gel fibres, plus rods of ettringite (SEM). (g) Interlocking of gel fibres (SEM). (h) Portlandite crystals in hardening cement pastes (SEM)

picks up again at the onset of hardening; nor does it explain why, if C_3S and C_3A are hydrating, there is a virtual absence of Si and Al in the liquid phase. Recent observations by transmission and scanning electron microscopy of setting cement pastes have helped to put together a mechanism that accounts for these, although many details are still contentious and the subject of active research.

The sequence of events is thought to be as follows:

1 When water is added to unreacted cement grains (Figure 8.7a), there is an initial period of very rapid hydration, mainly of C_3S and C_3A, lasting about one minute. This is accompanied by a high rate of heat evolution (Figure 8.5b), silicate polymerization, and a rapid increase in $[OH^-]$ (Figure 8.6), with a concomitant rise in the pH number

2 The reaction rate slows drastically as the surfaces of the grains become coated with silicate and aluminate **gels** (see ▼**Gels**▲ and Figure 8.7b). At the same time, the gypsum dissolves so that $[Ca^{2+}]$ and $[SO_4^{2-}]$ increase rapidly, peaking at about two minutes (Figure 8.6).

3 The subsequent fall in $[Ca^{2+}]$ and $[SO_4^{2-}]$ is due to the formation of very fine, needle-like crystals of a phase known as **ettringite**, which become visible within ten minutes (Figure 8.7c). This is a heavily hydrated calcium aluminosulphate, whose formation reaction can be written as

$$C_3A + 3C\bar{S}H_2 + 26H \rightarrow C_6A\bar{S}_3H_{32} \text{ (solid)}$$

4 The cohesion of the cement paste during the setting period is due to gel-to-gel contact on adjoining grains, aided increasingly by the interlocking of growing ettringite crystals. The fall in the hydration rate after the initial peak is due to the reaction becoming controlled by the rates at which the reacting species diffuse through the gel coatings. The gels are amorphous and of variable composition, so that the one formed on C_3S, for example, which is the main one present, is written in the shorthand nomenclature as C–S–H, the hyphens emphasizing its indeterminate nature.

5 Ca^{2+} and OH^- diffuse out through the gel causing a rise in $[Ca^{2+}]$ and $[OH^-]$ over the setting period (Figure 8.6), while water diffuses in and continues the hydration of the grain surface. The silicates and aluminates do not diffuse through the gel, but either add to it or build up in solution behind it, accounting for their absence from the liquid outside.

EXERCISE 8.4   What determines the direction of diffusion?

6 As the amount of water inside the gel coating increases, the pressure builds up, eventually leading to rupture of the gel. The ruptured gel peels away from the grain, forming gel foils and fibrils (Figure 8.7d) and, in the case of C_3A, also tubules (Figure 8.7e). This exposes the grain surface locally to further direct hydration and the process repeats.

▼Gels▲

Gels are swollen polymer networks of high viscosity. They are ubiquitous and variegated: vulcanized rubber forms a gel when it absorbs a liquid and swells; plasticized PVC (see Chapter 5), thixotropic paint, photographic emulsion, jam and mayonnaise are all based on gels; gel permeation chromatography (GPC), for measuring the molecular mass distribution in polymers, speaks for itself. Gels are used for fining beers and wines (e.g. isinglass and gelatin); silica gel is a well-known drying agent; gel spinning is an important route to highly oriented polymer fibres (see Chapter 6).

What these all have in common are properties intermediate between the liquid and the solid states. Thus, they deform elastically and recover, yet can often be induced to flow at higher stresses. They have extended three-dimensional network structures and are highly porous, so many gels contain a very high proportion of liquid to solid — one description of them is 'a highly concentrated solution of a liquid in a solid'. The networks can be permanent, as in the cases of swollen crosslinked rubber and GPC resins, or temporary as in many other cases. Polymeric molecules, however, are invariably involved.

One starting point for a gel is an emulsion, a suspension of a colloid, a collection of particles with sizes in the range 1 nm–0.1 µm, in a liquid. The stability of such suspensions is governed by interactions between the particles and between the particles and the liquid. Some, such as inks, can exist indefinitely, while others, such as milk, are unstable. Their stability is largely governed by the electric charges on the particles, which might be due to surface ionization in a polar liquid, or to absorption of dissolved ions. The type of ions present in the liquid determines whether the particles are positively or negatively charged. Thus, in a colloidal suspension in water, the charges on the particles change from positive to negative as the concentration of hydroxyl ions, OH^-, is increased (i.e. as the pH rises). The pH at which the changeover occurs is known as the **isoelectric point**.

The significance of this is that at the isoelectric point the particles are uncharged, therefore there is no electrostatic repulsion between them. Brownian motion can bring them close enough to combine through, for example, van der Waals' forces, forming larger and larger particles which fall out of the suspension, a process known as **coagulation**. If polymeric molecules are present, these can provide bridges between the particles, creating the open three-dimensional network required of a gel and this is known as **flocculation**.

7 Each grain sprouts a multitude of these fibres (Figure 8.7f) and, as they continue to grow and multiply, they start to interlock. This signals the end of the setting period and the paste starts to harden (Figure 8.7g). At about the same time $Ca(OH)_2$ starts to precipitate out of supersaturated solution although the trigger for this is not known. It forms plate-like hexagonal crystals known as **portlandite** (Figure 8.7h).

8 Continued hardening comes from the further multiplication, growth and interlocking of gel fibres and crystal species such as portlandite. At longer times these include the hydration products of C_2S, which are the same as C_3S, but develop more slowly (see Table 8.2). The continued, long-term polymerization of the silicate in the gel (Figure 8.5c) must also contribute to the hardening, as does the drying out of the gel to below its saturation value.

To summarize the roles of the ingredients:

C_3S Major ingredient, initial gel formation contributes to setting, hydration products C–S–H fibres and $Ca(OH)_2$ crystals make a major contribution to strength, particularly in the early stages of hardening.

C_2S Same hydration products as C_3S, contributes to increase in strength at later stages of hardening.

C_3A Contributes to setting through gel and ettringite formation, contributes little to mechanical strength (i.e. to hardening).

C_4AF Ferrite, like aluminate, acts as a flux in the cement kiln, hydration products play little part in setting or hardening, contributes colour to cement.

$C\bar{S}H_2$ Controls hydration rate of C_3A, constituent of ettringite which contributes to setting.

SAQ 8.1   (Objective 8.1)
Summarize the causes of: (a) setting; (b) the setting period; and (c) hardening in cement.

SAQ 8.2   (Objective 8.2)
Describe concisely the role of the CaO component in cement with reference to establishing and maintaining alkalinity.

SAQ 8.3   (Objective 8.3)
Outline the contributions of the four major mineral components of cement to: (a) setting and (b) hardening.

8.2.3 Water, pores and properties

The structure of hydrated and hardened cement is, as we have seen, an interlocked mass of crystals and amorphous fibres, whose two principal constituents are partially dried-out silicate gel in the fibres and portlandite in the crystals, together with any unhydrated cement particles.

EXERCISE 8.5   What sort of bonding would you expect in such a structure?

Water still has an important part to play in hardened cement, in addition to the part it plays in setting and hardening. The interstices between fibres and between fibres and crystals, as well as the larger spaces into which the gel did not grow, perhaps because of air bubbles or pockets of water, all constitute **pores**. The smallest pores are those associated with the gel and are a few nanometres in size. These invariably contain water and, as they are only a factor of ten or so larger than the water molecule, the water is fairly strongly bound to the pores by secondary bonding (it can be evaporated by heating). Typically this **gel water** constitutes some 15% by weight of the hydrated cement. This compares with the 25% by weight which is combined chemically with the hydrated compounds in cement and means that some 40% of water is required for complete hydration.

In terms of the properties of hydrated cement, the larger pores, whose sizes can range up to a millimetre or so (Figure 8.8), are much more important.

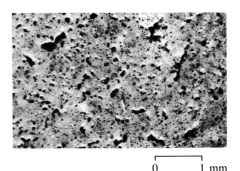

0 ⌐——⌐ 1 mm

Figure 8.8 Large pores in hydrated cement (optical)

The tensile strength of cement is only about 10% of its compressive strength, basically because of the presence of the large pores. As an example of their effect, Figure 8.9 shows the results from three-point bending tests (see ▼Flexural tests▲) on specimens of a hardened cement that were notched to different depths in the centre of the tensile surface. You can see that the longest notches reduced the strength the most, as you would expect from the Griffith equation (Chapter 2)

$$\sigma_t = \left(\frac{EG_c}{\pi a_c}\right)^{\frac{1}{2}} \tag{2.22}$$

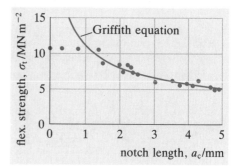

Figure 8.9 A graph of flexural strength against notch length for cement specimens

▼Flexural tests▲

Testing brittle materials in tension is made difficult by the problems of gripping specimens in a tensile machine without breaking them and of making test pieces in the 'dog-bone' shape of a tensile specimen without producing extra flaws in their surfaces. Because of this, brittle materials are frequently tested in bending. 'Roof tiles', discussed in Chapter 4, are one example of this. The specimens are usually parallel-sided bars, which are easier to make, and the two configurations used are three-point and four-point bending (Figure 8.10 and Chapter 1). Loads are transmitted to the specimens by knife-edges, or, if these cause too much local damage, by rollers.

In each case, the maximum tensile stress that the specimens can withstand is obtained by applying the equilibrium equation for bending that was given in Chapter 1: namely,

$$\frac{\sigma}{y} = \frac{M}{I} \tag{1.9}$$

The maximum stress, σ_{max}, occurs at $y = t/2$ and, from Table 1.1, the second moment of area, I, is given, in both cases, by

$$I = \frac{wt^3}{12}$$

From 'Bending moments from loads' in Chapter 1, you know that for three-point and four-point bending the maximum bending moment, M, is respectively

$$M = \frac{FL}{4} \quad \text{and} \quad M = \frac{F(L-s)}{4}$$

Thus for three-point and four-point bending

$$\sigma_{max} = \frac{1.5FL}{wt^2} \quad \text{(three-point)}$$

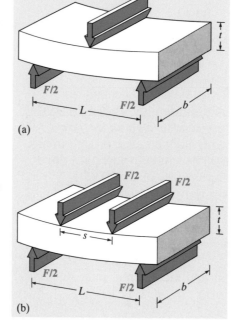

(a)

(b)

Figure 8.10 Schematic of (a) three-point and (b) four-point bending tests

and

$$\sigma_{max} = \frac{1.5F(L-s)}{wt^2} \quad \text{(four-point)}$$

Notice that the bending moment and, hence, the maximum tensile stress are constant between the two inner knife edges in four-point bending, while in three-point bending, the maximum tensile stress is located opposite the central knife edge at the middle of the span.

The interpretation of the results of flexural tests requires some care. Firstly, stress–strain data are not as unambiguous as in tensile tests, because of the gradient of stress and strain through the specimen, from compressive on one surface to tensile on the other. Unless the material is not only linearly elastic, but also has the same modulus in tension as in compression, you cannot obtain a meaningful stress–strain graph from a flexural test. Secondly, the apparent tensile strength values from a flexural test tend to be higher than those from a tensile test.

Because of this, rather than tensile strength, the strength values in bending are referred to as **flexural strength**. (They may also be referred to as **cross-breaking strength** or **modulus of rupture**.)

where σ_t is the tensile strength of a material containing crack-like flaws of critical length a_c and whose toughness is G_c. With reducing notch length, the strength rose, following the curve of the Griffith equation, until it reached a constant value of about $11\,\mathrm{MN\,m}^{-2}$ at $a_c \sim 1\,\mathrm{mm}$. This means that notch lengths less than 1 mm were *smaller* than the largest natural flaws in the material, so it is the latter which limit the tensile strength.

EXERCISE 8.8   If the cement used in the above tests had a Young's modulus, $E = 20\,\mathrm{GN\,m}^{-2}$, estimate its toughness, G_c.

Unreacted and unbound water is therefore at least partially responsible for the strength-impairing flaws in cement. What other effects does water have? Here you have to distinguish between the effect of the water content of the original mix — expressed as the **water–cement ratio** (in relative mass terms) — and the water content of the environment of the hardened cement.

You saw earlier that a totally hydrated cement required a minimum water–cement ratio of about 40%. Below this value the cement contains unreacted grains, while above it, the excess water helps to generate the larger capillary pores. The effect of the water–cement ratio on compressive strength is shown in Figure 8.11, while Figure 8.12 shows compressive stress–strain curves for cements with three different water–cement ratios at different hardening times. What emerges from these graphs is:

(a) Both the compressive strength and the stiffness decrease with increasing water–cement ratio.

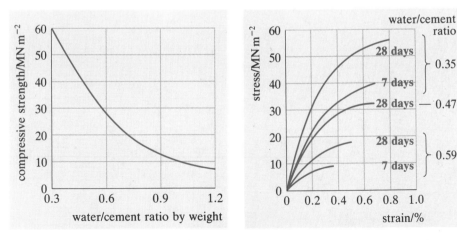

Figure 8.11 Compressive strength plotted against water-cement ratio

Figure 8.12 Compressive stress–strain curves for cement

367

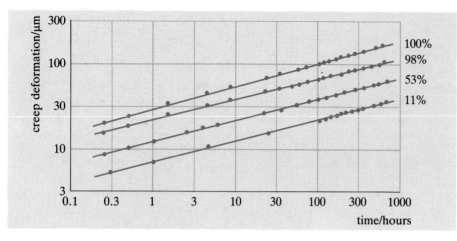

Figure 8.13 Compressive creep behaviour of cement loaded to half its failure load at different levels of relative humidity

(b) The stiffness increases with increasing hardening time.

(c) The stress–strain curves are nonlinear so that, as with polymers, there is no unique value of Young's modulus. Rather, a secant modulus or a tangent modulus should be used (see 'Creep in plastics' in Chapter 5).

There is a further similarity with polymeric materials — cement creeps under constant load and at least a part of the creep strain is recoverable on unloading. Cement therefore shows partially viscoelastic behaviour. An example of this is given in Figure 8.13, which shows the compressive creep in hardened cement at different levels of relative humidity. This demonstrates that environmental water plays a role in the creep of cement and it is thought to be due to movement of water in the pores under the action of the load.

Finally, there is the question: What can be done to overcome the deleterious effects of water and to modify the other properties of cement? This is considered in ▼Modifying the properties of cement▲

SAQ 8.4 (Objective 8.4)
Summarize the need for water in cement and its effect on the mechanical properties of hydrated cement.

▼Modifying the properties of cement▲

There are three main areas — cost, setting and hardening times and end use — that may be affected by modifying the properties of cement.

Low-cost, partial substitutes for cement are silicaceous industrial by-products that are pozzolanic: that is, they react with water in the presence of lime to harden in a similar way to cement. They include blast furnace slag, pulverized fly ash and rice-husk ash.

Additives to cement to modify properties are known as **admixtures**. In terms of setting and hardening, you have seen that gypsum controls the rate of hydration: calcium chloride does this as well, although chlorine ions can have a deleterious effect on the steel in reinforced concrete (see Section 8.4.4). Other admixtures are used to control such variables as the ability to work cement at low water–cement ratios and the viscosity of the mix.

End-use properties are affected by the distribution and size of air bubbles and by the permeability to water of the cement. Further admixtures are used to vary these properties. One such admixture is condensed silica fume, a by-product of the manufacture of silicon and silicon iron. This has very fine particles (average size 0.1–0.2 µm), which can effectively reduce the mean pore size in cement, thus increasing its strength and decreasing its permeability.

An important innovation has been the development of a process for eliminating the larger pores in cement, which limit its tensile strength. By subjecting a mixture of cement and about 5% of a polymer such as poly(vinyl alcohol) to a dough-type mixing, followed by compression at a pressure of about $5\,\mathrm{MN\,m^{-2}}$ to eliminate air, a cement is produced with a maximum pore size of about 15 µm and a flexural strength of about $60\,\mathrm{MN\,m^{-2}}$. This type of material was christened **macrodefect-free (MDF) cement**.

Another area of growing interest is the reinforcement of cement and of concrete with fibres to enhance tensile strength. These are usually discontinuous and include polymeric, inorganic and metallic materials.

8.3 Concrete

8.3.1 The nature of concrete

An optical micrograph of a typical piece of concrete is shown in Figure 8.14. In making concrete you are producing a composite and, as with all composites, its properties are dependent on the properties of the ingredients — their shape, size and the nature of the bonding between matrix and particle. The cement paste is the continuous phase that embeds particles of sand and aggregate. All the constituents can be regarded as brittle in a similar way to composite ceramics, such as porcelain and many glass-fibre reinforced plastics, especially those with thermosetting matrices.

Composites are produced so as to improve the behaviour of their ingredients: for example, in porcelain, the growth of mullite as fine needle-like crystals improves the toughness and, in GFRP, the resin matrix permits the high strength of the glass fibres to be exploited. In concrete you can look at it in two ways: either the sand and gravel serve to toughen the cement paste by introducing many weak interfaces, or the cement paste provides a means of binding together the low cost sand and gravel into a useful material. The truth lies somewhere between the two. Hydrated cement alone is not a useful building material, neither are raw sand and gravel; in combination they are.

Cement is expensive compared to the other ingredients of concrete, largely because of the energy required for kilning and grinding — typically about a tonne of coal or its equivalent for every five tonnes of cement. Despite this, cement and concrete, especially the latter, compare favourably with other materials in both money and energy costs — see ▼Weighing the costs▲. It still remains, however, an objective of specifiers of concrete mixes to produce a desired strength at lowest cost and this means minimizing the content of the most expensive ingredient, the cement. Because of its low tensile strength, concrete on its own can only be used in applications where there is little or no tensile loading. Its chief use is in foundations and it has been estimated that some 40% of all concrete goes into these, with a further 30% into slabs and paving for roads, runways and footpaths. In earlier times, it was used as a stone substitute in structures such as the dome of the Pantheon in Rome with its span of over 40 m. Before the use of reinforced concrete became widespread, it was used to build arched structures such as bridges, a fine example of which is shown in Figure 8.15.

8.3.2 The aggregate

The aggregate consists of a mixture of sand, which has a mean particle size of less than 2 mm, and crushed rock or gravel, in which the mean particle size is greater than 2 mm. If the concrete is to act as more than a loose pile of sand and gravel, the mix must contain sufficient cement paste to coat all the aggregate particles and fill the spaces between them.

0   10 mm

Figure 8.14 Optical micrograph of concrete

Figure 8.15 The twenty-one arch concrete Glenfinnan Viaduct, opened in 1901, on the Mallaig extension of the West Highland Railway

▼Weighing the costs▲

It is instructive to compare cement and concrete with the materials considered in the design of a bus shelter in Chapter 1. Here, as well as money costs, we are going to look at energy costs, which are becoming increasingly important. Table 8.3 sets out the relevant data, with the addition of the column headed Q, which gives the energy required to produce unit volume of the material.

What shows up immediately is how low both the cost per unit mass and the energy per unit volume are for cement and concrete compared with the other materials — reinforced concrete is only marginally more expensive and energy consuming.

> EXERCISE 8.9   Which material has the lowest cost per unit of production energy?

Of more interest perhaps, is to establish what it costs in money or energy terms to buy unit stiffness or unit strength. Here you have to decide the conditions under

> SAQ 8.5   (Objective 8.5)
> Suppose an application requires a short column of length L to carry a given compressive load, F. Which material in Table 8.3 offers:
>
> (a) the column with the lowest cost;
>
> (b) the column that requires the smallest energy to produce the material;
>
> (c) the column with the lowest cost per unit of stiffness under load?

which the loads are to be carried and these will depend on the application. Mass, for instance, is very important in aerospace applications, but not quite so critical in much of civil engineering.

In fact, because of the combination of low money cost and low energy requirement, concrete shows up very well in all such comparisons, despite its low value of tensile strength and its lowish value of Young's modulus. These, coupled with its mouldability, explain why it has become such an important construction material.

Table 8.3   Cost, property and energy data

Material	$C/£\,\mathrm{kg}^{-1}$	$\rho/\mathrm{kg\,m}^{-3}$	$E/\mathrm{GN\,m}^{-2}$	$\sigma_p/\mathrm{MN\,m}^{-2}$	$Q/\mathrm{GJ\,m}^{-3}$
cement	0.05	2400	30	16*	10
concrete	0.025	2400	30	26*	3.4
aluminium	1.3	2800	71	200	360
mild steel	0.3	7860	210	350	300
window glass	0.9	2500	71	50	50
UPVC	1.0	1450	3.0	50	70
GFRP (SMC)	1.7	1500	13	180	150

* Values in compression

The amount of cement paste needed can be reduced by using what's known as a **graded aggregate**. The usual ratio of aggregate to cement is about five volumes of graded aggregate to one volume of cement. The aggregate contains particles of a range of sizes. The small particles fit in the spaces left between the large particles and the cement is required only to flow into the remaining spaces (Figure 8.16). Typically the proportion of sand to gravel is about 60:40.

The sizes and size distributions of the aggregate particles are also determined by the surface finish required. A smooth finish needs a high proportion of fine aggregate. Most aggregates are obtained from

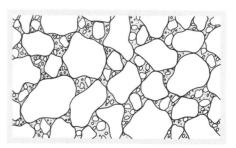

Figure 8.16 Schematic of graded aggregate

naturally occurring sand and stone deposits. The type of stone from which they are derived varies, as does the size and shape of the particles. Shape is particularly important in determining the strength and the ease with which the concrete mix can be worked. Angular stone particles, such as might be obtained from crushing quarried stone, produce a concrete of high strength, but one that is difficult to work with; the interlocking of the angular particles restricts the flow of the mix. River gravels comprise smoothly rounded stones which flow easily over each other, but produce a lower-strength concrete.

The requirement that the cement paste should both coat the aggregate and fill the spaces between the aggregate particles means that for a given aggregate there is a certain ratio of cement to aggregate that provides the optimum mix. Less cement and there is a certainty of voids, more cement than the optimum and the mix is overexpensive with no advantage to properties. This also knocks-on to the water–cement ratio and, hence, the workability of the mix. The practicalities of casting figure largely in designing and selecting a mix.

A pure cement paste made with a water–cement ratio of 0.4 will not flow easily. Hence a concrete mix containing cement with this ratio will be unworkable — it just can't be poured. In the same way as cement, concrete with a high water–cement ratio will yield a lower strength than concrete made from similar aggregate, but with a lower ratio. The extra water required to make the mix workable over and above that required to hydrate the cement increases the porosity of the concrete. It is possible that quite large holes can occur if the concrete is not properly compacted. One way round this is to use flow-enhancing admixtures (plasticizers), as mentioned in 'Modifying the properties of cement', and to keep the water–cement ratio low. Similar problems occur in selecting aggregate sizes. If thin sections are required, then large aggregate particles will hinder easy flow and can cause large voids.

Concrete is specified in terms of its compressive or crushing strength and this is discussed in ▼Cube strength▲.

▼Cube strength▲

In the United Kingdom, the standard quality control test for the strength of concrete involves measuring the load at which a 100 mm cube of the material is crushed. If a series of specimens is cast at the time the concrete is poured, they can be used to monitor the hardening with time of concrete in a structure.

The stress produced by dividing the crushing load by the cube's cross-sectional area is known as the **cube strength**. It is

not, however, an intrinsic material property and the in situ strength of cast concrete is typically only about 65% of its cube strength.

Concrete grades are specified in terms of their twenty-eight-day cube strength, measured in $MN\,m^{-2}$ (remember that this, too, is dependent on the water–cement ratio, see Figure 8.11). Grade 20, corresponding to a cube strength of $20\,MN\,m^{-2}$, is just a space-filling grade

and grades 30 and 40 tend to be the popular ones. Grades with strengths higher than $40\,MN\,m^{-2}$ are expensive and are normally only used in applications which justify the extra cost.

EXERCISE 8.10   What is the cube strength and grade of a concrete whose compressive failure load was 30.6 tonnes in a standard UK test?

8.3.3 Concrete as a composite

Concrete, with its matrix of cement binding the aggregate, has just the structure of a particulate composite. Thus its elastic properties should be described by the series model of composites (Chapter 7, Section 7.3.2). As it turns out, this is indeed so, as shown in Figure 8.17. Table 8.4 gives the density and Young's modulus for the components of concrete. Young's modulus is given because concrete behaves much more linearly than cement and Young's modulus is therefore appropriate.

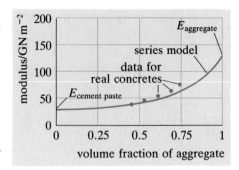

Figure 8.17 Application of series composite model to the elastic modulus of concrete

Table 8.4 Properties of cement and aggregate

Material	Density/kg m^{-3}	Young's modulus/GN m^{-2}
cement	2500	30
sand	2650	90
granite	2600	70
limestone	2700	74
sandstone	2600	130

The values for cement are those for a hypothetical pore-free material. If the porosity of a material is p, then its density ρ is related to p by,

$$p = \frac{\rho_0 - \rho}{\rho_0}$$

where ρ_0 is the pore-free density (see Exercise 7.4 in Chapter 7).

Empirically, the elastic modulus of cement of porosity p is

$$E = E_0(1 - p)^3$$

as noted in 'Stiffness of foams', Chapter 7.

SAQ 8.6 (Objective 8.6)
A concrete is made up of one part cement, two parts crushed limestone and three parts sand. If the porosity of the cement when set is 0.2, estimate the Young's modulus of the concrete.

While analyses of this type are useful for estimating Young's modulus, estimates of useful strength depend on the point at which the material ceases to behave elastically. This might occur because of the failure of one component by yielding or brittle failure, or by the breakdown of the interface between the components. The correlation between particle shape and strength perhaps gives some indication of the nature of the bonding between the cement paste and the aggregate. Mechanical interlocking must play a significant role in determining the strength of concrete, since the chemical bonding between aggregate and hydrated cement is weak relative to the bonding within the particles. The aggregate particles are themselves silicate and carbonate materials — the very type of mineral that forms the cement itself.

Figure 8.18 A steel-reinforcement assembly

Bonding between the cement paste and the aggregate will be largely via secondary bonding with the polar water molecules. As the interface is weak, it is found that the mechanical failure of concrete proceeds by cracking of the aggregate–cement interface. Concrete can be grouped with GFRP and timber as a composite that is toughened by the exploitation of weak interfaces between components. The presence of two brittle components of similar modulus and weak interfaces means that concrete can only be used safely in conditions where it is subjected to compressive stress. The successful use of GFRP in tension relies on the large difference in failure strain between the resin matrix and glass fibres. When a glass fibre breaks at a low strain, the resin retains its integrity and permits stress transfer to other fibres. In concrete both constituents fail at similar strains, so once the critical strain is reached in tension, a crack will propagate rapidly.

8.4 Reinforced concrete

8.4.1 How reinforcement works

The idea behind reinforced concrete is to offset the low tensile strength of concrete by combining the tensile strength of steel with the compressive strength of the concrete to give a durable, inexpensive structural material capable of withstanding bending loads. The steel used for the main reinforcing bars is a heavily cold-worked, mild steel. The complex assemblages of reinforcing steel are a common sight on building sites (Figure 8.18). They consist of the main reinforcing bars joined by links whose function is to hold the bars in place while the concrete sets. The links are made from a soft mild steel, with a yield stress of $250\,\mathrm{MN\,m^{-2}}$, which is easily bent to shape. The reason that the shapes are frequently complex is that the bars have to resist stresses other than the main tensile ones due to bending (see ▼Shear stresses in beams▲) and complex structures might have complicated loading patterns.

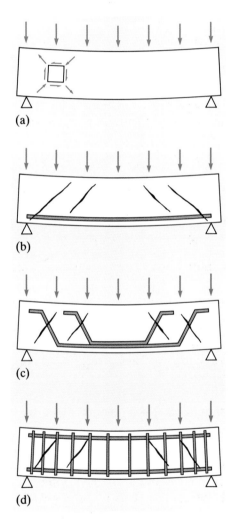

(a)

(b)

(c)

(d)

Figure 8.19 (a) Combined shear and bending stresses in beams. (b) Resulting cracks and (c) the placement of reinforcing bars to counteract this. (d) An alternative arrangement with vertical stirrups

▼Shear stresses in beams▲

When considering the bending of a beam, it was assumed that you only needed to worry about the longitudinal compressive and tensile stresses. However, as the depth of the beam in relation to its span increases, this assumption becomes invalid. Shear forces become significant when the ratio of the span of the beam to its depth is less than about ten and are also significant around the supports of longer beams.

The interaction between the shear stresses and the bending stresses produces resultant tensile forces that act diagonally (Figure 8.19a). These can lead to early cracking, as shown in Figure 8.19(b). To resist these stresses, the reinforcement bars need to be placed not as shown in Figure 8.19(b), but rather at right-angles to the anticipated cracks (Figure 8.19c). An alternative arrangement, using vertical steel stirrups, is shown in Figure 8.19(d).

Table 8.5   Design data for reinforced concrete

	Strength/MN m^{-2}	Design stress/MN m^{-2}	Young's modulus/GN m^{-2}	Strain to failure/%
concrete grade 40	26 (compression) 2.6 (tension)	16 (compression)	15 (long term)	0.35 (compression) 0.02 (tension)
cold-worked mild steel	425 (0.2% tensile proof stress)	370	210	0.2

Now, if you treat the reinforced concrete as a unidirectional fibre composite, loaded in the direction of the fibres, the appropriate model to apply would be the parallel, rule of mixtures model (Chapter 7). The starting assumption of this is that the strain in the concrete, ε_c, is the same as that in the steel, ε_s. If $\varepsilon_c = \varepsilon_s$, then, for linear elastic behaviour,

$$\frac{\sigma_c}{E_c} = \frac{\sigma_s}{E_s}$$

or

$$\sigma_s = \frac{\sigma_c E_s}{E_c}$$

Table 8.5 contains typical design data for the two materials and, using the values of Young's modulus given,

$$\sigma_s = 14\sigma_c$$

Thus, when $\sigma_c = 2.6\,\mathrm{MN\,m^{-2}}$, the tensile strength of concrete, $\sigma_s = 36\,\mathrm{MN\,m^{-2}}$. This is only about 8% of its proof stress and is a long way from taking full advantage of the high strength of steel. The only way round this is *to let the concrete crack* so that the steel carries *all* the tensile forces. This is the basis of the design of reinforced-concrete beams and obviously invalidates a rule of mixtures description, so a different approach is required.

First look at how a reinforced-concrete beam of straightforward geometry, such as that in Figure 8.20, behaves under bending load. Note that since you want to benefit from its tensile strength, the steel should clearly be incorporated on the tensile side of the beam.

As the loading on the beam is increased from zero up to failure, you can identify four different regions of behaviour. These are labelled I–IV on the load–deflection curve in Figure 8.21(a). In Region I, both the

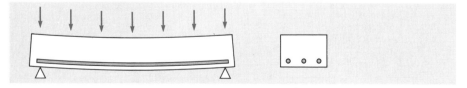

Figure 8.20 A simple rectangular beam in bending

steel and concrete behave elastically, the concrete in compression above the neutral axis and the steel and concrete in tension below it. This is reflected in the linear stress distribution through the thickness of the beam (Figure 8.21b), with a corresponding linear strain distribution.

Region I ends when the concrete on the tensile surface starts to fail at its fracture strain — for the grade 40 concrete in Table 8.5 this is about 2×10^{-4} (0.02%). In Region II, the deformation of the beam is still elastic, but, with the cracked region increasingly spreading up to the

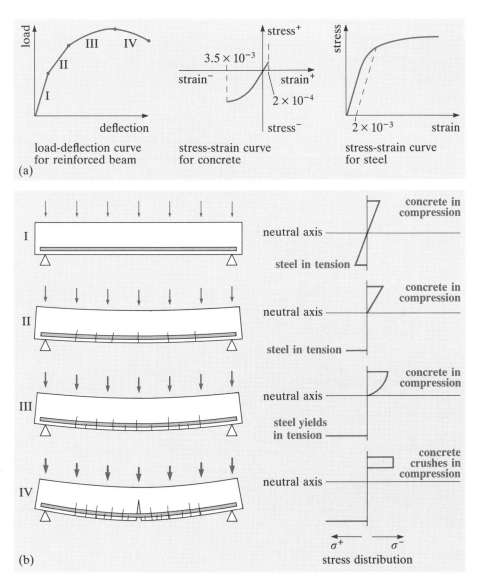

Figure 8.21 (a) Load–deflection and stress–strain characteristics of reinforced concrete, concrete and steel. (b) Loading to failure sequence and stress distributions for the beam

neutral axis, the principal tensile resistance is provided by the steel; the bending stiffness of the beam is lower.

The onset of yielding in the steel signals the start of Region III. The slope of the load–deflection curve decreases further and becomes nonlinear, accentuated by the increasingly nonlinear compressive deformation of the concrete. The collapse of the beam in Region IV, beyond the point of maximum load, is precipitated by failure of the concrete in compression at the upper surface of the beam.

EXERCISE 8.11 Describe the load–deformation behaviour if the beam were unloaded from each of Regions I–III and then reloaded.

Thus, to make the most efficient use of the two materials, a beam should be designed so that the steel starts to yield before the concrete fails in compression. This is also a safe strategy since tensile yielding of steel is progressive and the beam remains load-bearing, while the crushing of concrete is catastrophic.

8.4.2 A design example

We shall now take as an example a beam of a size that you might well see in a building such as a multistorey car park. It is 350 mm wide, 500 mm deep and has steel reinforcing bars of 32 mm diameter located 50 mm beneath its tensile surface. We shall assume that the concrete is grade 40 and that the design data of Table 8.5 apply. We want to know the maximum bending moment that the beam can sustain and how many reinforcing bars are required.

The design strategy considers the beam to be loaded to Region III of the load–deflection curve (Figure 8.21a), but to less than the maximum load. Thus, on the tensile side, the concrete has cracked and the steel has started to yield, while the concrete in compression has started to behave nonlinearly. The design loads usually include a factor of safety of, say, 1.5 times the intended load. The stress distribution for Region III is shown in Figure 8.21(b), and we shall assume that the concrete under compression has reached the maximum design stress for grade 40, which is given in Table 8.5 as $16 \, \mathrm{MN \, m^{-2}}$. Although this stress distribution is nonlinear, for mathematical convenience we shall assume that it is equivalent to a constant compressive stress of $14 \, \mathrm{MN \, m^{-2}}$ acting in the whole compressive region of the beam (i.e. above the neutral axis).

The combination of the distributed compressive stress in the concrete and the tensile stress in the steel results in an internal moment that is in equilibrium with the moment applied by the external loads on the beam and the self-weight of the beam.

The compressive stresses above the neutral axis of the beam can be replaced by a single compressive force, C, acting at the centre of the compressive zone. This together with the tensile force in the steel, T, form a couple which gives the maximum internal moment which the beam can be designed to resist (Figure 8.22a). However, since you no longer have a homogeneous beam, you can no longer assume that the neutral axis lies in the centre of the beam's cross-section. Therefore you need to check the location of the neutral axis. This, together with expressions for the bending moment and cross-sectional area of reinforcing bars, are derived in ▼Design calculations▲

Using the constant stress, $\sigma_c = 14\,\text{MN m}^{-2}$, $b = 0.35\,\text{m}$ and $d = 0.45\,\text{m}$, the maximum bending moment is

$$M_{\text{max}} = 0.44\sigma_c bd^2$$
$$= 0.437\,\text{MN m}$$

and, using $\sigma_s = 370\,\text{MN m}^{-2}$, the area of steel reinforcement is

$$A_s = \frac{0.64\sigma_c bd}{\sigma_s}$$
$$= 3814\,\text{mm}^2$$

A 32 mm bar has a cross section of 804 mm², so the A_s value would require five such bars.

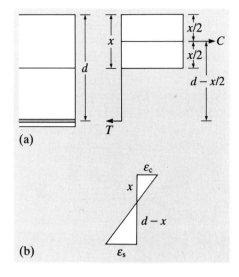

Figure 8.22 (a) Internal moment of the beam; (b) assumed strain distribution in the beam

▼Design calculations▲

1 Neutral axis Let the neutral axis lie x beneath the top surface of the beam and the steel d (Figure 8.22a). The simplest way of approximately locating the neutral axis is to consider an equivalent solid material with a compressive strain, ε_c, in its upper surface and a tensile strain, ε_s, in its lower one. From Equation (1.5)

$$\varepsilon = \frac{y}{R}$$

which shows that the strain varies linearly with the distance from the neutral axis, y. Thus you have the linear, but not symmetrical, strain distribution shown in Figure 8.22(b).

Then, by similar triangles

$$\frac{\varepsilon_c}{x} = \frac{\varepsilon_s}{(d - x)}$$

or

$$\frac{x}{d} = \frac{\varepsilon_c}{\varepsilon_s + \varepsilon_c}$$

If you now assume that the amount of steel is such that it just starts to yield as the concrete starts to fail in compression, then, using the failure strains given in Table 8.5,

$$x = 0.64d$$

2 Cross-sectional area of steel A_s If you now consider the compressive forces in the concrete to be replaced by C, then horizontal equilibrium requires that

$$C = T$$

or, if the breadth of the beam is b,

$$\sigma_c bx = \sigma_s A_s$$

Thus

$$A_s = \frac{\sigma_c bx}{\sigma_s}$$

which shows how the location of the neutral axis depends, among other things, on A_s. Substituting for x

$$A_s = \frac{0.64\sigma_c bd}{\sigma_s}$$

3 Bending moment M Taking moments about the line of action of T gives

$$M = C(d - x/2)$$
$$= \sigma_c bx(d - x/2)$$
$$= 0.44\sigma_c bd^2$$

SAQ 8.7   (Objective 8.7)

Explain the principles on which the design of reinforced-concrete beams is based.

If the beam just considered spanned 5 m between its fixing points, estimate the maximum distributed load it can safely carry per unit length. Use a safety factor of 1.5 and take $\rho_c = 2500\,\text{kg m}^{-3}$ and $\rho_s = 7860\,\text{kg m}^{-3}$; maximum bending moment in a beam with fixed (or 'built-in') ends of length L carrying a load w per unit length is

$$M_{\text{max}} = \frac{wL^2}{12}$$

8.4.3 Prestressed concrete

One of the problems with reinforced-concrete members is that the concrete on the tension side, which is cracked and therefore doing no work, represents a considerable penalty in weight. In addition, vibration from traffic, for example, can cause repeated opening and closing of the cracks, leading to a gradual deterioration and crumbling of the concrete. Also, the cracks expose the reinforcing rods to possible corrosion. These problems can be overcome if the concrete can be subjected to an overall compression, so that none of it is in tension when the structure is loaded. This is known as **prestressing**.

Two techniques of prestressing are used, **pretensioning** and **post-tensioning**. In both cases, high-tensile steel wires with a tensile strength of 1500 to 1800 MN m^{-2} are used to apply the compression to the concrete.

Pretensioning is used for the production of standard items, such as beams and floor slabs, in a factory. In this method the wires are pretensioned to about 1200 MN m^{-2} by hydraulic jacks and the concrete is cast around the wires. When the concrete has hardened sufficiently, the ends of the wire are cut. The elastically stretched wire is constrained from reverting to its unstressed length by its bond with the concrete. The contracting wire compresses the concrete and this inhibits cracking.

The stress distribution in a pretensioned beam under load is depicted in Figure 8.23. When the pretensioned wire shown in Figure 8.23(a) is released, the beam is compressed. An indication of the resulting distribution of the compressive stress through the depth of the beam is sketched in Figure 8.23(b); it is greater at the bottom of the beam than at the top. Note the convention of indicating a compressive stress on the right of the vertical line. Suppose an external load is imposed that produced the normal stress distribution of Figure 8.23(c). The beam will remain in compression at all points on its depth (see Figure 8.23d),

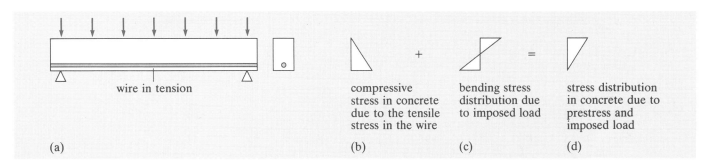

wire in tension

compressive
stress in concrete
due to the tensile
stress in the wire

bending stress
distribution due
to imposed load

stress distribution
in concrete due to
prestress and
imposed load

(a)   (b)   (c)   (d)

Figure 8.23 Stress distribution in a pretensioned beam under load

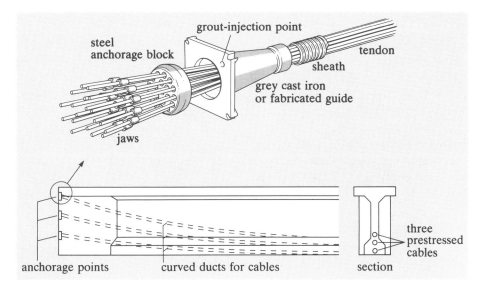

steel
anchorage block

grout-injection point

tendon

sheath

grey cast iron
or fabricated guide

jaws

anchorage points

curved ducts for cables

section

three
prestressed
cables

Figure 8.24 Post-tensioning

provided that the tensile stress due to the imposed load (the component to the left of the vertical line in Figure 8.23(c) does not exceed the compressive stress created by the pretensioning.

Post-tensioning is used for on-site construction of large sections. In this technique ducts are left when the structure is cast and wires are threaded loosely in the ducts. When the concrete has set, the wires are tensioned with hydraulic jacks acting on the end faces of the structure and the wires are wedged into the ends of the ducts (see Figure 8.24). Liquid grout is injected under pressure to fill the ducts and this, together with the wedges, transfers the tensile stress in the wires to the concrete, thus putting it safely into compression. Post-tensioning can be used to assemble precast units and in this way a long beam can be manufactured on site from a number of short segments manufactured in a factory. This technique is used extensively in building structures such as bridges and high rise buildings.

The Redheugh Bridge (Figure 8.1) is a post-tensioned structure, but it also incorporates another way of overcoming the inherent disadvantages of concrete by increasing the compressive stress on it. This is described in ▼Using stress concentrations▲

EXERCISE 8.13   The prestressing tendons used in the Redheugh Bridge each consisted of thirty-one steel wire strands of 12 mm diameter and were post-tensioned to a load of 400 tonnes. Estimate the stress in each strand.

▼Using stress concentrations▲

Although the presence of stress concentrations is frequently considered to be harmful in a stressed material, the harm usually comes from failing to take them into account properly. In fact, there are many instances of stress concentrations being used positively.

EXERCISE 8.12   Can you think of examples of things made from:
(a) paper, (b) plastic, (c) glass, (d) metal, which make deliberate use of stress concentrations?

One elegant example employed in concrete structures is the **Freyssinet hinge**. This is used, for example, in bridges, to allow some degree of rotation in a vertical plane of the superstructure relative to its supporting piers. This is particularly important during construction if the bridge is built by cantilevering the superstructure out from the piers.

The hinge works by using the stress-concentrating effect of notches cast into a pier to magnify the compressive stress on the pier due to the mass of the superstructure. Figure 8.25(a) shows the layout of a typical hinge, with reinforcing bars in the pier above and below the throat of the hinge. The resulting compressive stress distribution is shown in Figure 8.25(b), with maxima at either end of the throat. If rotation is applied to the hinge, a bending stress distribution such as that in Figure 8.25(c) will be generated. Adding this to the stresses due to the supported mass gives the net distribution shown in Figure 8.25(d). The important point about this is that the concrete in the hinge remains in compression despite the tensile component introduced by the rotation. Thus the incorporation of the stress-concentrating hinge permits a greater flexibility in the concrete than would otherwise be possible. Redheugh Bridge was built by cantilevering and has Freyssinet hinges at the tops of its piers.

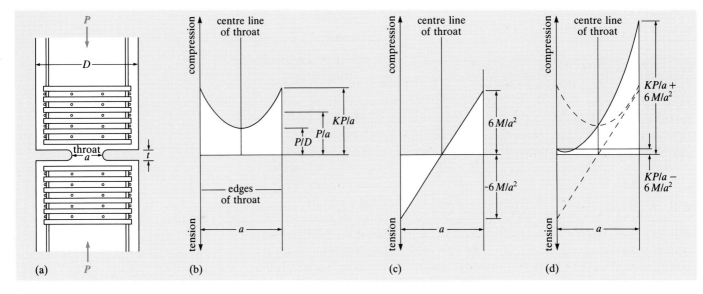

Figure 8.25 The Freyssinet hinge

Post-tensioning has the advantage over pretensioning in that the ducts can follow a curve and can thus be placed to counteract the line of maximum tension induced by both bending and shear in the loaded component. The wires have to run straight in a pretensioned component. In both cases, allowance has to be made when designing a prestressed component for the creep of the concrete, which partially relaxes the prestress, although creep in concrete is only about one-fifth of that shown for cement in Figure 8.13.

Compared with unstressed reinforced concrete, the use of prestressed concrete in, say, a bridge only requires about half the mass of concrete and a third the mass of steel, since the entire cross-section contributes to the bending stiffness — there is not much difference in their ultimate strengths. Thus, the potential economies are very large. Coupled with this are the improved aesthetic possibilities due to the more slender structural members.

One disadvantage that both unstressed and prestressed reinforced concrete share is the weight penalty of having to provide at least 50 mm of concrete between the steel and the outside world to protect the steel from corrosion. Thus there is considerable interest in developing ways of externally prestressing concrete, using high strength materials that do not need protecting from the environment in the same way.

EXERCISE 8.14   What materials are possible candidates for use in external prestressing tendons?

Of these materials, the two being worked on most actively are glass fibres encased in a polyester matrix (a bridge has been built in Düsseldorf incorporating these) and the aramid fibres (see Chapter 6), tendons of which require an outer sheath of, for example, polyethylene to protect the fibres from ultraviolet radiation.

SAQ 8.8   (Objective 8.8)
Describe briefly the distinction between pretensioning and post-tensioning techniques for reinforcing concrete. Why is steel with a yield stress of $1500\,\mathrm{MN\,m^{-2}}$ used as the reinforcement in prestressed concrete in preference to steel with a yield stress of $500\,\mathrm{MN\,m^{-2}}$, which is used for conventional reinforcement (non-prestressed)?

8.4.4 Degradation of reinforced concrete
'Rusting' in Chapter 3 described how iron and steel corrode easily in aqueous oxidizing environments. The porous structure of concrete means that water is available and that oxygen can readily diffuse into it.

Why doesn't the steel reinforcement corrode in the usual way?

The reason that it does not corrode arises from the high alkalinity of the concrete surrounding it. This high pH provides an environment in which iron has a natural tendency to passivate.

EXERCISE 8.15   What do you understand by the term 'passivate'?

Another way of looking at this is that, in contrast to the rusting process, where a mixture of iron oxides is produced, allowing the corrosion process to continue, the high alkalinity of concrete encourages iron to oxidize directly to its highest oxidation state, Fe_2O_3, without producing any of the intermediate oxides. Under these conditions the Fe_2O_3 forms a stable impermeable film on the steel surface, thus protecting the steel from further corrosion.

This means that any process which either significantly reduces the pH value in the vicinity of the steel, or which increases the permeability of the oxide film, will render the steel open to further corrosion. ▼Chloride attack▲ is a damaging example of the latter while ▼Carbonation▲ examines a prevalent example of the former.

EXERCISE 8.16   What are the likely effects of corrosion of the steel reinforcement in concrete?

▼A boring story!▲ speaks for itself.

SAQ 8.9   (Objective 8.9)
Describe the two classes of chemical attack on concrete that lead to corrosion of the steel reinforcement and one mechanism of physical degradation.

▼Chloride attack▲

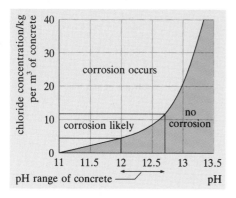

Figure 8.26 The concentration of chloride ions in concrete plotted against pH

Chloride ions react with the oxide film on steel, being incorporated into the oxide lattice and increasing the electrical conductivity of the film. This encourages corrosion. In effect, increasing $[Cl^-]$ shifts the corrosion–passivity boundary to higher values of pH, as Figure 8.26 shows. If this is operating in parallel with a process such as carbonation which is reducing the pH of the water in concrete, the concentration of chloride ions necessary for corrosive attack reduces.

Chloride ions are present, of course, in sea water, and in the salt used for deicing roads in winter. Also, if the aggregate contains gravel of marine or of desert origin, insufficient washing can lead to another source of chloride ion contamination of concrete. Although chloride was added to concrete to speed up the hydration rate, its use is no longer recommended in reinforced and prestressed concrete because of these corrosion problems.

▼Carbonation▲

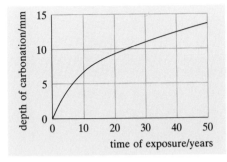

Figure 8.27 The depth of carbonated layer in concrete plotted against time

Carbon dioxide will dissolve in water until the solution reaches equilibrium at a pH value of about 3, the reaction being

$$CO_2 + H_2O \rightleftharpoons H_2CO_3 \rightleftharpoons 2H^+ + CO_3^{2-}$$

In concrete, atmospheric carbon dioxide can similarly dissolve in the water in the pores and the gel, but the situation is more complex because of the other ions in solution. Initially the carbonate ions will react with the Ca^{2+} to give calcium carbonate, which precipitates from solution thus

$$Ca^{2+} + CO_3^{2-} \rightleftharpoons CaCO_3 \text{ (solid)}$$

The hydrogen ions tend to combine with hydroxyl ions

$$H^+ + OH^- \rightleftharpoons H_2O$$

reducing the equilibrium pH to below 12.6. This is balanced by more portlandite — $Ca(OH)_2$ — going into solution so that the pH is maintained (see 'Mechanism of hydration' in Section 8.2.2). In comparison to the $Ca(OH)_2$ content, the supply of atmospheric carbon dioxide is virtually infinite, so a point will be reached where the $Ca(OH)_2$ is exhausted and the pH locally will begin to fall.

This carbonation is progressive from the surface of the concrete, its rate being governed by the rate at which carbon dioxide diffuses into the concrete (Figure 8.27).

EXERCISE 8.16   Estimate the diffusivity, D, of carbon dioxide in concrete from Figure 8.27 given that the depth of diffusion, x, at a time, t, is given by the equation

$$x = \sqrt{Dt}$$

Corrosion of the steel starts to occur when the pH falls to 8.7. The latest Code of Practice specifies a minimum depth of cover for concrete of 50 mm in exposed environments, which, from the data in Figure 8.27, should allow about 570 years before the steel is in any danger of corroding. Potentially, cracks could radically reduce this time by reducing the diffusion length required. However, it is found that in cracks less than 0.3 mm wide, the $CaCO_3$ precipitate acts as a seal. Further carbonation is slow since it relies on diffusion through it. Since prestressed concrete is not designed to crack, it is considerably more durable than reinforced concrete.

▼A boring story!▲

The article by Malcolm Smith, *Sponge that Eats Mortar*, reproduced below, reveals a rather unexpected mode of concrete degradation, despite the confusion of 'erosion' with 'corrosion'.

Sponge that eats mortar

BORING may put paid to concrete, at least in warm seas. For, in tropical and sub-tropical waters, where vast amounts of concrete are used to build harbour structures, retaining walls, oil installations, and many other shoreline facilities, such structures are more susceptible to the boring of marine animals than was previously thought possible. New research shows that, particularly in polluted coastal waters in the tropics, engineers and builders involved in fabricating submerged concrete structures need to be acutely aware that small boring worms and mussels — even sponges — can simply eat away all their aggregate and cement.

The first thorough study of the problem, by three Canadian geologists from McMaster University at Hamilton, Ontario is the result of a chance experiment on fish colonisation. In 1973 one of them built some artificial underwater reefs out of concrete blocks and limestone boulders in Discovery Bay, Jamaica to see what fish colonised them. In 1986 they were relocated and some of the blocks and limestone removed. To quantify the extent of bore holes, they were cut into small slabs, examined and the animals responsible for the damage identified.

What they found was that a veritable Pandora's box of marine organisms had bored their way into the concrete blocks — and even more into the limestone. Two animals, a boring sponge called *Cliona* and a boring mussel called *Lithophaga,* did the most damage to the concrete. the sponge concentrated its activities around the edges of the blocks while the mussels penetrated into their centres, creating large, ovoid cavities as they went on their way. Between them these two animals removed three per cent of the concrete.

One boring mussel, probably the most destructive — *Lithophaga antillarum* — is common around the world in tropical waters and can grow to lengths of 10 centimetres with a diameter of 2 centimetres. If it's abundant, as it is in Jamaica, its boreholes could quite easily compromise the structural stability of submerged concrete.

But boring into concrete isn't the sole preserve of certain sponges and mussels. Another bivalve, *Gastrochaena* and a burrowing mud shrimp, *Upogebia,* also managed to penetrate the outer edges of the blocks. Some of these creatures bore their way in physically; others secrete a chemical which softens the substrate; yet others use a combination of both methods.

Many of the factors that affect the susceptibility of concrete structures to attack by marine organisms are, as yet, unknown. But the amount of calcareous material in the structure — the amount of lime-based cement in the concrete for instance — is clearly an important one. So, too, it seems, is the amount of organic pollution in the sea offshore — which encourages the animals' growth — a particularly worrying finding because so many harbour and shoreline developments are adjacent to large conurbations which pour out their excrement into the nearest bay or estuary.

But the Jamaican study found less bioerosion than has recently been recorded in other seas. This is probably because the concrete block reefs became severely silted up as the result of storms, inhibiting the growth of most of the borers, and because they were located several hundred metres from the nearest source of the animals, a natural limestone reef in another part of Discovery Bay.

A recent episode in the Persian Gulf shows how serious concrete boring can become. A ten-kilometre length of submerged, very high strength, dense concrete was attacked by *Lithophaga* and *Cliona.* In just four years their burrows penetrated several centimetres and the creatures were living at densities of up to 2,000 per square metre of surface. The fear was they would penetrate in to the reinforcing steel, starting a steady process of dangerous erosion once seawater was in contact with it. Difficult — and expensive — repairs were needed urgently.

So engineers and builders constructing submerged concrete structures in warm, polluted seas should worry far more about tiny sponges and mussels than they ever have before — or boring may very well undo all their handywork.

(Marine Pollution Bulletin, vol. 19(5), p. 219-22)

Objectives for Chapter 8

After studying this chapter you should be able to:

8.1   Describe the phenomena of setting and hardening in cement and account for their origins (SAQ 8.1)

8.2   Show how the alkalinity in cement arises (SAQ 8.2)

8.3   Identify the major mineral constituents in cement and describe their roles (SAQ 8.3)

8.4   Discuss the function of water in cement and the effect it has on mechanical properties (SAQ 8.4)

8.5   Devise and apply simple merit indices to compare cement and concrete with other materials (SAQ 8.5)

8.6   Apply the appropriate composite model to concrete to estimate its elastic response (SAQ 8.6)

8.7   Explain the principles of the design of reinforced-concrete beams and perform simple design calculations (SAQ 8.7)

8.8   Describe and distinguish between the two techniques for prestressing concrete (SAQ 8.8)

8.9   Give an account of the physical and chemical degradation of concrete and reinforced concrete (SAQ 8.9)

8.10   Define and use the following terms:

admixture	gelwater
alite	graded aggregate
aluminate	gypsum
belite	hardening
carbonation	isoelectric point
cement	macrodefect-free cement
coagulation	modulus of rupture
concrete	pH
cross breaking strength	pore
cube strength	Portland cement
dimer	portlandite
ettringite	post-tensioning
ferrite	prestressing
flexural strength	pretensioning
flocculation	quicklime
Freyssinet hinge	reinforced concrete
gel	setting
	water–cement ratio

Answers to exercises

EXERCISE 8.1

specific surface area, $\alpha = \dfrac{\text{area, } A}{\text{mass, } M}$

density, $\rho = \dfrac{\text{mass, } M}{\text{volume, } V}$

For a sphere of radius r

volume $= \frac{4}{3}\pi r^3$

surface area $= 4\pi r^2$

The ratio of volume to surface area will be the same for the sample as a whole as for the individual grains. Thus

$$V = \tfrac{1}{3}rA$$

and

$$\rho = \frac{3M}{rA}$$

$$= \frac{3}{r\alpha}$$

Therefore

$$r = \frac{3}{\rho\alpha}$$

$$= 2.5 \times 10^{-6}\,\text{m}$$

Thus the diameter of the grains is $5\,\mu$m.

EXERCISE 8.2 The sketch below shows three silicate ions joined to form the trimeric silicate anion.

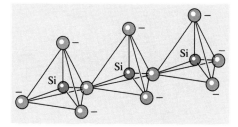

From this you can see that its formula is $Si_3O_{10}^{8-}$.

Since the formula for the dimer is $Si_2O_7^{6-}$ the equation for the reaction to produce the pentamer is

$$Si_3O_{10}^{8-} + Si_2O_7^{6-} \longrightarrow Si_5O_{16}^{12-} + O^{2-}$$

If the general formula is $Si_nO_x^{y-}$, you can now draw up a table based on the anions you know already.

Table 8.6

n	x	y
1	4	4
2	7	6
3	10	8
5	16	12

From this you can see that $x = (3n + 1)$ and $y = (2n + 2)$. Thus the general formula is

$$Si_nO_{(3n+1)}^{(2n+2)-}$$

EXERCISE 8.3 At 1 min, $[OH^-] \approx 2 \times 10^{-2}\,\text{mol l}^{-1}$. Therefore

$$[H^+] = \frac{1 \times 10^{-14}}{2 \times 10^{-2}}\,\text{mol l}^{-1}$$

$$= 5 \times 10^{-13}\,\text{mol l}^{-1}$$

Thus, at 1 min,

$$\text{pH} = \log_{10}\frac{1}{[H^+]}$$

$$= 12.3$$

At 10 min, $[OH^-] \approx 4 \times 10^{-2}\,\text{mol l}^{-1}$. Therefore, at 10 min, pH = 12.6.

At 3 h, $[OH^-] \approx 7.5 \times 10^{-2}\,\text{mol l}^{-1}$ and pH = 12.9.

EXERCISE 8.4 Diffusion occurs from regions of high concentration to those of low concentration, thus its direction is determined by the concentration gradient. Water is depleted within the gel layer as it takes part in the hydration reactions. Hence water from outside will tend to diffuse in. The converse applies to Ca^{2+} and OH^-.

EXERCISE 8.5 The answer is probably a bit of everything *except* metallic bonding. Covalent bonds in the silicates, ionic bonds in the portlandite and between Ca^{2+} in solution and the silicates, van der Waals' bonds between adjacent fibres and fibres and crystals, hydrogen bonding between residual water and the gel, are all likely to be present.

EXERCISE 8.6 There are two reasons. First, the larger the pore, the greater its effect as a stress-raising flaw, reducing the strength of the material in tension. Second, the pores can act as capillaries, providing a means of absorbing water and other chemicals. Water expands on freezing so this could generate stresses sufficient to cause spalling. Other chemicals can degrade the cement.

EXERCISE 8.7 The surface area of a specimen exposed to the maximum tension is smaller in bending than in tension, thus reducing the probability of it encompassing the worst flaw. Also the stress field ahead of a surface crack falls off more rapidly in bending than in tension.

EXERCISE 8.8 From the Griffith equation you have that the toughness is given by

$$G_c = \frac{\sigma_t^2 \pi a_c}{E}$$

From Figure 8.9, when $a_c = 1\,\text{mm}$, $\sigma_t \approx 11\,\text{MN m}^{-2}$ and hence

$$G_c = \frac{(11 \times 10^6)^2 \pi \times 1 \times 10^{-3}}{20 \times 10^9}\,\text{J m}^{-2}$$

$$G_c = 19\,\text{J m}^{-2}$$

EXERCISE 8.9 You want the answer in £ J^{-1}. From the table, C/Q has the units £ $m^3 J^{-1} kg^{-1}$, therefore you need to multiply by the density, ρ, to get the required quantity. Doing this gives the following values (in £ J^{-1}):

cement	1.2×10^{-8}
concrete	1.8×10^{-8}
aluminium	1.0×10^{-8}
mild steel	7.9×10^{-9}
window glass	4.5×10^{-8}
UPVC	2.1×10^{-8}
GFRP (SMC)	1.7×10^{-8}

Mild steel therefore has the lowest value.

385

EXERCISE 8.10 The breaking load of 30.6 tonnes corresponds to $30.6 \times 1000 \times 9.81\,\text{N}$. The compressive stress across the 100 mm cube faces was therefore

$$\frac{30.6 \times 1000 \times 9.81}{0.1 \times 0.1}\,\text{N m}^{-2}$$

$$= 3 \times 10^7\,\text{N m}^{-2}, \text{ or } 30\,\text{MN m}^{-2}$$

This corresponds to a grade 30 concrete.

EXERCISE 8.11

Region I Both components are still behaving elastically, unloading and reloading would follow the original curve.

Region II The beam would show partial elastic recovery on unloading, the cracks in the concrete being closed by the tensile force in steel. Reloading would be parallel to Region II as the cracked beam would be less stiff then before.

Region III The beam would have a permanent deformation on unloading. Reloading would initially be parallel to Region II.

EXERCISE 8.12 Some examples are
(a) Paper: perforations in stamps, toilet paper, etc.
(b) Plastic: break-off tabs on audio and video cassettes, the rubber phase in impact-modified thermoplastics.
(c) Glass: glass-cutting relies on a surface crack produced by a diamond scriber.
(d) Metals: ring-pull tops of drinking cans.

EXERCISE 8.13 The total cross-sectional area of steel in each tendon was

$$A = 31 \times (6 \times 10^{-3})^2 \times \pi\,\text{m}^2$$

Thus the stress was

$$\frac{4 \times 10^5 \times 9.81}{31\pi(6 \times 10^{-3})^2}\,\text{N m}^{-2}$$

$$= 1.12\,\text{GN m}^{-2}$$

EXERCISE 8.14 The materials are those listed in Table 3.4 of Chapter 3 and considered in Sections 6.2.4 and 6.2.5.

EXERCISE 8.15 When a metal passivates the corrosion process is prevented by the presence of a stable oxide film or the application of a potential that opposes the corrosion potential. The metal behaves in a manner more 'noble' than that indicated by its position in the electrochemical series.

Typical potentials of the steel used for reinforcement in concrete are $+200\,\text{mV}$ to $+400\,\text{mV}$ with respect to a standard hydrogen electrode. Comparing this with the electrochemical series shows that this steel behaves, electrochemically, as though it were copper!

EXERCISE 8.16

$$\text{If } x = \sqrt{Dt}$$

$$D = \frac{x^2}{t}$$

The carbonation depth reaches 10 mm after about 22 years. Thus

$$D = \frac{(10 \times 10^{-3})^2}{22 \times 365 \times 24 \times 3600}\,\text{m}^2\,\text{s}^{-1}$$

$$= 1.4 \times 10^{-13}\,\text{m}^2\,\text{s}^{-1}$$

EXERCISE 8.17 The two most serious effects are (a) loss of strength of the reinforcement — particularly acute in the case of prestressed concrete; (b) the increase in volume of the corrosion products leads to spalling of the concrete.

Answers to self-assessment questions

SAQ 8.1 Setting results from the initial hydration of C_3S and C_3A, producing a gel on the surface of the cement particles, together with needle-like crystals of ettringite. Gel-to-gel contact plus some interlocking of the ettringite crystals provide the cohesion of the paste during setting. Diffusion of water into the gel envelope causes an increase in pressure which eventually leads to rupture, with the formation of gel fibres and foils. As these multiply and grow, they begin to interlock, which, together with the precipitation of portlandite, signals the end of the setting period and the start of hardening. Accompanying this is an increasing degree of polymerization of the silicate anions forming the gel.

SAQ 8.2 Dissolution of the CaO component produces a very rapid increase in the pH of the liquid phase, reaching a pH of 12.3 in the first minute. The pH continues to rise during the setting period, with $[Ca^{2+}]$ and $[OH^-]$ reaching about double the level of a saturated solution before $Ca(OH)_2$ starts to precipitate when hardening starts. Thereafter the pH settles down to a steady level of pH 12.6, the equilibrium being maintained by

$$Ca^{2+} + 2OH^- \rightleftharpoons Ca(OH)_2 \text{ (solid)}$$

The high alkalinity provides the conditions for gel formation, as well as a passivating environment for steel reinforcement. The $Ca(OH)_2$ precipitate contributes to the hardening of cement.

SAQ 8.3 Alite, C_3S Initial hydration leads to high alkalinity plus gel formation. The latter provides setting cohesion. Subsequent rupture produces gel fibres and foils. Hardening occurs as these grow and interlock. Major contributor to the strength of cement.

Belite, C_2S Similar reactions to C_3S, but much slower, hence little contribution to setting. Substantial contribution to long-term hardening.

Aluminate, C_3A Initial hydration plays similar role in setting to that of C_3S plus the formation of ettringite crystals on reaction with gypsum. Minor contributor to strength of cement, but precursor helps the melting of the starting materials in the kiln.

Ferrite, C_4AF Very little role in either setting or hardening, but like C_3A precursor acts as a flux in the kiln.

SAQ 8.4 Water is necessary in cement to provide cohesion to the paste during setting and to provide the means whereby cement hardens to a stone-like solid. It does these through its hydration reactions with the mineral constituents of cement. About 40% by weight of water is required for complete hydration of cement.

Water can affect the mechanical properties largely through its role in creating pores in the hardened cement. These act as flaws that reduce the tensile strength and, since increasing porosity means reducing density, they also affect compressive strength and stiffness. Environmental water condensing in these pores increases the creep rate of cement and its susceptibility to frost damage, while also providing a medium for enhanced chemical attack to the interior of the cement.

SAQ 8.5

(a) In each case, the lowest material cost, c will be obtained by choosing the minimum volume of material that may be used without the stress in the column exceeding the strength σ_p.

Letting the sectional area be A

$$F = \sigma_p A$$

The cost of the material is

$$c = AL\rho C$$

Substituting for A

$$c = \frac{FL\rho C}{\sigma_p}$$

Since F and L are fixed, the lowest cost will come from the materials with the largest value of $\sigma_p/\rho C$.

(b) The energy required for the column is

$$q = ALQ$$

Substituting for A

$$q = \frac{FLQ}{\sigma_p}$$

The smallest requirement is that of the material with the largest value of σ_p/Q.

(c) Let the deflection or change in length under load F be ΔL. This corresponds to a strain ε of $\Delta L/L$. The stiffness of the column is

$$\frac{F}{\Delta L} = \frac{\sigma_p A}{\varepsilon L} = \frac{EA}{L}$$

The quantity sought, cost per unit stiffness, is then given by

$$\frac{c}{F/\Delta L} = \frac{AL\rho C}{EA/L} = \frac{L^2\rho C}{E}$$

The column with the lowest cost per unit stiffness is made from the material with the largest value of $E/\rho C$.

From Table 8.7, concrete is the preferred choice in all three cases.

SAQ 8.6 Young's modulus of the cement is given by

$$E_{cem} = 30(1 - 0.2)^3 \text{ GN m}^{-2}$$
$$= 15.4 \text{ GN m}^{-2}$$

From Chapter 7, the modulus of the concrete, E_c, is given by

$$\frac{1}{E_c} = \sum_i \frac{V_i}{E_i}$$

Table 8.7 Comparison of indices for various materials

Material	$\sigma_p/\rho C$ (MN m £$^{-1}$)	σ_p/Q (mN mJ^{-1})	$E/\rho C$ (GN m £$^{-1}$)
cement	0.133	1.60	0.250
concrete	0.433	7.65	0.500
aluminium	0.055	0.56	0.020
mild steel	0.148	1.17	0.089
window glass	0.022	1.00	0.032
UPVC	0.034	0.71	0.002
GFRP (SMC)	0.071	1.20	0.005

$$= \left(\frac{1}{6 \times 15.4} + \frac{2}{6 \times 74} \right.$$

$$\left. + \frac{3}{6 \times 90} \right) m^2 \, GN^{-1}$$

$$= 2.09 \times 10^{-2} \, m^2 \, GN^{-1}$$

Thus $E_c = 48 \, GN \, m^{-2}$.

SAQ 8.7 The steel is assumed to carry all tensile loads in the beam and so inhibits the catastrophic cracking that would occur if a concrete beam were subjected to significant bending moments. The concrete is assumed to carry all compressive loads. Some cracking of the concrete on the tensile side of the beam is inevitable, but is not damaging while the concrete on the compressive side of the beam resists the compressive loads without failure.

You have two contributions to the load per unit length: the weights of the beam, w_B, and the imposed load, w_L. Together with the safety factor, this gives

$$w = (1.5w_L + w_B)$$

The mass per unit length of the beam is $A_c \rho_c + A_s \rho_s$

$$A_s \rho_s = (5 \times \pi \times 0.016^2) \rho_s$$

$$= 31.6 \, kg \, m^{-1}$$

$$A_c \rho_c = (0.35 \times 0.5 - A_s) \rho_c$$

$$= 427 \, kg \, m^{-1}$$

Thus $w_B = 4.50 \, kN \, m^{-1}$.

M_{max} for the beam $= 437 \, kN \, m$ and $L = 5 \, m$, hence

$$M_{max} = \frac{(1.5w_L + w_B)L^2}{12}$$

$$w_L = \frac{12 M_{max}}{1.5 L^2} - \frac{w_B}{1.5}$$

$$= 137 \, kN \, m^{-1}$$

SAQ 8.8 In pretensioning, concrete is cast around the steel wires while they are under tension. The applied tension is released once the concrete has hardened and the bond between the steel and concrete ensures that the contracting steel puts the concrete into compression. In post-tensioning, the concrete is cast around ducts in which the steel wires are subsequently placed and tensioned. This tension acts on bearing faces at the ends of the component, thus placing the concrete in compression.

In prestressed concrete the steel is loaded elastically in tension and this tensile load is resisted by compressive loads in the concrete. When a prestressed concrete structure is loaded externally, an additional tensile load is exerted on the steel. Hence a steel with a higher yield stress is required if it is to supply the compressive load to inhibit cracking in the concrete and sustain additional tensile loads prior to yielding.

SAQ 8.9 The two main ways in which chemical attack can lead to corrosion of the steel are (a) by reducing the pH of the concrete, destroying the passivity of the oxide film on the steel (an example of this is carbonation) and (b) by the presence of species such as chloride ions which react with the oxide layer to make it more conducting, thus allowing corrosion to proceed.

A physical mechanism is the condensation of water in the pores, followed by freezing. The expansionary forces can lead to spalling of the concrete.

Chapter 9
Selection of Materials:
A Case Study

by Nick Reid

Chapter 9 Selection of materials: a case study

9.1 Introduction

The preceding chapters have covered a wide range of load-bearing materials — metals, ceramics and glasses, polymers and composites. We have focused on the mechanical properties of these materials and how these properties depend on microstructure, from the atomic scale upwards. This dependence clearly implies that changing the microstructure of a material, by using such stratagems as heat treatment or varying the chemical composition, alters its properties.

We have also compared the properties of different materials with a view to selecting the most appropriate material for a given application.

This chapter does not introduce any new principles or techniques. Rather it surveys the previous chapters in order to help *you* to select materials for a simple component and to select treatments that will improve their properties. We shall consider a fairly simple problem: how to improve the design of that ubiquitous product — the umbrella.

9.2 A problem with wind

If the only thing required of an umbrella was to cast a shadow, there would be no problem; the loads applied by the sun are light indeed! The loads due to rainfall are larger, but they are still of a modest magnitude. The maximum loading on an umbrella comes from high winds, which can occur under either sunny or rainy conditions. Unfortunately, an umbrella is often damaged by a violent gust of wind and what usually happens is that one of the ribs collapses. Therefore you need to look at the design of a rib and see whether you can modify it in order to improve its strength in bending, thereby improving its resistance to wind.

9.2.1 The conventional design

Consider a so-called 'gentleman's' umbrella (Figure 9.1). It consists of a fabric cover, usually a plain-weave, waterproofed nylon, supported on a folding framework which is attached to a central shaft. We shall focus on the design of the ribs of the frame because these are the critical members. A rib is hinged at one end to the shaft of the umbrella (Figure 9.2) and the other end is tied to the cover. Near the middle of

Figure 9.1 A gentleman's umbrella

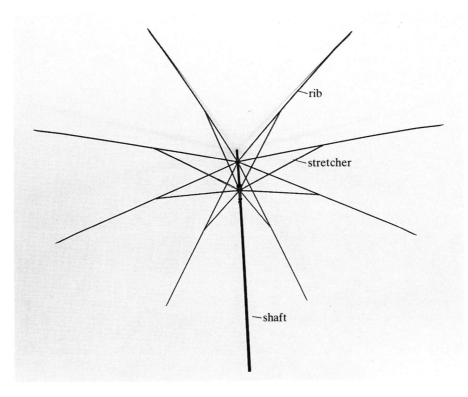

Figure 9.2 Components of an umbrella frame

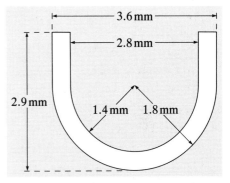

Figure 9.3 The cross-section of a roll-formed rib

the rib, a stretcher, similar in design to the rib, is attached by a hinge. The umbrella is erected by moving the stretchers in order to force the ribs against the cover, inducing a state of biaxial tension in the cover, while, in turn, the cover induces a state of elastic bending in the ribs. When a gust of wind blows, aerodynamic forces act on the cover, bending the ribs, culminating in failure if the forces are large enough.

Conventional designs of rib usually employ a rolled hypoeutectoid carbon steel, in lengths of 600 mm and with a constant U-shaped cross-section (Figure 9.3). A hardness test on one gave a value of $460 \pm 3.5\,H_V$, produced by quenching and tempering. To inhibit rusting, the ribs are either painted or metal plated. The bottom of the U-section bears against the cover.

The steel ribs fail when plastic deformation occurs and the rib is permanently strained. In bending, the compressed part of the section usually buckles: that is, the sides of the U splay inwards or outwards (Figure 9.4). Attempts to bend a buckled rib back to shape are rarely successful.

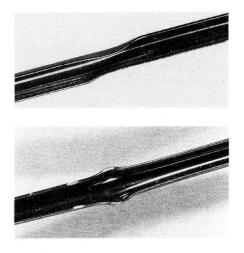

Figure 9.4 Buckled ribs

9.2.2 A specification for design

To make an umbrella more wind resistant, you want a rib that will withstand a larger bending load. To avoid having to redesign the cover as well, the bending stiffness of the rib should be unchanged; the existing cover will then stretch satisfactorily when the umbrella is erected. This rules out the option of strengthening the rib simply by increasing its cross-section, because this would also stiffen it. Furthermore, to avoid the ribs becoming too bulky, you should put an upper limit on their cross-section, say 6 mm.

In thinking about a new design, you need to draw on what you have learned in previous chapters, beginning with Chapter 1. This was primarily concerned with the deformation mechanics of elastic materials and therefore contains the means of quantifying stiffness and strength. First, you need to calculate the required value of the bending stiffness. Bending stiffness can be expressed as the product of Young's modulus, E, and the second moment of area I. This is appropriate because, for a given loading, the deflection produced is always proportional to EI. To evaluate this for the steel section, you must first recall the definition of I.

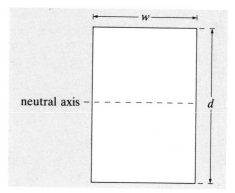

Figure 9.5 A rectangular section

> **SAQ 9.1 (Chapter 1)**
> Define I in terms of an element of cross-sectional area, dA, and the distance, y, between the element and the neutral axis.
>
> Consider a rectangular section of width w and depth d that is subjected to pure bending about a central axis parallel to the direction in which the width is measured (Figure 9.5). What is the formula for I about this axis? What is the name of the point on the cross-section through which the neutral axis passes?

Like the guttering in Chapter 5, the steel U-section is asymmetrical (the top and bottom differ) so the location of the neutral axis is not at half the sectional depth. Chapter 1 did not consider asymmetrical sections, but, as with symmetrical sections, the neutral axis passes through the **centroid** of the section: that is, the line about which the first moment of area is zero (in this case it is at a distance of 1.64 mm from the top of the U).

The results of a four-point bending test to evaluate the stiffness of a steel rib are shown in Figure 9.6. The spacing, a, of the adjacent loading point was 50 mm. The resulting graph of total bending force, $2F$, against central deflection, Δ is shown in Figure 9.7. The initial slope is a direct measure of the bending stiffness of the rib.

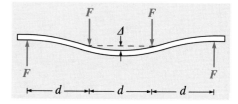

Figure 9.6 The geometry of the four-point bending test

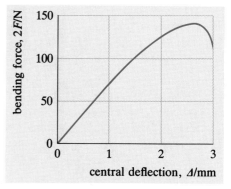

Figure 9.7 The graph of total bending force ($2F$) against central deflection Δ, (relative to the inner loading points), for a steel rib

SAQ 9.2   (Chapter 1)
What is the value of F/Δ for the rib?

Write down the expression from bending theory relating E, I, M and R. What is the value of the bending moment, M, in the centre of the test piece when a load F is applied at each loading point, spaced a apart?

Calculate the value of bending stiffness, EI, for the rib.

What would be the diameter of a solid circular-sectioned steel rib that had the same stiffness?

If circular sections are restricted to a maximum diameter of 6 mm, what does this imply for the value of modulus required?

You now have the value of the bending stiffness required for the redesigned rib. Note that, since we have imposed a limit of 6 mm on the diameter of the rib, the modulus of the material will have to exceed a critical value of $8.6\,\mathrm{GN\,m^{-2}}$ in order to provide the bending stiffness required.

The four-point bending test referred to in Figure 9.7 was continued until the rib collapsed by buckling — the appearance of the failed sample is shown in Figure 9.8. The maximum load sustained during the test provides a measure of the rib's resistance to collapse. This provides a target that the redesigned rib must beat.

SAQ 9.3   (Chapter 1)
What is the maximum value of the bending force ($2F$) carried prior to the collapse of the rib?

Calculate the value of the maximum bending moment carried by the rib.

Figure 9.8 The failed rib

9.2.3 A new design

The target is now clear — to design a rib with the required bending stiffness ($EI = 0.55\,\mathrm{N\,m^2}$) and with a value of the bending moment when the rib collapses that exceeds the value for the conventional design ($M_c = 3.5\,\mathrm{N\,m}$) by as much as possible. Of course, the stiffness and strength of a rib will depend both on the material chosen and on the size and shape of section used. If you are to compare the merits of differing materials on an equal basis, however, you must keep the shape constant while letting the size vary in order to achieve the required stiffness. Certain shapes (e.g. the U-shape) are difficult to produce in some materials, so we have adopted a solid circular section as the fixed shape. Using this, it should be possible to devise a merit index to help you to pick out the best material from a list of options.

Devising a merit index

In Chapter 2, you saw that a merit index is a combination of material properties that is proportional to the attribute sought in the product, as used under specified conditions. In this case the attribute sought is the maximum bending moment that can be carried by a section with a given value of bending stiffness. You now have to consider what combination of properties controls the maximum bending moment. Assume that the deformation of a rib is linearly elastic up to the point of collapse and that the bending stiffness, EI, has a specified value, k.

SAQ 9.4   (Chapters 1 and 2)
Write down an expression in terms of E and r for the bending stiffness, EI, of a beam with a circular section.

If the required value of bending stiffness is k, write down an expression for the radius of section that has this value when a material of modulus E is used. The parameters M, I, σ and y are related by a formula from engineers' bending theory. Write this formula down and explain what the symbols mean. Use this formula to obtain an expression for the maximum bending moment, $M_{\max}$, that, when applied to a section of radius r, produces a maximum stress of σ_c. Substitute for r in this expression the relation obtained earlier in terms of k. Write down a merit index, involving two material properties, that is proportional to $M_{\max}$.

You now have a merit index to guide you in your choice of material. What you seek is a material with a large strength and a relatively small Young's modulus. Note that the index has units $(\mathrm{N\,m^{-2}})^{1/4}$. Since we

have not laid down any conditions in respect of the weight of the
umbrella frame, the merit index does not contain the density of the
material. This is perhaps surprising for a product that is lifted and
carried around in use. The reason for ignoring weight is that the weights
of all designs are so small that they can be carried comfortably and the
absolute differences are not significant, although the relative weights
may differ considerably.

What is the value of the merit index for the steel currently used in
umbrellas?

Since this is a ferritic steel, you know that its Young's modulus has a
value of about $210\,\mathrm{GN\,m^{-2}}$. The tensile strength of the steel can be
estimated from its Vickers hardness number (460) by recalling from
Chapter 3 the section on 'Hardness measurements'.

SAQ 9.5 (Chapter 3)
For the umbrella steel, calculate the following:
(a) the tensile strength;
(b) the value of the merit index for a rib.

You now have a target value for the merit index. If you can find a
material with a value greater than $4.75\,(\mathrm{N\,m^{-2}})^{\frac{1}{4}}$, it will offer the
prospect of an umbrella frame of improved strength. Of course, there
are other requirements, too; perhaps the most important is that the
material should not degrade in service, for example, by exposure to
rainwater or sunlight. In order to look for promising materials, we shall
trawl through Chapters 2–8, thereby covering the major classes of
structural materials.

9.2.4 Nonferrous metals

You are looking for a material with a large strength relative to its
modulus. In Chapter 2 you saw that there are four stratagems for
strengthening metals: work hardening; the grain-size effect; solution
hardening; and age or precipitation hardening. In each case the
strengthening is accompanied by little, if any, increase in modulus.

SAQ 9.6 (Chapter 2)
Give a brief account of the mechanism of strengthening employed by
each of these four stratagems, commenting on the limiting factors.

In practice, the highest strengths are obtained by precipitation
hardening and you saw examples of this in aluminium-, magnesium-,
titanium- and nickel-based alloys, for which the properties were given in
Table 2.6.

SAQ 9.7 (Chapter 2)
Work out the values of the merit index for the alloys listed in Table 2.6. Which if any, of these alloys meets the target?

Describe qualitatively the type of treatment used to strengthen these alloys and how the accompanying changes in microstructure cause the strengthening to occur.

None of the alloys listed matches the steel in current use. There are, however, a few very-high-strength titanium alloys, such as Ti–4%Al–4%Sn–4%Mo–0.5%Si, with a merit index in the range 5–6, which are used by aeroengine manufacturers. These would offer a very modest increase in strength, but with an increase in cost of over ten times! You should also consider their resistance to environmental conditions, as well.

SAQ 9.8 (Chapter 2)
Comment on the resistance of aluminium-based and titanium-based alloys to atmospheric corrosion, with due reference to the nature of the surface.

9.2.5 Ferrous metals

A rolled carbon steel is the established material for umbrella ribs. Section 3.2.4 considers the range of strengths achievable in *drawn* eutectoid steel wire, (Fe–0.8%C). The strength varies with the diameter of wire produced, but a tensile strength of $2\,\text{GN}\,\text{m}^{-2}$ can be obtained in diameters up to 15 mm — higher than the strength of the umbrella steel inferred from the hardness ($1.47\,\text{GN}\,\text{m}^{-2}$). It appears that there may be some scope for improving the strength of the frame, by employing drawn eutectoid steel, but with attendant extra costs.

SAQ 9.9 (Chapter 3)
Describe the steps, and the accompanying microstructural changes, involved in producing 'patented' steel wire.

Why is this steel so strong?

You should consider whether any other ferrous alloy would be even better. One of the snags with a carbon steel is that it is prone to atmospheric corrosion (rusting) which may stain the cover if left unchecked. Painting or metal-plating is employed to prevent this, but rusting often occurs when the coating becomes damaged.

SAQ 9.10   (Chapter 3)
What is rust and what conditions are required for its formation on iron?

Why is rusting inhibited on a stainless steel, but not on a carbon steel?

In general, the ferritic and austenitic stainless steels have strength levels below that of the umbrella steel, but some of the martensitic steels can be hardened by heat treatment to higher levels. Fe–17%Cr–1%C, for example, can be quenched and tempered to give a yield strength of $1.8\,\mathrm{GN\,m^{-2}}$ and a modulus of $210\,\mathrm{GN\,m^{-2}}$ with a merit index of 5.8. This steel would provide both an increase in strength and freedom from rusting.

SAQ 9.11   (Chapter 3)
Describe the microstructural changes that occur during the formation and tempering of martensite in a plain carbon steel, relating them to changes in strength.

Give two reasons for adding chromium to martensitic stainless steel.

9.2.6 Ceramics and glasses
Here you have a class of materials held together by some of the strongest chemical bonds, giving a high chemical and thermal stability. Surely some of these materials will have the high mechanical strength that you are looking for, although they will also have large values of Young's modulus. Table 4.3 lists properties for a range of these materials from which you can work out values for the merit index.

SAQ 9.12   Evaluate the merit index for each of the glasses and ceramics listed in Table 4.3, noting which (if any) surpass that of the umbrella steel.

These materials have the opposite of what you want — low strengths and high moduli. While they can often be produced with high strength in the form of thin fibres (see Table 6.5), thicker sections are required and the strengths of these are invariably much lower.

SAQ 9.13   (Chapter 4)
Explain why the strengths of bulk ceramics are low in comparison with the values of their Young's moduli.

What could be done to improve their strength?

Use Table 6.5 to estimate values of the merit index for alumina and silicon carbide, supposing that the strengths of fibres could be reproduced in bulk material.

9.2.7 Polymeric materials

In Chapter 5 you saw that the values of tensile modulus of bulk, unreinforced, rigid polymers are normally small, of the order of 1 to $3\,\text{GN}\,\text{m}^{-2}$, well below the required minimum value mentioned earlier $(8.6\,\text{GN}\,\text{m}^{-2})$. You also saw in Sections 5.4.2 and 6.2.4 that this value can be increased considerably by increasing the molecular orientation, for instance, by drawing, a process that is used to produce high-strength fibres and 'Tensar' mesh. For extrusion grades of high-density polyethylene, cold drawing can be used to raise the short-term tensile modulus to $13\,\text{GN}\,\text{m}^{-2}$ and the tensile strength to about $500\,\text{MN}\,\text{m}^{-2}$, giving a value of the merit index of 13 — a substantial improvement on anything you have considered so far. Since the wind loading of umbrellas is due to gusts lasting about one second, the short-term modulus is appropriate.

SAQ 9.14 (Chapters 5 and 6)
Explain in terms of molecular changes why drawing increases the strength and tensile modulus of a partially crystalline polymer.

What are the effects of changing the average molecular mass on the improvements in tensile modulus and strength brought about by cold drawing high-density polyethylene?

Define the term creep modulus; explain qualitatively how and why this varies with the time under stress.

How are some polymers affected by water or sunlight?

9.2.8 Composite materials

In this particular design task you are concerned primarily with the properties of strength and modulus measured along the length of the rib. Thus you can consider a material with the maximum longitudinal strength, even if this is achieved at the expense of the transverse strength; the value of the properties in the transverse directions is of less importance. The task is clearly well-suited to a fibrous composite — a material in which the high strength of the fibres is exploited by binding them together with a matrix material to form a bulk product. The microstructure of such materials is characterized by the choice of fibre and matrix, and the orientations, length and volume fraction of the fibres.

SAQ 9.15 (Chapters 6 and 7)
Describe qualitatively the type of composite microstructure that would best suit this application.

Use the data in Tables 6.4 and 7.2 to predict the maximum value of the merit index that might theoretically be obtained from composites of fibres of E-glass and of Kevlar 29 in a matrix of polyester resin. State clearly any assumptions involved.

The theoretical values of merit index found in SAQ 9.15 are not achieved in practice in composites for several reasons: the packing factor achieved is no larger than about 70%; the fibres are not perfectly straight; failure occurs first by buckling of the compressed fibres at stresses below the tensile strength of the fibres. When the ingredients are formed into a uniaxial continuous-fibre composite by the process of pultrusion (a hybrid of drawing and extrusion, see Figure 9.9) a bending strength of $1450\,\mathrm{MN\,m^{-2}}$ can be achieved with a modulus of $45\,\mathrm{GN\,m^{-2}}$ in a glass–polyester composite. This material has a merit index of 14.8, surpassing even that of drawn polyethylene. A rod of diameter 4 mm would provide the stiffness required (Equation 9.4, SAQ 9.4), which is about the same size as the steel section currently used. It costs about 25 pence per metre.

You should remember that not all fibrous composites are synthetic. There are also natural composites, such as wood, that are used very successfully for structural purposes (see Section 7.2), and you should consider them too.

SAQ 9.16   (Chapter 7)
Use the data in Table 7.6 to investigate the suitability of each type of wood listed.

Describe briefly the microstructure of wood, indicating the component that controls the strength and modulus.

With the single exception of balsa, all the woods have moduli which exceed the minimum required value, but they all have small merit indices of about one. They are not serious contenders for this application because their strengths are low and their main virtue, low density, is not required.

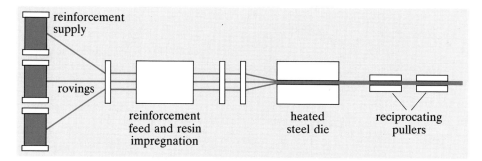

reinforcement supply

rovings

reinforcement feed and resin impregnation

heated steel die

reciprocating pullers

Figure 9.9 The pultrusion process. Fibres in the form of rovings are pulled through a tank in which they are coated with a liquid thermosetting resin. They are then drawn together and the continuous-fibre composite is shaped to a constant cross-section and cured in the heated steel die

Figure 9.10 A bamboo umbrella

In some designs of umbrella from the Far East, bamboo, which is more a grass than a wood, is used for the ribs and stretchers (see Figure 9.10). The section used is solid rectangular with rounded corners and it tapers towards the free end of the rib. Since these designs are usually hand made, there are significant variations in the dimensions of ribs, even within the same umbrella. Unlike steel, this material does not undergo significant plastic strain and its deformation is almost linearly elastic up to about half the failure load.

SAQ 9.17
Typically, solid bamboo has a bending strength of $300\,\mathrm{MN\,m^{-2}}$ and a Young's modulus of $10\,\mathrm{GN\,m^{-2}}$. Comment on its suitability for umbrella frames, bearing in mind that they get wet!

Bamboo is rather better than the woods considered above. Its modulus exceeds the minimum value and its merit index, at 4.1, is larger. This implies that a bamboo rib would be almost as strong as the steel rib, for a given stiffness. The bamboo frames made in the Far East tend to have a smaller span than the steel one under consideration and bamboo is chosen because it is an abundant, indigenous material with which local labour is familiar. Bamboo's favourable merit index plus its occurrence naturally as a hollow tube, together explain why this material was used widely for fishing rods and vaulting poles until displaced by synthetic fibre-composite materials. It is, however, still widely used in the Far East for scaffolding, even for high-rise buildings.

9.2.9 Cement and concrete

An immediate reaction to this class of materials might be to dismiss them scornfully for this application, but you must not be too hasty! Clearly, concretes containing aggregate with a size comparable to that of the section (6 mm) would be too inhomogeneous, but this does not apply to hydrated cement. This material has a modulus ($30\,\mathrm{GN\,m^{-2}}$), which comfortably exceeds the minimum value. It has, however, a very low strength in tension or bending ($\simeq 5\,\mathrm{MN\,m^{-2}}$), giving a very small value of the merit index (0.07).

> SAQ 9.18   (Chapter 8)
> Name the ingredients of Portland cement and describe briefly the chemical changes that occur during setting.
>
> Which microstructural feature controls the tensile strength of cement? How can the microstructure be changed in order to increase the strength of cement?

Section 8.2.3 describes how recent research has lead to the development of macrodefect-free cement. This can have a bending strength up to $200\,\mathrm{MN\,m^{-2}}$ and a modulus of $50\,\mathrm{GN\,m^{-2}}$, giving it a merit index of 1.9. This is below the target value, but better than that of mild steel. It is not a serious alternative to the umbrella steel as it stands, but it could probably be strengthened by reinforcement with continuous, parallel fibres made of a glass composition that is resistant to the highly alkaline cement. This is a materials development which has yet to happen!

9.3 Conclusions

We can conclude that for this application the best material is the one with the largest value of the failure stress in bending divided by the Young's modulus, raised to the power 3/4.

We have identified four materials with values of this merit index exceeding that of the steel in current use. In order of increasing index, they are a titanium alloy, a martensitic stainless steel, drawn high-density polyethylene and pultruded composites. The first would provide, at great extra cost, a marginal improvement in strength combined with excellent corrosion resistance, while the second would also be corrosion resistant, but stronger and cheaper, too. The last two materials would provide significant improvements on the steel rib by offering almost tripled collapse loads, a self-coloured finish and a freedom from environmental attack and rust stains on the cover.

Probably the best choice would be a pultruded glass or Kevlar 29 composite of 4 mm diameter. Its use, however, would significantly increase the cost of the product and would therefore be confined to 'up-market' umbrellas. Fishermen have shown themselves willing to buy GFRP and CFRP rods; perhaps it is time to offer them composite umbrellas to match!

Both the first and the last chapters of this book have considered structures subjected to loading by wind. Just in case you thought that these were mere academic exercises, Figure 9.11 provides dramatic evidence that bus shelters and umbrellas can be overloaded by wind!

Figure 9.11 In October 1989, winds of typhoon force — Typhoon Dan — swept Manila, capital of the Philippines, knocking over a bus shelter and turning one gentleman's umbrella inside out

Answers to self-assessment questions

SAQ 9.1 The second moment of area, I, is defined to be the integral, over the whole area of the cross-section, of the product of a small element of that area, dA, and the square of its distance from the neutral axis, y^2. Thus

$$I = \int_A y^2 \, dA$$

If the small element is a thin strip parallel to the neutral axis, $dA = w \, dy$ and the integral for a rectangle is

$$I = \int_{-\frac{d}{2}}^{\frac{d}{2}} y^2 w \, dy$$

$$= \frac{wd^3}{12}$$

The neutral axis passes through a point on the section called the centroid: that is, the point about which the first moment of area is zero (see 'Moments of area' in Chapter 1).

SAQ 9.2 The slope of the graph, $2F/\Delta$, is $70 \, \text{N mm}^{-1}$, so F/Δ is just half this, $35 \, \text{N mm}^{-1}$, or $35 \, \text{kN m}^{-1}$.

The formula relating E, I, M and R is Equation (1.10):

$$\frac{M}{I} = \frac{E}{R}$$

'Bending moments from loads' in Chapter 1 gave the bending moment between the central loading points in a four-point bending test as $M = Fd$, where d was the distance from the loading points to the outer supports (see Figure 1.17a). Here, in the bending test for the rib, the spacing between the loading points and the supports is a (Figure 9.6) and the bending moment between loading points is therefore Fa.

The value of the bending stiffness, EI, can be determined by using Equation (1.17)

$$\Delta_{\text{max}} = \frac{s^2 M}{8EI}$$

where s was the length of the central span. Substituting $M = Fa$ and $s = a$, this becomes

$$\Delta_{\text{max}} = \frac{Fa^3}{8EI}$$

This can be rearranged to give the bending stiffness

$$EI = \frac{F}{\Delta_{\text{max}}} \frac{a^3}{8} \qquad (9.1)$$

Putting in the values $a = 0.05 \, \text{m}$ and $F/\Delta_{\text{max}} = 35 \, \text{kN m}^{-1}$ gives the value required for an umbrella rib

$$EI = \frac{35000 \times (0.05)^3}{8}$$

$$= 0.55 \, \text{N m}^2$$

If the rib is of solid circular section, from Table 1.1

$$I = \frac{\pi r^4}{4}$$

For it to have the same stiffness as the one tested

$$\frac{E \pi r^4}{4} = 0.55 \, \text{N m}^2 \qquad (9.2)$$

Hence

$$r^4 = \frac{4 \times 0.55}{\pi \times 210 \times 10^9} \, \text{m}^4$$

$$r = 1.35 \times 10^{-3} \, \text{m}$$

Thus its diameter, $d = 2.7 \, \text{mm}$.

If the diameter $2r$ is to be kept to less than 6 mm, I cannot exceed

$$I = \pi r^4 / 4$$

$$= 6.36 \times 10^{-11} \, \text{m}^4$$

and the modulus must be at least

$$E = \frac{0.55}{6.36 \times 10^{-11}}$$

$$= 8.6 \, \text{GN m}^{-2}$$

SAQ 9.3
Collapse occurs at the maximum of the graph: namely, $2F = 140 \, \text{N}$. At this point the bending moment is

$$M_c = Fa \qquad (9.3)$$

$$= (140/2) \times 0.05$$

$$= 3.5 \, \text{N m}$$

SAQ 9.4 If the rib has a solid circular section, the expression for bending stiffness is

$$EI = \frac{E \pi r^4}{4}$$

Putting EI equal to k and rearranging

$$r = \left(\frac{4k}{E\pi}\right)^{1/4} \qquad (9.4)$$

The formula of engineer's bending theory is

$$\frac{M}{I} = \frac{\sigma}{y}$$

where σ is the bending stress at a point on the section a distance y from the neutral axis. Rearranging,

$$M_{\text{max}} = \frac{\sigma_c I}{y_{\text{max}}}$$

Since $y_{\text{max}} = r$

$$M_{\text{max}} = \frac{\sigma_c \pi r^4}{4r}$$

Substituting for r

$$M_{\text{max}} = \left[\frac{\sigma_c}{E^{3/4}}\right] k^{3/4} \left[\frac{\pi}{4}\right]^{1/4}$$

The merit index is the first term in brackets — the strength divided by the modulus raised to the power of 3/4.

SAQ 9.5 The Vickers hardness number H_V is defined as the load in kilograms divided by the area in square millimetres of the faces of the indentation. For high strength steels the tensile strength in MN m^{-2} is approximately equal to $3.2 \times$ the hardness, H_V. The hardness of the steel is $460 \, H_V$, implying that

$$\sigma_t = 460 \times 3.2$$

$$= 1472 \, \text{MN m}^{-2}$$

The merit index is

$$\frac{\sigma_t}{E^{3/4}} = \frac{1472 \times 10^6}{(210 \times 10^9)^{3/4}} \, (\text{N m}^{-2})^{1/4}$$

$$= 4.75 \, (\text{N m}^{-2})^{1/4}$$

SAQ 9.6

Work hardening This arises from the gradual accumulation of dislocations in the crystals that occurs during cold working. The strength is determined by the stress level that is required to move a dislocation through the 'forest' of other dislocations. This, in turn, depends on the square root of the dislocation density, $\rho^{1/2}$. The density increases with the plastic strain, thereby causing work hardening. The limit is set by the onset of cracking, which prevents further increases in plastic strain.

Grain-size hardening As a general rule, the yield or proof stress varies with the inverse square root of the average grain diameter $d^{-1/2}$, so polycrystalline metals can be strengthened by reducing the grain size. The limit is set by the smallest grain size that can be achieved by recrystallization — about $5\,\mu m$.

Solution hardening In general, the stress to move a dislocation and, hence, the strength increases with the concentration of solute in solution. The limit is set by the maximum concentration that the phase diagram will permit — only rarely (e.g. copper and nickel) do metals form solutions at all compositions.

Age hardening The stress to move a dislocation is increased by placing second phase particles in the dislocation's path. This can be done by 'precipitating' particles within a crystal during heat treatment. This stratagem is limited to the alloys with a suitable phase diagram containing a solvus composition which increases with temperature.

SAQ 9.7 None of these alloys meets the target of 4.75 $(N\,m^{-2})^{1/4}$ — see Table 9.1.

Heat treatment is used to cause age or precipitation hardening of these alloys. The first step is to heat the alloy to a temperature high enough to cause it to become a single-phase, solid solution. The second step is to quench the alloy rapidly, usually to room temperature, allowing insufficient time at high temperature to permit significant diffusion to occur. The solid-solution structure is thereby preserved. The final step is to age the alloy at an intermediate temperature, high enough to permit diffusion to cause the solid solution to break down into a mixture of phases. The new phase forms as an array of small, closely spaced particles (precipitates), which obstruct moving dislocations and thereby cause the strength of the alloy to increase.

SAQ 9.8 Under atmospheric conditions, aluminium and titanium do not corrode — they are passive. This arises from the formation on the surface of a continuous, adherent layer of oxide (Al_2O_3 and TiO_2, respectively). This layer separates the reacting elements (metal and oxygen) and since the rates of diffusion of these elements through the oxide layer are very slow at ambient temperature, it effectively arrests corrosion. The protective layer forms spontaneously on a clean surface of these metals, but thicker controlled layers are often grown deliberately by electrolytic processes such as anodizing, giving a good surface appearance with excellent corrosion resistance.

SAQ 9.9 The steel used has approximately the eutectoid composition ($\sim 0.8\,wt\%\,C$) and it is made to undergo heat treatment followed by cold working. The first step is to heat the steel at about $1150\,K$ in order to transform it to austenite, a face-centred cubic solid solution. It is then quenched into an oil or salt bath held at a temperature of about $850\,K$, whereupon the austenite transforms over a period of time into pearlite, a fine lamellar mixture of ferrite and cementite, Fe_3C. With this composition, little or no free ferrite is formed. When the transformation is complete, the steel is removed from the bath and cooled in air. The remaining treatment consists of repeated cold drawing of the steel through dies, which causes intense work hardening and the formation of a fibrous microstructure with a strongly preferred orientation.

The high strength of this structure is due to the accumulation of an exceedingly high dislocation density in the ferrite crystals.

SAQ 9.10 An oxygenated aqueous environment is required for the formation of rust on iron (Section 3.6.2). Rust is the name given to the mixture of hydrated iron oxides which forms on iron: that is, ferrous oxide (FeO) and ferric oxide (Fe_2O_3).

To form a protective layer, an oxide must be adherent, it must have a limited change in volume on formation from the metal and it must have a low diffusivity for oxygen. The oxides in rust fail on all three counts so rust is not protective. Stainless steels contain at least 12% of chromium which enables a protective layer of Cr_2O_3 to form on the surface.

SAQ 9.11 Martensite is the name given to the phase produced by rapidly cooling iron or steel from a temperature at which it is austenitic: that is, face-centred cubic. The austenite transforms by undergoing a spontaneous shear which forms elongated grains of martensite. In the absence of carbon, the martensite is simply body-centred cubic iron, which is soft and ductile. When carbon is present, however, the crystals are distorted to the body-centred tetragonal structure which is hardened by the carbon atoms trapped in solution. The hardness increases with the carbon content up to about $900\,H_V$ at 1% carbon (see Section 3.4.1).

Martensite is thermodynamically unstable and can be transformed by heating (tempering) in the range 300–$1000\,K$. This causes the carbon atoms to diffuse rapidly, forming small carbide particles. Some strength is lost by the removal of solute (carbon) from the body-centred tetragonal phase, but this is partially offset by the hardening effect of the particles. The net result is a decrease in strength, but an increase in ductility and toughness.

The presence of 17% chromium makes the steel 'stainless' and increases the 'hardenability': that is, slower cooling rates are required to form martensite.

Table 9.1

Alloy	Merit index/$(N\,m^{-2})^{1/4}$
2014-T6	3.0
6082-T6	1.9
7075-T6	3.7
Mg–2% Zn–1% Mn	1.2
Ti–2.5%Cu	3.0
Nimonic 263	2.0

SAQ 9.12
None of these materials attains the
required merit index — see Table 9.2.

SAQ 9.13 Since ceramics are held together
by strong primary bonds, they tend to
have large values of Young's modulus
($> 100\,\mathrm{GN\,m^{-2}}$). For the same reason, the
stress required to move dislocations (at a
given temperature) is also large, but this
does not confer high tensile strength. The
tensile strength is controlled by the
unstable growth of small, crack-like
defects, which exist on the surface and
within the microstructure. In tougher
materials such defects become blunted by
plastic deformation and do not grow.

To raise the strength these defects should
be eliminated or, failing that, reduced in
size. This must be done during the
processing of the ceramic by reducing the
porosity and the grain size.

Merit index for alumina

$$= \frac{1 \times 10^9}{(100 \times 10^9)^{3/4}} = 5.6$$

Merit index for silicon carbide

$$= \frac{4 \times 10^9}{(410 \times 10^9)^{3/4}} = 7.8$$

These materials have the potential to be
strong and light structural materials.

SAQ 9.14 Within a polymer in thermal
equilibrium, there is no direction along
which there is preferred alignment of the
molecules. In amorphous phases, the
molecules are coiled randomly into
spheres, while in crystalline phases the
lamellar crystals are arranged radially
within spherulites. In neither case is there
extra alignment of molecules in a
particular direction. This changes if the
polymer is deformed by drawing. The
molecules become straightened out in the
direction of drawing and stay in this
configuration. When the drawn polymer is
loaded in the direction of drawing, it is
the primary carbon-to-carbon bonds that
deform, rather than the secondary bonds,
which dominate in the unaligned case, and
both the modulus and strength increase.

Decreasing the molecular mass causes the
draw ratio to increase, which, in turn,
causes more molecular alignment to occur.
This increases the tensile modulus. The
strength, however, is sensitive to the

Table 9.2

Material	Merit index/$(\mathrm{N\,m^{-2}})^{1/4}$
silica glass	0.5
borosilicate glass	0.5
soda-lime silica glass (annealed)	0.5
(thermally toughened)	1.8
glass ceramic	0.7
RBSN	< 1.5
magnesium oxide	0.4
alumina	0.8
RBSC	0.9

molecular mass *per se*, so the strength
diminishes in spite of the extra alignment.

The creep modulus at any time t is the
constant applied stress divided by the
strain achieved at time t (Section 5.1.4).
Because of creep within the amorphous
phase, this strain increases with time, so
the modulus decreases with increasing
time.

Polymers may be affected by the
absorption of water causing swelling and a
reduction in the glass-transition
temperature (e.g. nylon, see Section 5.1.3).

They may also be affected by the
ultraviolet radiation in sunlight, which can
break bonds within the polymeric
molecules by the process of
photodecomposition. The molecular mass
is thereby reduced, allowing the free
radicals so produced to react with
atmospheric oxygen.

SAQ 9.15 For this application you want a
high strength compared with the modulus.
Looking at the potential fibres in Tables
7.2 and 7.2, the inorganic glasses and
Kevlar 29 appear to be best, due to their
lower moduli. Since you are concerned
only with longitudinal properties along the
length, the fibres should be aligned
lengthwise and, for optimum properties,
they should be continuous and should
occupy the maximum volume fraction.

According to Table 7.1, the theoretical
maximum volume fraction, V_f, is 0.91.
Assuming this value, the modulus of the
composite can be estimated from Equation
(7.9)

$$E_c = \sum_i E_i V_i$$

Substituting in this equation gives $E_c =$
$65.8\,\mathrm{GN\,m^{-2}}$ for E-glass and $E_c =$

$54.0\,\mathrm{GN\,m^{-2}}$ for Kevlar 29.

The composite's strength can be estimated
from the rule of mixtures for strength
(Section 7.4 and Exercise 7.7), in which σ_m
is assumed to be the stress in the matrix
when the strain is equal to the breaking
strain of the fibres. This strain is tensile
strength/modulus, which equals 3.4/72 for
E-glass and 2.8/59 for Kevlar 29.

Multiplying by the matrix modulus gives,
for both fibres,

$$\sigma_m = 0.047 \times 3.3$$
$$= 0.156\,\mathrm{GN\,m^{-2}}$$

Evaluating the expression for the rule of
mixtures gives, for E-glass,

$$\sigma^*_c = 3.1\,\mathrm{GN\,m^{-2}}$$

and for Kevlar 29

$$\sigma^*_c = 2.6\,\mathrm{GN\,m^{-2}}$$

The merit index for E-glass is

$$\frac{3.1 \times 10^9}{(65.8 \times 10^9)^{3/4}} = 23.9\,(\mathrm{N\,m^{-2}})^{1/4}$$

and for Kevlar 29 it is

$$\frac{2.6 \times 10^9}{(54.0 \times 10^9)^{3/4}} = 23.2\,(\mathrm{N\,m^{-2}})^{1/4}$$

SAQ 9.16 The values of the merit index
are shown in Table 9.3. None of these
woods is suitable.

The structure of wood consists of cells
(tracheids) elongated along the tree trunk,
with an aspect ratio of the order of 100,
with a smaller number of cells (rays)
oriented parallel to the radii of the tree.
The walls of the cells are composed of

Table 9.3

Material	Merit index$/(\mathrm{N\,m}^{-2})^{1/4}$
balsa	0.5
mahogany	1.4
ash	1.2
oak	1.3
beech	1.1
Sitka spruce	1.0
Scots pine	1.0
Douglas fir	1.1

'filament-wound' cellulose molecules (Section 7.2.4) and they contain small holes to interconnect neighbouring cells and allow the flow of sap. Annual-growth rings arise from differences in density of the wood grown at different seasons. The cells are held together with a thermoplastic called lignin and their walls constitute the main load-bearing component of wood.

SAQ 9.17 Bamboo has a merit index of 4.1 and this falls below your target. The value of its modulus, however, exceeds the minimum sought. It is a reasonable material to use for umbrella frames, but provides only 80% of the strength of a steel frame. Being a natural biological material, bamboo will ultimately decompose in a wet, outdoor environment, but an umbrella, dried after use, should last for years.

SAQ 9.18 This is answered in Sections 8.2.1. and 8.2.2. The ingredients are:

- alite
- belite
- aluminate
- ferrite (not to be confused with BCC Fe)
- gypsum.

The setting of cement arises from the hydration of alite and aluminate to produce a gel on the surface of the cement particles, together with needle-shaped crystals of ettringite. Contact between particles via the gel and interlocking of the ettringite crystals confers some cohesion on the paste during setting. The diffusion of water into the gel eventually causes the gel envelope to rupture, forming gel fibres and foils which interlock with one another. Together with the precipitation of portlandite, this marks the end of the setting period and the beginning of the hardening stage, involving an increasing degree of polymerization of the silicate ions within the gel.

The strength of cement, like that of ceramics, is controlled by the presence of crack-like features in the microstructure. These are associated with porosity.

To increase the strength, measures should be taken to reduce (or eliminate) the porosity.

Appendix 1
Units of Measurement

by George Weidmann

Appendix 2
Structural Formulae
for Polymers

by Peter Lewis

Appendix 3
The Periodic Table

Appendix 1 Units of measurement

A1.1 The International System of Units

The International System (SI) of units has seven **base units**, two **supplementary units** (the radian and the steradian) and a variety of **derived units**. The seven physical quantities on which the system is based, together with their dimensions, the SI base units and their symbols are listed in Table A1.1.

The SI derived units for other physical quantities are formed from the base units via the equation defining the quantity involved. Thus, for example, force = mass × acceleration and the unit of force, the newton, is equivalent to $kg\,m\,s^{-2}$. The two supplementary units and the principal derived units, together with some of their more important equivalents are shown in Table A1.2 opposite.

Table A1.1

Basic physical quantity	Dimension	SI base unit	Symbol
length	L	metre	m
mass	M	kilogram	kg
time	T	second	s
electric current	I	ampere	A
temperature	θ	kelvin	K
luminous intensity	J	candela	cd
amount of substance	–	mole	mol

Table A1.3

Multiplication factor		Prefix	Symbol
1 000 000 000 000 000 000 =	10^{18}	exa	E
1 000 000 000 000 000 =	10^{15}	peta	P
1 000 000 000 000 =	10^{12}	tera	T
1 000 000 000 =	10^{9}	giga	G
1 000 000 =	10^{6}	mega	M
1 000 =	10^{3}	kilo	k
100 =	10^{2}	hecto	h
10 =	10^{1}	deca	da
0.1 =	10^{-1}	deci	d
0.01 =	10^{-2}	centi	c
0.001 =	10^{-3}	milli	m
0.000 001 =	10^{-6}	micro	μ
0.000 000 001 =	10^{-9}	nano	n
0.000 000 000 001 =	10^{-12}	pico	p
0.000 000 000 000 001 =	10^{-15}	femto	f
0.000 000 000 000 000 001 =	10^{-18}	atto	a

Standard prefixes are used for multiples and submultiples of units (Table A1.3), the SI-preferred ones being multiples of 10^{3}. It should be noted that masses are still expressed as multiples of the gram, although the base unit is the kilogram. Thus $10^{-6}\,kg$ should be written as 1 mg.

Table A1.2

Quantity	Dimension	Unit	Symbol	Equivalent
plane angle	–	radian	rad	$(=180°/\pi)$
solid angle	–	steradian	sr	
density	ML^{-3}			$kg\,m^{-3}$
velocity or speed	LT^{-1}			$m\,s^{-1}$
acceleration	LT^{-2}			$m\,s^{-2}$
momentum	MLT^{-1}			$kg\,m\,s^{-1}$
moment of inertia	ML^2			$kg\,m^2$
force	MLT^{-2}	newton	N	$kg\,m\,s^{-2}$
pressure, stress	$ML^{-1}T^{-2}$	pascal	Pa	$N\,m^{-2}$
energy, work	ML^2T^{-2}	joule	J	$N\,m$
power	ML^2T^{-3}	watt	W	$J\,s^{-1}$
viscosity	$ML^{-1}T^{-1}$		$N\,s\,m^{-2}$	$kg\,m^{-1}\,s^{-1}$
frequency	T^{-1}	hertz	Hz	s^{-1}
electric conductance	$L^{-2}M^{-1}T^3I^2$	siemens	S	Ω^{-1}
electric charge	TI	coulomb	C	$A\,s$
electric potential difference	$L^2MT^{-3}I^{-1}$	volt	V	$W\,A^{-1}$
electric capacitance	$T^4I^2L^{-2}M^{-1}$	farad	F	$C\,V^{-1}$
electric resistance	$L^2MT^{-3}I^{-2}$	ohm	Ω	$V\,A^{-1}$
magnetic flux	$L^2MT^{-2}I^{-1}$	weber	Wb	$V\,s$
magnetic flux density (magnetic induction)	$MT^{-2}I^{-1}$	tesla	T	$V\,s\,m^{-2}$
inductance	$L^2MT^{-2}I^{-2}$	henry	H	$V\,s\,A^{-1}$
luminous flux	J	lumen	lm	$cd\,sr$
illuminance	JL^{-2}	lux	lx	$lm\,m^{-2}$
radioactivity	T^{-1}	becquerel	Bq	s^{-1}
absorbed dose of ionizing radiation	$L\,T^{-2}$	gray	Gy	$J\,kg^{-1}$

In any system of measurement in mechanics three fundamental units are required and in SI these are the metre, kilogram and second. Through common usage, certain multiples and submultiples of the three fundamental units have been given names. Some of these are given in Table A1.4, together with units of pressure and textile line density which are still in frequent use. None of them is a recognized SI unit.

Table A1.4

Physical quantity	Unit	Symbol	SI Equivalent
length	micron	μ	$10^{-6}\,m = 1\,\mu m$
length	ångström	Å	$10^{-10}\,m = 0.1\,nm$
mass	tonne	t	$10^3\,kg = 10^6\,g = 1\,Mg$
time	minute	min	$60\,s$
time	hour	h	$3600\,s$
time	day	d	$86\,400\,s$
time	year	a	$3.1578 \times 10^7\,s$
pressure	atmosphere	atm	$1.013 \times 10^5\,N\,m^{-2}$
pressure	bar	bar	$10^5\,N\,m^{-2}$
pressure	torr	Torr	$133.3\,N\,m^{-2}$
textile line density or yarn count	tex		$1\,g\,km^{-1} = 10^{-6}\,kg\,m^{-1}$
	denier		$0.1111\,g\,km^{-1} = 1.111 \times 10^{-7}\,kg\,m^{-1}$

A1.2 Other systems of units

After SI, the two most important systems of units are the cgs and fps systems, the latter being still very frequently used in the USA. Like SI, they are both based on the second as the unit of time, but differ in their base units for length (centimetre and foot) and for mass (gram and pound). Some of the more important conversion factors between SI units and the other two systems are set out in Table A1.5.

Temperatures are frequently expressed in degrees Celsius (or centigrade) or in degrees Fahrenheit. The relationship between these and the kelvin is

$$X\,°C = (X + 273.15)\,K$$

$$= (1.8X + 32)\,°F$$

Note that the symbol for the kelvin is K *not* °K.

In terms of temperature interval

$$1\,°C = 1\,K = 1.8\,°F$$

Temperature intervals are frequently differentiated from actual temperature in the Celsius and Fahrenheit systems by writing them as deg C and deg F, respectively.

Table A1.5

Physical quantity	cgs unit	fps unit
length	*centimetre* $1\,cm = 10^{-2}\,m$	*foot* $1\,ft = 0.3048\,m$ *inch* $1\,in = 2.54 \times 10^{-2}\,m$
mass	*gram* $1\,g = 10^{-3}\,kg$	*pound* $1\,lb = 0.4536\,kg$ *ounce* $1\,oz = 2.835 \times 10^{-2}\,kg$ *ton* $1\,ton = 1.016 \times 10^{3}\,kg$
area	$1\,cm^2 = 10^{-4}\,m^2$	$1\,ft^2 = 9.290 \times 10^{-2}\,m^2$ $1\,in^2 = 6.452 \times 10^{-4}\,m^2$
volume	$1\,cm^3 = 10^{-6}\,m^3$ $1\,litre = 10^{-3}\,m^3$	$1\,ft^3 = 2.832 \times 10^{-2}\,m^3$ $1\,in^3 = 1.639 \times 10^{-5}\,m^3$ *gallon* $1\,gal\,(UK) = 4.546 \times 10^{-3}\,m^3$ $1\,gal\,(USA) = 3.786 \times 10^{-3}\,m^3$ *fluid ounce* $1\,fl\,oz = 2.841 \times 10^{-5}\,m^3$
density	$1\,g\,cm^{-3} = 10^{3}\,kg\,m^{-3}$	$1\,lb\,ft^{-3} = 16.02\,kg\,m^{-3}$ $1\,lb\,in^{-3} = 2.768 \times 10^{4}\,kg\,m^{-3}$
velocity or speed	$1\,cm\,s^{-1} = 10^{-2}\,m\,s^{-1}$	$1\,ft\,s^{-1} = 0.3048\,m\,s^{-1}$ $1\,in\,s^{-1} = 2.54 \times 10^{-2}\,m\,s^{-1}$
momentum	$1\,g\,cm\,s^{-1} = 10^{-5}\,kg\,m\,s^{-1}$	$1\,lb\,ft\,s^{-1} = 0.1383\,kg\,m\,s^{-1}$
moment of inertia	$1\,g\,cm^2 = 10^{-7}\,kg\,m^2$	$1\,lb\,ft^2 = 4.214 \times 10^{-2}\,kg\,m^2$
force	*dyne* $1\,dyn = 10^{-5}\,N$	*poundal* $1\,pdl = 0.1383\,N$ *pound force* $1\,lbf = 4.448\,N$
pressure or stress	$1\,dyn\,cm^{-2} = 10^{-1}\,Pa$ *bar* $1\,bar = 10^{5}\,Pa$	$1\,lbf\,in^{-2}\,(psi) = 6.895 \times 10^{3}\,Pa$
energy or work	*erg* $1\,erg = 10^{-7}\,J$	$1\,ft\,pdl = 4.214 \times 10^{-2}\,J$ $1\,ft\,lbf = 1.356\,J$
power	$1\,erg\,s^{-1} = 10^{-7}\,W$	$1\,ft\,pdl\,s^{-1} = 4.214 \times 10^{-2}\,W$ $1\,ft\,lbf\,s^{-1} = 1.356\,W$ *horsepower* $1\,hp = 745.7\,W$
viscosity	*poise* $1\,P = 10^{-1}\,N\,s\,m^{-2}$	$1\,lbf\,s\,ft^{-2} = 47.88\,N\,s\,m^{-2}$
energy (thermal)	*calorie* $1\,cal = 4.187\,J$	*British thermal unit* $1\,BTU = 1.055 \times 10^{3}\,J$

Appendix 2 Structural formulae for polymers

A2.1 Thermoplastics and elastomers

Name	Abbreviation	Repeat unit	Comments
Polyolefins			
polyethylene: high-density medium-density low-density	HDPE MDPE LDPE	$-CH_2-CH_2-$	degree of crystallinity controlled by chain structure (branching)
polypropylene	PP	$-CH_2-CH-$ $\quad\quad\ \ \vert$ $\quad\quad\ CH_3$	usually isotactic
Vinyl polymers			
poly(vinyl chloride)	PVC	$-CH_2-CH-$ $\quad\quad\ \ \vert$ $\quad\quad\ Cl$	either plasticized or unplasticized (UPVC)
poly(vinylidene chloride)	PVDC	$\quad\quad\ Cl$ $\quad\quad\ \ \vert$ $-CH_2-C-$ $\quad\quad\ \ \vert$ $\quad\quad\ Cl$	crystalline polymer
polystyrene	PS	$-CH_2-CH-$	normally atactic, noncrystalline solids
polyacrylonitrile	PAN	$-CH_2-CH-$ $\quad\quad\ \ \vert$ $\quad\quad\ CN$	
poly(vinyl acetate)	PVA or PVAC	$-CH_2-CH-$ $\quad\quad\quad\ \ \vert$ $\quad\quad\quad O-CO-CH_3$	
poly(vinyl alcohol)	PVAL	$-CH_2-CH-$ $\quad\quad\ \ \vert$ $\quad\quad\ OH$	
poly(methyl methacrylate)	PMMA	$\quad\quad\ CH_3$ $\quad\quad\ \ \vert$ $-CH_2-C-$ $\quad\quad\ \ \vert$ $\quad\quad\ CO_2CH_3$	
polytetrafluoroethylene	PTFE	$-CF_2-CF_2-$	highly crystalline polymer

A2.1 Thermoplastics and elastomers

Name	Abbreviation	Repeat unit	Comments
Heteropolymers			
poly(methylene oxide)	POM	$-CH_2-O-$	'acetal' resin
poly(ethylene oxide)	PEO	$-CH_2-CH_2-O-$	
poly(phenylene oxide)	PPO	(2,6-dimethyl-1,4-phenylene ether; aromatic ring with CH_3 groups and $-O-$)	aromatic polymer often blended with PS
nylon 6	PA6	$-NH-(CH_2)_5-CO-$	fibre or bulk polymer
nylon 6,6	PA6,6	$-NH(CH_2)_6NHCO(CH_2)_4CO-$	
poly(ethylene terephthalate)	PET	$-OCH_2CH_2-O-CO-\langle\bigcirc\rangle-CO-$	aromatic polyester fibre-forming polymer
aramid	–	$-NH-\langle\bigcirc\rangle-NHCO-\langle\bigcirc\rangle-$	aromatic high-modulus fibre
polycarbonate	PC	$-O-\langle\bigcirc\rangle-C(CH_3)_2-\langle\bigcirc\rangle-O-CO-$	usually amorphous, transparent polymer which can crystallize when annealed or treated with solvent
polyimide	PI	(pyromellitic diimide unit linked to diphenyl ether)	thermally stable polymers
poly(etherether ketone)	PEEK	(aromatic rings linked by $C=O$ and $-O-$)	
polyethersulphone	PSu	(aromatic rings linked by SO_2 and $-O-$)	
Elastomers			
polybutadiene	BR	$-CH_2-CH=CH-CH_2-$	cis-, trans- or vinyl-configuration
polychloroprene (Neoprene)	CR	$-CH-C=CH-CH_2-$ with Cl	cis-configuration
polyisoprene (natural rubber)	IR NR	$-CH_2-C=CH-CH_2-$ with CH_3	

A2.1   Thermoplastics and elastomers

Name	Abbreviation	Repeat unit	Comments
polyisobutylene (butyl rubber)	PIB	$-CH_2-\underset{\underset{CH_3}{\mid}}{\overset{\overset{CH_3}{\mid}}{C}}-$	usually copolymerized with 1–2% diene
silicone rubber	VMQ	$-\underset{\underset{CH_3}{\mid}}{\overset{\overset{R}{\mid}}{Si}}-O-$	R = H, CH_3 or C_6H_5

Common copolymers

Name	Abbreviation		Comments
acrylonitrile butadiene styrene	ABS		SAN with BR
high-impact polystyrene	HIPS		PS with BR
styrene acrylonitrile	SAN		butadiene with acrylonitrile comonomers
styrene-butadiene rubber	SBR		random copolmer elastomer
styrene-butadiene-styrene rubber	SBS		block copolymer elastomer

A 2.2   Thermosets

Name	Abbreviation	Starting materials	Representative groups in prepreg or network	Crosslinking agent
phenolic resin	PF	phenol + formaldehyde OH (ring) + $CH_2=O$	$\sim OCH_2$ (ring with OH, $CH_2O\sim$, $\sim OCH_2$)	formaldehyde
urea formaldehyde resin	UF	urea $(NH_2)\,C=O$ + formaldehyde $CH_2=O$	$\sim HN-CO-NH-\underset{\underset{\wr}{\overset{\mid}{O}}}{CH_2}\sim$	formaldehyde
melamine resins	MF	melamine + formaldehyde	$\sim OCH_2NH$ ... $NHCH_2O\sim$ (triazine ring) $NHCH_2O$	formaldehyde

A2.2 Thermosets

Name	Abbreviation	Starting materials	Representative groups in prepreg or network	Crosslinking agent
epoxy resins	EP	bisphenol A + epichlorhydrin $Cl-CH_2-CH-CH_2$ (O)	$\sim O-\bigcirc-C(CH_3)(CH_3)-\bigcirc-O-CH_2-CH-CH_2\sim$ (O)	di- or tri-amines
polyester resin	UP	dicarboxylic acids + diols $\sim O-CO-CH=CH-CO-O\sim$ (styrene solvent)	polystyrene $\sim O-(CH_2)_4-O-CO-CH-CH-CO-O\sim$ polystyrene	benzoyl peroxide, producing free radicals
polyurethane resins, plastics or rubbers	PU	di-isocyanate + diol	$-N=C=O + HO-$ → $-NH-CO-O-$	triol or tri-isocyanate
ebonite	–	natural rubber	$-CH_2-C(CH_3)=CH-CH_2-$ → $-CH-C(CH_3)=CH-CH_2-$ S	sulphur (S_8) + accelerator

Appendix 3 The periodic table

A 3.1   Naturally occurring elements

Element	Symbol	Atomic number	Relative atomic mass	Element	Symbol	Atomic number	Relative atomic mass
actinum	Ac	89	227	neodymium	Nd	60	144.24
aluminium	Al	13	26.98	neon	Ne	10	20.18
antimony	Sb	51	121.75	nickel	Ni	28	58.71
argon	Ar	18	39.95	niobium	Nb	41	92.91
arsenic	As	33	74.92	nitrogen	N	7	14.01
astatine	At	85	210	osmium	Os	76	190.20
barium	Ba	56	137.34	oxygen	O	8	16.00
beryllium	Be	4	9.01	palladium	Pd	46	106.40
bismuth	Bi	83	208.98	phosphorus	P	15	30.97
boron	B	5	10.81	platinum	Pt	78	195.09
bromine	Br	35	79.90	polonium	Po	84	209
cadmium	Cd	48	112.40	potassium	K	19	39.10
caesium	Cs	55	132.90	praseodymium	Pr	59	140.91
calcium	Ca	20	40.08	promethium	Pm	61	145
carbon	C	6	12.01	protoactinium	Pa	91	231
cerium	Ce	58	140.12	radium	Ra	88	226
chlorine	Cl	17	35.45	radon	Rn	86	222
chromium	Cr	24	52.00	rhenium	Re	75	186.20
cobalt	Co	27	58.93	rhodium	Rh	45	102.91
copper	Cu	29	63.55	rubidium	Rb	37	85.47
dysprosium	Dy	66	162.50	ruthenium	Ru	44	101.07
erbium	Er	68	167.26	samarium	Sm	62	150.35
europium	Eu	63	151.96	scandium	Sc	21	44.96
fluorine	F	9	19.00	selenium	Se	34	78.96
francium	Fr	87	223	silicon	Si	14	28.09
gadolinium	Gd	64	157.25	silver	Ag	47	107.87
gallium	Ga	31	69.72	sodium	Na	11	22.99
germanium	Ge	32	72.59	strontium	Sr	38	87.62
gold	Au	79	196.97	sulphur	S	16	32.06
hafnium	Hf	72	178.49	tantalum	Ta	73	180.95
helium	He	2	4.003	technetium	Tc	43	98.91
holmium	Ho	67	164.93	tellurium	Te	52	127.60
hydrogen	H	1	1.008	terbium	Tb	65	158.92
indium	In	49	114.82	thallium	Tl	81	204.37
iodine	I	53	126.9	thorium	Th	90	232.04
iridium	Ir	77	192.2	thulium	Tm	69	168.93
iron	Fe	26	55.85	tin	Sn	50	118.69
krypton	Kr	36	83.80	titanium	Ti	22	47.90
lanthanum	La	57	138.91	tungsten	W	74	183.85
lead	Pb	82	207.19	uranium	U	92	238.03
lithium	Li	3	6.94	vanadium	V	23	50.94
lutetium	Lu	71	174.97	xenon	Xe	54	131.30
magnesium	Mg	12	24.31	ytterbium	Yb	70	173.04
manganese	Mn	25	54.94	yttrium	Y	39	88.91
mercury	Hg	80	200.59	zinc	Zn	30	65.37
molybdenum	Mo	42	95.94	zirconium	Zr	40	91.22

A 3.2 The periodic table

1a	2a	3b	4b	5b	6b	7b	8			1b	2b	3a	4a	5a	6a	7a	0
1 H $1s^1$						METALS						SEMI-METALS		NON-METALS			2 He $1s^2$
3 Li $-2s^1$	4 Be $2s^2$											5 B $-2s^2 2p^1$	6 C $-2s^2 2p^2$	7 N $-2s^2 2p^3$	8 O $-2s^2 2p^4$	9 F $-2s^2 2p^5$	10 Ne $-2s^2 2p^6$
11 Na $-3s^1$	12 Mg $-3s^2$											13 Al $-3s^2 3p^1$	14 Si $-3s^2 3p^2$	15 P $-3s^2 3p^3$	16 S $-3s^2 3p^4$	17 Cl $-3s^2 3p^5$	18 Ar $-3s^2 3p^6$
19 K $-4s^1$	20 Ca $-4s^2$	21 Sc $-4s^2 3d^1$	22 Ti $-4s^2 3d^2$	23 V $-4s^2 3d^3$	24 Cr $-4s^1 3d^5$	25 Mn $-4s^2 3d^5$	26 Fe $-4s^2 3d^6$	27 Co $-4s^2 3d^7$	28 Ni $-4s^2 3d^8$	29 Cu $-4s^1 3d^{10}$	30 Zn $-4s^2 3d^{10}$	31 Ga $-3d^{10} 4p^1$	32 Ge $-3d^{10} 4p^2$	33 As $-3d^{10} 4p^3$	34 Se $-3d^{10} 4p^4$	35 Br $-3d^{10} 4p^5$	36 Kr $-4s^2 3d^{10} 4p^6$
37 Rb $-5s^1$	38 Sr $-5s^2$	39 Y $-5s^2 4d^1$	40 Zr $-5s^2 4d^2$	41 Nb $-5s^1 4d^4$	42 Mo $-5s^1 4d^5$	43 Tc $-5s^1 4d^6$	44 Ru $-5s^1 4d^7$	45 Rh $-5s^1 4d^8$	46 Pd $-5s^0 4d^{10}$	47 Ag $-5s^1 4d^{10}$	48 Cd $-5s^2 4d^{10}$	49 In $-4d^{10} 5p^1$	50 Sn $-4d^{10} 5p^2$	51 Sb $-4d^{10} 5p^3$	52 Te $-4d^{10} 5p^4$	53 I $-4d^{10} 5p^5$	54 Xe $-5s^2 4d^{10} 5p^6$
55 Cs $-6s^1$	56 Ba $-6s^2$	57 to 71 *	72 Hf $-6s^2 4f^{14} 5d^2$	73 Ta $-6s^2 4f^{14} 5d^3$	74 W $-6s^2 4f^{14} 5d^4$	75 Re $-6s^2 4f^{14} 5d^5$	76 Os $-6s^2 4f^{14} 5d^6$	77 Ir $-6s^0 4f^{14} 5d^9$	78 Pt $-6s^1 4f^{14} 5d^9$	79 Au $-6s^1 4f^{14} 5d^{10}$	80 Hg $-6s^2 4f^{14} 5d^{10}$	81 Tl $-6p^1$	82 Pb $-6p^2$	83 Bi $-6p^3$	84 Po $-6p^4$	85 At $-6p^5$	86 Rn $-6s^2 6p^6$
87 Fr $-7s^1$	88 Ra $-7s^2$	89 to 92 †					METALS									NON-METALS	

* Lanthanides	57 La	58 Ce	59 Pr	60 Nd	61 Pm	62 Sm	63 Eu	64 Gd	65 Tb	66 Dy	67 Ho	68 Er	69 Tm	70 Yb	71 Lu
† Actinides	89 Ac	90 Th	91 Pa	92 U											

The entries give the atomic number, the symbol for the element and the configuration of the outer electrons. The dashes indicate that the lower levels are completely filled.

Acknowledgements

Grateful acknowledgement is made to the following sources for material reproduced in this book:

Figure 1.2(a), L. J. Gibson and M. F. Ashby *Cellular Solids: Structure and Properties*, © 1988 Pergamon Press plc. Figure 2.1(a–c), Courtesy of the Royal Mint. Figures 2.5 and 2.6, RHP Bearings Ltd. Figure 2.7, British Aerospace, Hatfield. Figure 2.23, J. E. Bailey and P. B. Hirsch *Proceedings of the Royal Society*, vol. A267, 11, © 1962 The Royal Society. Figures 2.35 and 2.37, H. J. Frost and M. F. Ashby *Deformation—Mechanism Maps*, © 1982, Pergamon Press plc. Figure 2.44, R. A. Smith 'An introduction to fracture mechanics for engineers', *Materials in Engineering Applications*, vol. 1, December 1978 (now *Materials & Design*) © 1978 Scientific & Technical Press Ltd. Figures 2.58 and 2.59, IMI Titanium Ltd. Figure 2.69, INCO Alloys Ltd (NIMONIC is a trademark of the INCO family of companies). Figure 3.47, E. L. Samuels *Optical Microscopy of Carbon Steel*, © 1980 American Society of Metals, ASM International. Figure 3.59, R. Cotterill *The Cambridge Guide to the Material World*, © 1985 Cambridge University Press. Figure 3.61, Musée Bartholdi, Colmar, France. Figure 4.8, A. C. Evans *et al. Journal of Materials Science*, Chapman & Hall. Figures 4.32, 4.39, 4.52 and 4.53, R. Morrell *Handbook of Properties of Technical and Engineering Ceramics*, © 1985 Crown copyright, reproduced with the permission of the Controller of HMSO. Figure 4.27, R. Gordon. Figure 4.28, IOP Roesler. Figure 4.47, Nissan Motor Co. Ltd. Figure 4.34, C. H. Brown. Figure 5.7, N. J. Mills *Plastics—Microstructure, Properties and Applications*, © 1986 Edward Arnold. Figure 5.31, R. P. Kambour 'Macromolecular reviews 7', *Journal of Polymer Science*, © 1973 John Wiley & Sons. Figure 5.32, P. Beahan, M. Bevis and D. Hull 'Fracture processes in polystyrene', *Proceedings of the Royal Society*, vol. A343, © 1975 The Royal Society. Figure 5.42, Netlon Limited. Figure 6.1, Reproduced by courtesy of the Trustees of the British Museum. Figures 6.3 and 6.4, J. P. Wild *Textiles in Archaeology*, Shire Publications Ltd, © 1988 John Peter Wild. Figures 6.5, 6.7 and 6.12(b,c), M. A. Taylor *Technology of Textile Properties* (2nd edn), © 1985 Forbes Publications Ltd. Figure 6.12(a), US Department of Agriculture. Figure 6.13, Courtesy of W. D. Cooke, Dept. of Textiles, UMIST. Figure 6.27, D. H. Page and P. A. Tydeman 'A new theory of the shrinkage, structure and properties of paper', in H. F. Rance (ed) *Handbook of Paper Science*, vol. 1, © 1980 Elsevier Science Publishers. Figure 6.35, B. J. Dunn (1980) *Ropes Made from Man-Made Fibres—B: Rope Properties*, Bridon Fibres Ltd. Figures 6.8, 7.5, 7.11 and 7.49, D. J. Hull *An Introduction to Composite Materials*, © 1981 Cambridge University Press. Figure 6.40, D. Kingston 'Development of parallel fibre tensile members', *Symposium on Engineering Applications of Parafil Ropes*, © 1988 Imperial College of Science and Technology. Figure 6.41, Linear Composites Ltd. Figures 7.1 and 7.3, Reproduced by courtesy of the Trustees of the British Museum. Figure 7.2, Milton Keynes Archaeology Unit. Figure 7.4, Vosper Thorneycroft U.K. Figure 7.12, Toyota Motor Corporation. Figure 7.13, Courtesy of General Dynamics. Figures 7.14 and 7.16, A. W. Birley, R. J. Heath and M. H. Scott *Plastics, Materials, Properties and Applications* (2nd edn), © 1988 Blackie and Son Ltd. Figure 7.20, Carello Lighting plc. Figure 7.21, London Underground Ltd and BP Chemicals. Figure 7.22, E.R.F. Ltd. Figures 7.27, 7.28 and 7.40, N. C. Hilyard (ed) *Mechanics of Cellular Plastics*, © 1982 Applied Science. Figures 7.29 and 7.30, L. J. Gibson and M. F. Ashby *Cellular Solids: Structure and Properties*, Pergamon Press plc, © 1980 Lorna J. Gibson and Michael F. Ashby. Figures 7.32 and 7.33, Permali Gloucester Ltd. Figure 7.34 (upper), Courtesy of Princes Risborough Laboratory, Building Research Establishment, © Crown copyright. Figure 7.34 (lower), B. G. Butterfield and B. A. Meylar *Three-dimensional Structure of Wood: an ultrastructural approach* (2nd edn), © 1980 Chapman and Hall. Figure 7.45, M. W. Darlington and M. Christie, Cranfield Institute of Technology. Figure 7.50, Courtesy of ICI Ltd. Figures 7.58 and 7.59, Courtesy of Dr T. J. Baker, Imperial College, London. Figure 8.1(a), North of Scotland Hydro-Electric Board. Figure 8.1(b), Edmund Nuttall Ltd. Figure 8.1(c), UPI/Bettmann Newsphotos. Figures 8.3, 8.4, 8.12 and 8.13, G. C. Bye *Portland Cement*, Pergamon Press plc, copyright © 1983 The Institute of Ceramics. Figures 8.6 and 8.7(b), D. D. Double *Proceedings of the Royal Society*, vol. 310, pp. 53–66, © 1983 The Royal Society. Figure 8.7(a,c,h), E. Breval, T. Knudsen and N. Tharlow, Aalborg Portland Betorforskings Laboratorium, Karlslunde, Denmark. Figure 8.7(d,f), Professor D. D. Double: from D. D. Double *et al.* (1978) 'The hydration of Portland cement', *Proceedings of the Royal Society*, vol. 359, pp. 435–51, The Royal Society. Figure 8.7(e), P. Barnes *et al.* (1980) 'Cement tubules—another look', *Cement and Concrete Research*, vol. 10, pp. 639–45, © 1980 Pergamon Press plc/G. C. Bye (1983) *Portland Cement*, Pergamon Press plc, © 1983 The Institute of Ceramics. Figure 8.7(g), Professor S. Diamond. from H. F. W. Taylor 'Portland cement: hydration', *Journal of Educational Modules for Materials Science and Engineering* (now *Journal of Materials Education*) vol. 3, no. 3, p. 439, © 1981 Materials Education Council, Pennsylvania State University. Figures 8.8 and 8.9, K. Kendall *et al.* 'The relation between porosity, microstructure and strength, and the approach to advanced cement-based materials', *Philosophical Transactions of the Royal Society*, vol. 310, pp. 139–53, © 1983 The Royal Society. Figure 8.15, Courtesy Gordon A. Buck. Figure 8.17, M. F. Ashby and R. H. Jones *Engineering Materials 2*, Pergamon Press plc, © 1986 M. F. Ashby and R. H. Jones. Figure 8.18, Cement and Concrete Association. Figure 8.25, D. Lee *Theory and Practice of Bearings and Expansion Joints for Bridges*, Cement and Concrete Association: Viewpoint Publication, © 1971. Figure 9.11, Paul Popper Ltd.

Index